Reaction Mechanisms in Organic Synthesis

Postgraduate Chemistry Series

A series designed to provide a broad understanding of selected growth areas of chemistry at postgraduate student and research level. Volumes concentrate on material in advance of a normal undergraduate text, although the relevant background to a subject is included. Key discoveries and trends in current research are highlighted, and volumes are extensively referenced and cross-referenced. Detailed and effective indexes are an important feature of the series. In some universities, the series will also serve as a valuable reference for final year honours students.

Editorial Board

Titles in the Series

Reaction Mechanisms in Organic Synthesis

Rakesh Kumar Parashar

Reader, Chemistry Department,
Kirori Mal College, University of Delhi, India

A John Wiley and Sons, Ltd., Publication

This edition first published 2009
© 2009 Rakesh Kumar Parashar

Blackwell Publishing was acquired by John Wiley & Sons in February 2007. Blackwell's publishing programme
has been merged with Wiley's global Scientific, Technical, and Medical business to form Wiley-Blackwell.

Registered office
John Wiley and Sons Ltd., The Atrium, Southern Gate, Chichester, West Sussex, PO19 8SQ, United Kingdom

Editorial offices
9600 Garsington Road, Oxford, OX4 2DQ, United Kingdom
2121 State Avenue, Ames, Iowa 50014-8300, USA

For details of our global editorial offices, for customer services and for information about how to apply for
permission to reuse the copyright material in this book please see our website at www.wiley.com/wiley-blackwell.

Library of Congress Cataloging-in-Publication Data
Parashar, R. K. (Rakesh Kumar)
[1st ed.]
Reaction mechanisms in organic synthesis / Rakesh K. Parashar
p. cm. — (Postgraduate chemistry series)
Includes bibliographical references and index.
ISBN 978-1-4051-5072-9 (hardback : alk. paper) — ISBN 978-1-4051-9089-3 (pbk. : alk. paper)
1. Organic compounds—synthesis. 2. Organic reaction mechanisms. 3. Physical organic chemistry. I. Title.
QD262.P34 2009
547'.2—dc22
2008034864

A catalogue record for this book is available from the British Library.

Set in 10/12 pt Minion by Aptara® Inc., New Delhi, India
Printed in Singapore by Utopia Press Pte Ltd

1   2009

To Riya, Manya and Indu with love and to
my parents with immense respect

Contents

Foreword

Exciting new methods and reagents are being discovered and used everyday in the synthesis of organic molecules. Knowing the mechanism of these reactions is very important, without which it is almost impossible to carry out the synthesis of important molecules in the laboratory or in industry. Thus, the importance of organic reaction mechanisms continues to increase, and this book is a welcome addition to the available sources on the subject.

While teaching organic synthesis and practicing it in the laboratory, a need is often felt of a handy book combining organic synthesis and mechanisms of reactions employed in synthesis instead of large volumes or monograms on synthesis. There are not many such books covering these two very essential aspects of organic chemistry.

Writing a textbook for any level is always a challenge. However, Dr Parashar deserves praise for undertaking this project and interlinking these two areas of organic chemistry so well throughout the book.

The book is designed to provide fundamental aspects of organic chemistry in a flexible way rather than presenting a traditional approach. The mechanisms and stereochemical features of common reactions used in organic synthesis are discussed in a qualitative and quantitative manner. Specific examples are taken from the latest literature.

The contents of the book give a general impression about what is dealt with. The selection of topics has been done very carefully and judiciously. The material is condensed to a manageable text of 363 pages and presented in a clear and logical fashion over eight chapters. This is done by focusing purely on the basics of the subject without going through exhaustive detail or repetitive examples.

This book would be of immense help to students at the postgraduate level as well as to research workers because of its contents and the way those have been dealt with. I sincerely hope that the book will go a long way to satisfy the long-felt need of students and teachers who inspire the students to take up synthetic organic chemistry as their research topic and career.

I hope practitioners and professionals will be benefited from the experience of learning reaction mechanisms of important synthetic reactions.

I am happy to recommend this book as a self-guide for students and professionals.

Virinder S. Parmar, PhD, FRSC
Professor and Head, Department of Chemistry, and
Chairman of the Board of Research Studies
University of Delhi, India

Preface

An organic chemist is primarily concerned with (a) the synthesis of organic molecules of particular interest to the pharmaceutical and agrochemical industries and (b) the way these molecules interact in biological pathways.

Synthesis involves a careful selection of reactions; new reactions are being developed everyday. Knowing how structure affects a reaction, a rational sequence of transformations can be used to synthesize target molecules. An understanding of organic reaction mechanisms is essential without which it is impossible to plan organic synthesis. It is also required to extend one's knowledge of different areas related to organic chemical reaction mechanisms. The vital importance of the organic synthesis processes is established by the fact that many Nobel laureates have been associated with this field.

Beginning with basic introductory course, this book covers all aspects of organic reaction mechanisms, expands on the foundation acquired in chemistry courses, and enables students and research workers to understand the mechanisms and then to plan syntheses. This book will help postgraduate students to write reasonable mechanisms for organic chemical transformations, which are arranged according to an ascending order of difficulty.

Established reactions are being subjected to both technical improvements and increasing number of applications. For example, intense efforts are made in industry and university laboratories to devise innovative ways to speed up reactions, to carry them out in a continuous fashion and to provide for separation of complex mixtures. For example, ultrasound can dramatically affect the rates of chemical reactions. Microwave-assisted protocols often result in high yields and time efficiency. Solid-phase synthesis allows for easy separation of the resulting products while providing for libraries of compounds to be made. Although these methods have been discussed in special monographs and review articles, there is no recent single book covering reactions (modern or newer) with latest procedural modifications and also simultaneously explaining reaction mechanism and covering stereospecificity and regiospecificity.

The book contains examples from recently published research work to illustrate the important steps involved in synthesis. The discussion is organized by the conditions under which the reaction is executed rather than by the types of mechanisms as is the case in most textbooks at the graduate level.

The author believes that students are well aware of the basic reaction pathways such as substitutions, additions, eliminations, aromatic substitutions, aliphatic nucleophilic substitutions and electrophilic substitutions. Students may follow undergraduate books on reaction mechanisms for basic knowledge of reactive intermediates and oxidation and reduction processes. *Reaction Mechanisms in Organic Synthesis* provides extensive coverage of various carbon–carbon bond forming reactions such as transition metal catalyzed reactions; use of stabilized carbanions, ylides and enamines for the carbon–carbon bond forming reactions; and advance level use of oxidation and reduction reagents in synthesis.

Thus, this book may prove to be an excellent primer for advanced postgraduates in chemistry. This book will be useful both for instructors and those who are preparing for examinations.

Following is a brief account of the contents of the eight chapters of this book.

Chapter 1 is devoted to exploring strategies involved in organic synthesis. It seeks to explain concepts like retrosynthetic analysis, atom economy, umpolung approach, click chemistry and asymmetric synthesis. On the basis of interesting and relevant examples, protection and deprotection of different functional groups are explained and the most probable mechanism is also mentioned for important reactions.

Chapter 2 includes complete discussion on reaction intermediates including carbocations, carbanions, free radicals, carbenes, nitrines and benzynes. The structure, methods of generation and important reactions of all the intermediates are discussed in this chapter. The author has emphasized on their applications in the asymmetric synthesis.

Chapter 3 discusses ylides and enamines, and also deals with the extended examples of carbanions.

Chapter 4 reviews the role of various reagents used in organic synthesis for the formation of carbon–carbon double bond. Specific examples are included at each stage to illustrate the mechanism under discussion.

Chapter 5 includes complete coverage of the transition metals-mediated carbon–carbon bond forming reactions. Pd-, Ni-, Cr-, Zr- and Cu-catalyzed reactions such as Heck, Negishi, Sonogashira, Suzuki, Hiyama, Stille, Kumada reactions are covered in adequate details including the applications of these reactions in organic synthesis.

Chapter 6 focuses on selected examples of reduction methods and their mechanisms in detail. The chapter gives a detailed account of reducing reagents and their applications in organic synthesis.

The oxidation examples in **Chapter 7** are arranged to elucidate key aspects of organic reaction mechanisms. The importance of oxidation reagents in synthesis and their mechanism of action have been explained in detail.

Chapter 8 covers extensively pericyclic reactions and also includes the aromatic transition state theory. Most of the examples are taken from latest literature and are useful for postgraduate and research students.

As an academic convenience to readers all reaction mechanisms leading to stereospecific products are highlighted. The book will also serve as an excellent reference book since references are offered at the end of each chapter.

The book seeks to cover the postgraduate syllabi of almost all the universities. Students will be spared the tedium of collecting all the information on the subject scattered in various books and journals. Even though a comprehensive effort was made to gather information from all sources, it is inevitable that some relevant papers and reviews may be left unscanned.

The author hopes that the book proves to be an easy-to-use general organic chemistry textbook and finds a place in libraries and personal bookshelves of the academic community.

All comments and suggestions will be received with gratitude.

Rakesh Kumar Parashar
Reader, Chemistry Department
Kirori Mal College
University of Delhi, India

About the Author

Dr Rakesh Kumar Parashar completed his PhD in 1990 from the University of Delhi, Delhi, in the field of synthetic organic chemistry. He is a Reader in Chemistry at Kirori Mal College, University of Delhi, Delhi. He has done his postdoctorate from the University of Barcelona, Spain. He has published 22 papers in various national and international journals and has delivered several lectures in India and abroad. He is also the author of several books. He is actively involved in teaching and research for the past 18 years.

Acknowledgements

I sincerely thank Prof. Jim Coxon who inspired me to take up this project. He also generously helped me to improve this book at the writing stage.

My special thanks are to Prof. Virinder S. Parmar, Head of Chemistry Department, University of Delhi, for writing foreword of this book. I acknowledge Prof. J. M. Khurana, University of Delhi, for his fruitful suggestions that helped me throughout the preparation of this manuscript. I also thank Dr S. Gera and Dr Geetanjali Pandey, Chemistry Department, Kirori Mal College, University of Delhi, for reviewing several chapters.

And, finally, I thank my wife, Indu, and daughters, Riya and Manya, for their love and encouragement during the lengthy, seemingly interminable period of writing this book.

Abbreviations

Ac	acetyl
Ac$_2$O	acetic anhydride
acac	acetylacetonate
AIBN	2,2′-azobisisobutyronitrile
All	allyloxycarbonyl
Ar	aryl
BBN	borabicyclo[3.3.1]nonane
BHT	butylated hydroxytoluene (2,6-di-*t*-butyl-*p*-cresol)
BINAL-H	2,2′-dihydroxy-1,1′-binaphthyllithium aluminum hydride
BINAP	2,2′-bis(diphenylphosphino)-1,1′-binaphthyl
BINOL	1,1′-bis-2,2-naphthol
bipy	2,2′-bipyridyl
Bn	benzyl
Boc	*t*-butoxycarbonyl
BOM	benzyloxymethyl
bp	boiling point
Bs	brosyl (4-bromobenzenesulfonyl)
BSA	*N,O*-bis(trimethylsilyl)acetamide
Bu	*n*-butyl
Bz	benzoyl
CAN	cerium(IV) ammonium nitrate
cat.	catalyst
Cbz	benzyloxycarbonyl
CHIRAPHOS	2,3-bis(diphenylphosphino)butane
CIP	Cahn–Ingold–Prelog priority rules
cod	cyclooctadiene
m-CPBA	*m*-chloroperbenzoic acid or *m*-chloroperoxybenzoic acid
CSA	10-camphorsulfonic acid
Cy	cyclohexyl
d	density
DABCO	1,4-diazabicyclo[2.2.2]octane
DAIPEN	1,1-dianisyl-2-isopropyl-1,2-ethylenediamine
DAST	*N,N*-diethylaminosulfur trifluoride
dba	dibenzylideneacetone
DBU	1,8-diazabicyclo[5.4.0]undec-7-ene
DCC	*N,N′*-dicyclohexylcarbodiimide
DCE	dichloroethane

DCM	dichloromethane
DDQ	2,3-dichloro-5,6-dicyano-1,4-benzoquinone
De	diastereomeric excess
DEG	diethylene glycol
DET	diethyl tartrate
(DHQ)$_2$PHAL	1,4-bis(9-*O*-dihydroquinine)phthalazine
(DHQD)$_2$PHAL	1,4-bis(9-*O*-dihydroquinidine)phthalazine
DIBAH or DIBAL-H	diisobutylaluminum hydride (*i*-Bu$_2$AlH)$_2$
DIEA	=DIPEA
DIOP	4,5-bis(diphenylphosphinomethyl)-2,2-dimethyl-1,3-dioxolane or 2,3-*O*-isopropylidene-2,3-dihydroxy-1,4-bis(diphenylphosphino) butane
DIPAMP	bis[(2-methoxyphenyl)phenylphosphino]ethane
DIPEA	diisopropylethylamine
DMA	dimethylacetamide
DMAP	4-(dimethylamino)pyridine
DME	1,2-dimethoxyethane, glyme or dimethyl glycol
DMEU	1,3-dimethylimidazolidin-2-one
DMF	dimethylformamide
DMPU	1,3-Dimethyltetrahydropyrimidin-2(1*H*)-one
DMS	dimethyl sulfide
DMSO	dimethyl sulfoxide
DPEN	diphenylethylenediamine
Dppe	1,2-bis(diphenylphosphino)ethane
DMMP	Dimethyl methylphosphonate
dppf	1,1′-bis(diphenylphosphino)ferrocene
dppm	1,1-bis(diphenylphosphino)methane
dppp	1,3-bis(diphenylphosphino)propane
Dod-S-Me	Dodecyl methyl sulfide
DTBP	di-*t*-butyl peroxide
E1cB	*e*limination *c*onjugate *b*ase
ee	enantiomeric excess
equiv.	equivalent(s)
Et	ethyl
EWG	electron-withdrawing group
Fmoc	9-fluorenylmethoxycarbonyl
h	hour(s)
HMDS	hexamethyldisilazane or 1,1,1,3,3,3,-hexamethyldisilazane
HMPA	hexamethylphosphoric triamide
HWE	Horner–Wadsworth–Emmons
i	iso
Ipc	isopinocampheyl
isc	intersystem crossing
IR	infrared
kcal	kilocalorie
KHDMS	potassium hexamethyldisilazide

LAH	lithium aluminum hydride
LDA	lithium diisopropylamide
LHMDS	LiHMDS
LiHMDS	lithium hexamethyldisilazide
LiTMP	lithium 2,2,6,6-tretramethylpiperidide
LTA	lead tetraacetate
LTEAH	lithium triethoxyaluminohydride
LVT	low-valent titanium
2,6-Lutidine	2,6-dimethylpyridine
M	metal; also molar
Me	methyl
MEM	(2-methoxyethoxy)methyl
min	minutes
mL	millilitre
MMPP	magnesium monoperoxyphthalate
MOM	methoxymethyl
mp	melting point
Ms	mesyl or methanesulfonyl
MS	molecular sieves
MTM	methylthiomethyl
MW	molecular weight; microwave
NaHMDS	sodium hexamethyldisilazide
NBA	*N*-bromoacetamide
NBS	*N*-bromosuccinimide
NCS	*N*-chlorosuccinimide
NIS	*N*-iodosuccinimide
NMO	*N*-methylmorpholine *N*-oxide
NMP	*N*-methyl-2-pyrrolidinone
NMR	nuclear magnetic resonance
Nu	nucleophile
OTf	Triflate or trifluoromethanesulfonate, functional group with the formula $CF_3SO_3^-$
PCC	pyridinium chlorochromate
PDC	pyridinium dichromate
Ph	phenyl
PhH	benzene
pent	pentyl
Piv	pivaloyl
PMB	*p*-methoxybenzyl
pmIm	1-methyl-3-pentylimidazolium
PMP	1,2,2,6,6-pentamethylpiperidine
PPTS	pyridinium *p*-toluenesulfonate
Pr	*n*-propyl
PTC	phase transfer catalyst/catalysis
PTSA	*p*-toluenesulfonic acid
py	pyridine
R	alkyl group

R	clockwise (R, for rectus)
rt	room temperature
S	counterclockwise (S, for sinister)
S_N1	nucleophilic substitution reaction unimolecular
S_N2	nucleophilic substitution reaction bimolecular
salen	bis(salicylidene)ethylenediamine
SET	single electron transfer
SMEAH	red-Al or sodium bis(2-methoxyethoxy)aluminum hydride
t	tertiary
TASF	tris(diethylamino)sulfonium difluorotrimethylsilicate
TBAB	tetrabutylammonium bromide
TBAF	tetrabutylammonium fluoride
TBAP	tetrabutylammonium perruthenate
TBDPS	*t*-butyldiphenylsilyl
TBHP	*t*-butyl hydroperoxide
TBS	*t*-butyldimethylsilyl
TEMPO	2,2,6,6-tetramethylpiperidinoxyl
TES	triethylsilyl
TFA	trifluoroacetic acid
TFAA	trifluoroacetic anhydride
tfp	tri-2-furylphosphine
THF	tetrahydrofuran
THP	tetrahydropyranyl
TIPS	triisopropylsilyl
TMEDA	$N,N,N'N'$-tetramethylethylenediamine
TMS	trimethylsilyl
TMSOTf	trimethylsilyl trifluoromethanesulfonate
Tol	*p*-tolyl
TPAP	tetrapropylammonium perruthenate
TPP	tetraphenylporphyrin
Ts	tosyl or *p*-toluenesulfonyl; also transition state
TSOH	*p*-toluenesulfonic acid (PTSA)
TTBS	tri-t-butylsilyl

Chapter 1
Synthetic Strategies

1.1 An introduction to organic synthesis

Organic synthesis is the construction of complex organic compounds from simple starting compounds by a series of chemical reactions. The compounds synthesized in nature are called **natural products**. Nature provides a plethora of organic compounds and many of these possess interesting chemical and pharmaceutical properties. Examples of natural products include cholesterol (**1.1**), a steroid found in most body tissues; limonene (**1.2**), a terpene found in lemon and orange oils; caffeine (**1.3**), a purine found in tea leaves and coffee beans; and morphine (**1.4**), an alkaloid found in opium.

The synthesis of organic molecules is the most important aspect of organic chemistry. There are two main areas of research in the field of organic synthesis, namely **total synthesis** and **methodology**. A total synthesis is the complete chemical synthesis of complex organic molecules from simple, commercially available or natural precursors. Methodology research usually involves three main stages, namely discovery, optimization and the study of scope and limitations. Some research groups may perform a total synthesis to showcase the new methodology and thereby demonstrate its application for the synthesis of other complex compounds.

The compound to be synthesized may have a small carbon framework such as vanillin (**1.5**) (vanilla flavouring) or have more complex carbon framework such as penicillin G (**1.6**) (an antibiotic) and taxol (**1.7**) (used for the treatment of certain types of cancer). However, three challenges must be met in devising a synthesis for a specific compound: (1) the carbon atom framework or skeleton that is found in the desired compound must be assembled;

(2) the functional groups that characterize the desired compound must be introduced or transformed from other groups at appropriate locations; and (3) if stereogenic centres are present, they must be fixed in a proper manner.

1.5

1.6

1.7

Thus, in order to understand the synthesis of a complex molecule, we need to understand the carbon–carbon bond forming reactions, **functional groups interconversions** and **stereochemistry** aspects.

Carbon–carbon bond forming reactions are the most important tool for the construction of organic molecules. The reaction in which one functional group is converted into another is known as functional group interconversion. The spatial arrangements of the substituents can have a significant impact on the reactivity and interaction towards other molecules. Many chiral drugs must be made with high enantiomeric purity because the other enantiomer may be inactive or has side effects. Thus, there is a need to develop methods to synthesize organic compounds as one pure enantiomer and the use of these techniques is referred to as **asymmetric synthesis** (section 1.5).

Therefore, carbon–carbon bond forming reactions, asymmetric synthesis, the design of new chiral ligands, environmental-friendly reactions and atom economical syntheses are the major aims of present-day research.

1.2   Retrosynthetic analysis (disconnection approach)

E. J. Corey[1,2] brought a more formal approach to synthesis design, known as retrosynthetic analysis. The analysis of synthesis in reverse manner is called **retrosynthetic analysis** or alternatively a **disconnection approach.** Retrosynthetic analysis or retrosynthesis is a technique for solving problems in synthesis planning, especially those presented by complex structures. In this approach, the synthesis is planned backwards starting from a relatively complex product to available simpler starting materials (Scheme 1.1). This approach requires construction of a carbon skeleton of the target molecule, placing the functional groups and appropriate control of stereochemistry.

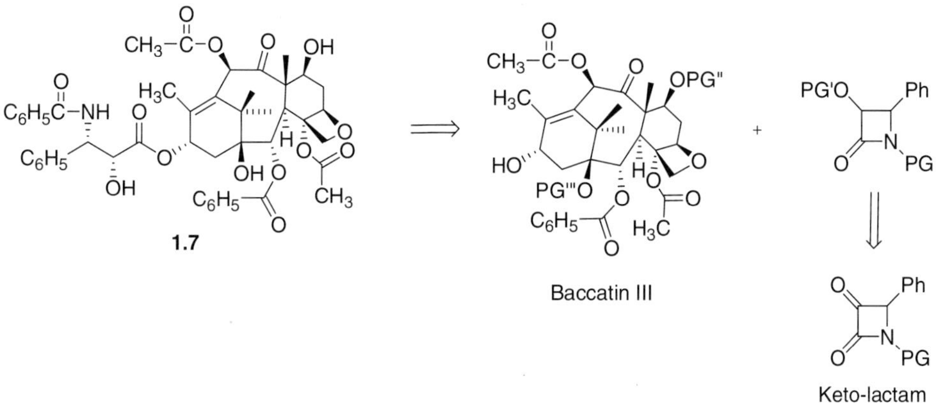

Scheme 1.1 Retrosynthetic analysis of taxol

Table 1.1 Synthetic versus retrosynthetic analysis

Direction	Synthetic	Retrosynthetic
Step	Reaction	Transform or retro-reaction
Arrow used in graphical depiction	$\longrightarrow$	$\Longrightarrow$
Starting structure	Reactant	Target
Resulting structure	Product	Precursor
Substructure required for operation	Reacting functionality	Retron

The terminology used in synthetic and retrosynthetic analysis is shown in Table 1.1.

A transform in the case of the retrosynthetic counterpart of the Wittig reaction is shown below:

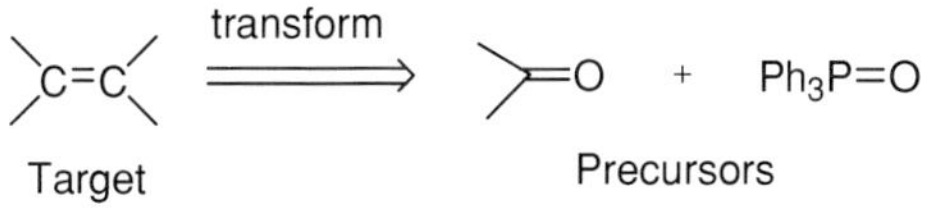

In a similar manner, the retrosynthetic analysis of the Diels–Alder reaction is represented below:

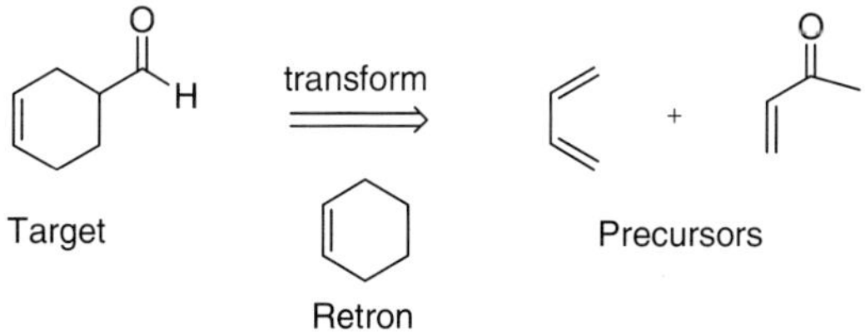

The retrosynthetic step involving the breaking of bond(s) to form two (or more) **synthons** is referred to as a **disconnection**. A synthon is an idealized fragment, usually a cation, anion or radical, resulting from a disconnection. One must select disconnections which correspond to the high yielding reactions.

Functional group interconversion is the process of the transformation of one functional group to another to help synthetic planning and to allow disconnections corresponding to appropriate reactions. In planning a synthetic strategy, apart from devising means of constructing the carbon skeleton with the required functionality, there are other factors which must be addressed including the control of regiochemistry and stereochemistry.

The above points are explained by discussing retrosynthetic analysis of cyclohexanol:

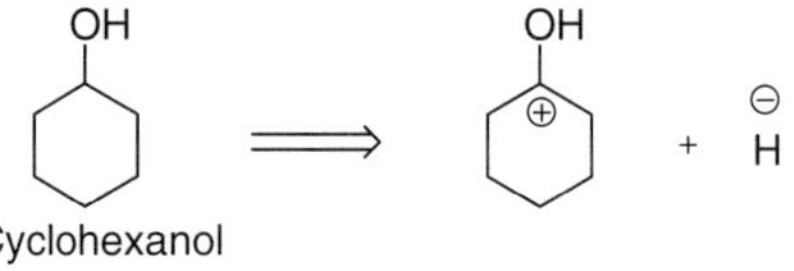

Cyclohexanol

The hydroxycarbocation and the hydride ion formed after disconnection of cyclohexanol are synthons. The synthetic equivalents of hydroxycarbocation and the hydride ion are cyclohexanone and sodium borohydride, respectively. Thus, the target molecule cyclohexanol can be prepared by treating cyclohexanone with sodium borohydride.

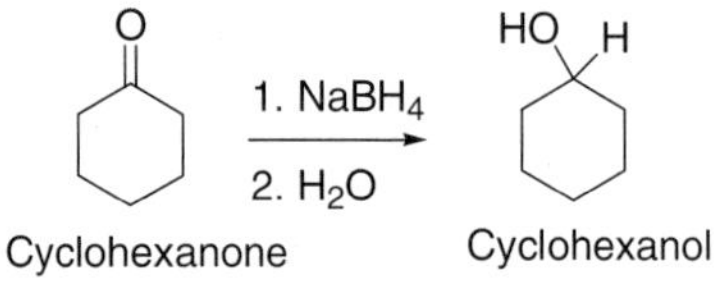

Cyclohexanone   Cyclohexanol

The C–C bond of cyclohexanol can also be disconnected as shown below:

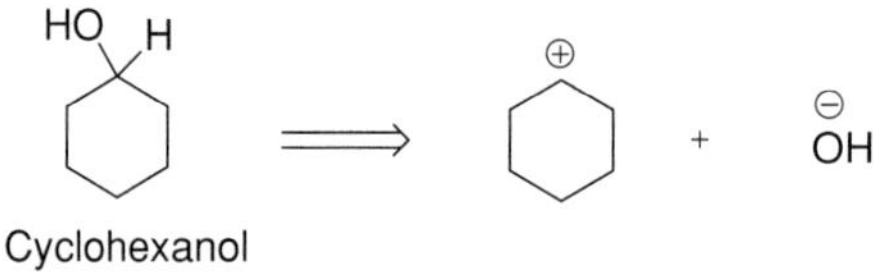

Cyclohexanol

The synthetic equivalent for the cyclohexyl carbocation is cyclohexyl bromide. Thus, cyclohexanol can be prepared by the reaction of cyclohexyl bromide with hydroxide ion.

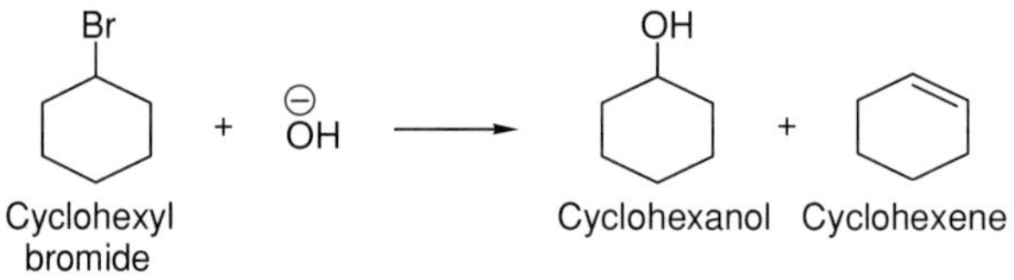

Cyclohexyl   Cyclohexanol   Cyclohexene
bromide

However, in this case cyclohexene is also formed; thus, this method may not be considered as effective as the previous one.

A **retrosynthetic tree** is a directed acyclic graph of several (or all) possible retrosyntheses of a single target. Retrosynthetic analysis, then, consists of applying transforms to a given target, thereby generating all precursors from which that target can be made in a single step. The analysis can be repeated for each precursor, generating a second level of precursors. Each precursor molecule so generated is in some way simpler than the target from which it was derived and then considered to be a target and analyzed similarly. The analysis terminates when precursors are elaborated, which are considered to be relatively simple or readily available, generating a tree of synthetic intermediates.

The final result is a complete retrosynthetic tree that will contain all possible syntheses of the given target – reasonable and unreasonable, efficient and cumbersome. Of course, such a tree would be unmanageably large both for humans and computers, even when the number of precursor levels is limited. To keep the size of the retrosynthetic tree under control, examine all possible disconnections – check which are chemically sound (corresponding to known reactions, reagents, directing effects). The guiding principles for this selection are called **strategies**.

Some **guidelines for retrosynthesis** are given below:

1. It is better to use convergent approach rather than divergent for many complex molecules.
2. Use only disconnections corresponding to disconnect C–C bonds and C–X bonds wherever possible.
3. Disconnect to readily recognizable synthons by using only known reactions (transform).
4. The synthesis must be short.
5. It is better to use those reactions which do not form mixtures.
6. The focus is on the removal of stereocentres under stereocontrol. Stereocontrol can be achieved through either mechanistic control or substrate control.

The computer-assisted synthetic analysis designated **OCSS** (*organic chemical simulation of synthesis*) and **LHASA** (*logic and heuristics applied to synthetic analysis*) were designed to assist chemists in synthetic analysis by Corey *et al.*[3,4]. LHASA generates trees of synthetic intermediates from a target molecule by analysis in the retrosynthetic direction.

Click chemistry is a modular synthetic approach towards the assembly of new molecular entities. The nature has overall preference for carbon–heteroatom bonds over carbon–carbon bonds; e.g. all the proteins are created from 20 building blocks that are joined via reversible heteroatom links. Thus following nature's lead, the term 'click chemistry'[5] was coined by Kolb, Finn and Sharpless in 2001 for synthesis restricted to molecules that are easy to make. The click chemistry as defined by Sharpless is reactions that are modular, wide in scope, high yielding, create only inoffensive products, are stereospecific, simple to perform and require the use of only benign solvent. Of all the reactions which fall under the umbrella of click chemistry, the Huisgen 1,3-dipolar cycloaddition of alkynes and azides to yield 1,2,3-triazoles is undoubtedly the premier example of a click reaction. The reaction is accelerated under copper(I) catalysis, requires no protecting groups, and almost complete conversion takes place. The reaction is selective, as only 1,4-disubstituted 1,2,3-triazole is the only product formed and there is no formation of 1,5-disubstituted triazole, which is also formed in the thermally induced Huisgen cycloaddition (Scheme 1.2).

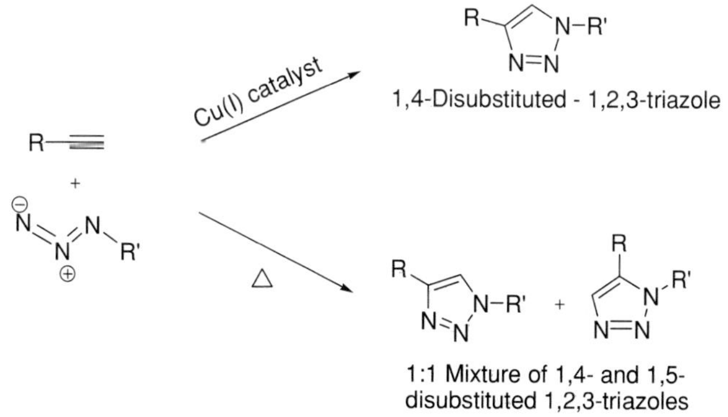

Scheme 1.2

Due to the reliability, specificity and biocompatibility of **click chemistry**, its application is found in nearly all areas of modern chemistry from drug discovery to material science.

1.3 Umpolung strategy

Umpolung is a general class of reactions in which the characteristic reactivity of a group or an atom is temporarily reversed. The concept of umpolung is helpful especially with carbonyl groups. But to understand this concept, it is important to understand the normal reactivity of the carbonyl group. For example, under normal conditions carbonyl carbon is electrophilic and the α-carbon is nucleophilic because of the resonance, as shown below:

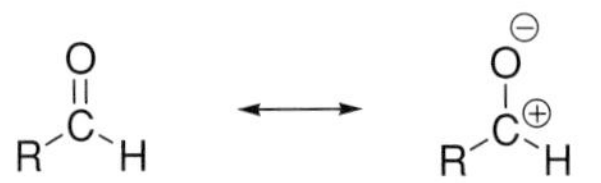

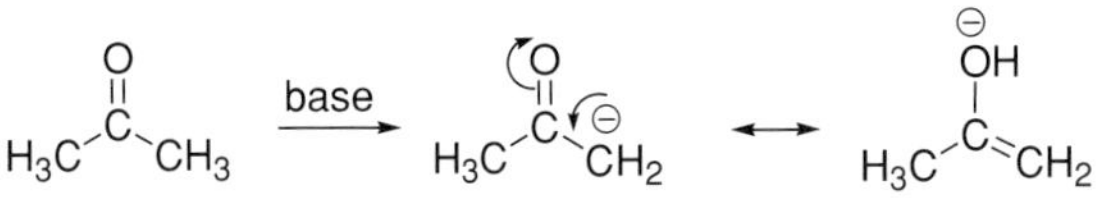

But if the polarity of a carbonyl compound is reversed, the acyl carbon becomes nucleophilic. This is achieved by first converting the carbonyl group into dithianes **1.8**, and then the carbon becomes nucleophilic. The strong base can remove the hydrogen adjacent to the sulfur in the dithiane to give 2-lithio-1,3-dithiane **1.9**. The acyl anion equivalent **1.9** generated in this manner reacts with an alkyl halide to give the alkylated product **1.10**. Finally, the carbonyl group is regenerated by unmasking the dithiane (Scheme 1.3). Thus, this type of inversion of the normal polarization of a functional group atom is known as umpolung.

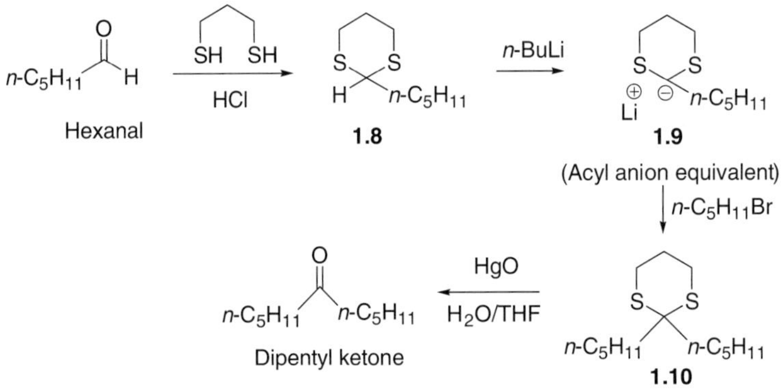

Scheme 1.3 Conversion of hexanal into dipentyl ketone (corey-seebach reaction)

In Scheme 1.3, hexanal on reaction with 1,3-propanedithiol gives the 1,3-dithiane derivative **1.8**. A strong base such as *n*-butyllithium abstracts the proton to give the corresponding 2-lithio-1,3-dithiane **1.9**, which reacts with 1-bromopentane to give alkylated product **1.10**. Treatment of **1.10** with HgO and BF_3 (boron trifluoride) in aqueous THF (tetrahydrofuran) yields the dipentyl ketone (the corey-seebach reaction[6]). Thus, dithianyllithium (2-lithio-1,3-dithiane) **1.9** is an 'acyl anion' synthetic equivalent.

The dithiane anion **1.9** also reacts with acyl halides, ketones and aldehydes to give the corresponding dioxygenated compounds. Schemes 1.4 and 1.5 show the reaction of dithiane anions **1.11** and **1.12** with ketones. The most common example of umpolung reactivity of a carbonyl group is the benzoin condensation (Scheme 1.6).

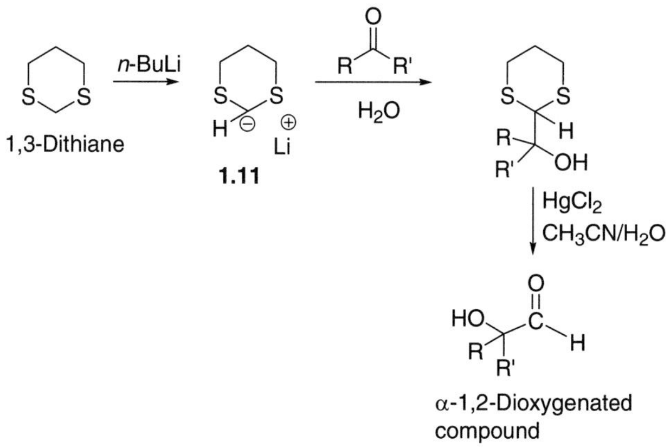

Scheme 1.4

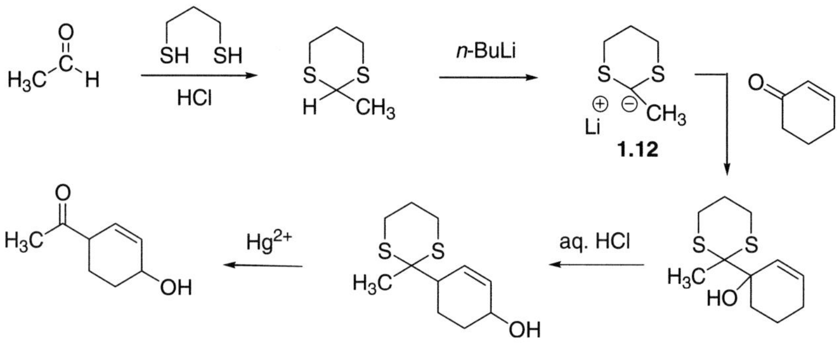

Scheme 1.5

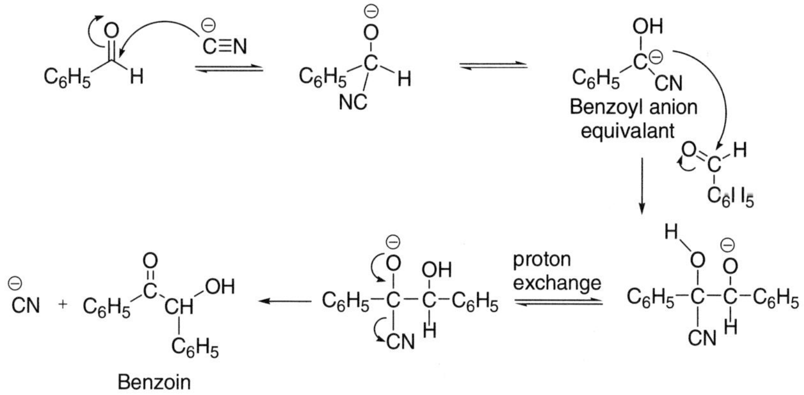

Scheme 1.6 Mechanism of benzoin condensation

A synthetic route for the synthesis of 2-deoxy-*C*-aryl glycosides using an umpolung strategy has been reported by Aidhen and co-worker[7] (Scheme 1.7). The synthetic endeavour led to a versatile intermediate aryl ketone **1.13**, which has paved the way for two important classes of *C*-glycosides, i.e. *C*-alkyl furanosides **1.14** and methyl 2-deoxy-*C*-aryl pyranosides **1.15**.

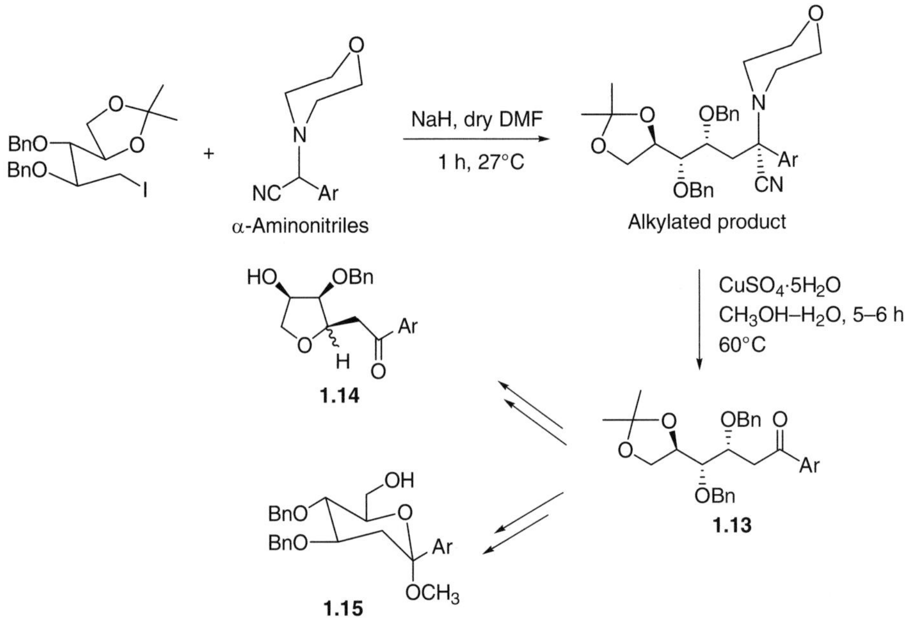

Scheme 1.7 Synthesis of C-aryl glycosides

1.4 Atom economy

The concept of atom economy was developed by B. M. Trost[8,9] which deals with chemical reactions that do not waste atoms. Atom economy describes the conversion efficiency of a chemical process in terms of all atoms involved. It is widely used to focus on the need to improve the efficiency of chemical reactions.

A logical extension[10] of B. M. Trost's concept of atom economy is to calculate the **percentage atom economy**. This can be done by taking the ratio of the mass of the utilized atoms to the total mass of the atoms of all the reactants and multiplying by 100.

$$\text{Percentage atom economy} = \frac{\text{Mass of atoms in the final product}}{\text{Mass of atoms in reactants}} \times 100$$

R. A. Sheldon[11] has developed a similar concept called **percentage atom utilization**. For instance, the percentage atom economy and percentage atom utilization calculation for the oxidation reaction of benzene to maleic anhydride is given below:

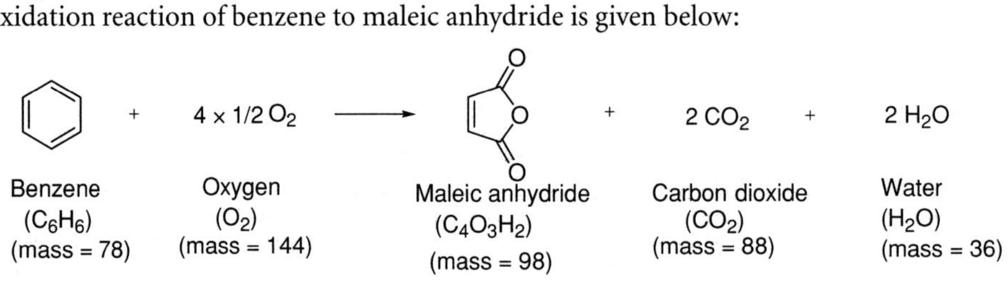

| Benzene (C_6H_6) (mass = 78) | Oxygen (O_2) (mass = 144) | Maleic anhydride $(C_4O_3H_2)$ (mass = 98) | Carbon dioxide (CO_2) (mass = 88) | Water (H_2O) (mass = 36) |

$$\text{Percentage atom utilization} = \frac{\text{MFW of maleic anhydride}}{\text{MFW (maleic anhydride} + 2 \text{ carbon dioxide} + 2 \text{ water)}} \times 100$$

Even if the reaction were to proceed with 100% yield, only 44.14% (by weight) of the atoms of the reactants are incorporated into the desired product, with 55.86% of the reactant atoms ending up as unwanted by-products.

It is often difficult to know the structures of all the by-products; therefore, the percentage atom economy may be determined by dividing the molecular formula weight (MFW) of the desired product by the sum of the MFWs of all the reactants and multiplying by 100.

$$\text{Percentage atom economy} = \frac{98}{78 + 144} \times 100 = 44.14$$

The percentage atom economy of this reaction is 44.14. This means that 44.14% of the mass of the reactants ends up in the desired product.

Recent developments including the advent of green chemistry and high raw material (oil) prices increasingly demand high atom economy. In a chemical process involving simple additions, with anything else needed only catalytically, the amount of starting materials or reactants equals the amount of all products generated and no atom is wasted. The Diels–Alder reaction is an example of a potentially atom efficient reaction. Since so few of the existing used reactions are additions, synthesis of complex molecules requires the development of new atom economic methodology. Atom economy can be improved on by careful selection of starting materials and catalyst system.

A classic example of improving the route to a commercial product is **ibuprofen (1.16)**, which is an analgesic (a pain reliever) and is also effective as a **non-steroidal anti-inflammatory drug**. Ibuprofen was produced using six steps (Scheme 1.8) by the Boots Company, with an overall atom economy of just 40%.

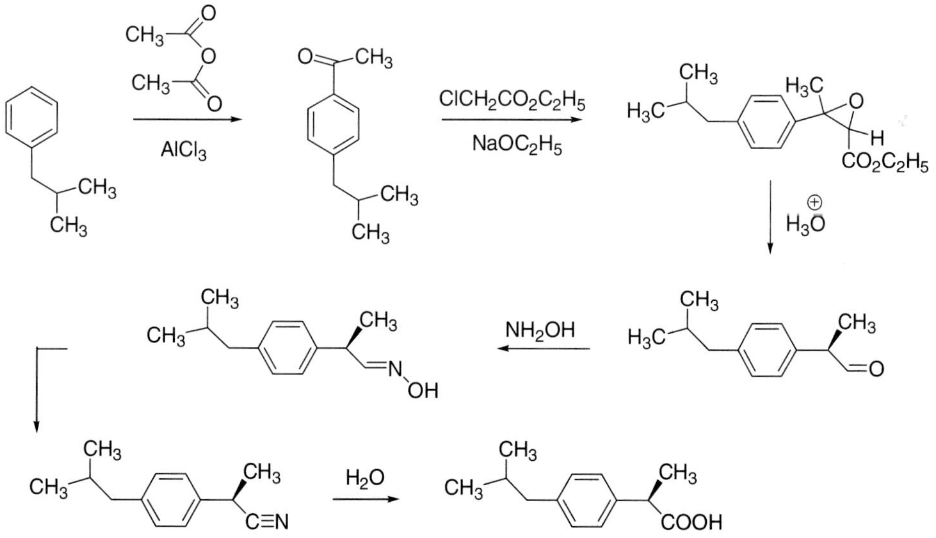

Scheme 1.8   Synthesis of ibuprofen by the Boots Company

The total MFW of all the reactants used is 514.5 ($C_{20}H_{42}NO_{10}ClN_9$) and the total MFW of atoms utilized is 206 (ibuprofen; $C_{13}H_{18}O_2$).

$$\text{Percentage atom economy} = \frac{\text{MFW of atoms utilized}}{\text{MFW of all reactants}} \times 100 = \frac{206}{514.5} \times 100 = 40$$

In the 1990s the Hoechst Celanese Corporation (together with the Boots company they formed the BHC process to prepare and market ibuprofen, **1.16**) developed a new three-stage process (Scheme 1.9), with an atom economy of 77.4%.

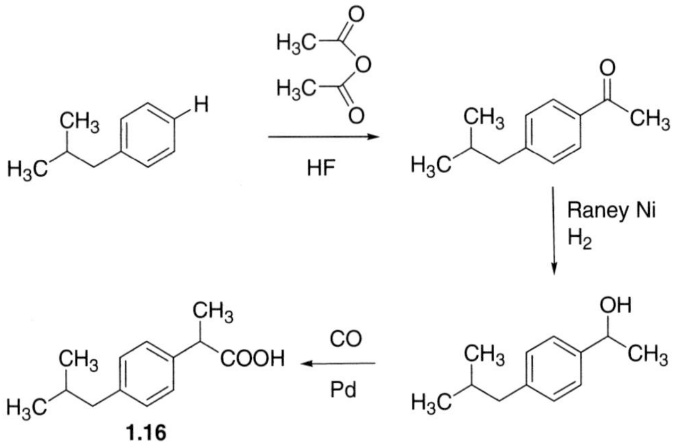

Scheme 1.9 Synthesis of ibuprofen by the BHC process

The total MFW of all the reactants used is 266 (all the reagents; $C_{15}H_{22}O_4$) and the total MFW of atoms utilized is 206 (ibuprofen; $C_{13}H_{18}O_2$).

$$\text{Percentage atom economy} = \frac{\text{MFW of atoms utilized}}{\text{MFW of all reactants}} \times 100 = \frac{206}{266} \times 100 = 77.4$$

In addition to higher atom economy, in the BHC process, HF is used in catalytic amounts and is recovered and reused. However, in the first step in the Boots process, the AlCl$_3$ (aluminium chloride) hydrate is produced in large amounts as a by-product because AlCl$_3$ is used in stoichiometric amounts. Thus, there is a significant improvement in the BHC process over the Boots process.

1.5   Selectivity

B. M. Trost has enunciated a set of criteria by which chemical processes can be evaluated. They fall under two categories: selectivity and atom economy. The atom economy has already been discussed in section 1.4. The issues of selectivity can be categorized under the following headings: chemoselectivity, regioselectivity, diastereoselectivity and enantioselectivity.

1.5.1 Chemoselectivity

Chemoselectivity is the differentiation among various functional groups in a polyfunctional molecule by preferential reactivity of one functional group over another. For example, chemoselective reduction of the aldehyde group in **1.17** occurs with $NaBH_4$ (sodium borohydride) in methanol at low temperature to give **1.18**. However, in the presence of $CeCl_3$ (ceric chloride), the keto group is reduced with $NaBH_4$ to give **1.19**.

Selective monoprotection[12] of 1,4-butanediol (**1.20**) with TBDPSCl (*tert*-butyldiphenylsilyl chloride) (see Table 1.2) is another example of chemoselectivity. Monoprotected alcohol **1.21** on oxidation with PDC (pyridinium dichromate) in DMF (dimethylformamide) afforded the corresponding carboxylic acid derivative **1.22** in 75% yield.

The chemoselective 1,2-reduction of αβ-unsaturated carbonyl compounds has been carried out with metal hydride or by hydrogenation. However, chemoselective 1,4-reduction of αβ-unsaturated carbonyl compounds is challenging. Recently, αβ-unsaturated carbonyl compounds **1.23**, **1.25** and **1.27** were selectively reduced to the corresponding saturated carbonyl compounds **1.24**, **1.26** and **1.28**, respectively, by cobalt octacarbonyl and water $[Co_2(CO)_8$–H_2O system][13].

Table 1.2 Various trialkylsilyl chlorides (R′$_3$SiCl) used for the protection of R–OH as ROSiR′$_3$

R′$_3$SiCl Trialkylsilylchloride	ROSiR′$_3$ Trialkylsilylether
(CH$_3$)$_3$SiCl Trimethysilyl chloride (TMSCl)	$\begin{array}{c} CH_3 \\ \| \\ R-O-Si-CH_3 \\ \| \\ CH_3 \end{array}$ (TMSOR)
(C$_2$H$_5$)$_3$SiCl Triethylsilyl chloride (TESCl)	$\begin{array}{c} C_2H_5 \\ \| \\ R-O-Si-C_2H_5 \\ \| \\ C_2H_5 \end{array}$ (TESOR)
(CH$_3$)$_3$CSi(CH$_3$)$_2$Cl t-Butyldimethylsilyl chloride (TBSCl)	$\begin{array}{c} CH_3 \\ \| \\ R-O-Si-C(CH_3)_3 \\ \| \\ CH_3 \end{array}$ (TBSOR)
(CH$_3$)$_3$CSi(C$_6$H$_5$)$_2$Cl t-Butyldiphenylsilyl chloride (TBDPSCl)	$\begin{array}{c} C_6H_5 \\ \| \\ R-O-Si-C(CH_3)_3 \\ \| \\ C_6H_5 \end{array}$ (TBDPSOR)
[(CH$_3$)$_2$CH]$_3$SiCl Triisopropylsilyl chloride (TIPSCl)	$\begin{array}{c} CH(CH_3)_2 \\ \| \\ R-O-Si-CH(CH_3)_2 \\ \| \\ CH(CH_3)_2 \end{array}$ (TIPSOR)
[(CH$_3$)$_3$C]$_3$ SiCl Tri-t-butylsilylchloride (TTBSCl)	$\begin{array}{c} C(CH_3)_3 \\ \| \\ R-O-Si-C(CH_3)_3 \\ \| \\ C(CH_3)_3 \end{array}$ (TTBSOR)

1.5.2 Regioselectivity

Regioselectivity (orientational control) is the formation of one constitutional isomer as the major product in which two or more constitutional isomers could be obtained. For example, addition of HBr (hydrogen bromide) to 1-methylcyclohexene (**1.29**) gives 1-bromo-1-methylcyclohexane (**1.30**) as the main product and 1-bromo-2-methylcyclohexane (**1.31**) is formed as the minor product.

1.29   + HBr   ⟶   **1.30**   +   **1.31**
                          (Major product)   (Minor product)

The LiAlH$_4$ (lithium aluminium hydride) attacks the epoxides on the sterically less hindered C–O bond to give the corresponding alcohol.

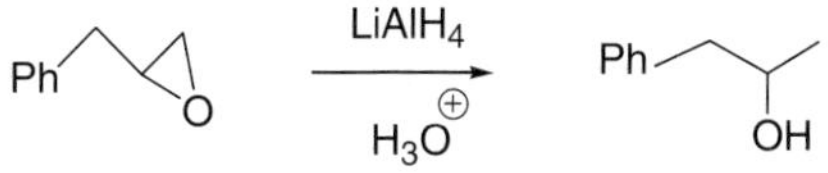

Regioselectivity is also observed in Diels–Alder reactions (also see Chapter 8, section 8.3.1).

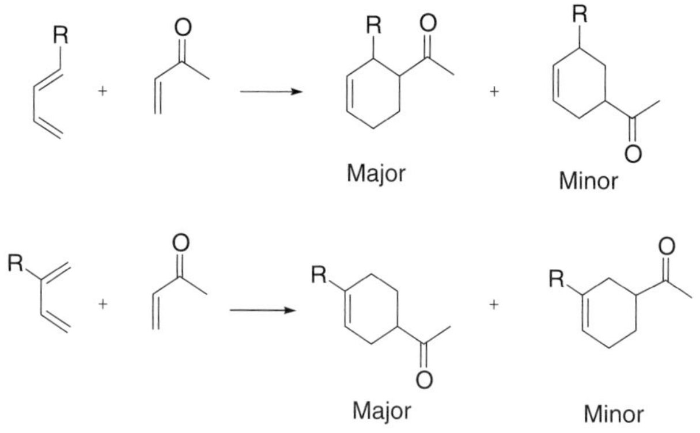

1.5.3 Stereoselectivity

Stereoisomerism is the arrangement of atoms in molecules whose connectivity remains the same but their arrangement in space is different in each isomer (atoms are connected in the same sequence). The two main types of stereoisomerism are ***cis–trans*** or ***Z–E* isomerism** and **optical isomerism**.

cis–trans **or *Z–E* isomerism:** The *cis*- and *trans*-1,2-dibromoethenes isomers cannot be easily interconverted because of restriction of carbon–carbon double bond. Both have the same molecular formula $C_2H_2Br_2$, but the arrangement of their atoms in space is different. The *cis*- and *trans*-1,2-dibromoethenes are not mirror images of each other; thus, they are not enantiomers and are therefore diastereomers.

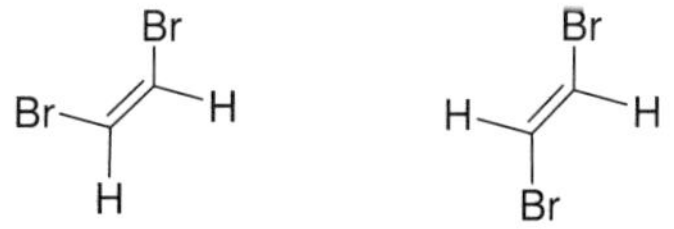

cis-1,2-Dibromoethene   *trans*-1,2-Dibromoethene

If the alkene is trisubstituted or tetrasubstituted, the terms *cis* and *trans* are normally not applied but the *E–Z* system of nomenclature is applied to alkene diastereomers. Generally, **Cahn-Ingold-Prelog (CIP) rules** are used to assign priorities to each end of the double bond. If $R^1 > R^2$ and $R^3 > R^4$ then alkene with R^1 and R^3 on the same side is designated *Z* from the German word *Zusammen*, meaning together. If R^1 and R^3 are on the opposite side then alkene is designated *E* from the German word *Entgegen*, meaning opposite. For an example, (*Z*)- and (*E*)-3-chloro-2-pentenes are shown below:

$$Z \qquad R^1 > R^2$$
$$R^3 > R^4$$

$$Z \qquad CH_3 > H$$
$$Cl > C_2H_5$$

$$E \qquad\qquad\qquad E$$

Optical isomers: Optical isomers are stereoisomers that can be formed around asymmetrical carbon(s) also known as **chiral** carbons. A **stereocentre** or chiral centre in organic chemistry generally refers to a carbon atom in a chemical compound that is an **asymmetric carbon atom** or a **chiral** carbon. A compound is chiral if it is non-superimposable on its mirror image. **Enantiomers** are two optical isomers that are reflections of each other. They have the same physical properties, except for their ability to rotate plane-polarized light, which they do in equal magnitudes but in opposite direction. A mixture of equal amounts of both enantiomers is said to be a **racemic mixture**. A racemic mixture does not rotate plane-polarized light. The assignment of each stereocentre as either *R* or *S* follows from the **CIP sequence rules**. Details of CIP sequence rules can be found in any undergraduate textbook on organic chemistry. **Diastereoisomers** are two optical isomers that are neither reflections of each other nor superimposable. Diastereomers can have different physical properties and different reactivities. A compound can have 2^n stereoisomers, where *n* is the number of stereocentres. Tartaric acid contains two asymmetric centres, but two of the configurations are equivalent and together are called ***meso* compounds**. A *meso* compound is optically inactive (or achiral) because it contains an internal plane of symmetry. The remaining two configurations are (+)- and (−)- mirror images, thus enantiomers. The *meso* form is a diastereomer of the other forms.

L(+)-Tartaric acid
(laevo-tartaric acid)

D-(−)-Tartaric acid
(dextro-tartaric acid)

meso-Tartaric acid

In a **stereoselective reaction**, one stereoisomer is formed in a major amount than another. When the stereoisomers are enantiomers the selectivity is known as **enantioselectivity**. The degree of enantiomeric purity of a solution is measured by its **enantiomeric excess, or ee.** The percentage enantiomeric excess is found by dividing the observed optical rotation by the optical rotation of pure enantiomer in excess and multiplying by 100.

$$\text{Optical purity} = \text{Percentage enantiomeric excess} = \frac{[\alpha]_{\text{mixture}}}{[\alpha]_{\text{pure sample}}} \times 100$$

For example, the observed specific rotation of the racemic mixture is +8.52 degrees of rotation. The specific rotation of the pure *S*-enantiomer is −15.00 degrees of rotation. Since the pure *S*-enantiomer has −15.00° and the specific rotation of mixture is +8.52°, then the *R*-configured isomer with specific rotation +15.00° is in excess. The percentage enantiomeric excess for the *R*-isomer is given as:

$$\text{Percenatge enantiomeric excess for } R\text{-isomer} = \frac{+8.52}{+15.00} \times 100 = 56.8$$

The 0% ee means 50:50 racemic mixture, while 50% ee means 75:25 mixture. Thus, **enantiomeric excess** or **ee** is a measure for how much of one enantiomer is present compared to the other. For example, in a sample with 40% ee in *R*, the remaining 60% is racemic with 30% of *R* and 30% of *S*, and so the total amount of *R* is 70%. Thus, the percentage enantiomeric excess is also written as:

$$\text{Percentage enantiomeric excess} = \frac{\text{Major enantiomer} - \text{Minor enantiomer}}{\text{Major enantiomer} + \text{Minor enantiomer}} \times 100$$

Recently, the enantiomeric excess of α-amino acid ester hydrochlorides has been determined directly by using FAB (fast atom bombardment) mass spectrometry without chromatographic separation of the enantiomers[14].

When each stereoisomeric reactant forms a different stereoisomeric product the reaction is known as **stereospecific reaction**. For example, the addition of :CBr$_2$ (dibromocarbene, prepared from bromoform and base) to *cis*-2-butene gives *cis*-2,3-dimethyl-1,1-dibromocyclopropane (**1.32**), whereas addition of :CBr$_2$ to the *trans*-isomer exclusively yields the *trans*-cyclopropane **1.33**.

Bromination of alkenes is also a stereospecific reaction.

Therefore, all the stereospecific reactions are also stereoselective reactions. However, all the stereoselective reactions are not necessarily stereospecific.

When a molecule which already contains at least one stereocentre undergoes reaction in which a new stereocentre is created, there is a possibility of formation of two (or more) stereoisomeric products. For example, reduction of **1.34** leads to **diastereoisomeric** products **1.35** and **1.36** with diastereoselectivity (de) 83% (*si* face addition) (**1.35**). For more details about *re* or *si* face additions, see Chapter 6, section 6.4.2.

de 83% *si*-face addition)
91.5: 8.5% (**1.35:1.36**)

1.5.4   Asymmetric synthesis or chiral synthesis

A chiral substance is **enantiopure** or **homochiral** when only one of two possible enantiomers is present. A chiral substance is **enantioenriched** or **heterochiral** when an excess of one enantiomer is present but not to the exclusion of the other. If the desired product is an enantiomer, the reaction needs to be sufficiently stereoselective even when atom economy is 100%. For the biological usage we almost need one enantiomer and in high purity. This is because when biologically active chiral compounds interact with its receptor site which is chiral, the two enantiomers of the chiral molecule interact differently and can lead to different chemistry. For example, one enantiomer of asparagines (**1.37**) is bitter while the other is sweet. As far as medicinal applications are concerned, a given enantiomer of a drug may be effective while the other is inactive or potentially harmful. For example, one enantiomer of ethanbutol (**1.38**) is used as antibiotic and the other causes blindness.

Bitter   Sweet

Asparagine (**1.37**)

Antibiotic   Causes blindness

Ethanbutol (**1.38**)

Despite its importance, the ability to obtain chiral molecules in enantiopure form is a difficult challenge. One strategy to make a pure enantiomer is to produce the racemic mixture and then separate both enantiomers and effectively throw away the undesired enantiomer. Separation of enantiomers is a very difficult endeavour, and destroying half the reaction product at every stereogenic step is not viable as yields in multi-step synthesis decrease exponentially.

   Chiral synthesis, also called **asymmetric synthesis**, is synthesis which preserves or introduces a desired chirality. Principally, there are three different methods to induce asymmetry in reactions. There can be either one or several stereogenic centres embedded in the substrate inducing chirality in the reaction (i.e. **substrate control**) or an external source providing the chiral induction (i.e. **reagent control**). In both cases the obtained stereoselectivity reflects the energy difference between the diastereomeric transition states.

   The obvious approach for chiral synthesis would be to find a chiral starting material, such as a natural amino acid, carbohydrates, carboxylic acids or terpene. The major source of these chiral starting materials sometimes called **chirons** is nature itself. The synthesis of a complex enantiopure chemical compound from a readily available enantiopure substance such as natural amino acids is known as **chiral pool synthesis**. For example, chiral lithium amides[15a] **1.39** that are used for several types of enantioselective asymmetric syntheses can be prepared in both enantiomeric forms starting from the corresponding optically active amino acids, and these are often available commercially.

1.39

X = Li or H

Y =

or

R' = CH$_2$Bu^t, CH$_2$CF$_3$

R = Ph, *t*-Bu

However, **chiral pool synthesis** is restricted by the number of possible starting enantiomeric pure compounds and requires a stoichiometric amount of the starting material, which may be scarce and expensive.

Chiral auxiliaries are optically active compounds which are used to direct asymmetric synthesis. The chiral auxiliary is temporarily incorporated into an organic synthesis which introduces chirality in otherwise racemic compounds. This temporary stereocentre then forces the asymmetric formation of a second stereocentre. The synthesis is thus **diastereoselective**, rather than enantioselective. After the creation of the second stereocentre the original auxiliary can be removed in a third step and recycled. E. J. Corey in 1975, B. M. Trost in 1980 and J. K. Whitesell in 1985 introduced the chiral auxiliaries 8-phenylmenthol[15b] (**1.40**), chiral mandelic acid[15c] (**1.41**) and *trans*-2-phenyl-1-cyclohexanol[15c] (**1.42**), respectively.

1.40   **1.41**   **1.42**

In order to maximize the diastereoselectivity observed for an auxiliary, it would appear reasonable that the stereocontrolling functional group is in a position in space as close as possible to the newly forming stereogenic centre. Chiral imide auxiliaries such as Evans' *N*-acyloxazolidinones (**1.43**) are used for asymmetric alkylation and asymmetric aldol condensation (Scheme 1.10).

Many structural variants of *N*-acyloxazolidinones have been reported and exhibit different cleavage reactivity or complimentary diastereoselectivity compared to *N*-acyloxazolidinone (**1.43**).

Several examples of chiral auxiliaries[16] that rely on relatively remote stereogenic centres to control diastereoselectivity are known. For example, alkylation of the enolates of **1.44** and **1.46** to **1.45** and **1.47** is controlled via 1,4- and 1,3-asymmetric induction, respectively.

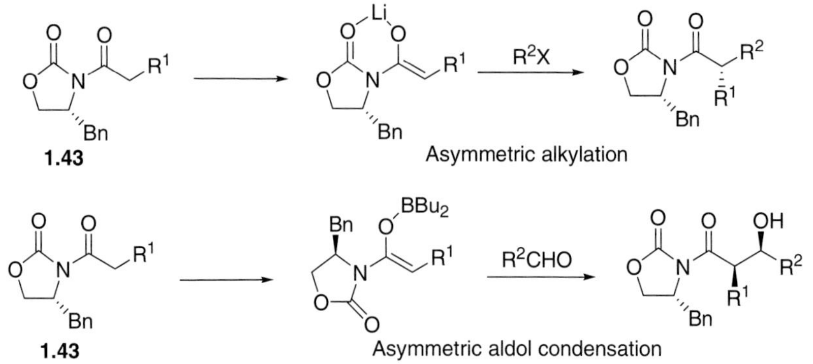

1.43   Asymmetric alkylation

1.43   Asymmetric aldol condensation

Scheme 1.10

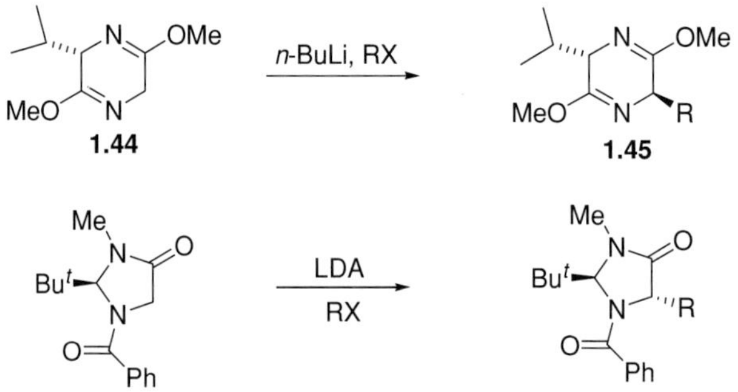

1.44   **1.45**

1.46   **1.47**

The steric and electronic factors combine to transfer or relay stereochemical information from the stereogenic centre to the site of reaction. Small changes in bond angles or heteroatom hybridization can result in large changes in diastereoselectivities. For example, changing the nitrogen protecting group in imidazolidinone-derived auxiliaries **1.48** results in significant improvements in the observed diastereoselectivities of **1.49** during enolate alkylation.

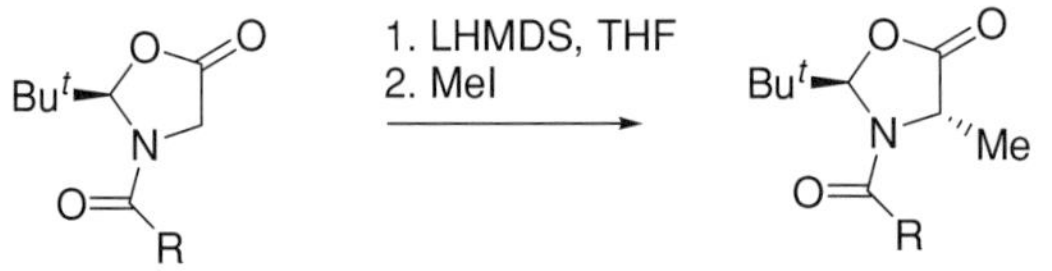

1.48   **1.49**

R	de
t-Bu	50%
Ph	88%
OPh	96%

The conformationally flexible group serves to both relay and amplify the stereochemical information of the existing stereogenic centre, thus enabling efficient control of diastereoselectivity (Scheme 1.11).

The chiral auxiliary method may be used for the synthesis of all-carbon quaternary stereocentres[17]. Thus, chiral bicyclic thioglycolate lactam **1.50** is alkylated three times and

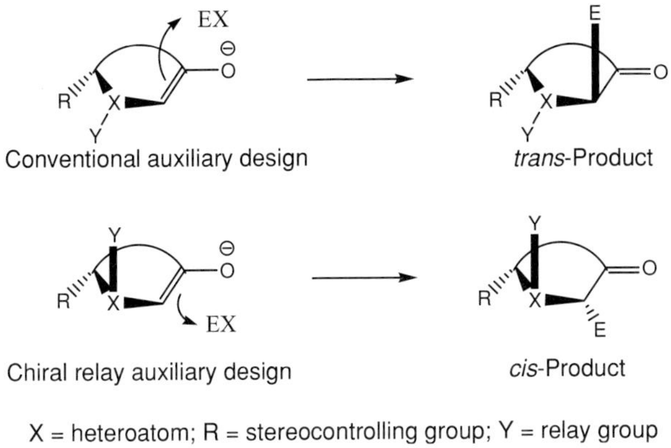
EX
O
R
X
Y
Conventional auxiliary design
E
O
trans-Product
Y
O
EX
Chiral relay auxiliary design
cis-Product
X = heteroatom; R = stereocontrolling group; Y = relay group

Scheme 1.11

the alkylated products **1.51** may be cleaved under either acidic or reductive conditions to furnish either carboxylic acids **1.52** or primary alcohols **1.53**, respectively (Scheme 1.12).

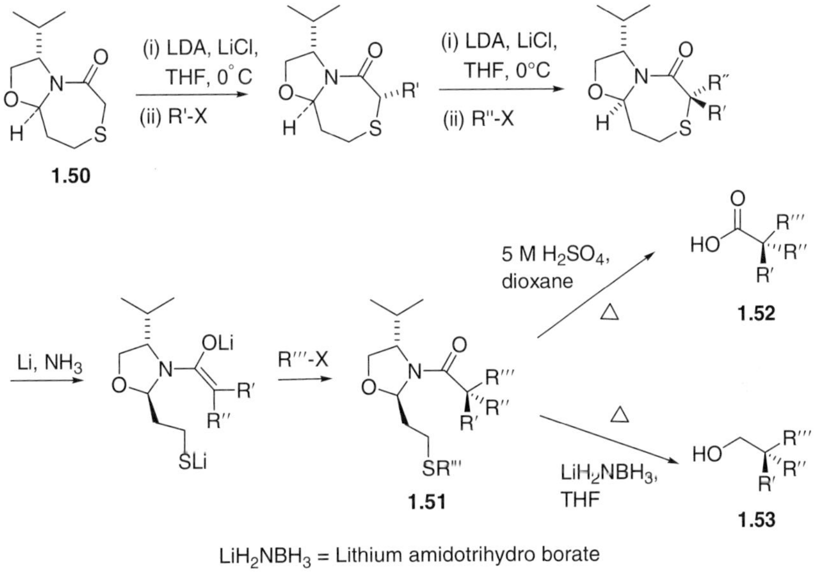
1.50
(i) LDA, LiCl, THF, 0°C
(ii) R'-X
(i) LDA, LiCl, THF, 0°C
(ii) R''-X
Li, NH3
OLi
SLi
R'''-X
SR'''
1.51
5 M H2SO4, dioxane
Δ
1.52
LiH2NBH3, THF
1.53
LiH2NBH3 = Lithium amidotrihydro borate

Scheme 1.12

The problem in the use of **chiral reagents** in chiral synthesis is that the chiral reagents are used up in the reaction. Thus, the most economic and convenient chiral synthesis is by the use of chiral catalysts where a small amount of chiral catalyst can produce large amounts of enantiomerically enriched product.

Asymmetric catalysis, the introduction of chirality into non-chiral reactants through usage of a chiral catalysts, is an important aspect of asymmetric synthesis. The most extensively studied asymmetric catalysis reaction is that of hydrogenation of alkenes. In addition to hydrogenation reactions, platinum group metal complexes can effectively be used for the asymmetric hydrosilations, allylic alkylations, isomerizations, hydroformylations and carbonylations. All members of the platinum group metals have been successfully used. The reaction of carbonate **1.54** with sodium dimethyl malonate with catalyst $[Mo(CO)_3C_7H_8]$ and chiral ligand **1.55** gave branched product **1.56** in 95% yield and 95% ee[18].

chiral ligand (*S*,*S*)
(1.55)

$Mo(CO)_3C_7H_8$,

$\oplus \ \ominus$
$NaCH(COOCH_3)_2$

1.54

1.56
95% yield, 95% ee

Chiral ligands once attached to the starting materials physically dictate trajectory for attack, leaving only the desired trajectory open. One chiral ligand that is widely used for introducing chirality in combination with ruthenium or rhodium is **BINAP** (2,2′-bis(diphenylphosphino)-1,1′-binaphthyl) (**1.57**). Both (*S*)-BINAP and (*R*)-BINAP are commercially available. It consists of two naphthyl groups linked by a single bond with diphenylphosphino groups at the end of each naphthyl group. Rotation about the single bond binding the two naphthyl groups is restricted because of the rigidity of their p system. Therefore, the angle made by the two p planes is fixed to approximately 90° and two separate enantiomers exist. BINAP is prepared from BINOL (1,1′-bis-2-naphthol) (see Chapter 6, section 6.1.1, Scheme 6.3).

(*S*)-(–)-BINAP

(*S*)-(–)-2,2′-Bis(diphenyl phosphino)-1,1′-binaphthyl

(*R*)-(+)-BINAP

(*R*)-(+)-2,2′-Bis(diphenyl phosphino)-1,1′-binaphthyl

1.57

The chiral structure of BINAP enables highly enantioselective reactions in **organic synthesis.** The Ru– and Rh–BINAP complexes catalyze the hydrogenation of functionalized alkenes and carbonyls on only one face of the molecule. For example, asymmetric hydrogenation of methyl 3-oxobutanoate (**1.58**) using (*R*)-BINAP–ruthenium complex yields (*R*)-(–)-methyl 3-hydroxybutanoate (**1.59**) in 99.5% ee[19,20].

1.58

H_2, cat.

CH_3OH

1.59
99.5% ee

Cat. = (*R*)-BINAP-RuCl$_2$

Similarly, catalytic hydrogenation of geraniol (**1.60**) in the presence of Ru–(R)-BINAP complex yields (S)-(−)-citronellol (**1.61**) in high yield[21,22].

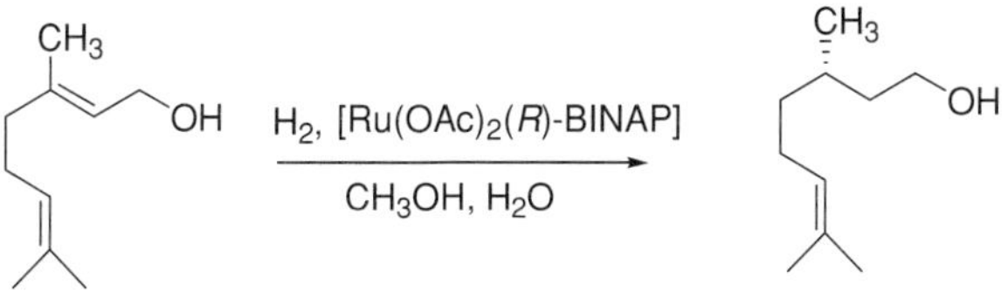

1.60   **1.61**

The use of chiral rhodium–BINAP complexes for the asymmetric isomerization of alkenes has been utilized in the industrial synthesis of menthol by Ryoji Noyori (winner of the 2001 Nobel Prize in Chemistry). This synthetic method was industrialized by Takasago International Corporation and provides (−)-menthol to pharmaceutical and food companies worldwide. In this case the catalyst [(S-BINAP)-Rh(COD)] or [(S-BINAP)$_2$-RuClO$_4^-$] is used for the asymmetric isomerization of diethylgeranylamine (**1.62**) to 3-(R)-citronellalenamine (**1.63**) (Scheme 1.13).

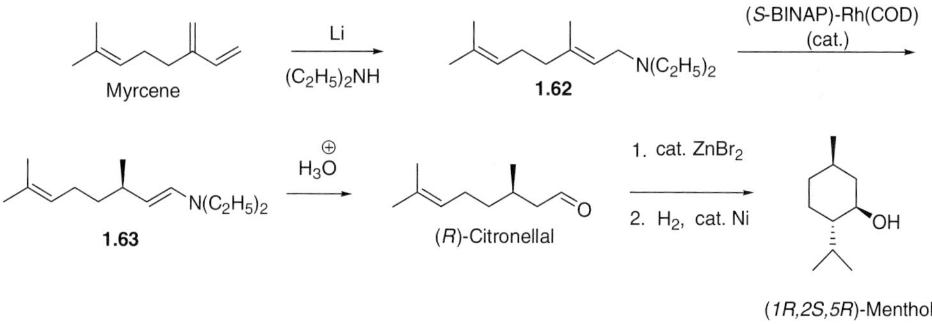

Scheme 1.13 Asymmetric synthesis of menthol by Takasago International Corporation

Several novel catalysts in which borohydride is complexed with a difunctional chiral ligand have been developed and used for the enantioselective reduction of prochiral ketones to chiral alcohols. **Corey–Bakshi–Shibata reduction**[23,24] (**CBS reduction**) is an organic reaction which reduces ketones enantioselectively into alcohols by using chiral oxazaborolidines and BH$_3$·THF or catecholborane as stoichiometric reductants (**CBS reagent, 1.64**) (also see Chapter 6, section 6.4.2).

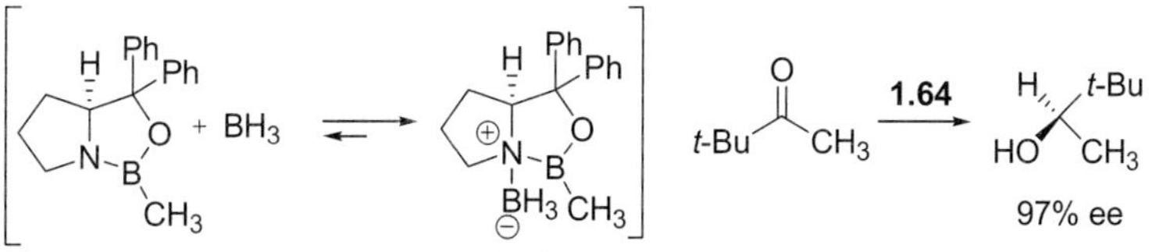

1.64

For example, (S)-2-methyl-CBS-oxazaborolidine binds reversibly with diborane to form the reactive reducing species **1.64**. Coordination of the ketone oxygen with the Lewis acidic boron orients and activates the carbonyl group for hydride transfer to its *si* face. The intramolecular hydride transfer mechanism serves as a model for achieving enantioselective reduction (see Scheme 6.20).

Naproxen, an anti-inflammatory drug, is synthesized by utilizing an asymmetric enantioselective hydrocyanation of vinylnaphthalene **1.65** utilizing a chiral ligand **1.66**. Since the S-enantiomer is medicinally desirable whereas the R-enantiomer produces harmful health effects, the enantioselectivity of this reaction is important. The synthesis of naproxen nitrile (**1.67**) shown below produces the S-(−)-enantiomer with 75% ee.

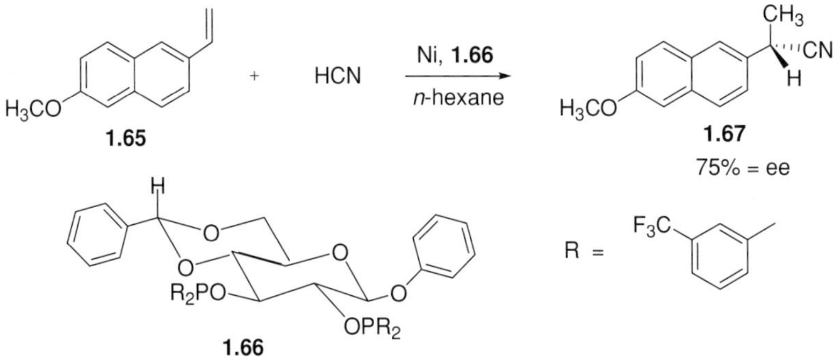

In 2001, K. B. Sharpless won the Nobel Prize in Chemistry for his work on **asymmetric aminohydroxylation**[25–27] and **asymmetric epoxidation**[28–30]. These stereoselective oxidation reactions are powerful catalytic asymmetric methods that have revolutionized synthetic organic chemistry.

Sharpless asymmetric epoxidation[28–30] is an enantioselective epoxidation of an allylic alcohol with *tert*-butyl hydroperoxide (*t*-BuOOH), titanium tetraisopropoxide [Ti(O-*i*Pr)$_4$] and (+)- or (−)-diethyl tartrate [(+)- or (−)-DET] to produce optically active epoxide from achiral allylic alcohol. The reaction is diastereoselective for α-substituted allylic alcohols. Formation of chiral epoxides is an important step in the synthesis of natural products because epoxides can be easily converted into diols and ethers.

For example, asymmetric epoxidation of geraniol (**1.60**) gives (2S,3S)-epoxygeranial (**1.68**) in 77% yield and 95% ee.

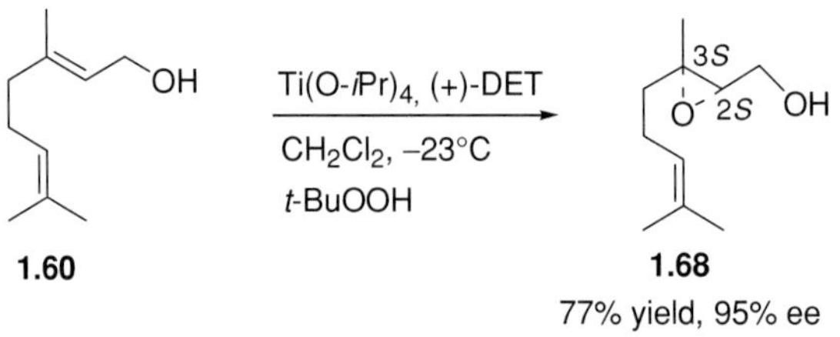

Epoxidation of alkenes with complex of a chiral **salen ligand** and manganese(III), **1.69** or **1.70**, is known as **Jacobsen epoxidation**[31,32].

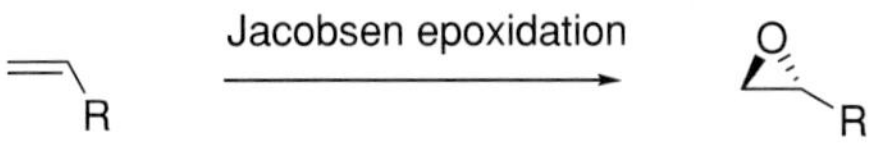

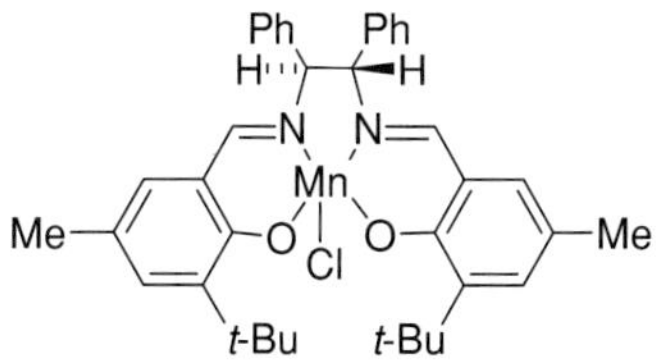

1.69

(Jacobsen catalysts)

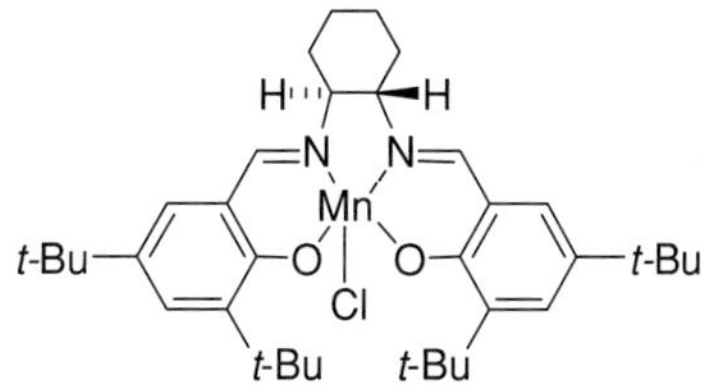

1.70

(More selective Jacobsen catalyst)

Mechanism of epoxidation:[33a] The oxygen transfer occurs by a two-step catalytic cycle (Scheme 1.14). In the first step oxygen is transferred to the Mn(III) by an oxidant. The oxygen coordinates to the metal. In the second step the activated oxygen is delivered to the alkene.

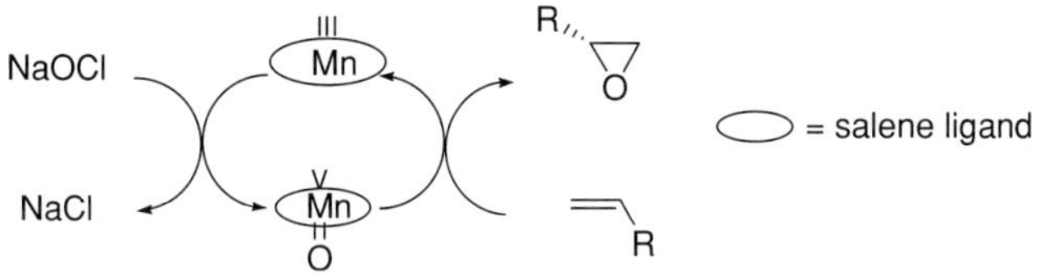

Scheme 1.14 The epoxidation of alkenes

The transfer of oxygen to alkene may occur by several different mechanisms (Scheme 1.15). Oxygen radicals may be formed as intermediates when radical stabilizing groups are attached. This mechanism is supported by the fact that *cis*-alkene gives both *cis*- and *trans*-epoxide. The formation of the metallaoxetane as an intermediate is also proposed by Norrby *et al.*[33b] However, concerted oxygen delivery has also been proposed.

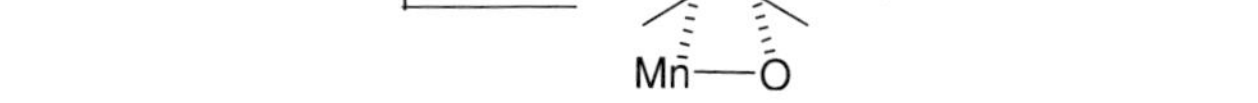

Scheme 1.15 Possible modes of oxygen transfer from Mn to double bond

The selectivity of oxygen transfer from oxo–manganese complex to the alkene depends on the relative orientation of the activated catalyst and the alkene. The alkene approaches the salen ligand in such a manner so that it avoids the raised butyl groups and keeps its substituents as far from the **salen ligand** as possible.

$R^1 = -(CH_2)_4-; R^2 = t\text{-Bu}$
$R^1 = Ph; R^2 = CH_3$

The **dihydroxylation of alkenes** with catalytic amount of osmium tetroxide in the presence of stoichiometric oxidizing reagent, like barium chlorate, *tert*-butyl hydroperoxide (*t*-BuOOH), N-methyl-N-oxo-morpholine (NMO), sodium peroxodisulphate ($Na_2S_2O_8$), iodine (I_2) or potassium ferricyanide [$K_3Fe(CN)_6$], is an important method for the production of diols. The reaction is stereospecific as *syn*-diols are obtained. However, the stereoselectivity varies depending on the structure of the alkene. **Sharpless** provided a method of asymmetric dihydroxylation[34] of alkene. When osmium tetroxide and an oxidizing agent like NMO or $K_3Fe(CN)_6$ are used in the presence of a chiral *cinchona alkaloid*, high enantiomeric excess (%ee) of the diol product is formed. It is the chiral cinchona alkaloid that provides the optically active component to the catalyst. The reaction is carried out in a buffered solution to ensure a stable pH. NMO or $K_3Fe(CN)_6$ is used for the regeneration of osmium(VIII).

These reagents are commercially available premixed: asymmetric dihydroxylation mix (AD-mix). **AD-mix** is available in two variations: **AD-mix α** is $(DHQ)_2PHAL + K_2OsO_2(OH)_4 + K_3Fe(CN)_6$ and **AD-mix β** is $(DHQD)_2PHAL + K_2OsO_2(OH)_4 + K_3Fe(CN)_6$. Ligand $(DHQ)_2PHAL$ is 1,4-bis(9-O-dihydroquinine)phthalazine (**1.71**) and ligand $(DHQD)_2PHAL$ is 1,4-bis(9-O-dihydroquinidine)phthalazine (**1.72**).

1.71

AD-mix α = $K_2OsO_2(OH)_4$ (cat.), K_2CO_3, $K_3Fe(CN)_6$, $(DHQ)_2PHAL$ (cat.)

1.72

AD-mix β = $K_2OsO_2(OH)_4$ (cat.), K_2CO_3, $K_3Fe(CN)_6$, $(DHQD)_2PHAL$ (cat.)

$K_2OsO_2(OH)_4$ = potassium osmate dihydrate

Sharpless AD reaction is extremely useful and efficient for the asymmetric dihydroxylation of alkenes.

The reaction mechanism of the Sharpless dihydyroxylation is given in Chapter 7, section 7.5.

Ruthenium catalysis allows dihydroxylation providing an easy access to *syn*-diols, but over-oxidation is a common side reaction. The improved protocol for the Ru-catalyzed *syn*-dihydroxylation uses only 0.5 mol% catalyst under acidic conditions that gave products in high yields with only minor formation of side products[35].

Sharpless and co-workers first reported the aminohydroxylation[25-27] of alkenes in 1975 and have subsequently extended the reaction into an efficient one-step catalytic **asymmetric aminohydroxylation**. This reaction uses an osmium catalyst [$K_2OsO_2(OH)_4$], chloramine salt (such as chloramine T; see Chapter 7, section 7.6) as the oxidant and cinchona alkaloid **1.71** or **1.72** as the chiral ligand. For example, asymmetric aminohydroxylation of styrene (**1.73**) could produce two regioisomeric amino alcohols **1.74** and **1.75**. Using Sharpless asymmetric aminohydroxylation, (1R)-N-ethoxycarbonyl-1-phenyl-2-hydroxyethylamine (**1.74**) was obtained by O'Brien *et al.*[36] as the major product and with high enantiomeric excess than its regioisomeric counterpart (R)-N-ethoxycarbonyl-2-phenyl-2-hydroxyethylamine (**1.75**). The corresponding free amino alcohols were obtained by deprotection of ethyl carbamate (urethane) derivatives.

Synthesis of the taxol side chain[37] involves the asymmetric aminohydroxylation reaction (Scheme 1.16).

The ability to select reaction conditions to provide an enantiomerically pure compound is important in industry. Methods to increase yields of specific enantiomers at low cost are in high demand, and this has resulted in research to develop new chiral building blocks, chiral ligands and catalysts. An additional problem encountered in industry is that catalysts are often expensive and difficult to recover, and research on catalyst recovery is important.

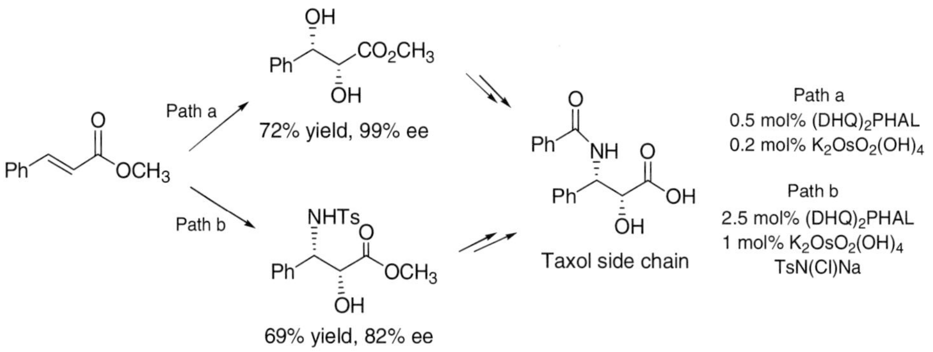

Scheme 1.16

1.6 Protecting groups

When a chemical reaction is to be carried out selectively at one reaction site in a multi-functional organic compound (organic molecule contains two or more than two reactive groups) and we want reaction at one reactive site, then other reactive sites must be temporarily blocked or protected. Many protecting groups have been, and are being, developed for this purpose. Whenever a protecting group is used to facilitate a synthetic operation, it normally must be removed once the operation is complete. This step is called **deprotection**. Protection and deprotection of functional groups have received attention in recent years not only because of their fundamental importance, but also for their role in multi-step synthesis.

The preparation of complex organic molecules demands the availability of different protecting groups to allow the survival of reactive functional groups during the various synthetic operations, finally resulting in the selective production of the target molecule. For example, in the conversion of ethyl 5-oxo-hexanoate (**1.76**) into 6-hydroxy-2-hexanone (**1.77**), it is required to block the ketone group first and then the ester group is reduced with LiAlH$_4$. The ketone group is protected as an acetal because an acetal group does not react with the reducing agent LiAlH$_4$. In the final step the acetal group is removed by treatment with acid. The overall scheme for this transformation is given in Scheme 1.17.

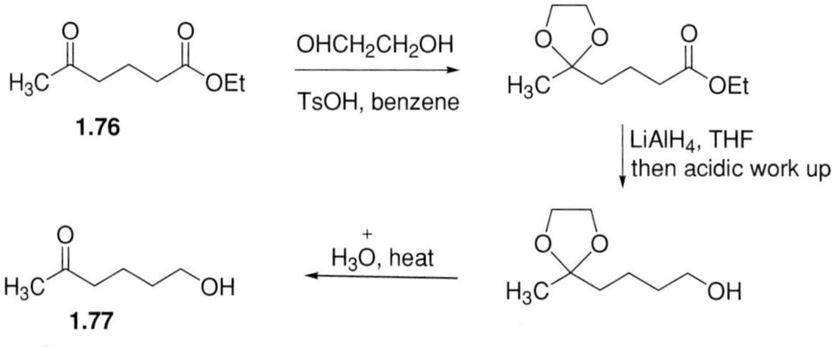

Scheme 1.17

A protecting group must fulfil a number of requirements. A good protecting group should be easy to put on, without the generation of new stereogenic centres, and easy to remove. The protecting group should have a minimum of additional functionality to avoid further sites of reaction. The protecting group should form a crystalline derivative with high reaction yields that can easily be separated from side products. The protecting group should not interfere with the reactions performed before it is removed.

Protecting groups can be cleaved under various conditions including basic solvolysis, acids, heavy metals, fluoride ions, reductive elimination, β-elimination, hydrogenolysis, oxidation, dissolving metal reduction, nucleophilic substitution, transition metal catalysis, light and enzymes. The assisted electrolytic and photolytic methods are the important methods for the removal of protecting groups. Photolabile groups are called **caged compounds** or phototriggers, deprotected on irradiation at wavelengths 254–350 nm with high quantum yield.

The protecting group must remain attached throughout the synthesis and may be removed after completion of synthesis. However, these protecting groups are not incorporated into the final product; thus, their use makes a reaction less atom economical. In other words, the use of protecting group should be avoided whenever possible.

A wide range of protecting groups is currently available for the different functional groups. A very short overview of the most commonly used protecting groups is given in this chapter. They are classified according to the functional group they protect. Conditions for their lability are also discussed.

1.6.1 Common hydroxy protecting groups

Hydroxyl group must be protected during **oxidation, acylation, halogenation, dehydration** and other reactions to which it is susceptible. Hydroxyl groups are protected by forming their alkyl ethers, alkoxyalkyl ethers, silyl ethers and esters. However, ethers are preferred over esters because of their stability in acetic acid and basic conditions.

Alkyl and alkoxyalkyl ethers

Alkyl ethers are generally prepared by acid-catalyzed addition of alcohols to an alkene or by **Williamson ether synthesis** (Scheme 1.18).

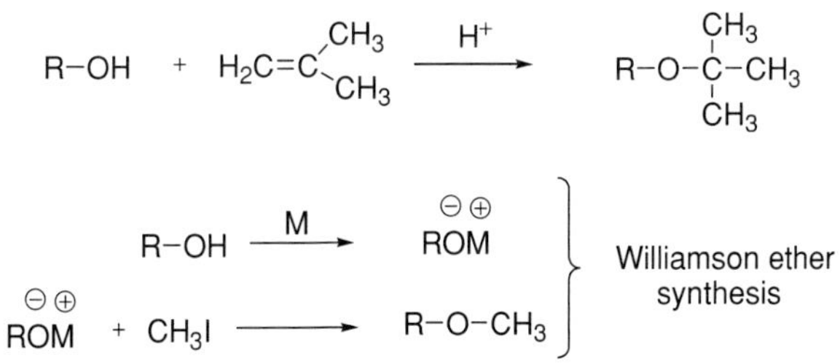

Scheme 1.18

Tetrahydropyranyl (THP) ethers **1.79** can be prepared from 3,4-dihydro-2H-pyran (DHP) **1.78** by an acid-catalyzed reaction[38].

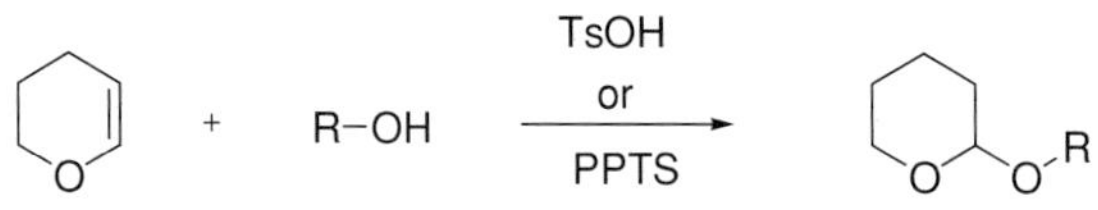

1.78   **1.79**

Tetrahydropyranyl ethers are stable to bases and the protection is removed by acid-catalyzed hydrolysis. For example, geraniol (**1.60**) is protected as geraniol tetrahydropyranyl ether (**1.80**) in the presence of pyridinium *p*-toluenesulfonate (PPTS) reagent. These ethers are cleaved with PPTS in warm ethanol[39] (Scheme 1.19).

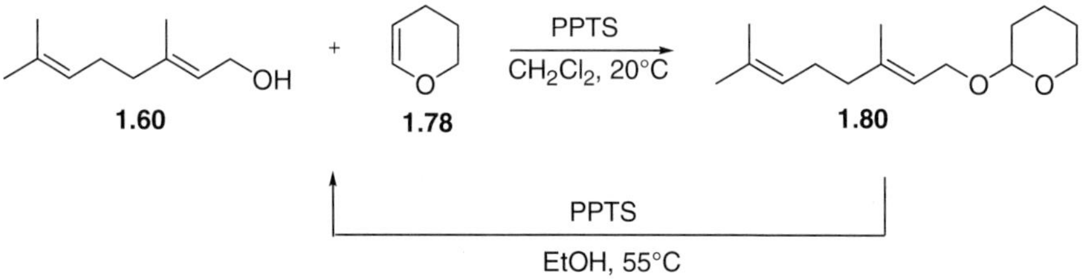

Scheme 1.19

However, formation of THP ether introduces a new stereogenic centre. The introduction of the THP ether onto a chiral molecule therefore results in the formation of diastereoisomers.

Phenols are protected as their methyl ethers[40,41], *tert*-butyl ethers, allyl ethers and benzyl ethers.

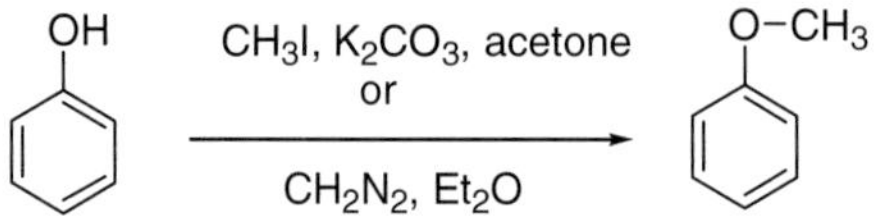

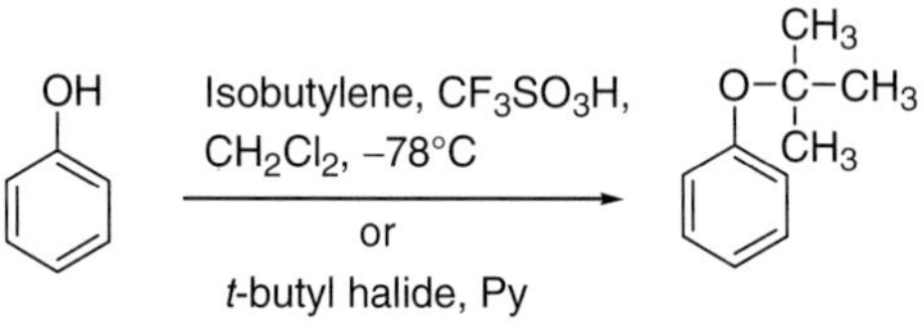

Miura and co-workers[42] reported the protection of phenols by allyl alcohols in the presence of catalytic amounts of palladium(II) acetate and titanium(IV) isopropoxide. The reaction is remarkably general; however, it fails in the case of 3,5-dimethoxyphenol because of the exclusive formation of a C-allylated product.

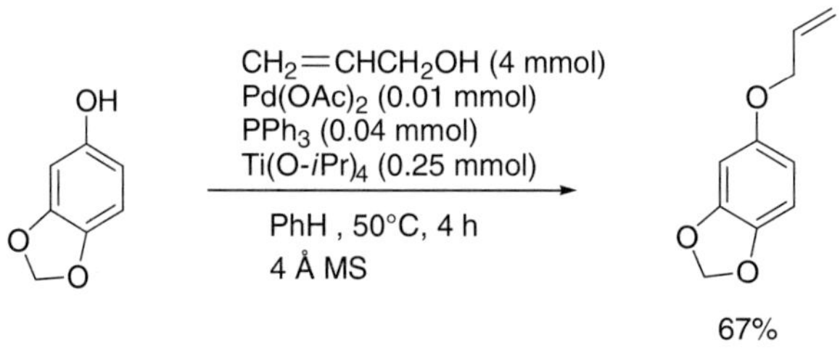

67%

Ethers may be removed commonly by acid, with the THP derivative **1.79** reacting more rapidly than the *tert*-butyl ether. Benzyl ethers may be removed under a variety of conditions such as hydrogenolysis, dissolving metal reduction (Na in NH$_3$) and HBr (mild). Methyl ethers are cleaved[43] by refluxing with EtSNa and DMF. *tert*-Butyl ethers can be cleaved with trifluroacetic acid (CF$_3$COOH) at 25°C.

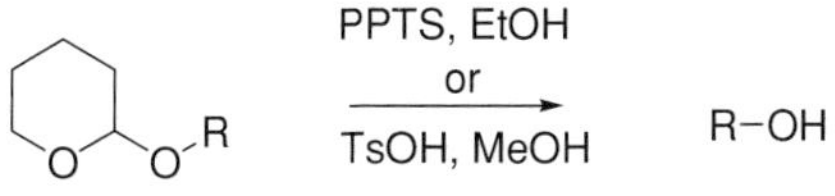

1.79

The nucleophilic cleavage of aryl alkyl ethers gives the corresponding phenol with only 1 equiv. of thiophenol in the presence of N-methyl-2-pyrrolidinone (NMP) in a catalytic amount of potassium carbonate[44]. The aromatic nitro and chloro substituents which are displaced with stoichiometric thiolates are preserved by this method. Moreover, αβ-unsaturated carbonyl compounds do not undergo Michael addition of thiolate under these conditions.

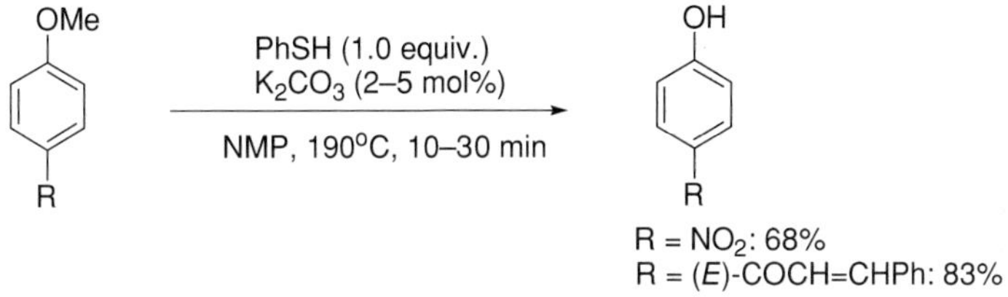

R = NO$_2$: 68%
R = (*E*)-COCH=CHPh: 83%

The tetrahydropyranylation of alcohols under solvent-free conditions is efficiently catalyzed by bismuth triflate (0.1 mol%). The experimental procedure is simple and works well with a variety of alcohols and phenols. The catalyst is insensitive to air and small amounts of moisture, easy to handle and relatively non-toxic. The deprotection of THP ethers is also catalyzed by bismuth triflate[45] (1.0 mol%).

The benzyl (Bn) or *p*-methoxybenzyl groups (PMB) can be removed under reducing cleavage conditions (Scheme 1.20).

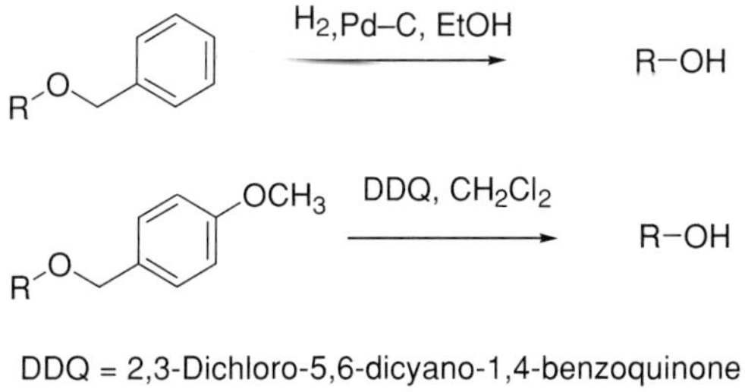

DDQ = 2,3-Dichloro-5,6-dicyano-1,4-benzoquinone

Scheme 1.20

Alkali metals (such as Li) in liquid ammonia are commonly applied for the deprotection of benzyl (Bn) ethers[46]. Lithium naphthalenide prepared from lithium and naphthalene in stoichiometric amount[47] or catalytic amount[48] is often used to deprotect benzyl ethers.

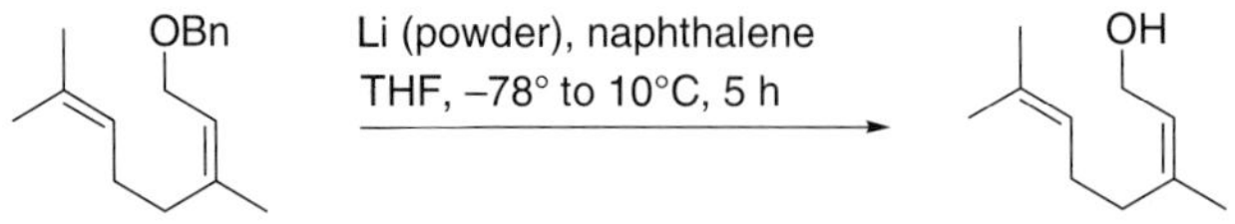

1.60

Hwu and co-workers[49] reported selective cleavage of a benzyl (Bn) ether with lithium di-isopropylamide (LDA) in the presence of a methoxy group; however, by using sodium bis(trimethylsilyl)amide [NaN(SiMe₃)₂], the dimethoxybenzene undergoes selectively mono-O-demethylation (Scheme 1.21).

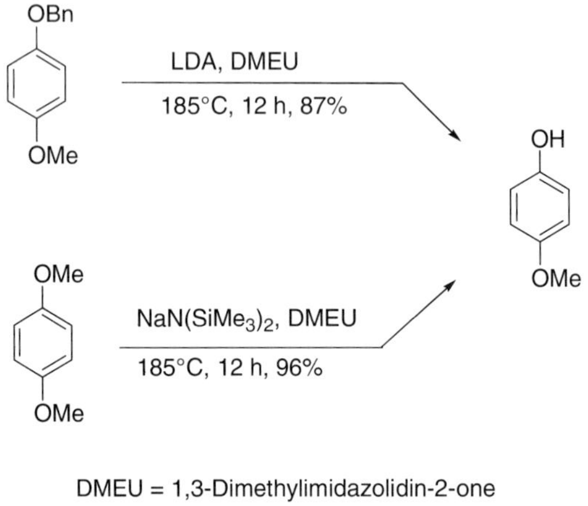

DMEU = 1,3-Dimethylimidazolidin-2-one

Scheme 1.21

Hirota and co-workers[50] reported the selective removal of the benzyl (Bn) group with Pd–C-catalyzed hydrogenolysis of PMB protected phenols. The removal of the PMB group is inhibited by the presence of pyridine.

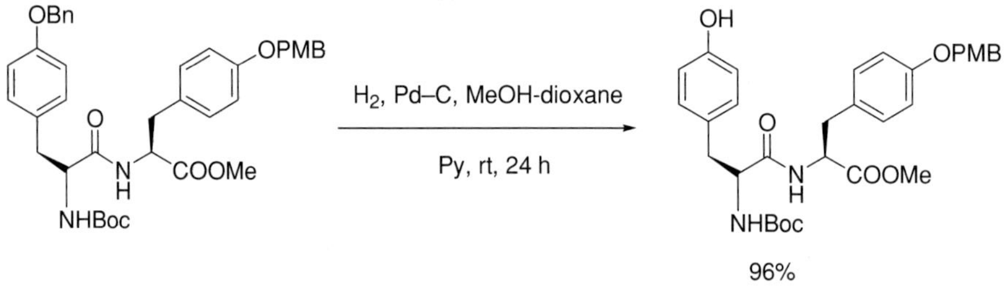

96%

The classical procedure for the removal of the allyl group involves a two-step sequence, where the allyl group is first isomerized to the corresponding propenyl function with a strong base such as potassium *tert*-butoxide (*t*-BuOK) or a metal catalyst such as Pd–C, followed by conversion of the propenyl group into free alcohol. However, recently several other methods were reported in the literature for the removal of the allyl group using various reagents such as DDQ, CeCl₃·7H₂O/NaI, Ti(O-*i*Pr)₄ or *p*-TsOH. Another efficient method for allyl deprotection is by using DMSO (dimethylsulfoxide)/NaI reagent. The benzyl, ethyl and *tert*-butyl protecting groups were quite stable under these reaction conditions[51].

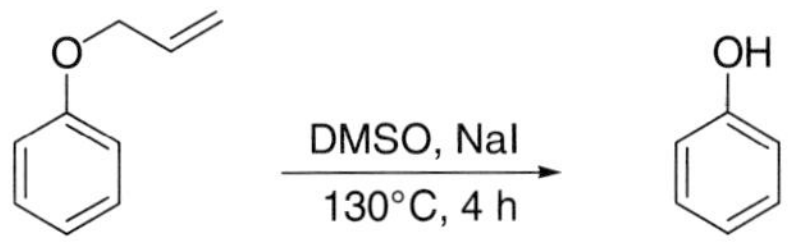

Propargylic aryl ethers (also esters) are cleaved by benzyltriethylammonium tetrathiomolyb-date in acetonitrile at room temperature[52]. Allyl esters are not cleaved under these conditions. Electroreduction in the presence of Ni–bipyridine complex as catalyst is another method to affect the deprotection of propargyl ethers[53] (Scheme 1.22).

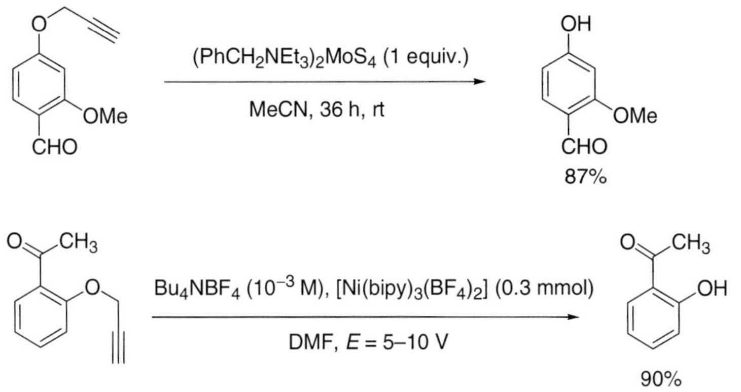

Scheme 1.22

A general method for forming acyclic mixed acetals (alkoxyalkyl ethers) is given below:

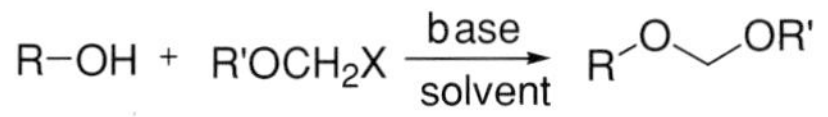

$$R{-}OH\ +\ R'OCH_2X \xrightarrow[\text{solvent}]{\text{base}} R{\diagup}O{\diagdown}OR'$$

The following alkoxyalkyl ethers are commonly formed from hydroxy compounds:

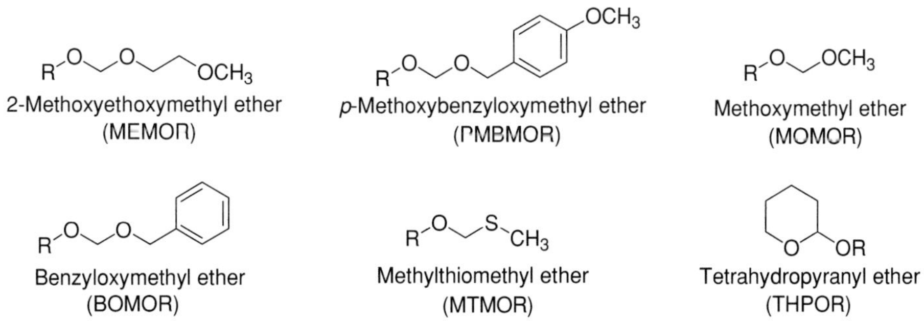

2-Methoxyethoxymethyl ether (MEMOR)

p-Methoxybenzyloxymethyl ether (PMBMOR)

Methoxymethyl ether (MOMOR)

Benzyloxymethyl ether (BOMOR)

Methylthiomethyl ether (MTMOR)

Tetrahydropyranyl ether (THPOR)

A **2-methoxyethoxymethyl ether** (**MEMOR**) is normally prepared under non-acidic conditions in methylene chloride solution or under basic conditions. The MEM ether group can be removed in excellent yield with trifluoroacetic acid (TFA) in dichloromethane (1:1). The MEM group can also be removed by treatment with zinc bromide ($ZnBr_2$), titanium chloride ($TiCl_4$) or bromocatechol borane. When MEM-protected diols are treated with zinc bromide ($ZnBr_2$) in ethyl acetate, 1,3-dioxane is formed and a mechanism of this reaction is given in Scheme 1.23.

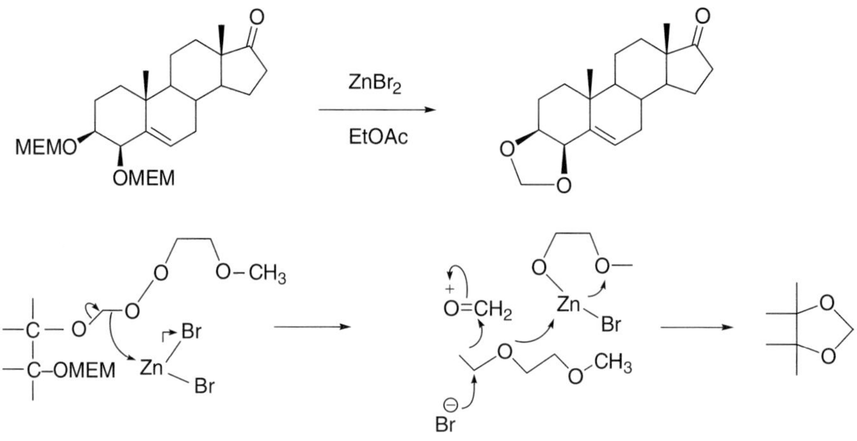

Scheme 1.23

The **MEM group** can be selectively removed with trimethylsilyl iodide in acetonitrile without affecting a methyl ether or an ester group.

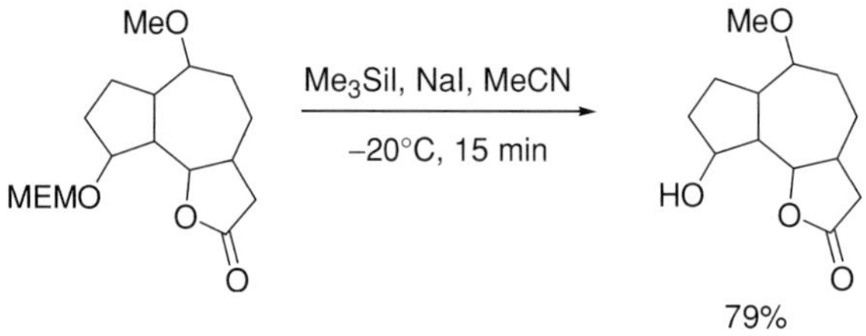

The **methoxymethyl (MOM) group** is one of the best protecting groups for alcohols and phenols[54]. The MOM ethers can be prepared by treating either alcohols or phenols with MOMCl (methoxymethyl chloride) or MOMOAc (methoxymethyl acetate) (Scheme 1.24). It is stable to a variety of commonly used reagents, such as strong bases, Grignard reagents, alkyllithiums and lithium aluminium hydride.

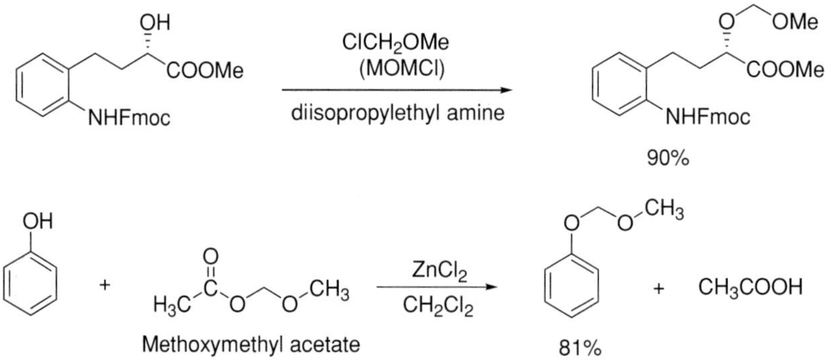

Scheme 1.24

Acid-catalyzed hydrolysis removes the MOM group.

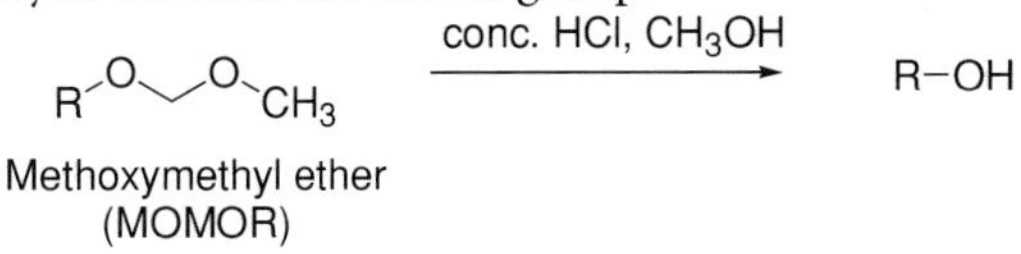

$$R{-}O{-}CH_2{-}O{-}CH_3 \xrightarrow{\text{conc. HCl, CH}_3\text{OH}} R{-}OH$$

Methoxymethyl ether
(MOMOR)

Bromocatechol borane or LiBF$_4$ (lithium fluoroborate) in acetonitrile (CH$_3$CN) and water has also been used for the deprotection of the MOM group.

Benzyloxymethyl ethers (BOMOR) are usually prepared by reacting BOMCl (benzyloxymethyl chloride) with alcohols under basic conditions and can be selectively removed with H$_2$, Pd–C or Na/NH$_3$.

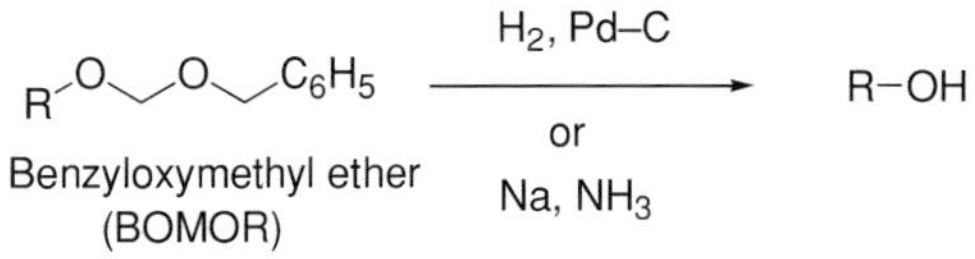

$$R{-}O{-}CH_2{-}O{-}CH_2{-}C_6H_5 \xrightarrow[\text{or} \atop \text{Na, NH}_3]{\text{H}_2, \text{Pd–C}} R{-}OH$$

Benzyloxymethyl ether
(BOMOR)

A **p-methoxybenzyloxymethyl (PMBM) group** is removed by acid hydrolysis or by reduction[55].

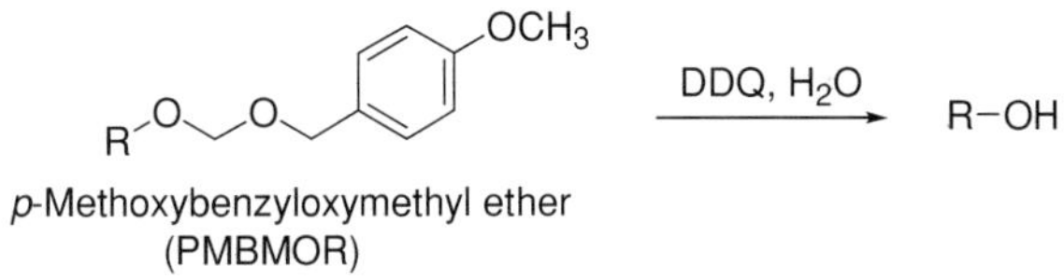

$$R{-}O{-}CH_2{-}O{-}CH_2{-}C_6H_4{-}OCH_3 \xrightarrow{\text{DDQ, H}_2\text{O}} R{-}OH$$

p-Methoxybenzyloxymethyl ether
(PMBMOR)

The methylthiomethyl ether (MTMOR): Tertiary hydroxyl groups, which are susceptible to acid-catalyzed dehydration, can be easily protected as MTM **ethers** and recovered in good yield. The MTM ether of a hydroxyl group can be formed either by a typical **Williamson ether synthesis** or on reaction with dimethylsulfoxide (DMSO) and acetic anhydride (Ac$_2$O). In the latter case, the reaction proceeds with the **Pummerer rearrangement**[56–58] (Scheme 1.25).

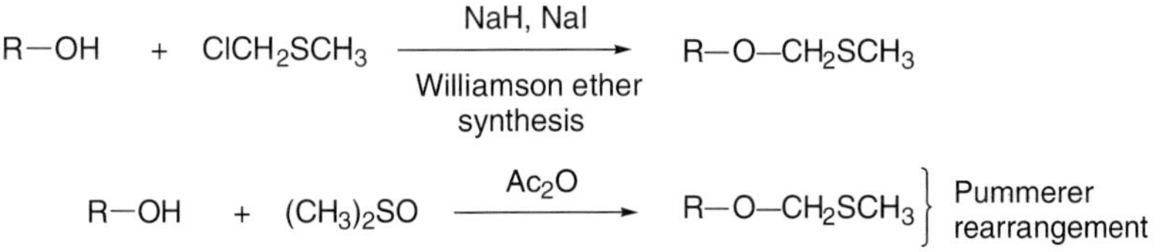

$$R{-}OH \ + \ ClCH_2SCH_3 \xrightarrow[\text{Williamson ether} \atop \text{synthesis}]{\text{NaH, NaI}} R{-}O{-}CH_2SCH_3$$

$$R{-}OH \ + \ (CH_3)_2SO \xrightarrow{\text{Ac}_2\text{O}} R{-}O{-}CH_2SCH_3 \ \bigg\}\ \begin{array}{l}\text{Pummerer}\\\text{rearrangement}\end{array}$$

Scheme 1.25

The mechanism of the Pummerer rearrangement is given in Scheme 1.26.

A **methylthiomethyl (MTM) group** is removed by acid or can be cleaved by mild treatment with aqueous silver or mercury salts (neutral mercuric chloride) to which most other ethers are stable; as a result, the selective deprotection of polyfunctional molecules becomes possible using MTM ethers for the hydroxy groups.

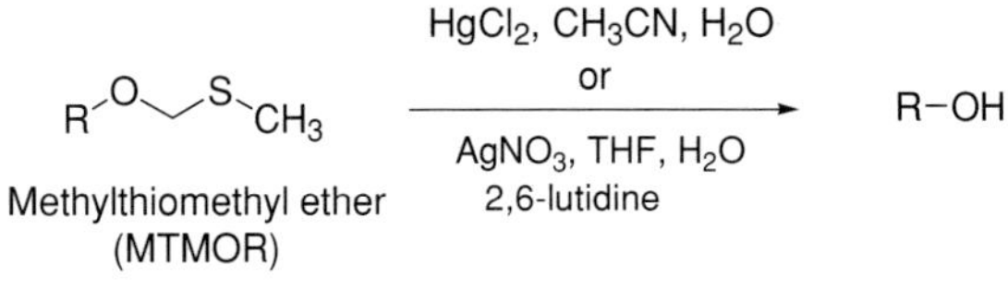

$$R{-}O{-}CH_2{-}S{-}CH_3 \xrightarrow[\text{AgNO}_3, \text{THF, H}_2\text{O} \atop \text{2,6-lutidine}]{\text{HgCl}_2, \text{CH}_3\text{CN, H}_2\text{O} \atop \text{or}} R{-}OH$$

Methylthiomethyl ether
(MTMOR)

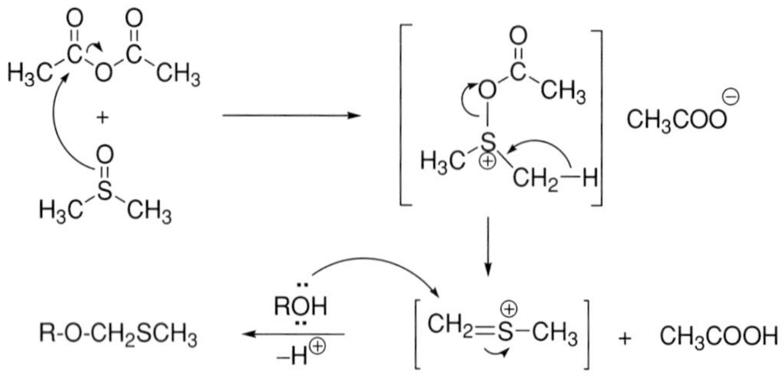

Scheme 1.26   Mechanism of the Pummerer rearrangement

Silyl ethers

Protection of hydroxyl groups through the formation of silyl ethers has been extensively utilized in organic synthesis. Silyl ethers are resistant to oxidation, have good thermal stability, low viscosity and are easily recoverable from their starting compounds.

Numerous methods can be used for the synthesis of trialkylsilyl ethers (Scheme 1.27). Alcohols react rapidly with trialkylsilyl chloride (R'_3SiCl) to give trialkylsilyl ethers[59] ($ROSiR'_3$) in the presence of an amine base like triethylamine, pyridine, imidazole or 2,6-lutidine (Table 1.2).

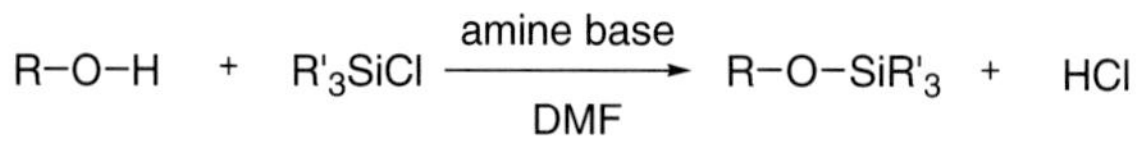

$$R-O-H \quad + \quad R'_3SiCl \quad \xrightarrow[\text{DMF}]{\text{amine base}} \quad R-O-SiR'_3 \quad + \quad HCl$$

Scheme 1.27

Unlike 3°-alkyl halides, **trialkylsilyl chlorides** (R'_3SiCl) undergo nucleophilic substitution by a mechanism similar to the S_N2. Enolate anions obtained from alcohols react with trialkylsilyl chlorides (R'_3SiCl), generating trialkylsilyl ethers (R'_3SiOR) by substitution at oxygen. The exceptional strength of the Si–O bond combined with longer C–Si bond lengths (less steric crowding) serve to stabilize such transition states as shown in Scheme 1.28.

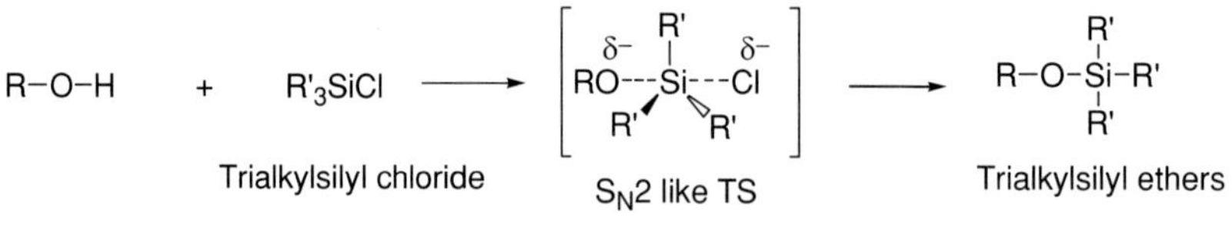

Scheme 1.28

With bulkier groups, such as TBS, it is possible to distinguish between primary and secondary alcohols. This is an example of regiocontrol (see section 1.5).

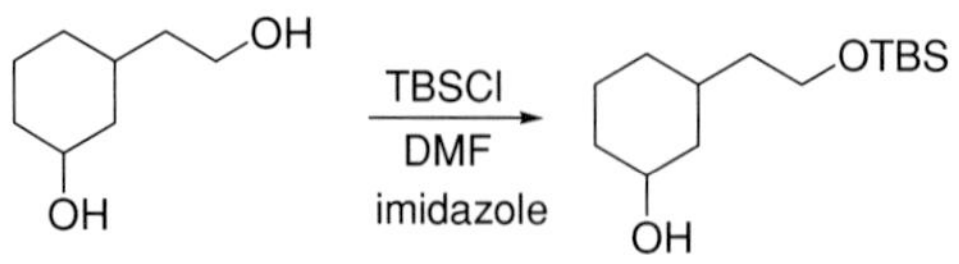

Selective monoprotection of 1,4-butanediol (**1.20**) with TBDPSCl gave a 90% yield of the corresponding alcohol **1.21**[12].

The removal of the TMS group is commonly carried out in the presence of a catalyst including iron(III) and tin(II) chlorides, copper(II) nitrate, cerium(III) nitrate, citric acid and sodium hydroxide or various fluoro derivatives. TMS derivatives are rather easily hydrolyzed to their alcohol precursors, but the bulkier silyl ethers are more resistant and are stable over a wide pH range. These protective groups are readily cleaved by fluoride anion, often introduced as a tetraalkylammonium salt such as tetrabutylammonium fluoride (TBAF).

Maiti and Roy[60] reported a selective method for deprotection of primary allylic, benzylic, homoallylic and aryl TBS ethers using aqueous DMSO at 90°C. All other TBS-protected groups, benzyl ethers, THP ethers as well as methyl ethers remain unaffected.

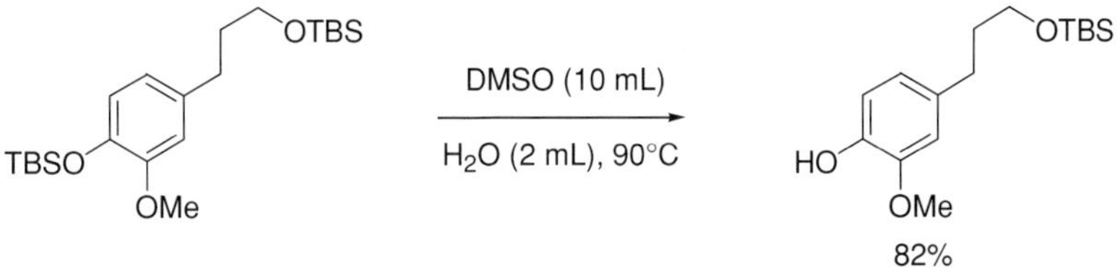

Esters

The acylation of alcohols represents an important reaction for the synthetic organic chemist; it was historically used for derivatization and characterization of alcohols. The acylation is usually performed by using acyl chlorides or the corresponding anhydrides in the presence of a base such as triethylamine or pyridine (Scheme 1.29). Faster reaction rates can be achieved by adding 4-(dimethylamino)pyridine (DMAP) as a co-catalyst.

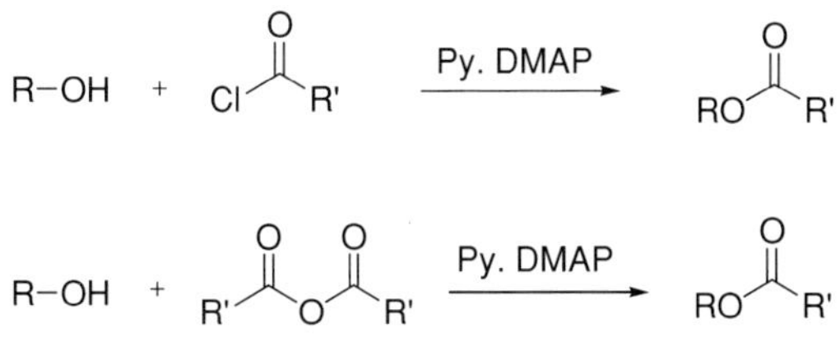

Scheme 1.29

Under these conditions, base-sensitive substrates may undergo decomposition. To avoid this drawback, protic and Lewis acid can be utilized, such as *p*-toluenesulfonic acid, zinc chloride, cobalt chloride or scandium triflate.

Acetate, chloroacetate, benzoate, *p*-methoxy benzoate, benzyl carbonate (Cbz), *tert*-butyl carbonate (Boc) and 9-(fluorenylmethyl) carbonate (Fmoc) are commonly prepared to protect the hydroxyl group (Table 1.3).

Methyl carbonates are cleaved under basic conditions (K_2CO_3/MeOH). Fmoc can be cleaved with base like Et_3N, Py, morpholine or diisopropylethylamine. Allyl carbonates can be

Table 1.3

X	Ester	X	Carbonate
CH_3	Acetate	$-OCH_3$	Methyl carbonate
CH_2-Cl	Chloroacetate		Allyl carbonate (Alloc)
	Benzoate		
	p-Methoxy benzoate		Benzyl carbonate (Cbz)
			t-Butyl carbonate (Boc)
			9-(Fluorenylmethyl) carbonate (Fmoc)

cleaved by $Pd_2(dba)_3$/dppe/Et_2NH/THF. Benzyl carbonates can be cleaved with H_2, Pd–C, EtOH. The regeneration of alcohols from their esters can also be carried out under heterogeneous conditions by using some solid catalysts, such as zeolites, silica, alumina or acidic resins.

1.6.2 Common diols protecting groups

Diols (1,2 and 1,3) are commonly protected as their O,O-acetals and ketals. Acetals are compounds with general structure $RR^1 C (OR^2) (OR^3)$, where R and R^1 may be H (but not necessary), but R^2 and $R^3 \neq H$. Ketals are subclass of acetals where neither R nor R^1 is H.

In a similar manner, ethylidene acetals, cyclopentylidene acetals, cyclohexylidene acetals, arylidene acetals and cyclic carbonates can be prepared.

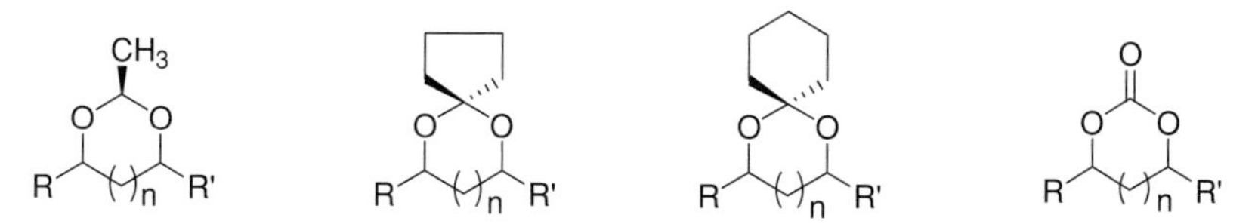

Ethylidene acetal   Cyclopentylidene acetal   Cyclohexylidene acetal   Cyclic carbonate

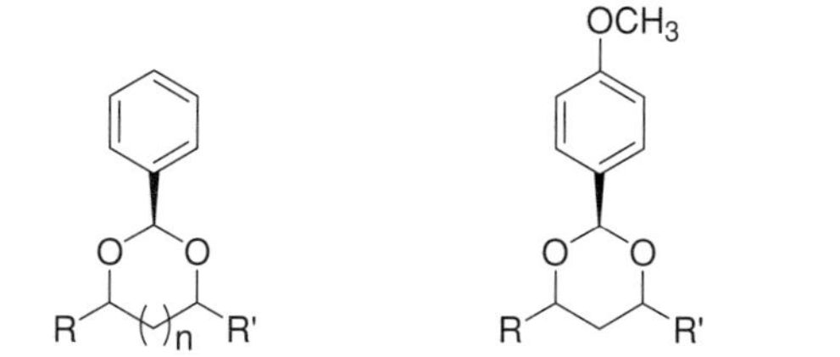

Benzylidene acetal   *p*-Methoxybenzylidene acetal

These acetals and ketals can be cleaved either under acidic conditions or by reduction (Scheme 1.30).

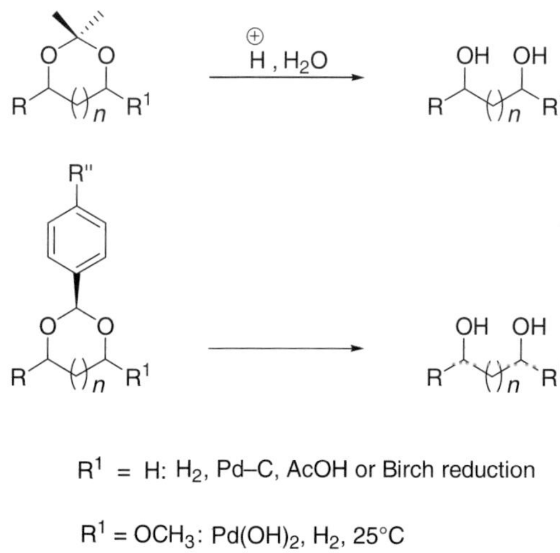

R^1 = H: H_2, Pd–C, AcOH or Birch reduction

R^1 = OCH_3: Pd(OH)$_2$, H_2, 25°C

Scheme 1.30 Deprotection of acetals and ketals

Acetals are also cleaved by ferric chloride either absorbed or not absorbed on silica gel. The TBS group (see Table 1.2) is not deprotected under these conditions. Depending on the number of equivalents of ferric chloride used, the selective deprotection of either one or both acetal groups can be achieved[61–63] (Scheme 1.31).

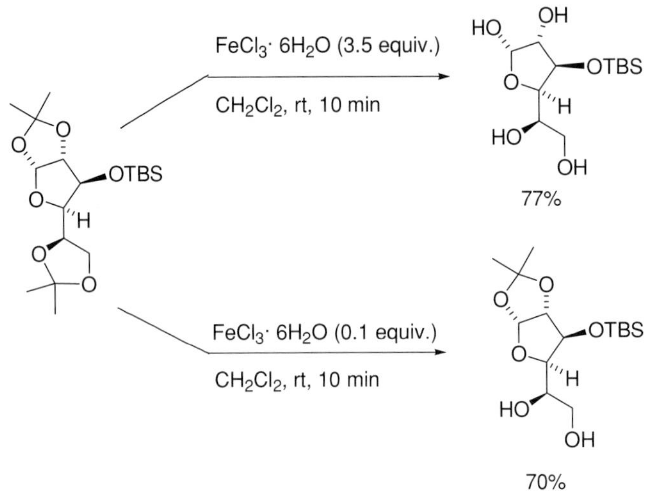

Scheme 1.31

1.6.3 Common amine protecting groups

Nitrogen protection continues to attract a great deal of attention in a wide range of chemical fields, such as peptide, nucleoside, polymer and ligand synthesis. Moreover, in recent years, a number of nitrogen protecting groups have been used as chiral auxiliaries. Thus, the design of new, milder and more effective methods for nitrogen protection is still an active topic in synthetic chemistry.

Imide and amide protecting groups: The phthalimide group has been successfully used to protect the amino group. Cleavage of *N*-alkylphthalimide (**1.81**) is easily carried out with hydrazine, in a hot solution or in the cold for a prolonged period to give **1.82** and the amine. Base-catalyzed hydrolysis of *N*-alkylphthalimide **1.81** also gives the corresponding amine (Scheme 1.32).

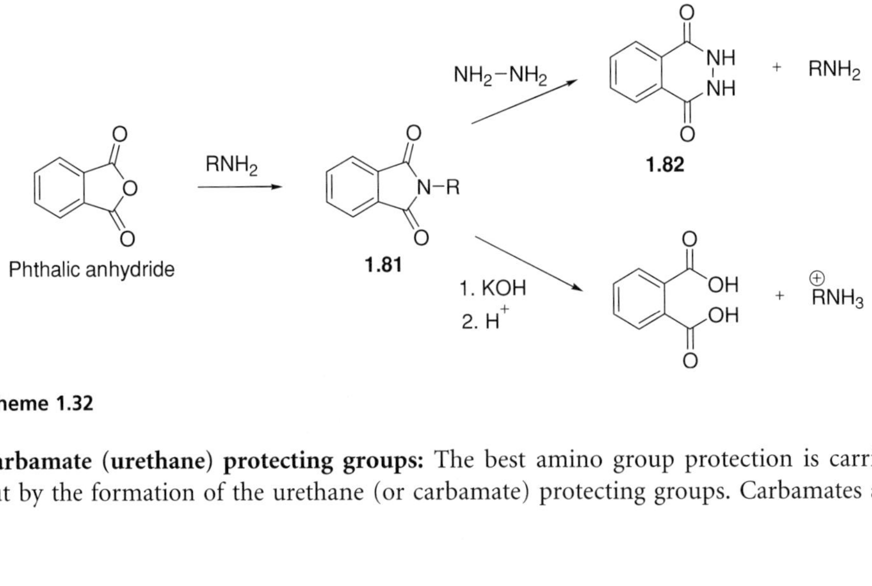

Scheme 1.32

Carbamate (urethane) protecting groups: The best amino group protection is carried out by the formation of the urethane (or carbamate) protecting groups. Carbamates are prepared from amines by the following methods (Scheme 1.33):

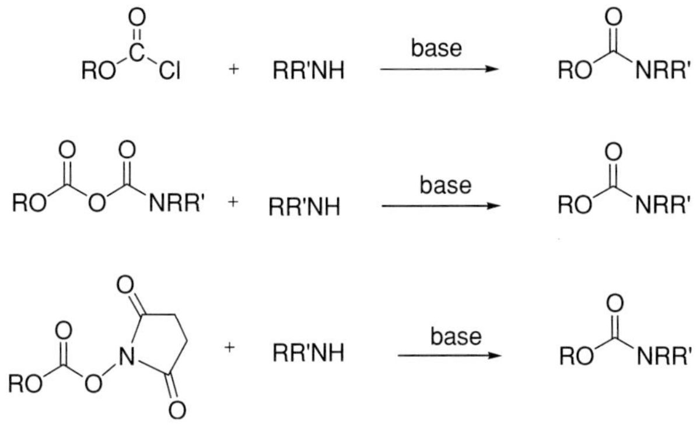

Scheme 1.33

For example, urethane protecting groups such as benzyloxycarbonyl (Cbz), *tert*-butoxycarbonyl (Boc) and (fluorenylmethoxy) carbonyl (Fmoc) are easily introduced as shown in Scheme 1.34:

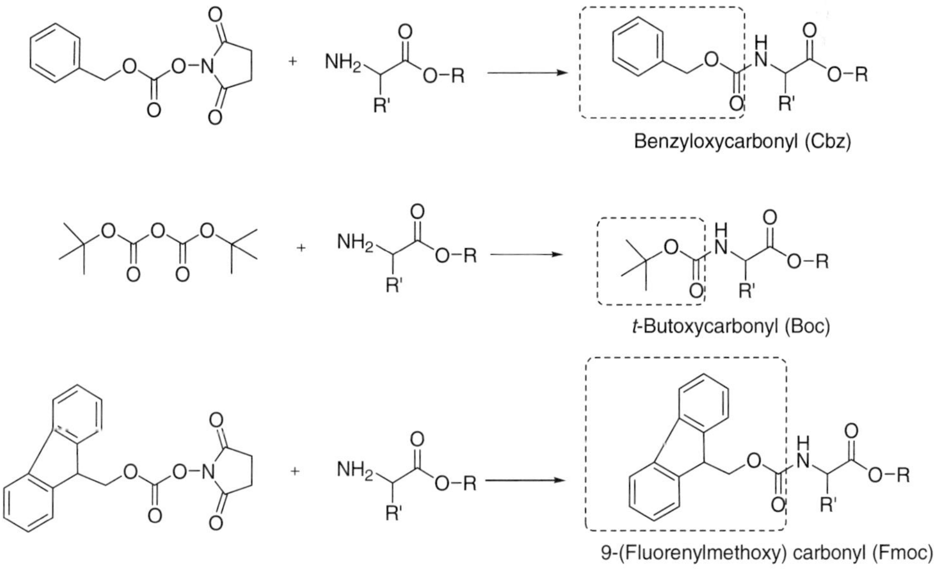

Benzyloxycarbonyl (Cbz)

t-Butoxycarbonyl (Boc)

9-(Fluorenylmethoxy) carbonyl (Fmoc)

Scheme 1.34

These protecting groups withstand a variety of harsh reaction conditions.

Boc is a good labile group because it is stable at room temperature and easily removed with dilute solution of TFA either neat or in dichloromethane. Other mineral acids or Lewis acids have also been used, although less frequently.

Fmoc is a base labile protecting group which is easily removed by reaction with concentrated solutions of amines. Both Cbz and the acid labile *t*Boc are commonly used. These owe their lability to the stability of the carbocation produced on deprotection (Scheme 1.35). The Boc group, which generates a stable tertiary cation on deprotection, is more prone to deprotection by weaker acids than is Cbz.

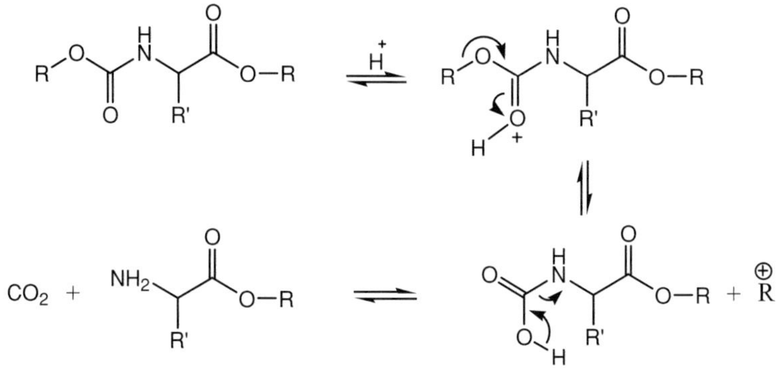

Scheme 1.35

The Fmoc group protection is common in solid-phase peptide synthesis. Fmoc is resistant to acidic conditions and easily deprotected by weak bases, particularly secondary amines. Deprotection occurs through base-catalyzed abstraction of the β-proton of the protecting group with elimination leading to formation of dibenzofulvene (**1.83**) (Scheme 1.36).

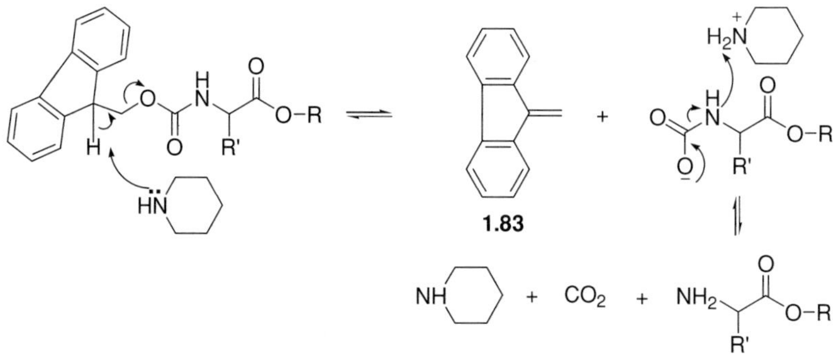

Scheme 1.36   Mechanism of deprotection of the Fmoc group

The different cleavage conditions for the above urethane protecting groups have enabled so-called *orthogonal* protection strategies to be developed, which in turn enable selective deprotection to be performed on different amines present in the same molecule. For example in peptide synthesis[64], the N-Boc group could be cleaved selectively using TMSOTf, followed by aqueous work up.

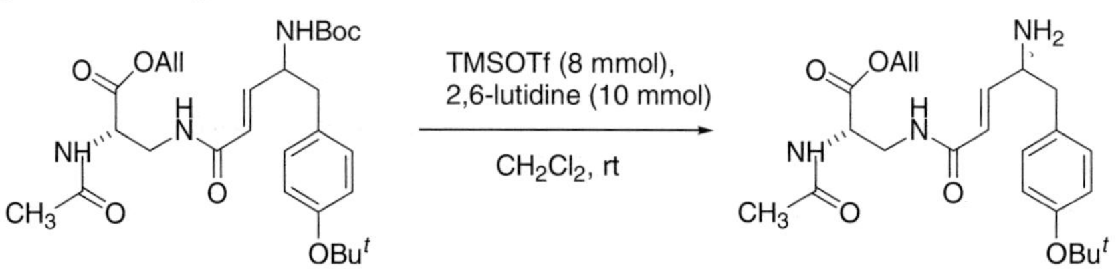

Treatment of the adenine derivative **1.84** with 1-(benzyloxycarbonyl)-3-ethylimidazolium tetrafluoroborate (Rapoport's reagent; **1.85**) forms **1.86** in 82% yield in which the NH$_2$ group is protected with Cbz.

Both the Boc groups in substituted guanidine **1.87** can be removed with stannic chloride in ethyl acetate. The reagent is milder than the TFA and gives a high yield of deprotected product **1.88** in 88% yield.

The *p*-toluenesulfonyl (Ts) group from *N*-arylsulfonylcarbamates and *N*-acylsulfonamides can be removed by the use of magnesium in anhydrous methanol under ultrasonic conditions.

The amino group can be protected by forming its sulfonyl [such as arylsulfonyl or 2-(trimethylsilyl) ethyl sulfonyl], sulfenyl and silyl derivatives. The 2- or 4-nitrophenylsulfonamide derivatives of amino acids are useful substrates for mono-N-alkylation using only cesium carbonate (Cs$_2$CO$_3$) as the base. The sulfonamide group can be removed in **1.89** by potassium phenyl thiolate (PhSH and K$_2$CO$_3$) in acetonitrile to give the N-alkylated α-amino esters **1.90** and the reaction occurs without racemization.

Benzyl and allyl amines: The amines can also be protected as benzylamines and allyl amines (Scheme 1.37).

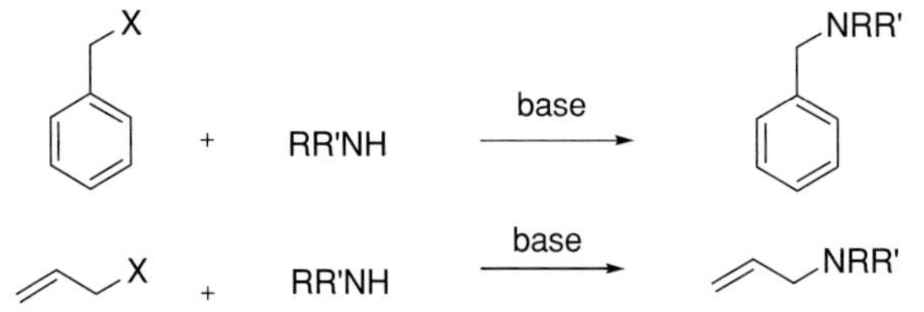

Scheme 1.37

These amines can be deprotected under reduction conditions (Pd–C/ROH/HCO$_2$NH$_4$ or Na/NH$_3$). The allyl amines can be deprotected by oxidative cleavage with ozone (dimethyl-sulfide work up) or with KMnO$_4$ in acetone.

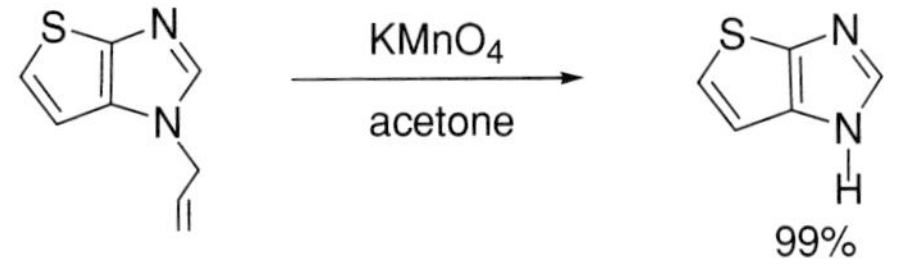

1.6.4 Common carbonyl protecting groups

Carbonyls can be protected as acyclic or cyclic acetals, S,S'-dialkyl acetals, oxathiolanes, 1,1-diacetates and nitrogenous derivatives.

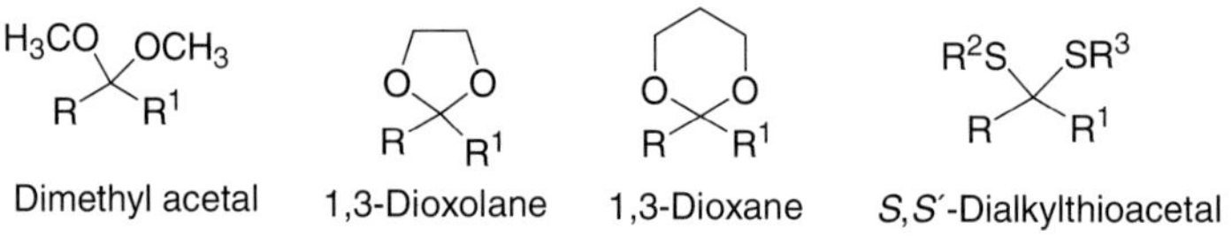

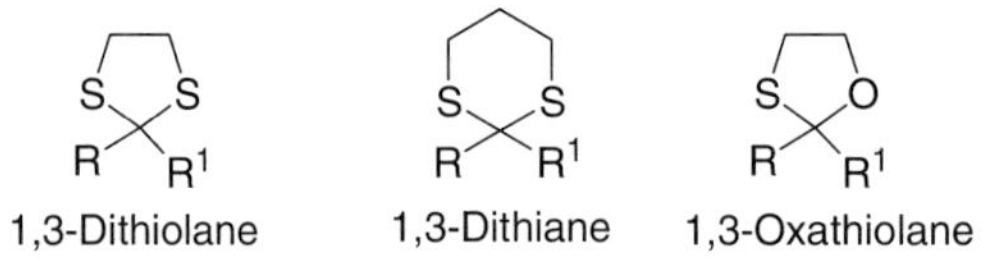

Acyclic and cyclic acetals are stable to base but removed with acid. Aliphatic aldehydes are more reactive than aromatic aldehydes, which in turn are more reactive than ketones.

Dimethyl acetals can be prepared under different conditions from aldehydes and ketones[65,66] as shown below:

These acetals can be cleaved with HCl–water or dioxane–water. Cyclic acetals are cleaved more slowly than their open-chain analogues; for example, dimethyl acetals can be cleaved in the presence of a 1,3-dithiane and a dioxolane acetal by using TFA/CHCl$_3$/H$_2$O. p-Toluenesulfonic acid (TsOH)/acetone or 70% H$_2$O$_2$/Cl$_3$CCOOH/CH$_2$Cl$_2$/t-BuOH/Me$_2$S can also be used for the cleavage of dimethyl acetals.

Formation of cyclic acetals of αβ-unsaturated carbonyls is usually slower than for the saturated carbonyls. Thus, saturated ketones can be selectively protected in the presence of αβ-unsaturated ketones with ethylene glycol and a stoichiometric amount of *p*-TsOH and water[67].

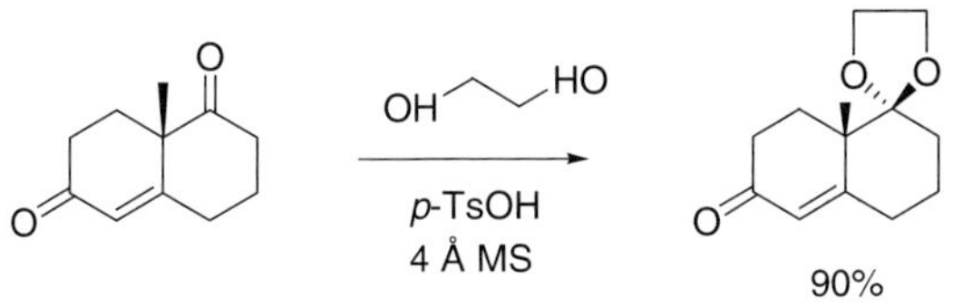

90%

However, αβ-unsaturated ketones are also selectively protected as shown below[68]:

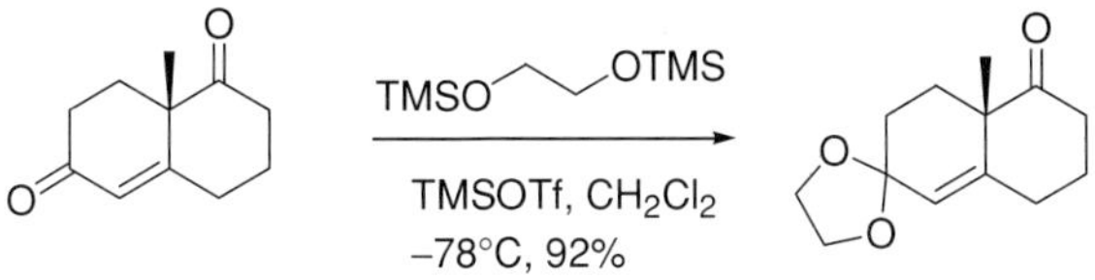

The selective cleavage[69] of cyclic acetal of αβ-unsaturated ketone over cyclic acetal of saturated ketone can be achieved by using NaI and $CeCl_3 \cdot 7H_2O$ in acetonitrile.

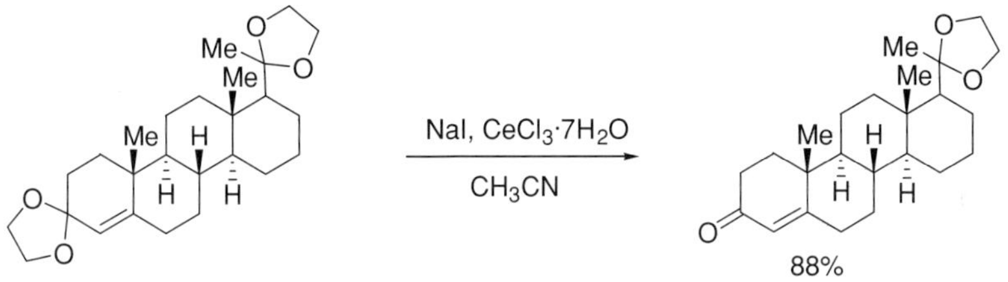

88%

Acetals can be deprotected[70] to the corresponding carbonyl compounds using a catalytic amount of carbon tetrabromide (CBr_4) in the acetonitrile–water mixture under thermal or ultrasound conditions.

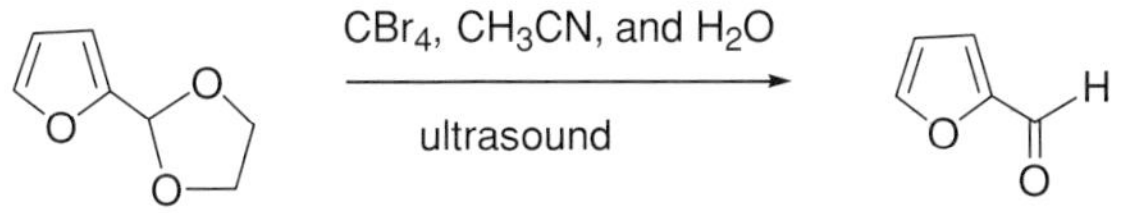

The selective deprotection[71] of acetal to a carbonyl group can be achieved with lithium tetrafluoroborate $LiBF_4$ in THF.

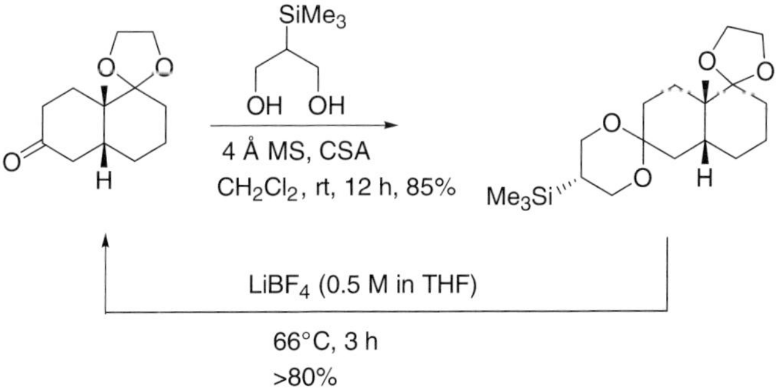

CSA = Camphorsulfonic acid

Open chain and cyclic thioacetals: 1,3-Dithiolane and 1,3-dithiane derivatives are versatile intermediates in the synthesis and interconversion of monocarbonyl and 1,2-dicarbonyl compounds. Protection of carbonyl groups as their open-chain and cyclic thioacetals is an important method in the synthesis of organic molecules[72]. Thioacetals are stable towards ordinary acidic and basic conditions and can act as acyl synthetic equivalent groups (see section 1.3). Although many procedures are available for preparing thioacetals, their deprotection is not a very easy process.

S,S'-Dialkyl thioacetals can be formed by reacting aldehydes with RSH or $RSSi(CH_3)_3$.

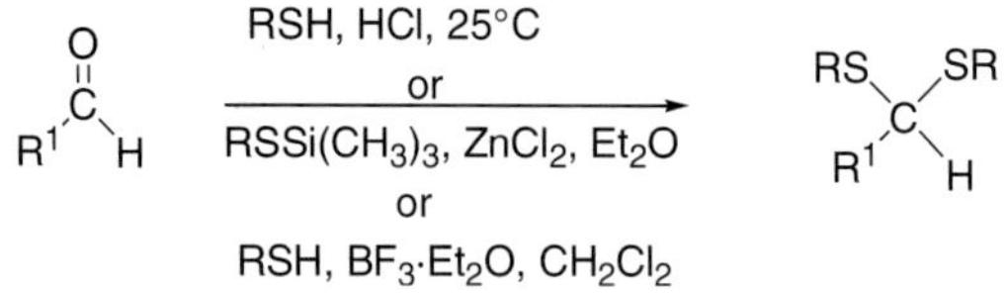

1,3-Dithiane works well for aromatic aldehydes and aromatic and aliphatic ketones, while 2-phenyl-1,3-dithiane should be used for aliphatic aldehydes[73].

Many procedures[74] are available in the literature for the deprotection of S,S'-dialkyl thioacetals to their carbonyl compounds such as clay supported ammonium ion, ferric or cupric nitrates, zirconium sulfonyl phosphonate, oxides of nitrogen, DDQ, $SeO_2/AcOH$, $DMSO/HCl/H_2O$, TMSI(Br), $LiN(i-C_3H_7)_2/THF$, ceric ammonium nitrate in aqueous CH_3CN, $CuCl_2/CuO$/acetone and reflux, $Hg(ClO_4)_2$/chloroform and m-$CPBA/Et_3N/Ac_2O/H_2O$.

Recently new methods have been introduced using mercury(II) nitrate trihydrate[75], MnO_2, $KMnO_4$ and $BaMnO_4$ catalyzed with $AlCl_3$ in CH_3CN at room temperature for the deprotection of benzylic dithioacetals of aldehydes and non-enolizable ketones[76].

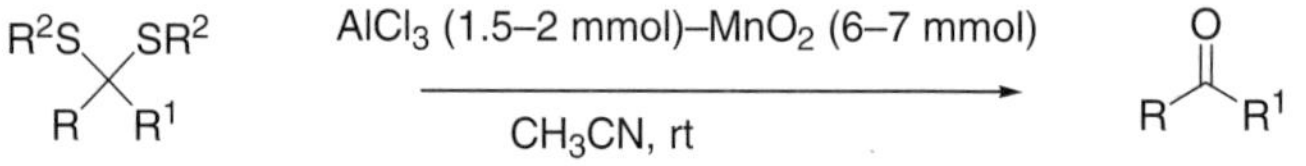

$R^1 = H$; R = aryl
$R^2 = -(CH_2)_3, -(CH_2)_2-$, n-Bu

A mechanism of deprotection of S,S'-thioacetals under non-aqueous reaction conditions has been suggested in which oxygen of MnO_2 acts as a nucleophilic species (Scheme 1.38).

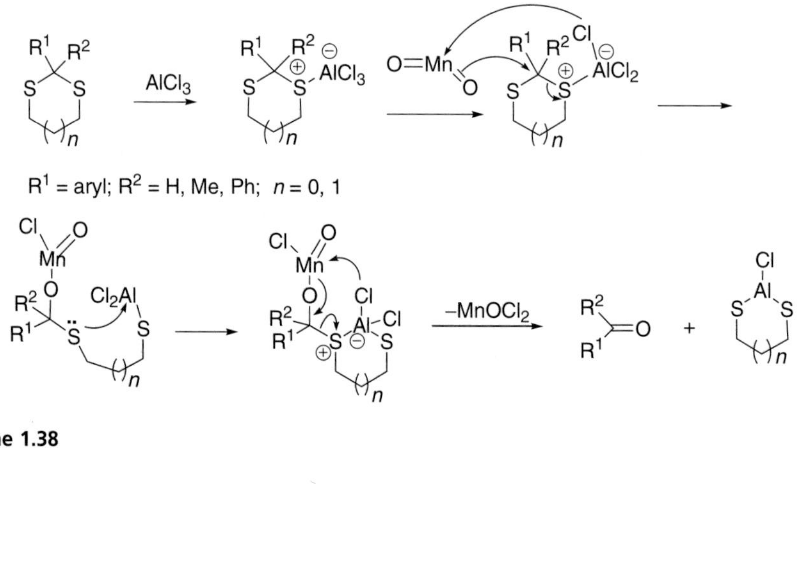

R^1 = aryl; R^2 = H, Me, Ph; n = 0, 1

Scheme 1.38

A similar mechanism has also been proposed for deprotection of *S,S'*-thioacetals using seleninic anhydride.[77]

1.6.5 Common carboxylic acid protecting groups

Carboxylic acids are protected as their esters such as methyl esters, *tert*-butyl esters, allyl esters, benzyl esters, phenacyl esters and alkoxyalkyl esters. The esters are formed by the reaction of carboxylic acid with alcohol, and the reaction is known as esterification.

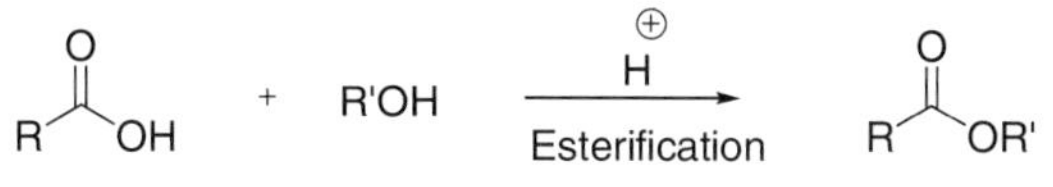

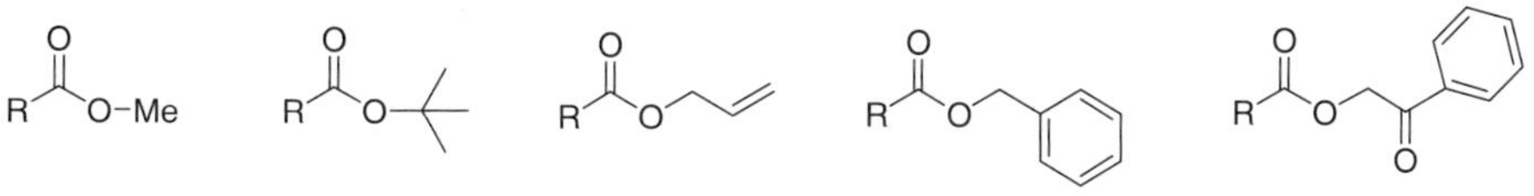

| Methyl ester | *t*-Butyl ester | Allyl ester | Benzyl ester | Phenacyl ester |

Although many useful and reliable methods for esterification of carboxylic acids are usually employed, there is need to find versatile processes to replace the classical methodologies with more benign alternatives, characterized by general applicability. Two new methods for the formation of esters from carboxylic acids are shown in Scheme 1.39.

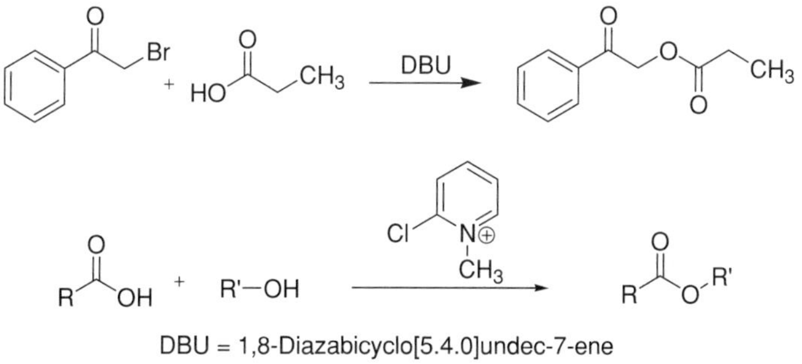

DBU = 1,8-Diazabicyclo[5.4.0]undec-7-ene

Scheme 1.39

The mechanism of Mukaiyama esterification is given in Scheme 1.40.

The *tert*-butyl esters are usually prepared by the reaction of an acid with isobutylene in the presence of an acidic catalyst. In the modified procedure instead of isobutylene, *tert*-butyl alcohol (*t*-BuOH) is used in the presence of heterogeneous acid catalyst[78].

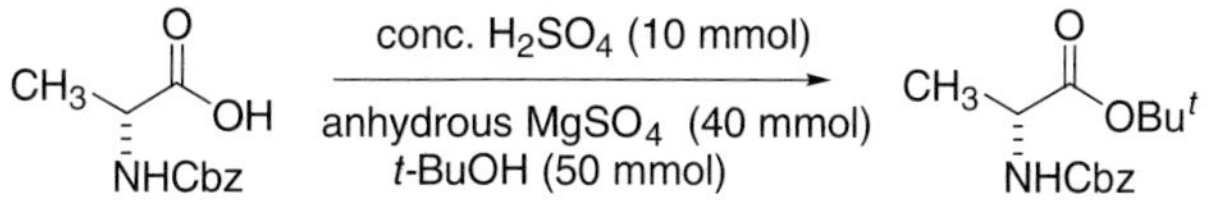

Allyl esters can be prepared by the reaction of carboxylic acid with allyl bromide in the presence of Cs_2CO_3 in DMF or allyl alcohol in the presence of TsOH in benzene.

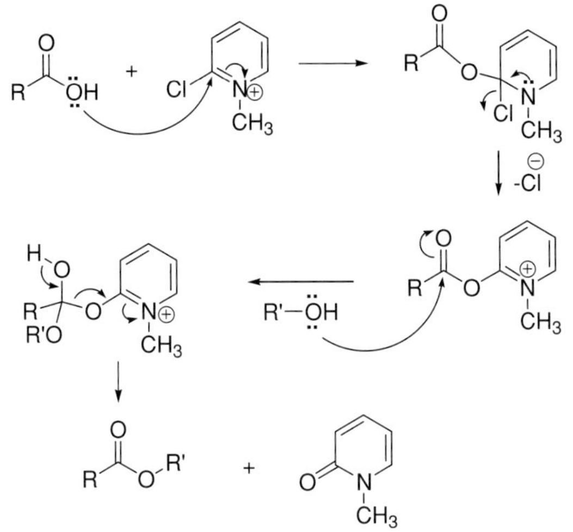

Scheme 1.40 Mechanism of Mukaiyama esterification

Alkoxyalkyl esters and silyl esters are also easily prepared and cleaved. For example, 2-(trimethylsilyl)ethoxymethyl esters are usually cleaved with HF in acetonitrile by fluoride ion.

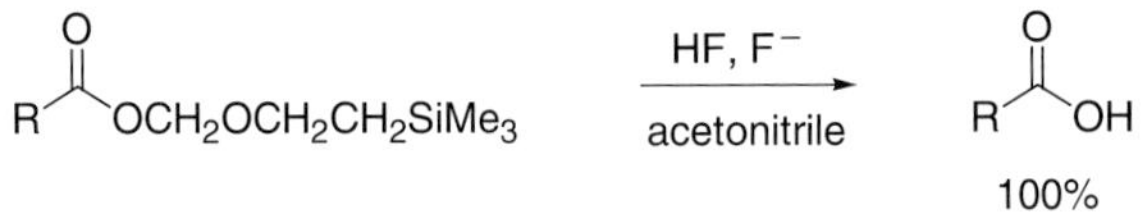

Methyl esters are removed by an acid or a base. The lithium hydroxide also cleaved the methyl ester group while the Boc group remained intact.

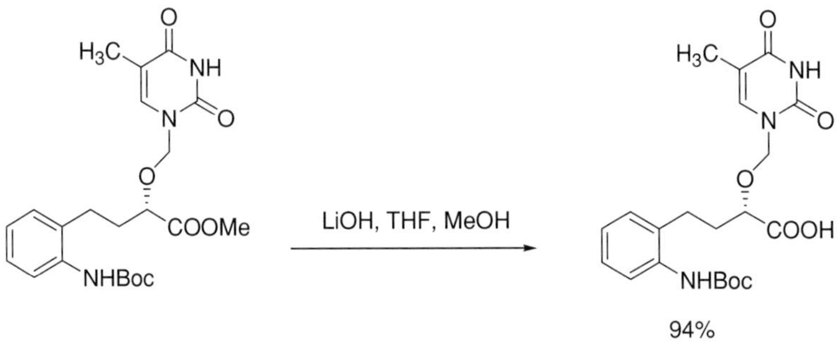

94%

Benzyl esters can be removed by hydrogenolysis. *tert*-Butyl esters can be cleaved by CF$_3$COOH in CH$_2$Cl$_2$. Selective deprotection of *tert*-butyl esters in the presence of *N*-Boc protecting groups for several amino acids can be carried out with CeCl$_3$·7H$_2$O–NaI. The main advantage of this method is the low cost of the reagents and mild nature of the ceric chloride interaction in comparison to other Lewis acids.

Phenacyl esters can be removed by light at wavelengths 308–313 nm with >70% yield. The mechanism of the photodeprotection of the phenacyl group is shown in Scheme 1.41:

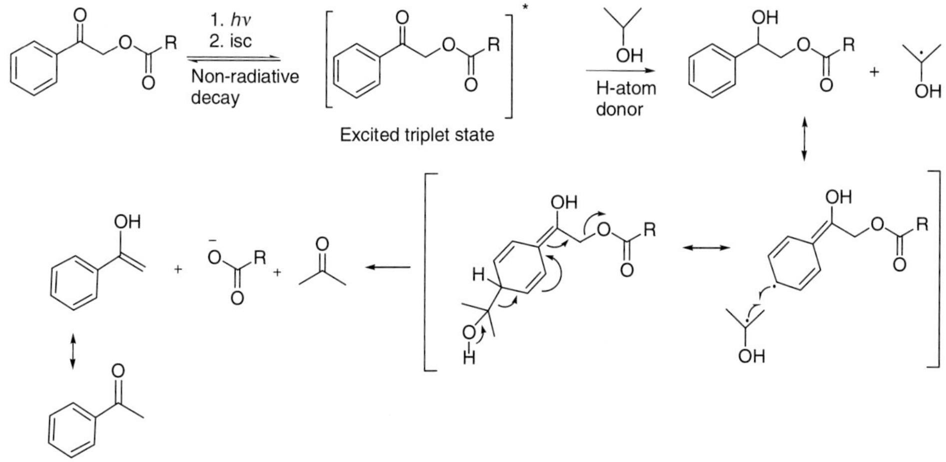

Scheme 1.41   Mechanism of photodeprotection of the phenacyl group

However, irradiation of buffered solutions of esters of the *p*-hydroxy phenacyl at room temperature leads to the rapid release of carboxylate anion with the formation of *p*-hydroxyphenyl acetic acid[79,80] (**1.91**) (Scheme 1.42).

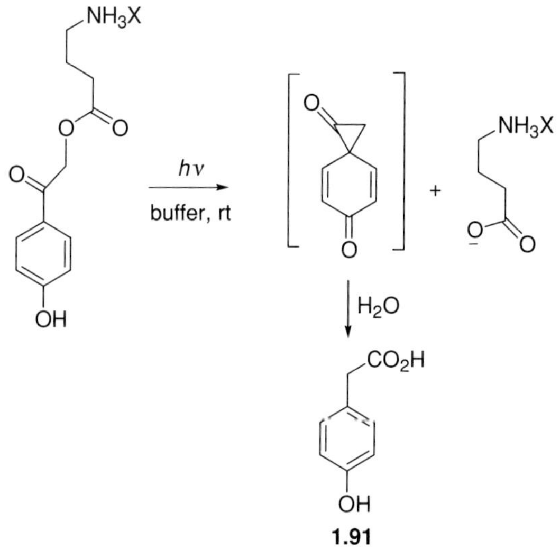

1.91

Scheme 1.42   Mechanism of photodeprotection of the *p*-hydroxyphenacyl esters

1.6.6   Common arenesulfonic acid protecting groups

Roberts and co-workers[81] prepared neopentyl (2,2-dimethylpropyl) esters of arenesulfonic acids which are compatible with a wide range of standard organic synthesis methodologies. These esters withstand several reagents such as *tert*-butyllithium, vinylmagnesium bromide, CrO_3, NBS (*N*-bromosuccinimide)–benzoyl peroxide, H_2–Raney Ni, DIBAL-H, NaI,

$HONH_2$, NaH, aqueous HBr and NaOH. Deprotection of these esters can be accomplished by heating the ester with excess of tetramethylammonium chloride in DMF (Scheme 1.43).

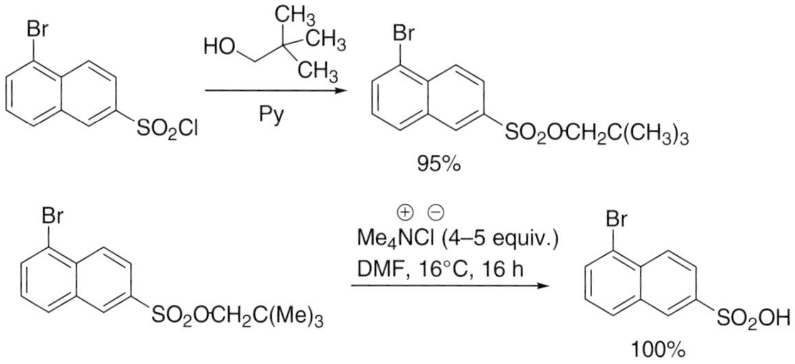

Scheme 1.43

1.6.7 Common alkyne protecting groups

Alkynes can be protected as their silyl derivatives and the most common silyl groups TMS, TES, TIPS and TBS are introduced by reacting alkyne with the corresponding trialkylsilyl chlorides (see Table 1.2 for the structures of R'_3SiCl).

$$R\text{---}\equiv\text{---}M \xrightarrow{R'_3SiX} R\text{---}\equiv\text{---}SiR'_3$$

M = Li, Mg; X = Cl, OTf

Cleavage of trialkylsilylalkynes can be achieved by using TBAF in the presence of THF.

$$R\text{---}\equiv\text{---}SiR'_3 \xrightarrow[\text{THF}]{\text{TBAF}} R\text{---}\equiv\text{---}H$$

Cleavage[82] of trimethylsilylalkynes can also be carried out by using KF/MeOH, $AgNO_3$/2,6-lutidine or K_2CO_3/MeOH.

References

1. Corey, E. J. and Cheng, X.-M., *The Logic of Chemical Synthesis*, Wiley, New York, **1995.**
2. Corey, E. J., *Angew. Chem., Int. Ed. Engl.*, **1991**, *30*, 455.
3. Corey, E. J., Wipke, W. T., Cramer, R. D., III and Howe, W. J., *J. Am. Chem. Soc.*, **1972**, *94*, 421.
4. Corey, E. J., Howe, W. J. and Pensak, D. A., *J. Am. Chem. Soc.*, **1974**, *96*, 7724.
5. Moses, J. E. and Moorhouse, A. D., *Chem. Soc. Rev.*, **2007**, *36*, 1249; Kolb, H. C., Finn, M. G. and Sharpless, K. B., *Angew. Chem., Int. Ed. Engl.*, **2001**, *40*, 2004.
6. Gröbel, B. T. and Seebach, D., Synthesis, **1977**, 357.
7. Vijayasaradhi, S. and Aidhen, I. S., *Org Lett.*, **2002**, *4* (10), 1739.
8. Trost, B. M., *Science*, **1991**, *254*, 1471.
9. Trost, B. M., *Angew. Chem., Int. Ed. Engl.*, **1995**, *34*, 259.
10. Cann, M. C. and Connelly, M. E., *Real-World Cases in Green Chemistry*, American Chemical Society, Washington, DC, **2000.**
11. Sheldon, R. A., *Chem. Ind. (Lond.)*, **1992**, 903.
12. Jacobi, P. A., Murphree, S., Rupprecht, F. and Zheng, W., *J. Org. Chem.*, **1996**, *61*(7), 2413.
13. Lee, H. Y. and An, M., *Tetrahedron Lett.*, **2003**, *44*, 2775.

14. Sawada, M., Takai, Y., Yamada, H., Sawada, M., Yamaoka, H., Azuma, T., Fujioka, T., Kawai, Y. and Tanaka, T., *Chem. Commun.*, **1998**, *15*, 1569.
15a. Koga, K., *Pure Appl. Chem.*, **1994**, *66* (7), 1487.
15b. Corey, E. J. and Ensley, H. E., *J. Am. Chem. Soc.*, **1975**, *97*, 6908.
15c. Trost, B. K., O'Krongly, D. and Belletire, J. E., *J. Am. Chem. Soc.*, **1980**, *102*, 7595.
15d. Whitsell, J. K., Chen, H. H. and Lawrence, R. M., *J. Org. Chem.*, **1985**, *50*(23), 4663.
16. Bull, S. D., Davies, S. G., Fox, D. J., Garner, A. C. and Sellers, T. G. R., *Pure Appl. Chem.*, **1998**, *70*(8), 1501.
17. Arpin, A., Manthorpe, J. M. and Gleason, J. L., *Org. Lett.*, **2006**, *8*(7), 1359.
18. Trost, B. M. and Dogra, K., *Org. Lett.*, **2007**, *9*(5), 861.
19. Kitamura, M., Tokunaga, M., Ohkuma, T. and Noyori, R., *Org. Synth. Coll.*, **1998**, *9*, 589.
20. Kitamura, M., Tokunaga, M., Ohkuma, T. and Noyori, R., *Org. Synth. Coll.*, **1993**, *71*, 1.
21. Takaya, H., Ohta, T., Inoue, S., Tokunaga, M., Kitamura, M. and Noyori. R., *Org. Synth. Coll.*, **1998**, *9*, 169.
22. Takaya, H., Ohta, T., Inoue, S., Tokunaga, M., Kitamura, M. and Noyori. R., *Org. Synth. Coll.*, **1995**, *72*, 74. (Also see References 3 and 4 of Chapter 1 and Reference 14 of Chapter 5.)
23. Corey, E. J., Shibata, S. and Bakshi, R. K., *J. Org. Chem.*, **1988**, *53*, 2861.
24. Corey, E. J., Shibata, T. and Lee, T. W., *J. Am. Chem. Soc.*, **2002**, *124* (15), 3808.
25. Li, G., Chang, H. T. and Sharpless, K. B., *Angew. Chem., Int. Ed. Engl.*, **1996**, *35*, 451.
26. Kolb, H. C., VanNieuwenhze, M. S. and Sharpless, K. B., *Chem. Rev.*, **1994**, *94*, 2483.
27. Sharpless, K. B., *Org. Synth.*, **1991**, *70*, 47.
28. Sharpless, K. B., *J. Am. Chem. Soc.*, **1987**, *109*, 5765.
29. Katsuki, T. and Sharpless, K. B., *J. Am. Chem. Soc.*, **1980**, *102*(18), 5974.
30. Martin, V. S., Woodard, S. S., Katsuki, T., Yamada, Y., Ikeda, M. and Sharpless, K. B., *J. Am. Chem. Soc.*, **1981**, *103* (20), 6237.
31. Zhang, W., Loebach, J. L., Wilson, S. R. and Jacobsen, E. N., *J. Am. Chem. Soc.*, **1990**, *112*, 2801.
32. Jacobsen, E. N., Deng, L., Furukawa, Y. and Martinez, L. E., *Tetrahedron*, **1994**, *50*, 4323.
33a. Linber, T., *Angew. Chem., Int. Ed. Engl.*, **1997**, *36*(19), 2004.
33b. Linde, C., Akermark, B., Norrby, P.-O. and Svensson, M., *J. Am. Chem. Soc.*, **1999**, *121*, 5083.
34. Wang, Z. M., Kolb, K. C. and Sharpless, K. B., *J. Org. Chem.*, **1994**, *59*, 5104.
35. Shimada, T., Mukaide, K., Shinohara, A., Han, J. W. and Hayashi, T., *J. Am. Chem. Soc.*, **2002**, *124*, 1584.
36. O'Brien, P., Osborne, S. A. and Parker, D. D., *J. Chem. Soc., Perkin Trans. 1*, **1998**, 2519.
37. Li, G. and Sharpless, K. B., *Acta Chem. Scand.*, **1996**, *50*, 649.
38. Parham, W. E. and Anderson, E. L., *J. Am. Chem. Soc.*, **1948**, *70*, 4187.
39. Miyashita, M., Yoshikoshi, A., and Grieco, P. A., *J. Org. Chem.*, **1977**, *42*(23), 3772.
40. Vyas, G. N. and Shah, N. M., *Org. Synth. Coll.*, **1963**, *4*, 836.
41. Bracher, F. and Schulte, B., *J. Chem. Soc., Perkin Trans 1*, **1966**, 2619.
42. Satoh, T., Ikeda, M., Miura, M. and Nomura, M., *J. Org. Chem.*, **1997**, *62*, 4877.
43. Ahmad, R., Saa, J. M. and Cava, M. P., *J. Org. Chem.*, **1997**, *42*, 1228.
44. Nayak, M. K. and Chakraborti, A. K., *Tetrahedron lett.*, **1997**, *38*, 8749.
45. Stephens, J. R., Butler, P. L., Clow, C. H., Oswald, M. C., Smith, R. C. and Mohan, R. S., *Eur. J. Org. Chem.*, **2003**, *19*, 3827.
46. Kocienski, P. J., *Protecting Groups*, George Thieme Verlag, Stuttgart, Germany, **1994.**
47. Liu, H. J., Yip, J. and Shia, K.S., *Tetrahedron Lett.*, **1997**, *38*, 2253.
48. Alonso, E., Ramon, J. and Yus, M., *Tetrahedron Lett.*, **1997**, *53*, 14355.
49. Hwu, J. R., Wong, F. F., Haung, J. J. and Tasy, S. C., *J. Org. Chem.*, **1997**, *62*, 4097.
50. Sajiki, H., Kuno, H. and Hirota, K., *Tetrahedron Lett.*, **1997**, *38*, 399.
51. Nagaraju, M., Krishnaiah, A. and Mereyala, H. B., *Synth. Commun.*, **2007**, *35*, 2467.
52. Swamy, V. M., Hankumaran, P. and Chandrasekaran, S., *Synlett*, **1997**, 513.
53. Olivero, S. and Dunach, E., *Tetrahedron Lett.*, **1997**, *38*, 6193.
54. Lee, A. S. Y., Hu, Y. J. and Chu, S. F., *Tetrahedron*, **2001**, *57*, 2121.
55. Kozikowaski, A. P. and Wu, J.-P., *Tetrahedron Lett.*, **1987**, *28*, 5125.
56. Pummerer, R., *Chem. Ber.* **1909**, *42*, 2282.
57. Pummerer, R., *Chem. Ber.* **1910**, *43*, 1401.
58. Laleu, B., Machado, M. S. and Lacour, J., *Chem. Commun.*, **2006**, 2786–2788.
59. Corey, E. J. and Venkateswarlu, A., *J. Am. Chem. Soc.*, **1972**, *94*, 6190.
60. Maiti, G. and Roy, S. C., *Tetrahedron Lett.*, **1997**, *38*, 495.
61. Fadel, A., Yefsah, R. and Salaun, J., *Synthesis*, **1987**, 37.
62. Kim, K. S., Song, Y. H., Lee, B. H. and Hahn, C. S., *J. Org. Chem.*, **1986**, *51*, 104.
63. Sen, S. E., Roach, S. L., Boggs, J. K., Ewing, G. J. and Magrath, J., *J. Org. Chem.*, **1997**, *62*, 6684.
64. Bastiaana, H. M., Van DerBaan, J. L. and Ottenheijm, H. C. J., *J. Org. Chem.*, **1997**, *62*, 3880.
65. Bégué, J. P., M'Bida, A., Bonnet-Delpon, D., Novo, B. and Resnati, G., *Synthesis*, **1996**, 399.
66. Cameron, A. F. B., Hunt, J. S., Oughton, J. F., Wilkinson, P. A. and Wilson, B. M., *J. Chem. Soc.*, **1953**, 3864.
67. Bosch, M. P., Camps, F., Coll, J., Guessero, T., Tatsuoka, T. and Meinwald, T., *J. Org. Chem.*, **1986**, *51*, 773.

68. Tsunoda, T., Suzuki, M. and Noyori, R., *Tetrahedron Lett.*, **1980**, *21*, 1357.
69. Marcantoni, E., Nobili, F., Bartoli, G., Bosco, M. and Sambri, L., *J. Org. Chem.*, **1997**, *62*(12), 4183.
70. Lee, A. S. Y. and Cheng, C. L., *Tetrahedron*, **1997**, *53*, 14255.
71. Lipshutz, B. H., Mollard, P., Lindshy, C. and Chang, V., *Tetrahedron Lett.*, **1997**, *38*, 1873.
72. Groblel, B. T. and Seebach, D., *Synthesis*, **1977**, 357, and the references cited therein.
73. McHale, W. and Kutateladze, A., *J. Org. Chem.*, **1998**, *63*(26), 9924.
74. Vedjes, E. and Fuchs, P. L., *J. Org. Chem.*, **1971**, *36*, 366, and the references cited therein.
75. Habibi, M. H., Tangestaninejad, M., Montazerozohori, M., and Baltork, I. M., *Molecules*, **2003**, *8*, 663.
76. Firouzabadi, H., Hazarkhani, H., Karimi, B, Niroumand, U. and Ghassamipour, S., *Fourth International Electronic Conference on Synthetic Organic Chemistry (ECSOC-4)*, **2000.**
77. Cussans, N. J. and Ley, S. V., *J. Chem. Soc., Perkin Trans. 1*, **1980**, 1654.; Barton, D. H. R., Lester, D. J. and Ley, S. V., *J. Chem. Soc. Perkin Trans.* I, **1980**, 1.
78. Wright, S. W., Hageman, D. L., Wright, A. S. and McClure, L. D., *Tetrahedron Lett.*, **1997**, *38*, 7345.
79. Givens, R. S., Jung, A., Park, C. H., Weber, J. and Barlett, W., *J. Am. Chem. Soc.*, **1997**, *119*, 8369.
80. Khanbabaee, K. and Lotzerich, K., *J. Org. Chem.*, **1998**, *63*, 8723.
81. Roberts, J. C., Gao, H., Gopalsamy, A., Kongsjahju, A. and Patch, R. J., *Tetrahedron Lett.*, **1997**, *38*, 355.
82. Carreira, E. M. and Bois, D. J., *J. Am. Chem. Soc.*, **1995**, *117*, 8106.

Chapter 2
Reactive Intermediates

Reactive intermediates[1-4] are believed to be transient intermediates in the majority of reactions. The main types of reactive intermediates of interest to organic chemists are carbocations, carbanions, radicals, radical ions, carbenes, nitrenes, arynes, nitrenium ions and diradicals.

Reactive intermediate chemistry assists chemists in the design of new reactions for the efficient synthesis of pharmaceuticals, fine chemicals and agricultural products. Reactive intermediates are usually short lived, very reactive and are rarely isolated under normal reaction conditions. However, their structures are established by indirect means either by chemical trapping, spectroscopically or sometimes by isolating them at very low temperature. The shapes of these intermediates become important while considering the stereochemistry of reactions in which they play a role.

Carbocations are electrophiles and carbanions are nucleophiles. Reactions of these intermediates involving, at some stage, the bonding of a nucleophile to an electrophile are sometimes called ionic reactions.

2.1   Carbocations

Carbocation has a positively charged carbon atom which has only six electrons in its outer valence shell instead of the eight valence electrons (octet rule).

2.1.1   Structure and stability of carbocations

The heterolytic fission of a C—X bond in an organic molecule, in which X is more electronegative than carbon, generates the negatively charged anion (X$^-$) and positive charged species known as carbocations (called carbonium ions in the older literature).

(X is more electronegative than C)

The carbon atom in a typical carbocation is sp^2 hybridized. The p_z orbital is empty and is perpendicular to the plane of the other three bonds. Thus, carbocation adopts a trigonal planar shape.

Side view   Top view

Carbocation planar (sp^2)

Because carbocation assumes a planar structure, its formation is inhibited in compounds which do not permit attainment of a planar geometry as in the case of bridge head compounds. Also, on the basis of quantum mechanical calculations for simple alkyl carbocations, it has been found that the planar (sp^2) configuration is more stable than the pyramidal (sp^3) configuration by about 84 kJ (20 kcal) mol^{-1}. Thus, the difficulty in the formation of carbocations increases as the attainment of planarity is inhibited. The planar configuration of simple carbocations has also been confirmed by NMR and IR spectra. Ned Arnett measured carbocation stabilities directly by measuring the enthalpy of reaction for the ionization of the RX process in hypermedia SbF$_5$/FSO$_3$H/SO$_2$ClF at $-40°$C.

$$R_3C{-}X \longrightarrow R_3\overset{\oplus}{C} + \overset{\ominus}{X}$$

The electron-donating groups attached to positively charged carbons in carbocations increase the stability of the carbocations by inductive effect and/or hyperconjugation (no bond resonance). Thus, a tertiary carbocation is more stable than a secondary carbocation, which in turn is more stable than a primary carbocation.

However, the presence of electron-withdrawing groups adjacent to the carbon atom bearing positive charge makes the carbocation less stable (by $-$I and/or $-$M effects).

Resonance effect further stabilizes the carbocations when present. By resonance the positive charge on the central carbon atom gets dispersed over other carbon atoms and this renders stability to the carbocation. The more the canonical (resonating) structures for a carbocation, the more stable it will be. For example, benzyl and allyl carbocations are very stable because of resonance.

Benzyl carbocation
(Canonical structures)

Allyl carbocation
(Canonical structures)

Thus, the order of the stability of the benzyl, allyl and *n*-propyl carbocations is as follows:

$$C_6H_5\overset{\oplus}{C}H_2 > CH_2{=}CH{-}\overset{\oplus}{C}H_2 > CH_3CH_2\overset{\oplus}{C}H_2$$

In certain cases, the carbocations are so stable that their solid salts have been isolated. For example, triphenylmethyl perchlorate (**2.1**) exists as a red crystalline solid and tropylium bromide (**2.2**) has been isolated as a yellow solid. Tropylium bromide (**2.2**) is stabilized by aromatization as the tropylium cation is planar and has 6π electrons like benzene.

2.1   **2.2**

The order of the stability of the tropylium, triphenylmethyl, benzyl and allyl carbocations is as follows:

$$\text{Tropylium cation} \quad > \quad (C_6H_5)_3\overset{\oplus}{C} \quad > \quad C_6H_5\overset{\oplus}{C}H_2 \quad > \quad H_2C=CH-\overset{\oplus}{C}H_2$$

Triphenylmethyl   Benzyl carbocation   Allyl carbocation
carbocation

The order of the stability of the alkyl (1°, 2° and 3°), allyl and benzyl carbocations is as follows:

$$\overset{\oplus}{C}H_3 \; < \; CH_3\overset{\oplus}{C}H_2 \; < \; (CH_3)_2\overset{\oplus}{C}H \; = \; CH_2=CH-\overset{\oplus}{C}H_2 \; < \; C_6H_5\overset{\oplus}{C}H_2 \; < \; (CH_3)_3\overset{\oplus}{C}$$

The stability of the carbocation is also increased due to the presence of heteroatom having an unshared pair of electrons, e.g. oxygen, nitrogen or halogen, adjacent to the cationic centre. Such carbocations are stabilized by resonance. The methoxymethyl cation is obtained as a stable solid, $MeOCH_2{}^+SbF_4{}^-$.

$$\underset{\oplus}{\overset{H}{H-C-\ddot{O}-Me}} \quad \longleftrightarrow \quad \overset{H}{H-C=\overset{\oplus}{\ddot{O}}-Me}$$

The stability of the carbocation increases by the presence of each additional cyclopropyl group.

2.1.2 Generation of carbocations

From alcohols
Alcohols on treatment with concentrated acid get protonated and then may lose a molecule of water to form carbocations.

$$R-OH \; + \; \overset{\oplus}{H} \; \longrightarrow \; R\overset{\oplus}{O}H_2 \; \xrightarrow{-H_2O} \; \overset{\oplus}{R}$$

From alkyl halides
Ionization of alkyl halides gives carbocations.

$$R-X \; \xrightarrow{\text{polar solvent}} \; \overset{\oplus}{R} \; + \; \overset{\ominus}{X}$$

X = I, Br, Cl

The process is accelerated by the presence of powerful ion-solvating medium and metal ions such as Ag^+ ions or Lewis acid. In place of alkyl halides, alkyl tosylates and alkyl mesylates can also be used.

Friedel–Crafts alkylation of aromatic compounds involves the formation of a carbocation that acts as electrophile (see section 2.1.3).

From alkenes

Addition of proton to alkenes give carbocations. The addition is regioselective and follows **Markovnikov's addition rule** (also spelt as Markownikoff's). The rule states that with the addition of H–X to an alkene, the acid hydrogen (H) becomes attached to the carbon with the greatest number of hydrogens and the halide (X) group becomes attached to the carbon with the fewest number of hydrogens. The addition of the hydrogen to one carbon atom in the alkene creates a positive charge on the other carbon, forming a carbocation intermediate. The more substituted the carbocation the more stable it is, due to +I effect and hyperconjugation.

$$(CH_3)_2C{=}CH_2 \ + \ \overset{\oplus}{H} \longrightarrow (CH_3)_2\overset{\oplus}{C}{-}CH_3$$
$$3^0\text{-carbocation}$$

From diazonium ions

The alkyl diazonium ions (in contrast to aryl diazonium ions) are unstable and decompose at room temperature to give carbocations.

$$R{-}\overset{\oplus}{N}{\equiv}N \longrightarrow \overset{\oplus}{R} \ + \ N_2$$

From acyl halides

The acyl halides (RCOX) on treatment with anhydrous aluminium chloride ($AlCl_3$) give a complex, which decomposes to give acyl electrophile, an acylium ion (RCO^+). Friedel–Crafts acylation of aromatic compounds involves the formation of a carbocation that acts as an electrophile (see section 2.1.3).

$$\underset{R}{\overset{O}{\|}}{\overset{}{C}}{-}Cl \ \xrightarrow{AlCl_3} \ \left[R{-}\overset{\oplus}{C}{=}\ddot{O}{:} \longleftrightarrow R{-}C{\equiv}\overset{\oplus}{O}{:} \right] \ + \ \overset{\ominus}{AlCl_4}$$
$$\text{Acylium ion}$$

2.1.3   Reactions of carbocations

Carbocations usually undergo elimination reactions, addition reactions, reactions with nucleophiles and rearrangements.

Elimination of a proton

A carbocation may lose a proton to form an alkene. For example, 1-propyl carbocation generated from diazonium salt may lose a proton (H^+) to form an alkene (propene). Alternatively, 1-propyl carbocation may rearrange to more stable secondary carbocation, which may also lose a proton to give propene (Scheme 2.1).

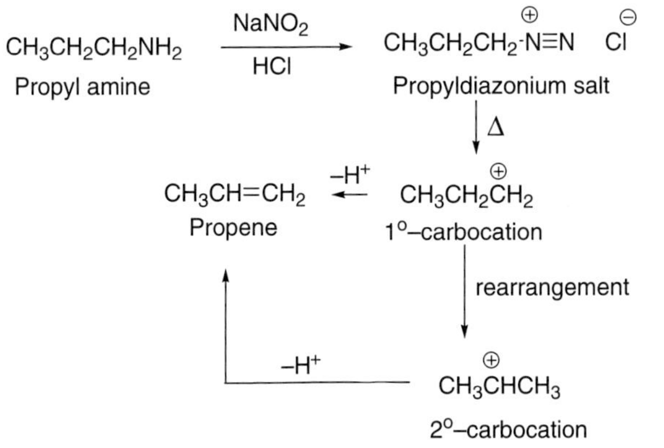

Scheme 2.1

Reaction with alkenes and aromatic systems

A carbocation may react with an alkene to produce another carbocation.

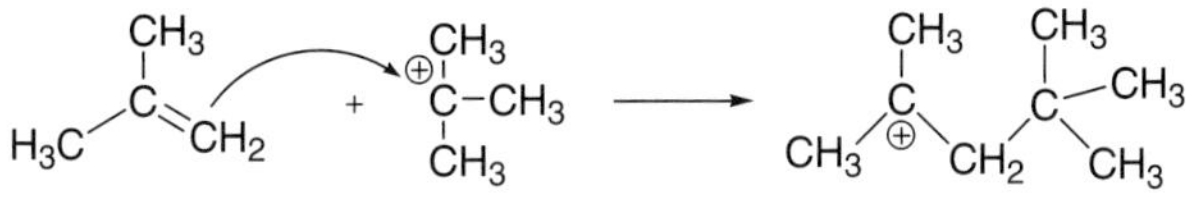

2-Methylpropene

The alkyl carbocations formed from alkyl halides, alkenes and alcohols act as an electrophile in **Friedel–Crafts alkylation reactions**[5].

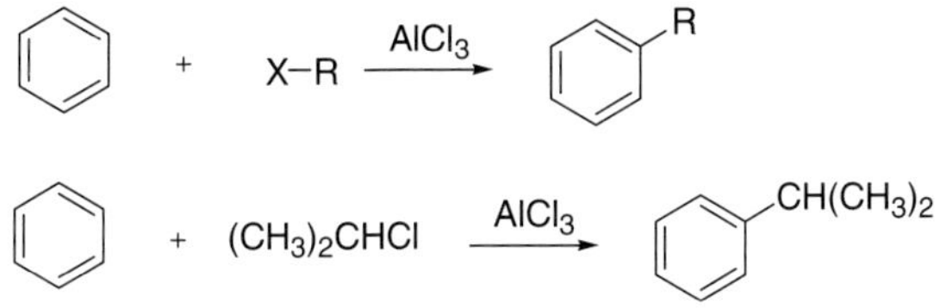

The Friedel–Crafts alkylation mechanism involves the generation of an electrophile by adding an alkyl halide to the Lewis acid aluminium trichloride, which results in the formation of an organometallic complex. In this complex the carbon attached to the chlorine has a great deal of positive charge character (in fact, for practical purposes it is considered as a carbocation).

$$R-Cl \ + \ AlCl_3 \ \longrightarrow \ \overset{\delta+}{R}---\overset{\delta-}{Cl}--AlCl_3$$

After the generation of electrophile, π-electrons in a benzene ring attack the positive carbon to create a non-aromatic intermediate, which has several resonance structures. Elimination of a proton reestablishes the aromaticity of the ring, and the aluminium trichloride catalyst is regenerated along with a molecule of hydrochloric acid (Scheme 2.2).

In a similar manner, acyl carbocations formed from acyl halides act as an electrophile in **Friedel–Crafts acylation reactions**[5] (Scheme 2.3).

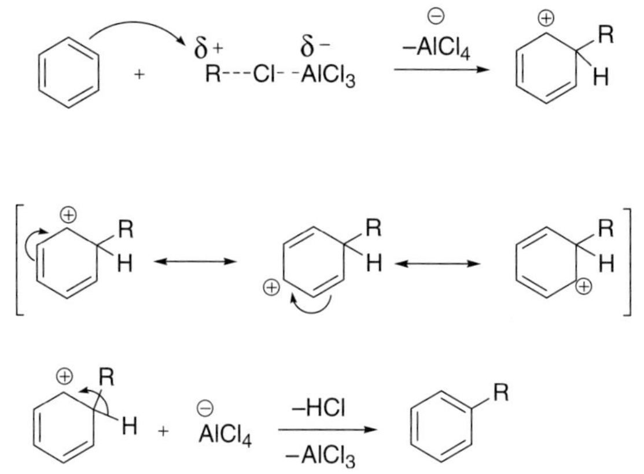

Scheme 2.2

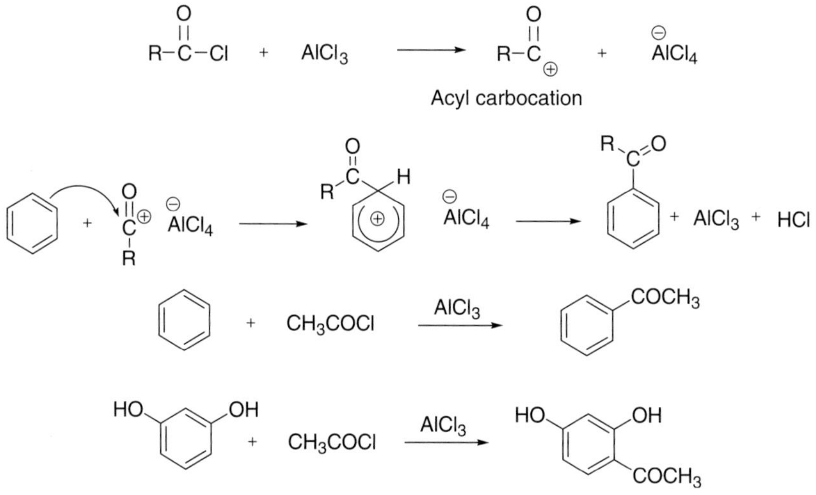

Scheme 2.3

Reaction with nucleophiles

A carbocation may combine with a nucleophile to form a new bond. For example, addition of the electrophile H^+ to a propene produces the most stable product (secondary carbocation in this case according to Markovnikov's addition rule). The second step involves nucleophilic attack by Cl^- to give isopropyl chloride (Scheme 2.4).

The reaction of a carbocation with a neutral nucleophile such as water gives a protonated alcohol. Tertiary butyl carbocation, for example, reacts with water (neutral nucleophile) to give protonated *tert*-butyl alcohol, which eliminates a proton to give *tert*-butyl alcohol (Scheme 2.5).

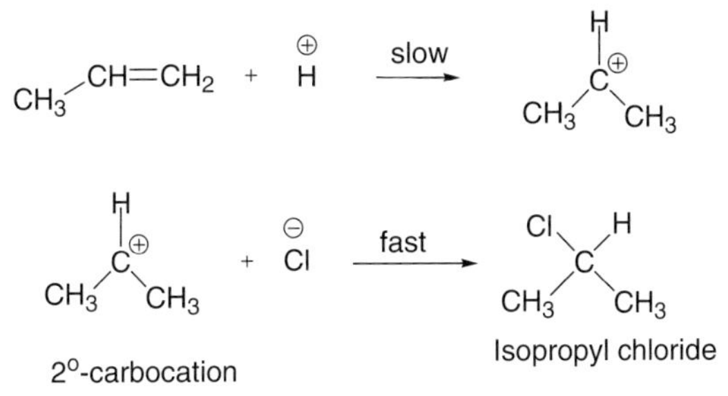

Scheme 2.4

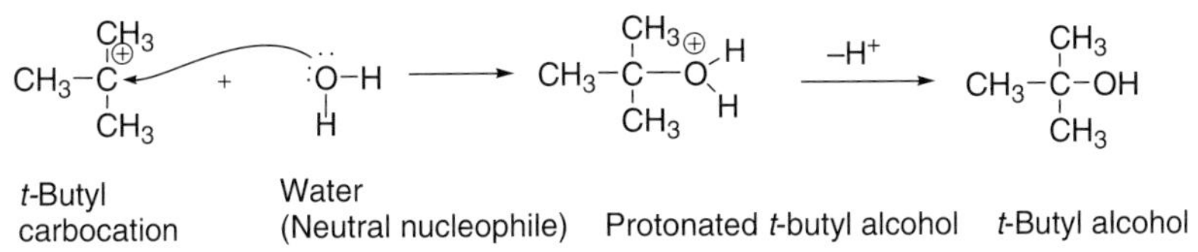

Scheme 2.5

Rearrangement of carbocations

Molecular rearrangements involving carbocations as reactive intermediates are very common in organic chemistry. The first-formed carbocation, which is less stable, can rearrange by 1,2-shift of either H or alkyl group to more stable carbocation.

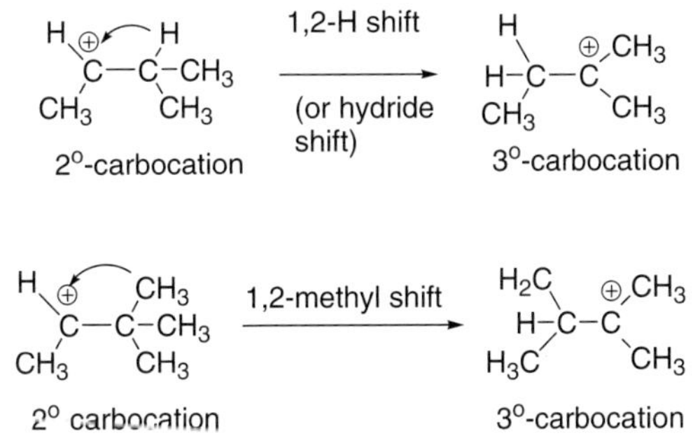

For example, hydrolysis of a secondary alkyl halide 2-iodo-3-methylbutane (**2.3**) gives rearranged tertiary alcohol 2-methyl-2-butanol (**2.4**) (Scheme 2.6).

Wagner–Meerwein rearrangements[6–8]

The low S_N2 reactivity of 1°-alkyl bromide, 2,2-dimethyl-1-bromopropane (neopentyl bromide, **2.5**), is explained by steric hindrance to the required 180° alignment of reacting orbitals. However, under S_N1 conditions, neopentyl bromide (**2.5**) reacts at roughly the same rate as other 1°-alkyl halides such as ethyl bromide. Ionization of alkyl halides to carbocation in S_N1 is the rate-determining step. Although the product from ethyl bromide is ethanol as expected, neopentyl bromide (**2.5**) yields 2-methyl-2-butanol (**2.6**) instead of the expected 2,2-dimethyl-1-propanol (neopentyl alcohol) (**2.7**). This is because once formed the ethyl carbocation can only be transformed by a substitution or elimination process. In the case of the neopentyl carbocation, however, the initially formed 1°-carbocation may be converted

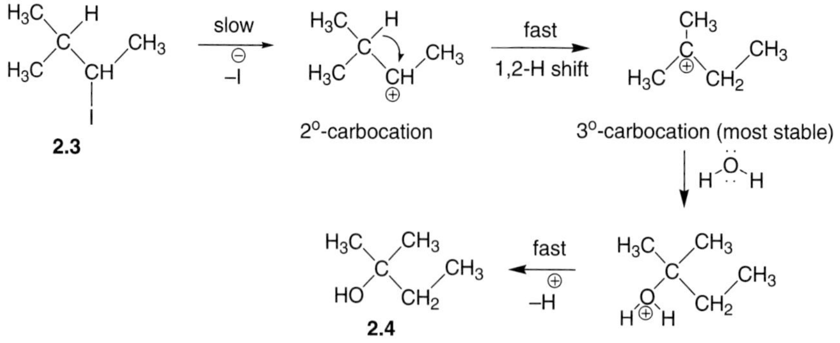

Scheme 2.6

into a more stable 3°-carbocation by the 1,2-shift of an adjacent methyl group with its bonding electrons (Scheme 2.7). The rearrangement that involves migration of the methyl group is often referred to as **Nametkin rearrangement**.

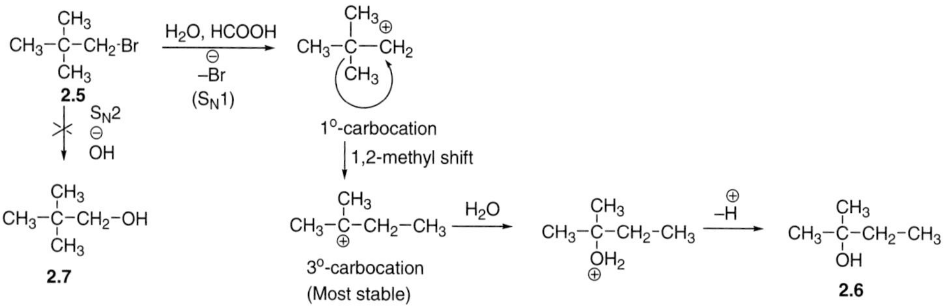

Scheme 2.7   Rearrangement of the neopentyl system

Increasing the stability of the carbocation intermediates is not the only factor that leads to molecular rearrangement. If angle strain, torsional strain or steric crowding in the reactant structure is relieved by an alkyl or aryl shift to a carbocation site, such a rearrangement is commonly observed. Thus, in some cases molecular rearrangements involving carbocations may lead to ring expansion, as shown in Scheme 2.8.

Many of the most interesting rearrangements involving 1,2-shifts were discovered during structural studies of naturally occurring compounds such as terpenes by H. Meerwein and G. Wagner.

In the conversion of α-pinene (**2.8**) into bornyl chloride (**2.9**) (*endo* isomer), the rearrangement to a 2° carbocation is favoured by relief of small-ring strain (Scheme 2.9). In a similar manner the conversion of camphene hydrochloride (**2.10**) into isobornyl chloride (**2.11**) involves rearrangement known as the Wagner–Meerwein rearrangement (Scheme 2.10).

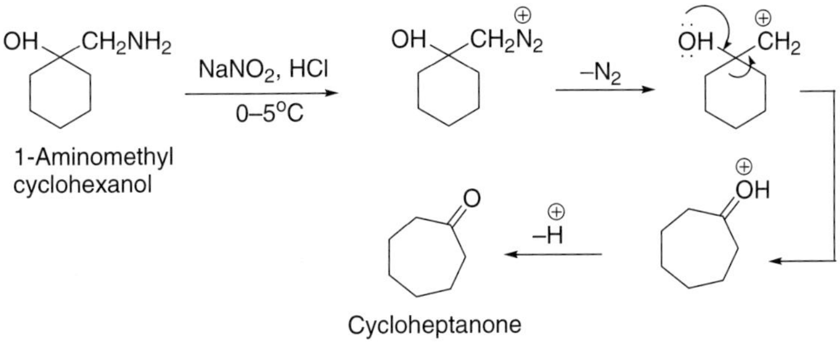

1-Aminomethyl
cyclohexanol

NaNO$_2$, HCl
0–5°C

–N$_2$

–H

Cycloheptanone

Scheme 2.8

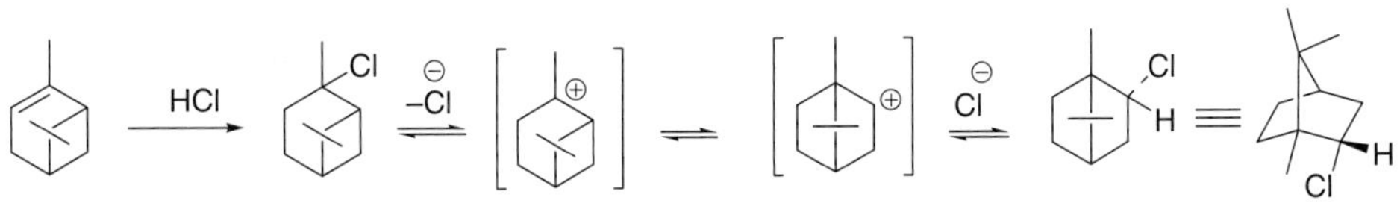

HCl

–Cl

2.8   **2.9**

Scheme 2.9

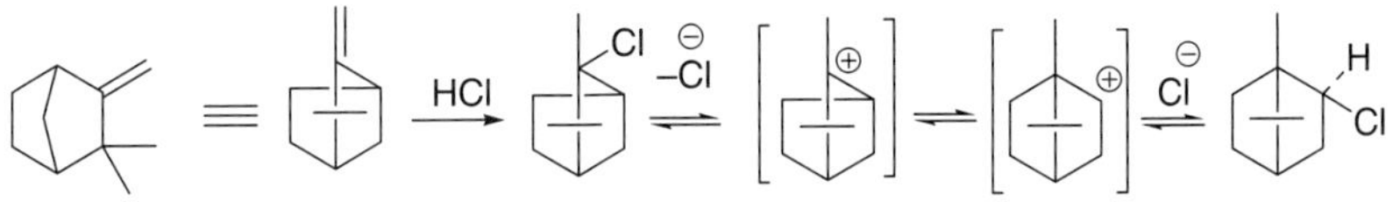

HCl

–Cl

2.10   **2.11**

Scheme 2.10

Pinacol–pinacolone rearrangement

2,3-Dimethyl-2,3-butanediol (pinacol) (**2.12**) on treatment with H$_2$SO$_4$ generates 3,3-dimethyl-2-butanone, commonly known as pinacolone (**2.13**). Pinacol itself is produced by magnesium reduction of acetone, probably by way of a ketyl intermediate.

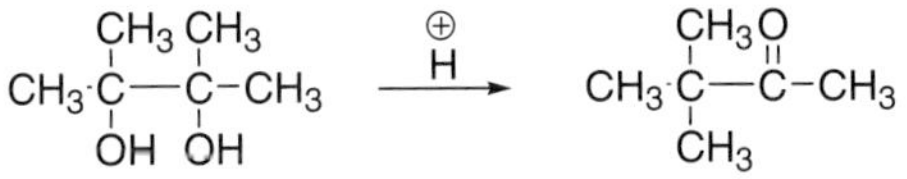

$$CH_3\text{-}C\text{—}C\text{-}CH_3 \xrightarrow{\overset{\oplus}{H}} CH_3\text{-}C\text{—}C\text{-}CH_3$$

2.12   **2.13**

The mechanism involves the loss of water molecule from a protonated diol, followed by 1,2-nucleophilic shift of a group. Since the diol is symmetrical, protonation and loss of water take place with equal probability at either hydroxyl group. The resulting 3°-carbocation is relatively stable but 1,2-methyl shift generates an even more stable carbocation in which the charge is delocalized by heteroatom resonance (Scheme 2.11).

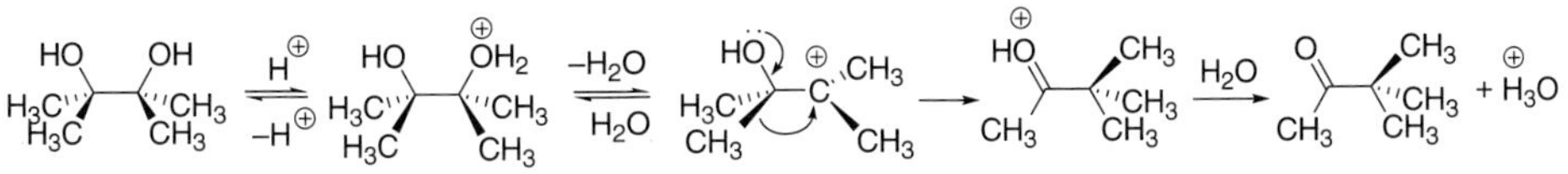

$\xrightarrow[-H^{\oplus}]{H^{\oplus}}$ $\xrightarrow[H_2O]{-H_2O}$ $\xrightarrow{}$ $\xrightarrow{H_2O}$ + H$_3$O$^{\oplus}$

2.12   **2.13**

Scheme 2.11   Mechanism of the pinacol–pinacolone rearrangement

In the case of unsymmetrical glycols the product formation depends mainly on which OH group is lost to leave behind the most stable carbocation, and thereafter which group migrates. The order of migrating aptitude is Ar > H > R.

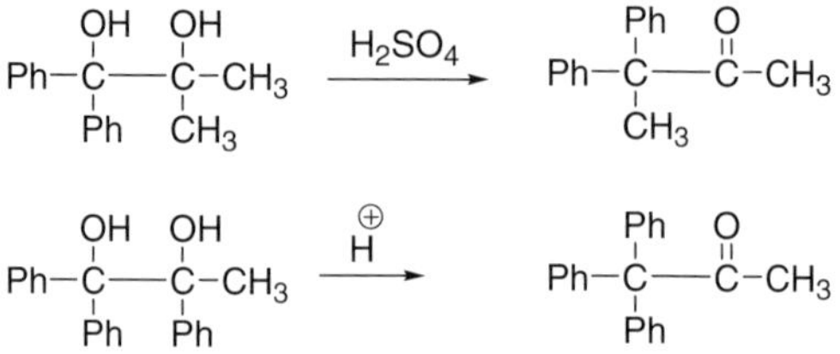

The clear preference for a methylene group shift over a methyl group shift explains the predominant ring contraction in Scheme 2.12.

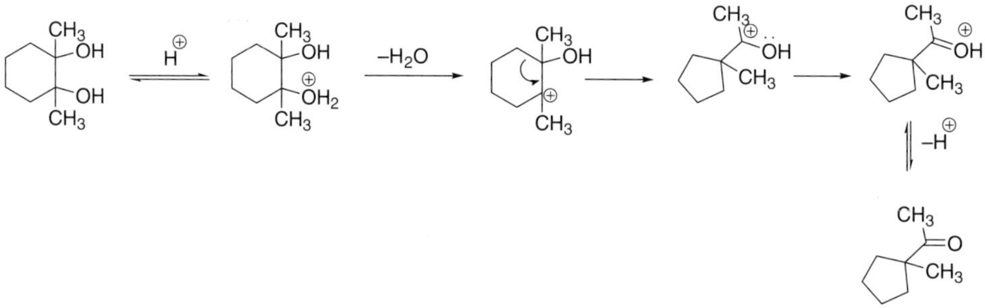

Scheme 2.12

Strained ring contracts to an even smaller ring. Phenyl groups generally have a high migratory aptitude, so the failure to obtain 2,2-diphenylcyclobutanone (**2.15**) as a product from **2.14** might seem surprising. However, the carbocation resulting from a phenyl shift would be just as strained as its precursor, whereas the shift of a ring methylene group generates **2.16** in which an unstrained carbocation is stabilized by phenyl and oxygen substituents. Conjugative stabilization of the phenyl ketone and absence of sp^2 hybridized carbon atoms in the small ring may also contribute to the stability of the observed product.

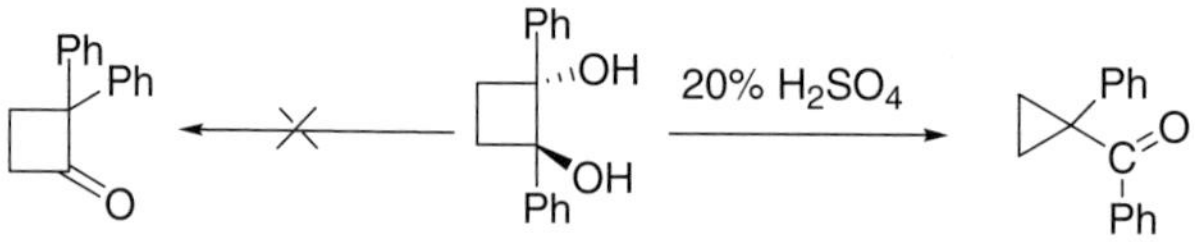

2.15   **2.14**   **2.16**

2.1.4   Non-classical carbocations

A non-classical carbocation is a cyclic, bridged structure with delocalized σ-bond in 3-centre-2-electron bonds of bridged system.

It is well known that acetolysis of both *exo*-2-norbornyl brosylate (**2.17**) and *endo*-2-norbornyl brosylate (**2.18**) produces exclusively *exo*-2-norbornyl acetate (**2.19**). However, **2.17** is 350 times more reactive than **2.18** (Scheme 2.13). Further, optically active **2.17** gives completely racemic *exo*-acetate **2.19**, whereas **2.18** is substituted with slight inversion of configuration (93% racemic *exo*) (Scheme 2.14).

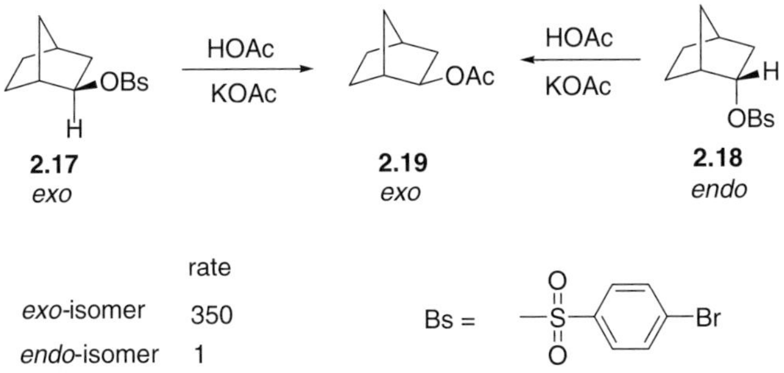

2.17
exo

2.19
exo

2.18
endo

	rate
exo-isomer	350
endo-isomer	1

Bs =

Scheme 2.13

2.17
exo

50%   50%

2.19, *exo*
(Complete racemization)

2.18
endo

46%   54%

2.19, *exo*

Optical activity retained but
slight inversion of configuration)

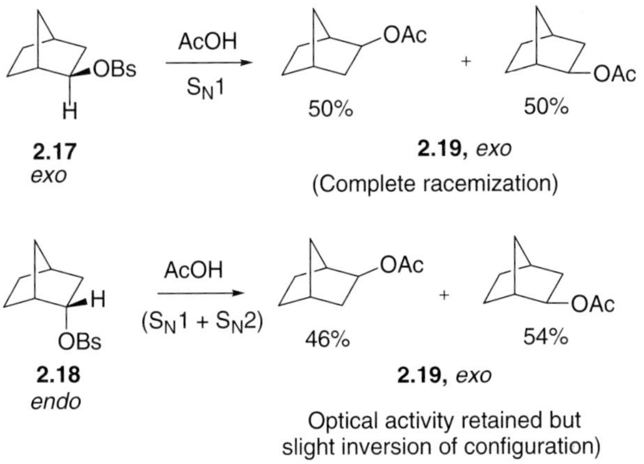

Scheme 2.14

The rate of substitution of the *endo*-brosylate **2.18** was considered normal, since its reactivity is comparable to that of cyclohexyl brosylate. Ionization of the *exo*-brosylate **2.17** is assisted by the neighbouring C1–C6 bonding electrons participation with the expulsion of the leaving group. The non-classical carbocation **2.20** is formed as an intermediate in which positive charge residing on C1 is delocalized on C2 as well (Scheme 2.15).

The stereochemical outcome is due to a symmetric intermediate **2.20** which is achiral and has a plane of symmetry passing through C4, C5, C6 and the midpoint of C1–C2 bond. C6 is pentacoordinate and serves as a bridging atom in the cation.

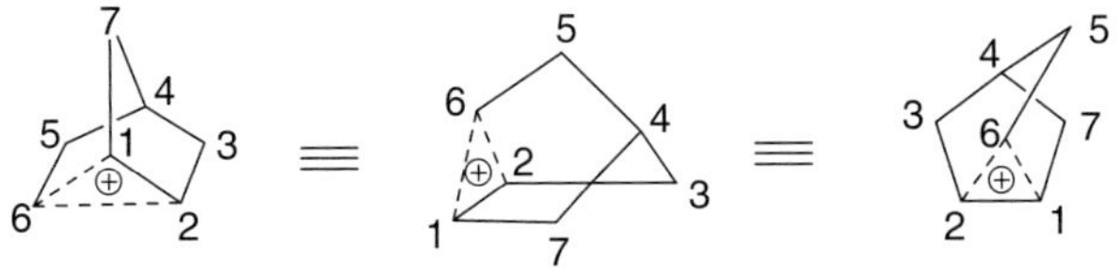

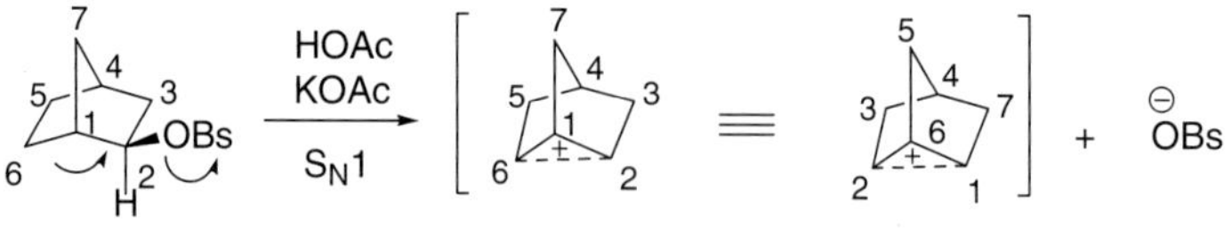

2.17   **2.20**
                                   Norbornyl carbocation
                                   (Non-classial carbocation)

Scheme 2.15

Attack of acetate at C1 or C2 would be equally likely and would result in equal amounts of enantiomeric acetates (Scheme 2.16). The acetate ester would be *exo*, **2.19**, since the reaction must occur from the direction opposite that of the bridging interaction. The non-classical carbocation **2.20** can be formed only from the *exo*-brosylate **2.17** because it has the proper *anti* relationship between C1–C6 bond and the leaving group. The bridged ion **2.20** can be formed from the *endo*-brosylate **2.18** only after an unassisted ionization, which is then converted into a non-classical carbocation, **2.20**.

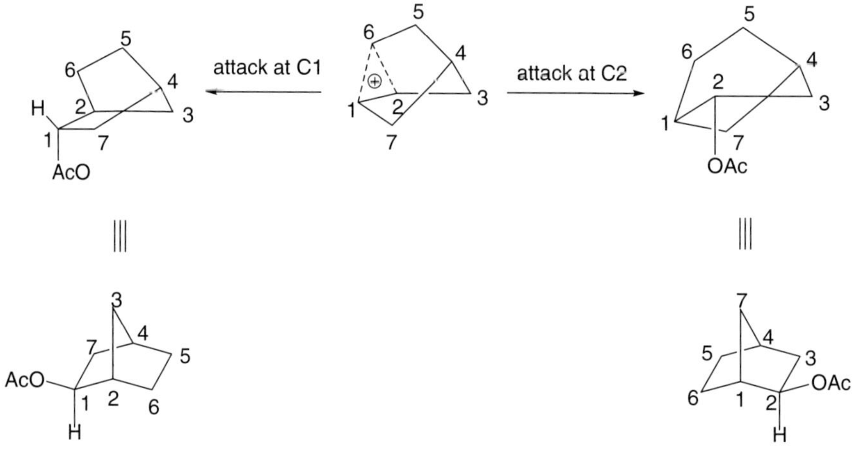

Scheme 2.16

However, not everyone was convinced by the existence of the non-classical carbocation. H. C. Brown[9] 1977 pointed out that the norbornyl compounds are compared with cyclopentyl rather than with cyclohexyl analogues, **2.21** (eclipsing strain), and in such a comparison the *endo*-isomer is abnormally slow, the *exo*-isomer being only 14 times faster than cyclopentyl analogues. He also pointed out that the formation of racemic product is due to two rapidly equilibrating classical carbocation species (Scheme 2.17). The interconversion of enantiomeric classical carbocation species must be very rapid on the reaction timescale.

Despite Brown's classical realization to solve this controversy, several experiments including tracer studies provided additional support for the non-classical view. Although there was some damage to the theory of non-classical carbocations because of contentious debates between Winstein and Brown, the extraordinary detailing of mechanisms and structure was a major achievement of human intellect.

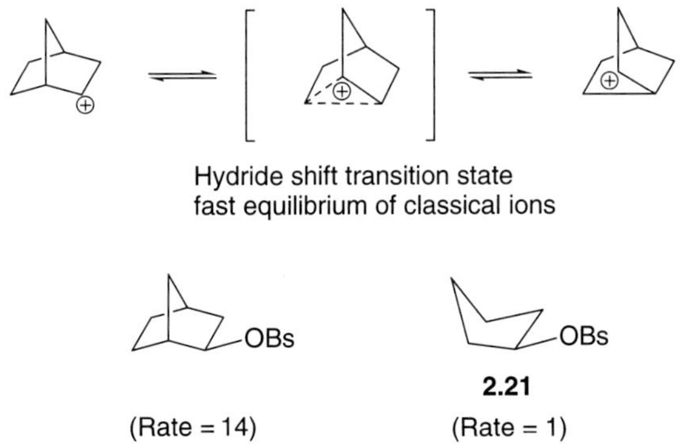

Hydride shift transition state
fast equilibrium of classical ions

2.21

(Rate = 14)   (Rate = 1)

Scheme 2.17

2.2 Carbanions

A carbanion can be considered as a species containing a trivalent carbon with a lone pair of electrons and is therefore negatively charged.

2.2.1 Structure and stability of carbanions

Carbanions are considered to be derived by the heterolytic fission of the C−X bond in an organic molecule in which carbon is more electronegative than X.

(C is more electronegative than X)   Carbanion

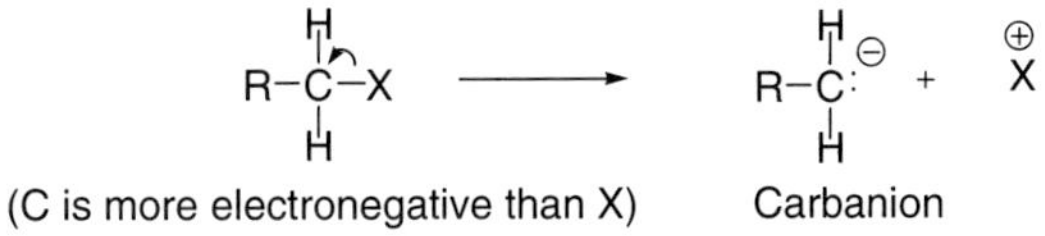

The shape of simple carbanions as determined on the basis of a number of experiments is found to be pyramidal, similar to that of amines. The central carbon atom is sp^3 hybridized with the unshared pair occupying one apex of the tetrahedron (if the electron pair is viewed as a substituent). These species invert rapidly at room temperature, passing through a higher energy planar form in which the electron pair occupies a p-orbital. However, when the carbanion is stabilized by delocalization, it assumes sp^2 hybridization for effective resonance.

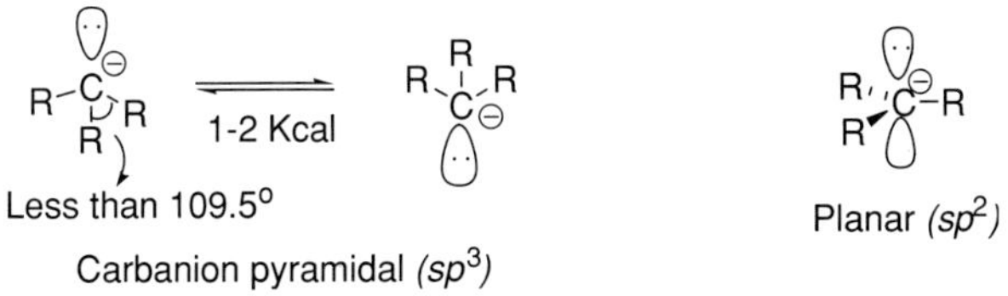

1-2 Kcal

Less than 109.5°   Planar (sp^2)

Carbanion pyramidal (sp^3)

In practice, any organic compound having a C−H bond (all such compounds can be considered to be an acid in the classical sense) can donate a proton to a suitable base; the species obtained as a result is the carbanion.

$$R_3C\!-\!H \ + \ \overset{\ominus}{B}: \ \rightleftharpoons \ R_3\overset{\ominus}{C} \ + \ BH$$

Table 2.1 Relative stability of the carbanions

Acid	Base	Approximate pK_a
RCH_2CN	$R\overline{C}HCN$	25
$HC\equiv CH$	$HC\equiv C^-$	25
Ar_3CH	Ar_3C^-	31.5
Ar_2CH_2	Ar_2CH^-	33.5
$PhCH_3$	$PhCH_2^-$	38
$CH_2{=}CH_2$	$CH_2{=}CH^-$	44
Cyclopropane (Δ)	Δ^-	46
$(CH_3)_2CH_2$	$(CH_3)_2CH^-$	51

A carbanion possesses an unshared pair of electrons and is therefore a base. Thus, a carbanion may accept a proton to give its conjugate acid. In fact, the stability of a carbanion depends on the strength of the conjugate acid. The weaker is the acid (higher pK_a value), the greater is the strength of a conjugate base and therefore lower will be the stability of the carbanion. Table 2.1 shows the conjugate bases obtained from the corresponding acids along with pK_a values.

The carbanions being electron rich are very reactive intermediates and are readily attacked by electrophiles (electron-deficient reagents). The stability of the carbanion is increased by electron-withdrawing groups ($-I$, $-M$). However, the stability is decreased by electron-donating groups ($+I$, $+M$). Thus, 3° carbanion is less stable than 2°, which is less stable than 1° in solution due to destabilization of electron-donating alkyl groups.

Like carbocations, carbanions are also stabilized by resonance. Thus, benzyl carbanion and allyl carbanion are more stable than ethyl carbanion. The stabilization by resonance is due to the delocalization of the negative charge, which is distributed over other carbon atoms. The canonical (resonating) forms of benzyl carbanion and allyl carbanion are given below:

Canonical forms of benzyl carbanion

Canonical forms of allyl carbanion

When a functional group X is present at the α-position to the carbon having negative charge, it may increase or decrease the stability of the carbanion. The effect of various groups (X) on the stability of the carbanion is of the order:

$$X = NO_2 > RCO > COOR > SO_2 > CN \approx CONH_2 > \text{halogens} > H > R$$

When there is an increase in s character of the carbanionic carbon the carbanion becomes stable.

$$CH_3CH_2^- < CH_2{=}CH^- < HC\equiv C^-$$
$$sp^3 \qquad\quad sp^2 \qquad\quad sp$$

If the unshared pair of electrons is involved in ring current and system becomes aromatic, carbanions become greatly stabilized.

The stability of various carbanions is found to be in the following order:

Benzyl > vinyl > phenyl > cyclopropyl > ethyl > *n*-propyl > isobutyl > neopentyl > cyclobutyl > cyclopentyl.

2.2.2 Generation of carbanions

Following methods are generally used for the generation of carbanions.

Abstraction of H by a base

An appropriate organic substrate having a C−H bond on treatment with a suitable base results in the abstraction of hydrogen to generate a carbanion.

From unsaturated compounds

Addition of a nucleophile to carbon–carbon double bond generates a carbanion.

From alkyl halides

Reduction of carbon–halogen bond by metal yields carbanion. Reaction of alkyl halide with Mg in the presence of anhydrous ether as solvent generates Grignard reagent. The Grignard reagent behaves like a carbanion. Alkyllithiums are also obtained from alkyl halides and behave as carbanions.

$$R-X + Mg \xrightarrow{\text{dry ether}} RMgX$$

$$R-X + 2\,Li \xrightarrow{\text{dry pentane}} RLi + LiX$$

2.2.3 Reactions of carbanions

Carbanions take part in the usual types of reactions, viz., addition, elimination, displacement, oxidation and rearrangement.

Addition reactions

The carbanions being electron rich behave like nucleophiles and add to the carbonyl group, for example, aldol condensation (see Chapter 3 for the addition reactions of carbanion to the carbonyl group) (Scheme 2.18).

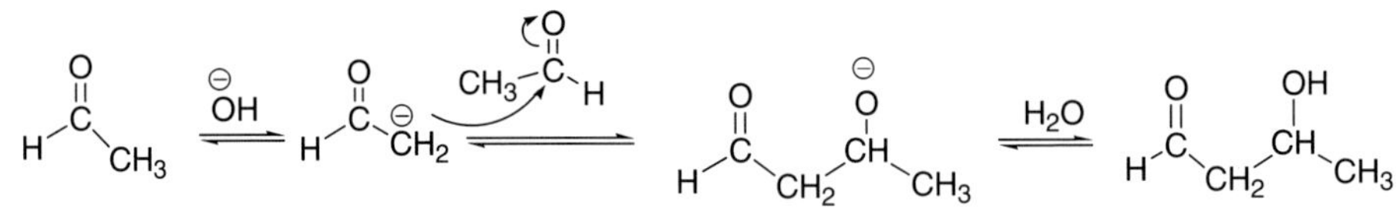

Scheme 2.18

Elimination reactions

Carbanions are involved as intermediates in E1cB elimination reactions.

$$Cl_2CHCF_3 \xrightarrow{\text{base}} Cl_2\overset{\ominus}{C}CF_3 \xrightarrow{-F^{\ominus}} Cl_2C=CF_2$$

In decarboxylation, the loss of CO_2 from the carboxylate anion is believed to involve a carbanion intermediate, which acquires a proton from solvent or other sources. The anion of β-ketoacids can undergo facile decarboxylation (Scheme 2.19).

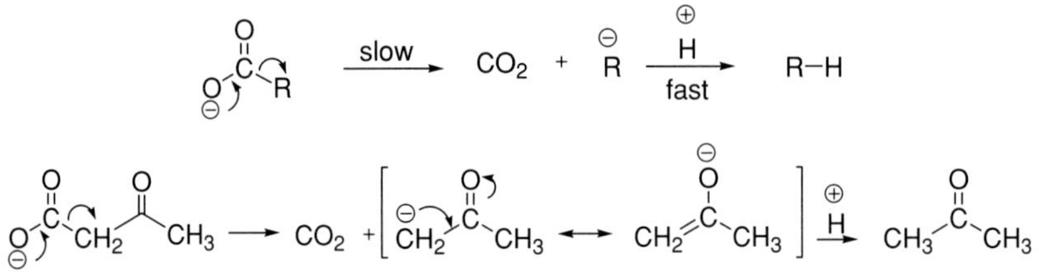

Scheme 2.19

Displacement reactions

The carbanions are involved in a number of displacement reactions. The synthetic applications of diethyl malonate (**2.22**) and acetoacetic ester (ethyl acetoacetate) (**2.23**) are due to the formation of the corresponding carbanions (Scheme 2.20) (also see Chapter 3).

The alkynyl (or acetylide) carbanions undergo alkylation in a similar manner.

$$RC{\equiv}CH \xrightarrow[]{\overset{\oplus\ominus}{NaNH_2}} RC{\equiv}C\overset{\ominus\oplus}{Na} \xrightarrow{RBr} RC{\equiv}C-R + NaBr$$

Alkyne   Alkynyl anion   Alkyne

Oxidation of carbanions

Under suitable conditions the carbanions can be oxidized. Thus, triphenylmethyl carbanion is oxidized to triphenylmethyl radical slowly by air. The triphenylmethyl radical so obtained can be reduced back to the carbanion by shaking with sodium amalgam.

$$Ph_3C\overset{\ominus\oplus}{Na} \underset{Na/Hg}{\overset{O_2}{\rightleftharpoons}} Ph_3\overset{.}{C}$$

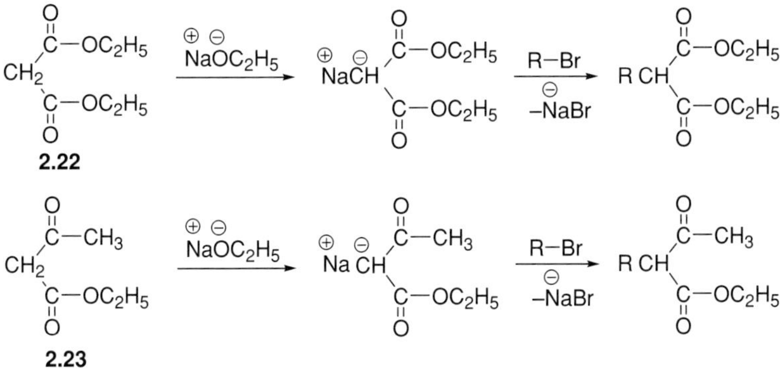

Scheme 2.20

In certain cases, the carbanions can be oxidized with one-electron oxidizing agents like iodine (Scheme 2.21).

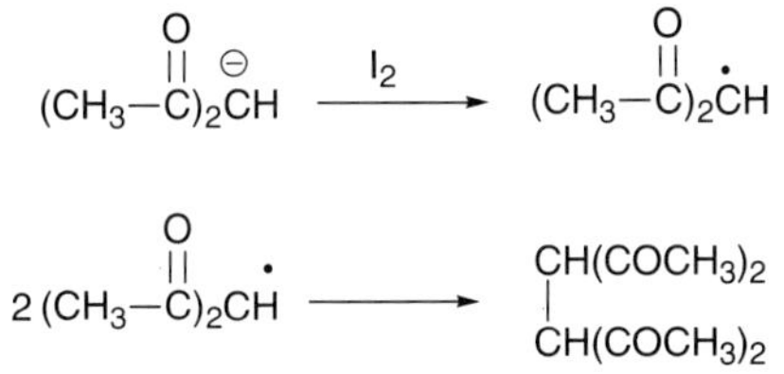

Scheme 2.21

This is a useful reaction for the formation of a carbon–carbon bond via dimerization of the radicals formed by the oxidation of carbanions.

Rearrangement reactions

Rearrangements involving carbanions are much less common than those involving carbo-cations. One such example is given in Scheme 2.22.

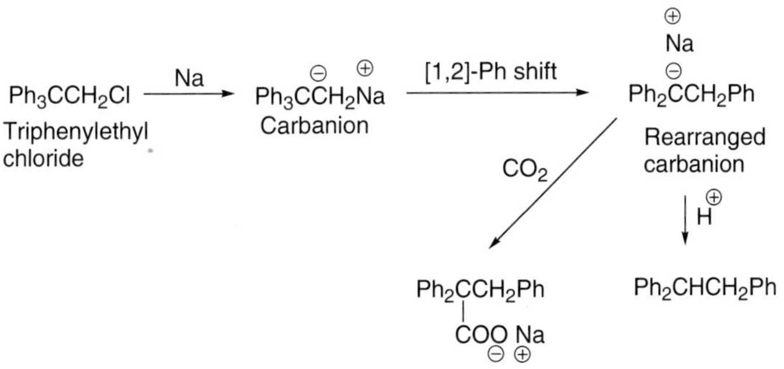

Scheme 2.22

The migrating group shifts to the carbanion carbon without an electron pair as compared to rearrangements involving carbocations in which the migrating group migrates with an electron pair. Simple 1,2-shifts of alkyl from carbon to carbon are not many. However, examples are recorded in literature in which the alkyl is involved in the 1,2-shift from N or S to a carbanion carbon, for example **Stevens rearrangement**.

Stevens rearrangement[10–12]

A quaternary ammonium salt, in which none of the alkyl group is having a β-hydrogen atom but one of the alkyl group has an electron-withdrawing group β to the nitrogen atom, undergoes base-catalyzed rearrangement to yield a tertiary amine. The rearrangement involves migration of a group, without pair of electrons, from nitrogen to carbon having negative charge. For example, phenacylbenzyldimethylammonium bromide (**2.24**) gives α-dimethylamino-β-phenylpropiophenone (**2.25**) on treatment with aqueous hydroxide.

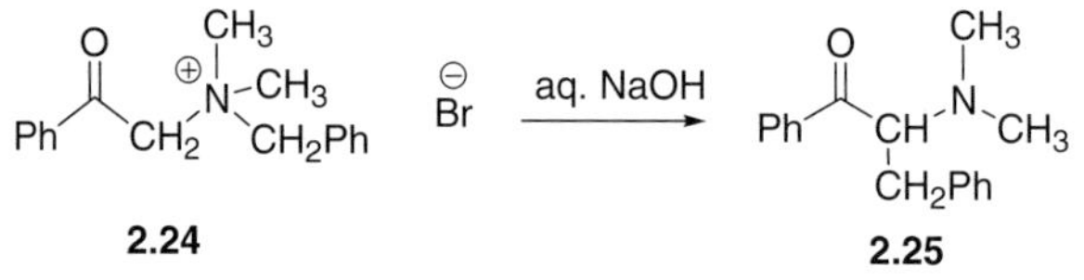

However, quaternary ammonium salts having a β-hydrogen atom undergo elimination with base to give alkene and tertiary amine. The reaction is known as the **Hofmann elimination**[13–15].

The role of the electron-withdrawing group, such as a carbonyl group, in the Stevens rearrangement is to stabilize the ylide. Benzyl group migrates in preference to alkyl group. Crossover experiments showed the mechanism of the Stevens rearrangement to be intramolecular (Scheme 2.23). It was thought earlier that this rearrangement occurred in a concerted manner. The base abstracts the hydrogen from the ammonium salt **2.26** to give the ylide **2.27**, which rearranges to give tertiary amine **2.28**. It was further observed that the rearrangement occurs with retention of absolute stereochemistry at the migrating centre.

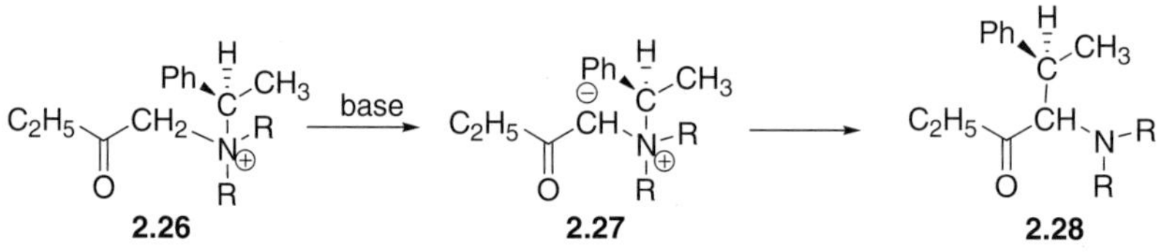

Scheme 2.23

Concerted rearrangement would not be an allowed reaction according to orbital symmetry rules. Thus a concerted mechanism is ruled out and a radical pair mechanism is proposed. The radical pair mechanism involves deprotonation followed by homolytic fragmentation of ylide **A** to produce a pair of radicals **B**. The rapid recombination of pair of radicals, which remain together in a tight solvent cage, gives the final product (Scheme 2.24). Formation of small amounts of coupling product R–R supports this mechanism.

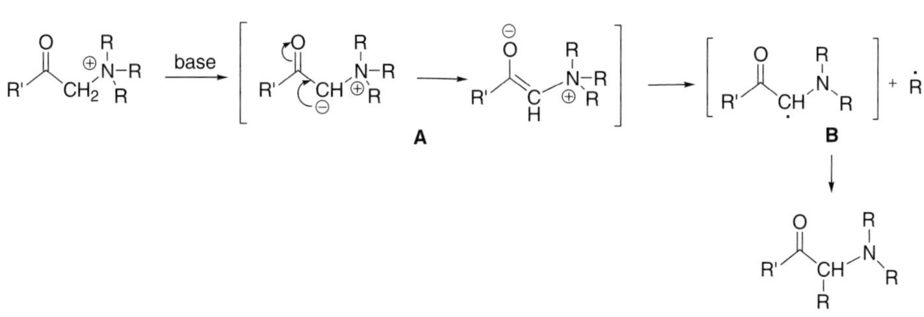

Scheme 2.24 Free radical dissociation–recombination mechanism

A third mechanism is also proposed in which ion pairs are formed instead of radical pairs. A variant of the Stevens rearrangement is the rearrangement of sulfur ylides.

Favorskii rearrangement[16–22]

The base-catalyzed rearrangement of α-haloketones (chloro and bromo) to carboxylic acid derivatives is called **Favorskii rearrangement**.

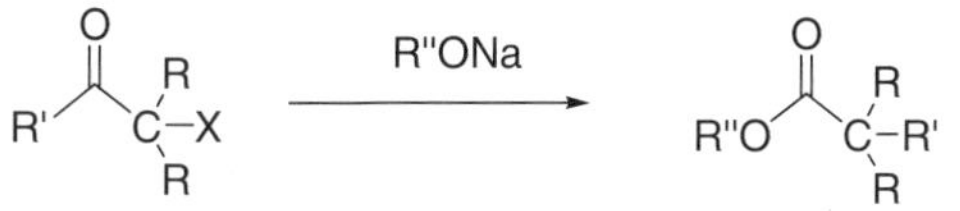

The rearrangement of cyclic ketones involves ring contraction. Thus, 2-chlorocyclobutanone (**2.29**) on treatment with sodium methoxide followed by hydrolysis is converted into cyclopropanecarboxylic acid (**2.30**). In a similar manner, 2-bromocyclohexanone (**2.31**) on treatment with sodium methoxide gives methyl ester of cyclopentanecarboxylic acid (**2.32**).

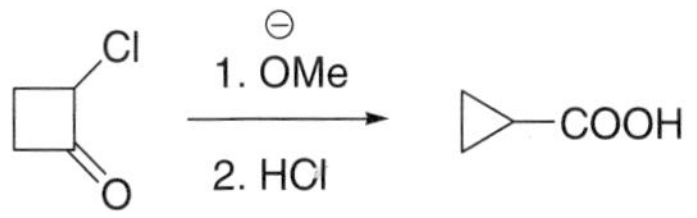

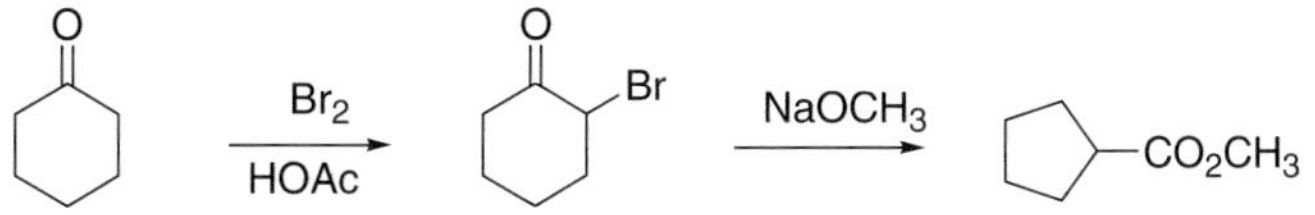

The generally accepted mechanism for the Favorskii rearrangement involves the formation of reactive cyclopropanone intermediate **C**. Base abstracts the α-hydrogen from **A** to give the carbanion **B**, which undergoes intramolecular S_N2 displacement of the halide ion. The resulting cyclopropanone intermediate **C** is opened under the reaction conditions to give the more stable carbanion **D**, which takes proton from solvent to furnish the final product, an ester **E** (Scheme 2.25).

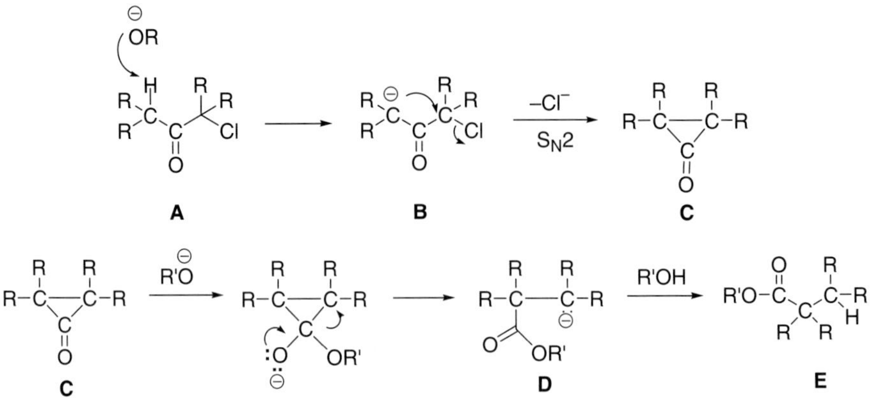

Scheme 2.25

The above mechanism is supported by the fact that the same ester **2.35** is formed from the isomeric haloketones **2.33** and **2.34** (Scheme 2.26).

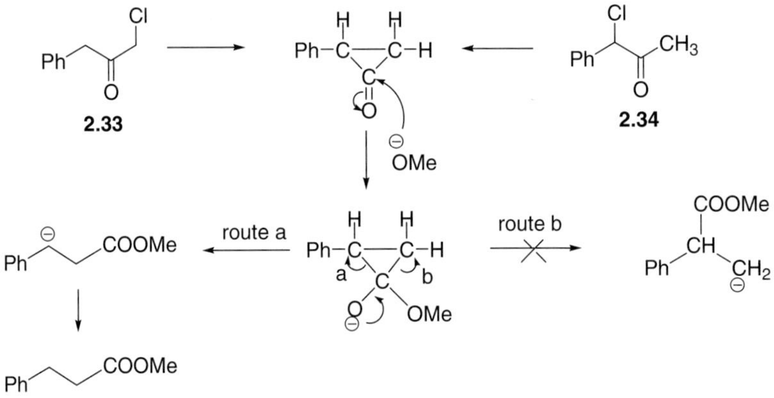

Scheme 2.26

In the symmetrical cyclopropanone intermediate, the two α-carbons are equivalent and ring opening via route a or route b gives the same carbanion. However, unsymmetrical cyclopropanone ring opens in such a way so that the more stable carbanion of the two possible carbanions is formed.

2.3   Free radicals

A radical (often called a free radical) is an atom or a group of atoms that have one or more unpaired electrons. Thus, carbon radicals have only seven valence electrons, and may be considered electron deficient of one electron. However, they do not in general bond to nucleophilic electron pairs, so their chemistry exhibits unique differences from that of conventional electrophiles. A prominent feature of radicals is that they are highly reactive. In fact, the high reactivity of free radicals is due to the tendency of the odd electrons to 'pair up' with another available electron. They readily react with O_2; thus, their reactions must be carried out under an inert atmosphere.

Radicals are often uncharged (neutral) but radical cations and radical anions also exist.

2.3.1   Structure and stability of free radicals

A radical is paramagnetic and so can be observed by electron spin resonance (esr) spectroscopy. Radicals are planar in configuration, but the energy difference between pyramidal and planar forms is very small.

Free radical planar (sp^2)   Pyramidal (sp^3)

The stability of the radicals depends on the nature of the atom that is the radical centre and on the electronic properties of the groups attached to the radical. As in the case of carbocations, the order of stability of the free radicals is tertiary > secondary > primary > methyl. This can be explained on the basis of hyperconjugation as in the case of carbocations. The stability of the free radicals also increases by resonance possibilities. Thus, benzylic and allylic free radicals are more stable and less reactive than the simple alkyl radicals. This is due to the delocalization of the unpaired electron over the π-orbital system in each case.

Canonical forms of benzyl free radical

$$CH_2^-CH-\overset{\bullet}{C}H_2 \quad \longleftrightarrow \quad \overset{\bullet}{C}H_2-CH=CH_2$$

Canonical forms of allyl free radical

The decreasing order of the stability of various radicals is as follows:

Benzyl   Allyl   Tertiary   Secondary   Primary   Methyl

Vinyl   Alkynyl   Phenyl

The stability of a radical increases as the extent of potential delocalization increases. Therefore, $Ph_2CH^{\bullet}$ is more stable than $PhCH_2^{\bullet}$, and $Ph_3C^{\bullet}$ is a reasonably stable radical. Adjacent functional groups, electron withdrawing or electron donating, both seem to stabilize radicals.

Certain radicals have rigid molecular structures with fixed bond angles and dihedral angles. These are known as **bridge head radicals** and have pyramidal structure. This has been supported on the basis of physical and chemical evidences.

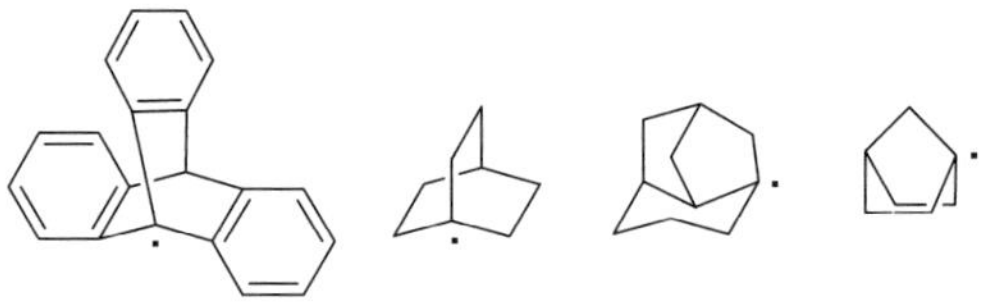

Bridge head radicals

2.3.2 Generation of free radicals

The homolytic cleavage of a covalent bond generates a pair of free radicals (Scheme 2.27). Energy in the form of ultraviolet–visible light, heat or some other form is needed to break a covalent bond.

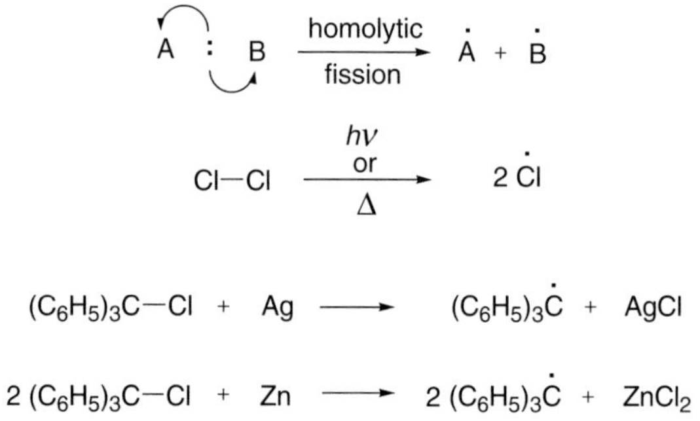

Scheme 2.27

The dissociation energies of C–C, C–H, C–O and C–X bonds are quite high, but the very weak O–O bonds of peroxides are cleaved at relatively low temperatures (Table 2.2).

Organic azo compounds (R–N=N–R) also easily decompose to alkyl radicals and nitrogen. The thermodynamic stability of nitrogen provides an overall driving force for this decomposition. Homolysis of several weaker bonds initiates the carbon radical reactions

Table 2.2 Standard bond energies and approximate homolysis temperature

Bond	D (kcal mol^{-1})	T (°C)	Bond	D (kcal mol^{-1})	T (°C)	Bond	D (kcal mol^{-1})	T (°C)
C–C	85	670	O–O	34	160	O–Cl	49	280
C–H	99	850	N–N	39	230	C–I	51	350
C–O	84	680	S–S	55	440	C–Br	67	480

and then subsequent transfer of radical to carbon occurs. Typical radical initiators are dibenzoyl peroxide (**2.36**) and azobisisobutyronitrile (AIBN) (**2.37**).

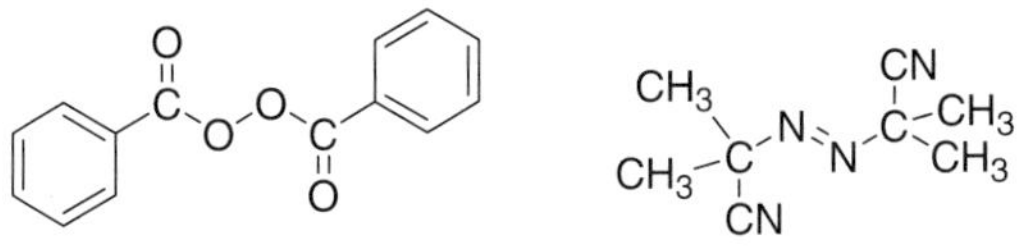

2.36   **2.37**

Following are some of the methods used for the generation of free radicals.

Thermolysis

The radicals are formed by heating the appropriate substrate at suitable temperature (Scheme 2.28). For example, on heating, chlorine forms chlorine radicals. Dialkyl peroxides form alkoxy radicals. Peroxy esters such as **2.36** and **2.38** fragment to acyl radicals, which eliminate carbon dioxide to give the corresponding alkyl radicals. Azo compounds eliminate nitrogen to give a pair of alkyl radicals. The cleavage of bonds can be achieved by heating in non-polar solvents or in vapors phase.

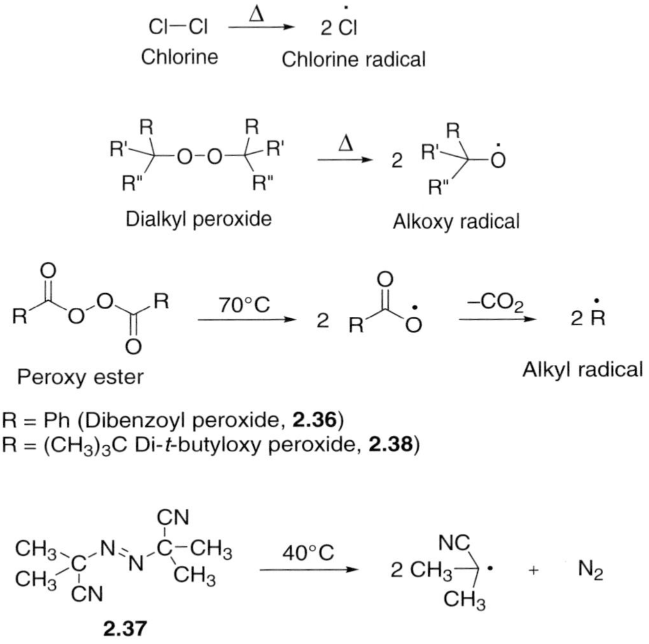

$$Cl-Cl \xrightarrow{\Delta} 2\ \dot{C}l$$
Chlorine   Chlorine radical

Dialkyl peroxide   Alkoxy radical

Peroxy ester   Alkyl radical

R = Ph (Dibenzoyl peroxide, **2.36**)
R = (CH$_3$)$_3$C Di-*t*-butyloxy peroxide, **2.38**)

2.37

Scheme 2.28

Photolysis

Compounds having absorption bands in the visible or near-ultraviolet spectrum may be electronically excited to such a degree that weak covalent bonds undergo homolysis (Scheme 2.29). Photolysis of the peroxides gives alkoxy radicals. Many azo compounds are important source of alkyl radicals. Acetone in vapour phase is decomposed by light having a wavelength of about 320 nm (3200 Å). In this reaction, two molecules of methyl free radicals are generated.

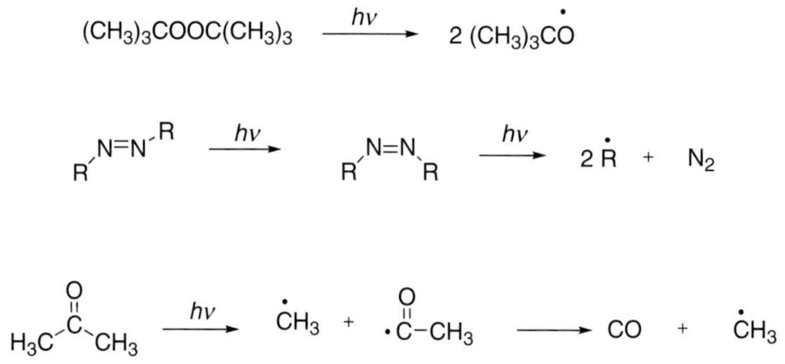

$$(CH_3)_3COOC(CH_3)_3 \xrightarrow{h\nu} 2\ (CH_3)_3C\overset{\bullet}{O}$$

$$\underset{R}{\overset{R}{\diagdown}}N{=}N{\diagup}^R \xrightarrow{h\nu} \underset{R}{\overset{R}{\diagdown}}N{=}N_{\diagdown R} \xrightarrow{h\nu} 2\ \overset{\bullet}{R}\ +\ N_2$$

$$\underset{H_3C}{\overset{\overset{O}{\|}}{\diagdown}}\overset{}{C}_{\diagdown CH_3} \xrightarrow{h\nu} \overset{\bullet}{C}H_3\ +\ \underset{}{\overset{\overset{O}{\|}}{\bullet C}}{-}CH_3 \longrightarrow CO\ +\ \overset{\bullet}{C}H_3$$

Scheme 2.29

Halogens, alkyl nitrites and alkyl hypochlorites also undergo photolysis easily (Scheme 2.30). Alkyl nitrites and alkyl hypochlorites generate alkoxy radicals.

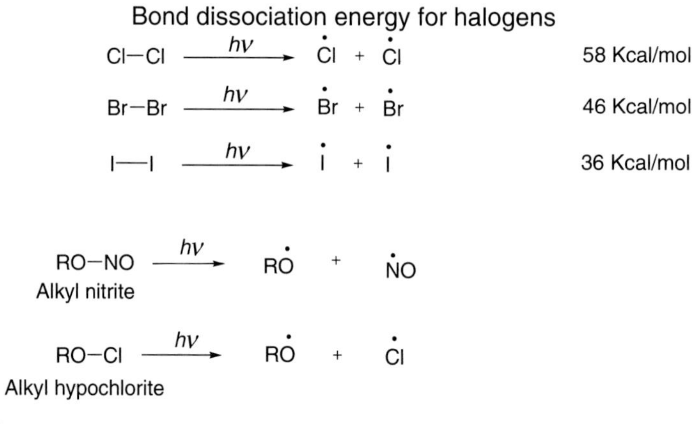

Bond dissociation energy for halogens

$$Cl{-}Cl \xrightarrow{h\nu} \overset{\bullet}{Cl}\ +\ \overset{\bullet}{Cl} \qquad\qquad 58\ \text{Kcal/mol}$$

$$Br{-}Br \xrightarrow{h\nu} \overset{\bullet}{Br}\ +\ \overset{\bullet}{Br} \qquad\qquad 46\ \text{Kcal/mol}$$

$$I{-}I \xrightarrow{h\nu} \overset{\bullet}{I}\ +\ \overset{\bullet}{I} \qquad\qquad 36\ \text{Kcal/mol}$$

$$RO{-}NO \xrightarrow{h\nu} R\overset{\bullet}{O}\ +\ \overset{\bullet}{N}O$$
Alkyl nitrite

$$RO{-}Cl \xrightarrow{h\nu} R\overset{\bullet}{O}\ +\ \overset{\bullet}{Cl}$$
Alkyl hypochlorite

Scheme 2.30

Redox reactions

An electron can be removed from an anion, and the process is known as the oxidation process. For example, the phenoxide ion is oxidized by Fe^{3+} to give the phenoxy radical and the Fe^{3+} is co-reduced to Fe^{2+}. This is known as the **single-electron-transfer (SET) oxidation** (Scheme 2.31).

An electron can be given to a cation and the process is known as the **single-electron-transfer (SET) reduction**. The source of one-electron transfer is the metal ion. For example, Cu^+ ions are used for the decomposition of acyl peroxides (Scheme 2.32). This is a convenient method for the generation of the ArCOO˙ radicals, especially because in thermolysis the ArCOO˙ radicals further decompose to Ar˙ and CO_2.

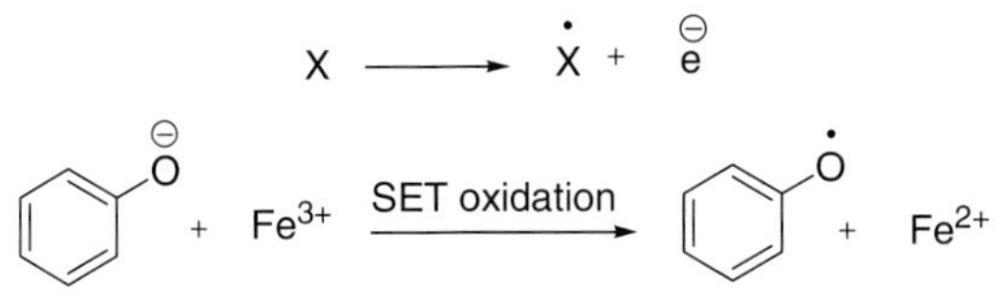

$$X \longrightarrow \overset{\bullet}{X} + \overset{\ominus}{e}$$

Scheme 2.31

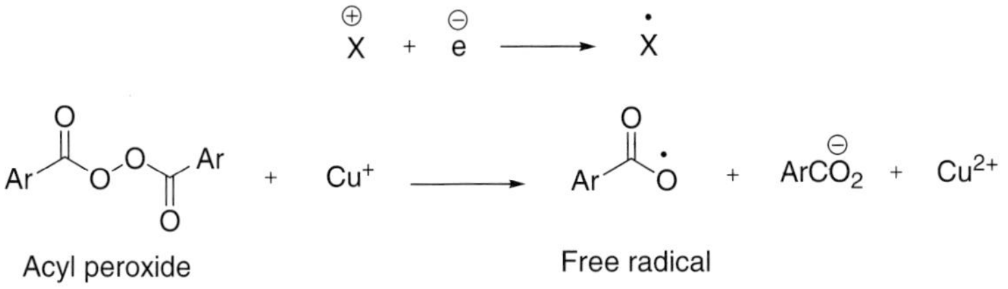

$$\overset{\oplus}{X} + \overset{\ominus}{e} \longrightarrow \overset{\bullet}{X}$$

Acyl peroxide   Free radical

Scheme 2.32

Cu^+ also finds application in the **Sandmeyer reaction**[23–26] involving the decomposition of diazonium salts. In this reaction the free radical $Ar^{\bullet}$ is formed as an intermediate.

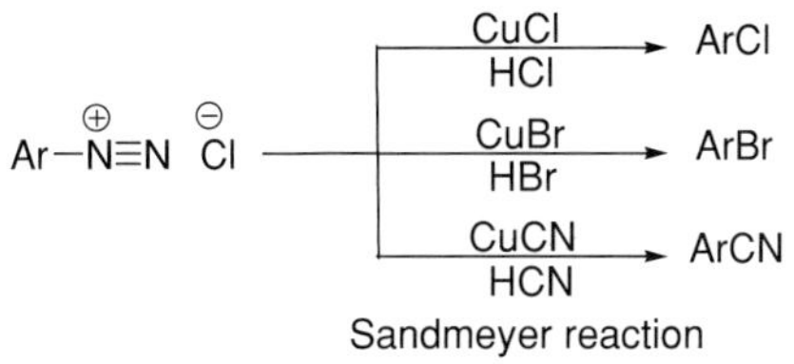

Sandmeyer reaction

The mechanism of the **Sandmeyer reaction** involves the transfer of electron from Cu^+ to diazonium salt. Elimination of nitrogen generates aryl free radical, which reacts with CuXCl to give aryl halide via either path a or path b (Scheme 2.33).

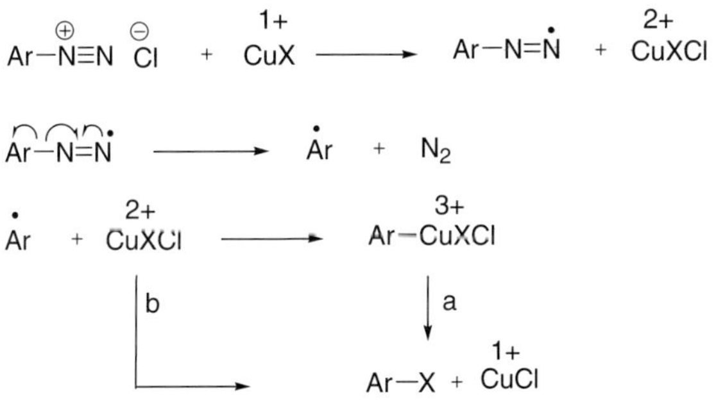

Scheme 2.33

The iron(II) ion (Fe^{2+}) reduces hydrogen peroxide to hydroxyl radical and hydroxide ion. The mixture of H_2O_2 and Fe^{2+} is known as **Fenton's reagent**[27]. It was developed in the 1890s by Henry John Horstman Fenton. The effective oxidizing agent is the hydroxyl radical ($HO^{\bullet}$).

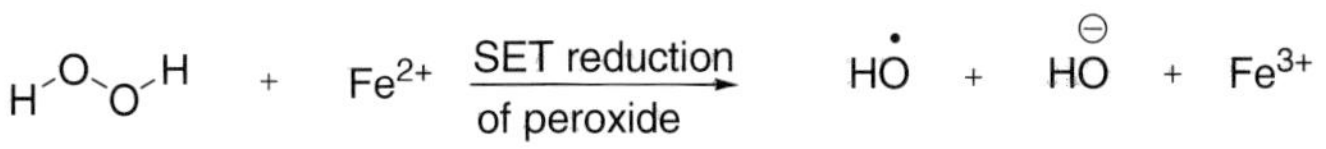

Iron(III) is then reduced back to iron(II), a peroxide radical and a proton by the same hydrogen peroxide (disproportionation).

Stable phenoxy radical (**2.40**) can also be generated by one-electron oxidation of **2.39** with $K_3Fe(CN)_6$.

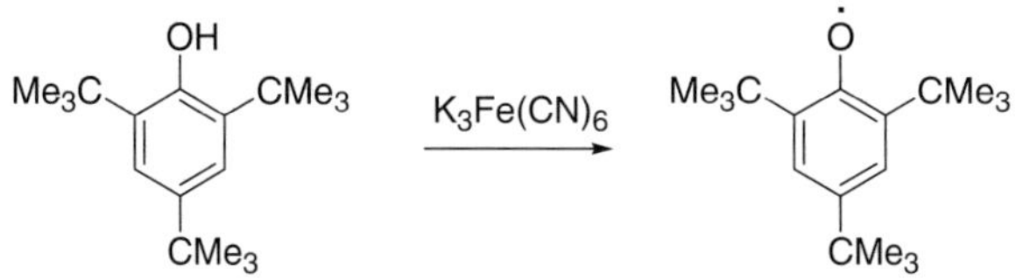

2.39   **2.40**

The dimerization of carbanions with iodine also takes place via a free radical (see Scheme 2.21).

Kolbe electrolytic reaction[28-30] for the synthesis of alkanes also involves the radicals as intermediates. For example, when a solution of diphenylacetic acid (**2.41**) is electrolyzed in DMF (dimethylformamide), the product **2.42** is obtained in 24% yield.

$$2\ Ph_2CHCOOH \xrightarrow[\text{DMF, 17 h}]{\text{electrolysis}} Ph_2CHCHPh_2$$

2.41   **2.42**

The mechanism of the Kolbe reaction involves electrochemical decarboxylation–dimerization via radicals (Scheme 2.34).

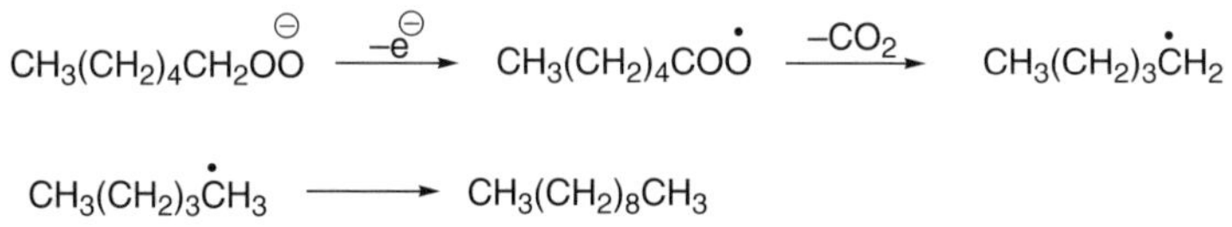

$$CH_3(CH_2)_4CH_2O\overset{\ominus}{O} \xrightarrow{-\overset{\ominus}{e}} CH_3(CH_2)_4C\overset{\bullet}{O}O \xrightarrow{-CO_2} CH_3(CH_2)_3\overset{\bullet}{C}H_2$$

$$CH_3(CH_2)_3\overset{\bullet}{C}H_3 \longrightarrow CH_3(CH_2)_8CH_3$$

Scheme 2.34

Radicals are also formed as intermediates in the reductions of the carbonyl group (see sections 6.4.3 and 6.5.3).

2.3.3   Radical ions

A **radical ion** is a free radical species that carries a negative charge (**radical anion**) or a positive charge (**radical cation**). When a neutral, spin-paired species gains a single electron it becomes a **radical anion**. Likewise, when a neutral, spin-paired species loses an electron it becomes a **radical cation**.

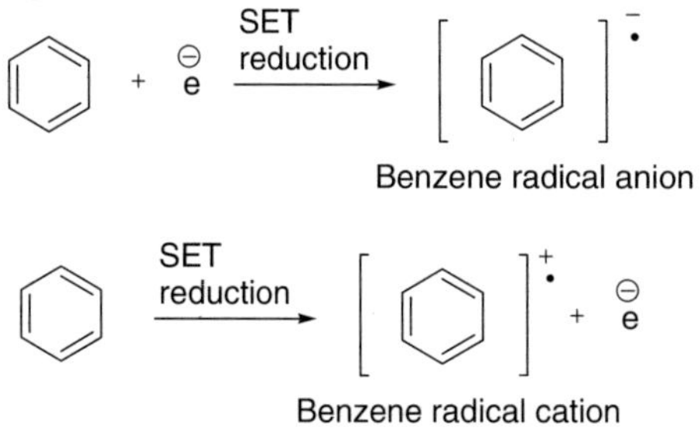

Benzene radical anion

Benzene radical cation

Radical cations and radical anions are known in the gas phase. They are routinely generated and studied in the complementary techniques of mass spectrometry and negative ion mass spectrometry.

Radical anions

Many aromatic compounds can undergo one-electron reduction by alkali metals, such as Na and Li. For example, the reaction of naphthalene with sodium in an aprotic solvent gives the naphthalene radical anion – sodium ion salt.

Radical cations

Cationic radicals are much less stable and noticed prominently in mass spectroscopy. When a molecule in gas phase is subjected to electron ionization, one electron is abstracted by the electron beam to create a **radical cation**. This species represents the molecular ion or parent ion, which on fragmentation gives a complex mixture of ions and uncharged radical species. For example, the methanol radical cation fragments into a methyl cation CH_3^+ and a hydroxyl radical. Secondary species are also generated by proton gain $(M + 1)$ and proton loss $(M - 1)$.

$$CH_3OH^{+\bullet} \longrightarrow CH_3^+ + \overset{\bullet}{O}H$$

2.3.4 Reactions of radicals

Radicals are very reactive reaction intermediates and their half-life period is very short. Some of the reactions of radicals are given below.

Halogenation of alkanes

The reaction of methane with chlorine in the presence of ultraviolet light gives a mixture of methyl chloride, methylene chloride, chloroform and carbon tetrachloride. When excess of chlorine is used and the time of the reaction is prolonged, the final product is predominantly carbon tetrachloride.

$$CH_4 + Cl_2 \xrightarrow{UV\ light} CH_3Cl + CH_2Cl_2 + CHCl_3 + CCl_4$$

The free radical halogenation of alkanes takes place in three steps: initiation, propagation and termination (Scheme 2.35).

The substitution of a hydrogen atom in the benzylic position by a bromine atom on reaction with *N*-bromosuccinimide in the presence of catalytic amounts of AIBN (2.37) is known as the **Wohl–Ziegler process**[31].

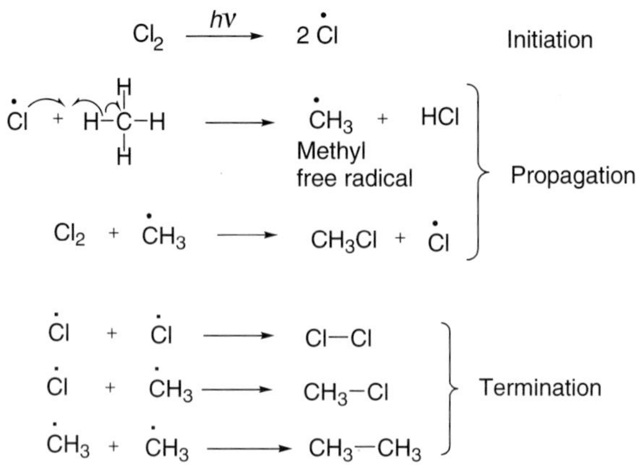

Scheme 2.35

The mechanism of the allylic Wohl–Ziegler bromination involves free radicals, as shown in Scheme 2.36.

Radical defunctionalization reactions

Dehalogenation: Dehalogenation of haloalkanes[32,33] (R–X) is often carried out with tributyltin hydride (**2.43**) in the presence of AIBN (**2.37**). The reactivity of R–X is in the order of R–I > R–Br > R–Cl (R–F being inert); tertiary > secondary > primary > aryl or vinyl.

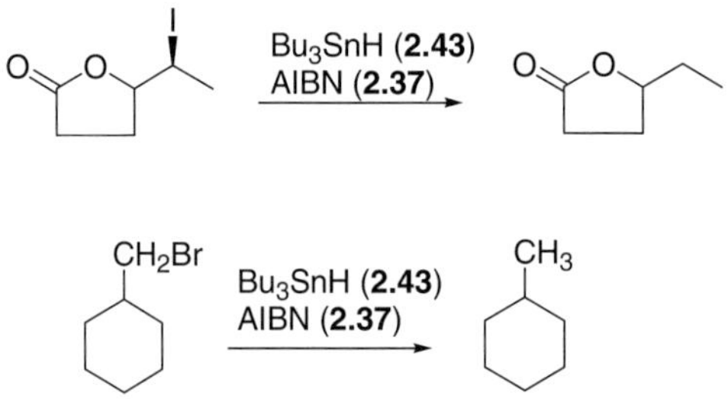

The mechanism of dehalogenation of haloalkanes is shown in Scheme 2.37.

     The overall reaction results in the reduction of carbon–halogen bonds to carbon–hydrogen bonds.

     Deoxygenation of alcohols: The **Barton–McCombie deoxygenation**[34,35] is a useful procedure for converting alcohols to alkanes. For hindered and polar alcohols the traditional methods of deoxygenation, for example conversion of the alcohol into tosylate and reduction with lithium aluminium hydride, are ineffective. Sir Derek Barton developed a highly effective alternative which involves the formation of a thiocarbonyl derivative, a xanthate ester, from primary and secondary alcohols and other thiocarbonyl derivatives from tertiary alcohols (xanthates of tertiary alcohols can undergo **Chugaev elimination**; see section 4.2.2), followed by reaction with tributyltin hydride. This procedure of deoxygenation of alcohols is known as the **Barton–McCombie deoxygenation reaction.**

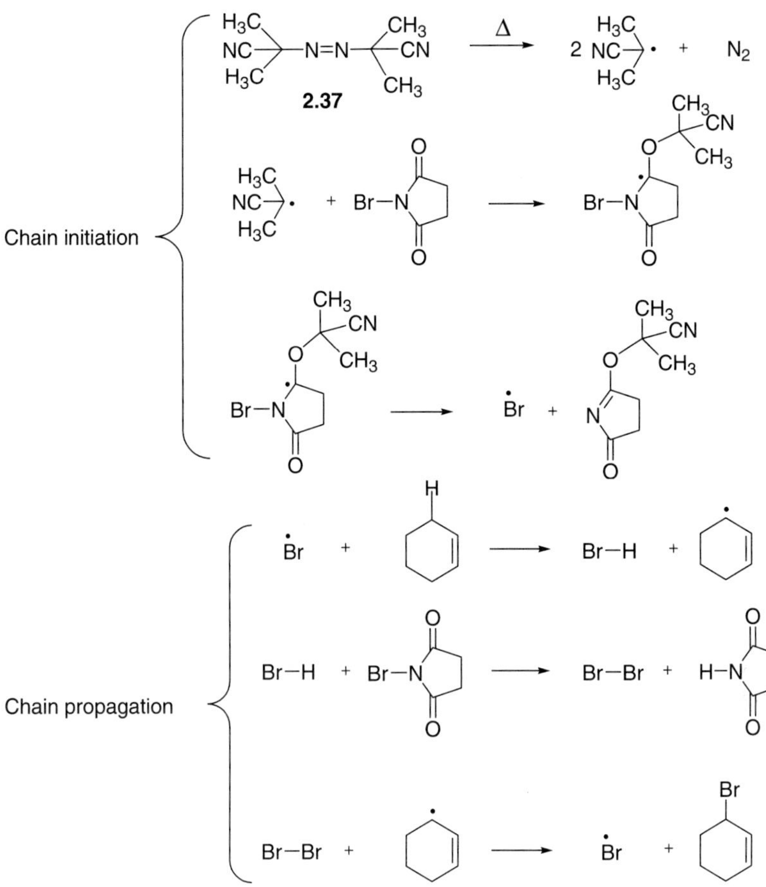

Scheme 2.36

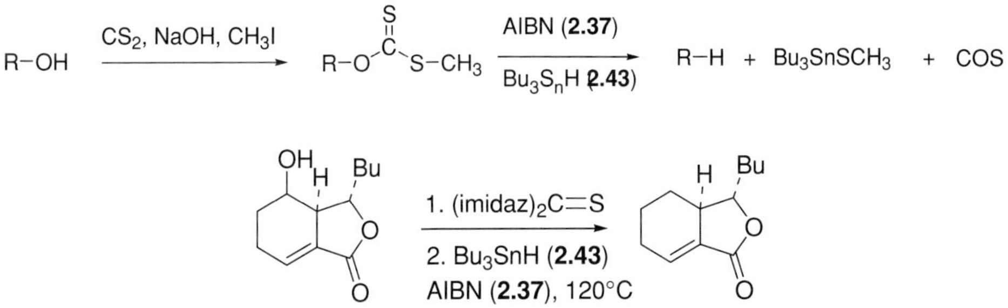

A thionoester **A**, such as a xanthate, is first prepared from the alcohol. Then tributyltin radical adds to the sulfur of the thiocarbonyl (C=S) function. This is followed by the fragmentation of the intermediate species to give **B** and **C**, and Bu$_3$SnH (**2.43**) reacts with **B** to give alkane (Scheme 2.38).

The tin–hydrogen bond of tributyltin hydride (**2.43**) is relatively weak and undergoes fission in the presence of AIBN (**2.37**). Since tin reagents are highly toxic, other methods[36,37]

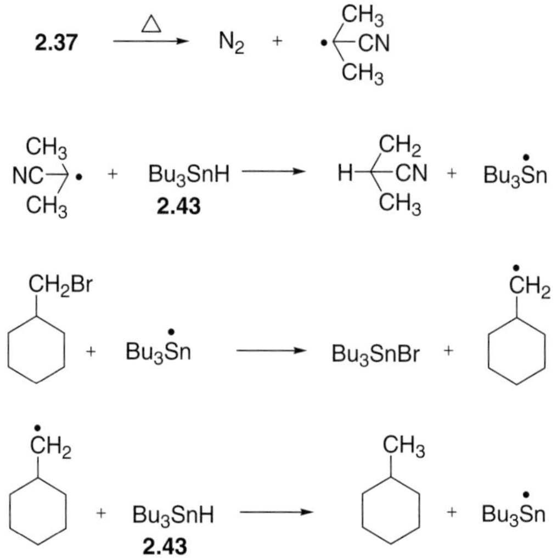

Scheme 2.37

that use tin compounds in catalytic amounts are developed: for example, (i) catalytic hexabutylditin [(Bu₃Sn)₂] or tributyltin chloride (Bu₃SnCl) can be used as alternative conditions, (ii) it is also possible to use stoichiometric amounts of sodium cyanoborohydride (NaBH₃CN) to prepare **2.43** *in situ* and (iii) catalytic amounts of tris(trimethylsilyl)silane [(Me₃Si)₃SiH] can also be used, which has similar properties to **2.43**.

The Barton–McCombie reaction is extremely useful in the preparation of deoxysugars in carbohydrate chemistry.

Barton decarboxylation: A variation of the Barton–McCombie reaction is called the Barton decarboxylation[38].

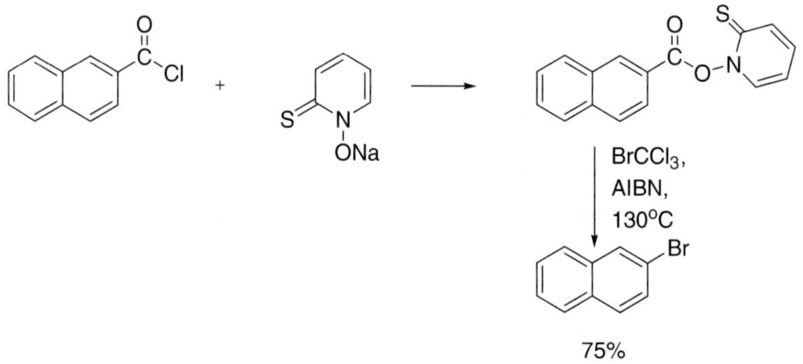

75%

The Barton decarboxylation involves the preparation of *O*-acyl thiohydroxamates **A** from acid chlorides and sodium salt of *N*-hydroxy-2-thiopyridine (**2.44**), which undergoes

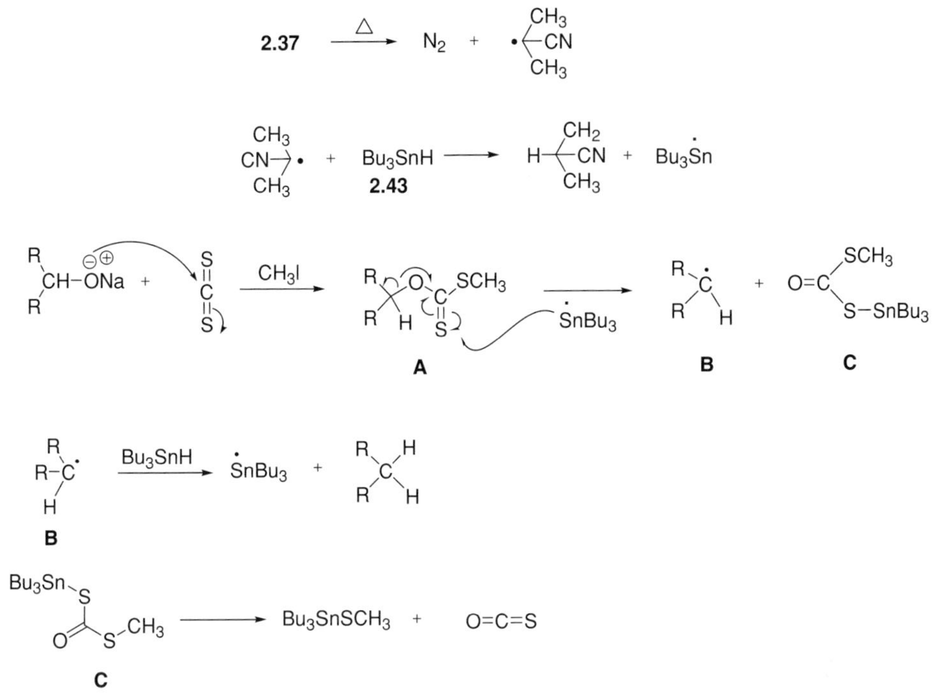

2.37 —Δ→ N₂ + •C(CH₃)(CH₃)CN

(CH₃)(CH₃)C•CN + Bu₃SnH (2.43) ⟶ H–C(CH₂)(CH₃)CN + Bu₃Sn•

R₂CH–ONa⁻⊕ + S=C(=S)SCH₃ →CH₃I→ A →→ B + C

A: R₂C(H)(O–C(=S)SCH₃) with SnBu₃

B: R₂C•H

C: O=C(SCH₃)(S–SnBu₃)

R–C•(R)(H) + Bu₃SnH ⟶ Bu₃Sn• + R–CH(R)(H)

B

Bu₃Sn–S–C(=O)–S–CH₃ (**C**) ⟶ Bu₃SnSCH₃ + O=C=S

Scheme 2.38

reaction with **2.37** and **2.43** to give **C** and carboxyl radical **D** via an intermediate radical **B**. Decarboxylation of **D** gives alkyl radicals, which are finally trapped by **2.43** to give the alkane (Scheme 2.39).

It is also possible to decarboxylate the carboxylic acids with other reagents, leading to useful functionalized products such as sulfide, selenide and bromo derivatives. For example, decarboxylative halogenation of **A** on treatment with CCl_4, $BrCCl_3$ or CH_2I_2 gives the corresponding alkyl halide[39].

In a similar manner, deamination, denitration, desulfurization and deselenation can also be carried out by using tributyltin hydride (**2.43**).

Barton reaction: In the Barton reaction[40–43] a methyl group in the γ-position to an OH group is converted into an oxime group. For example, corticosterone acetate (**2.45**) on reaction with nitrosyl chloride (NOCl) in pyridine forms its 11β-nitrite **2.46**, which on photolysis in toluene gives aldosterone acetate oxime (**2.47**) in 21% yield. Treatment of **2.47** with nitrous acid (HNO_2) gives aldosterone acetate (**2.48**) (Scheme 2.40).

Photohomolysis of the weak N–O bond of a nitrite ester **A** forms a pair of radicals **B** and NO˙. The oxy radical **B** abstracts a hydrogen atom from a nearby carbon and the resulting radical **C** and NO˙ couples to give a nitroso compound **D**. Tautomerism of the nitroso product **D** followed by treatment with nitrous acid converts this to the carbonyl group (Scheme 2.41).

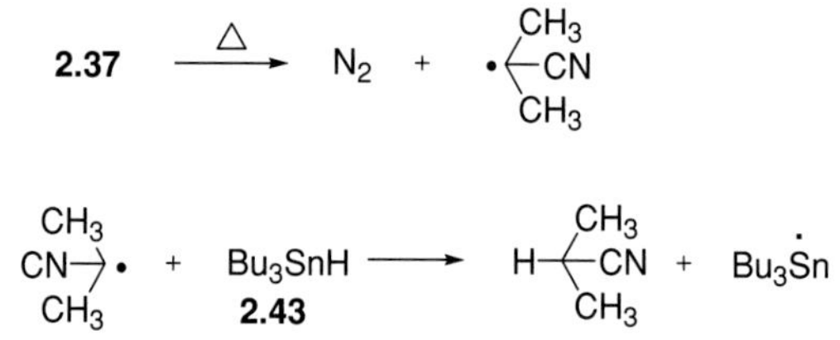

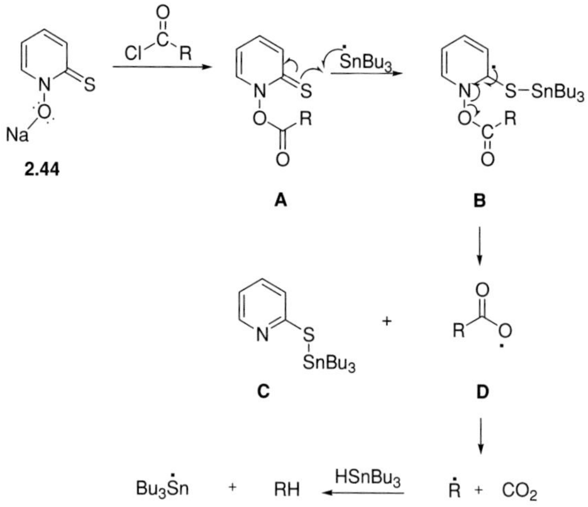

Scheme 2.39

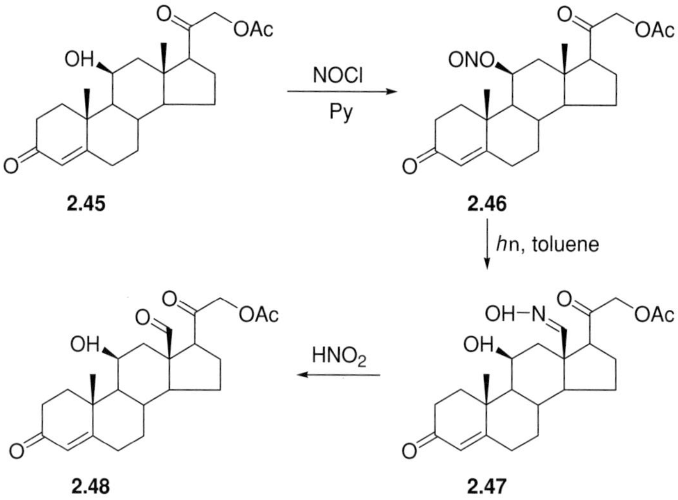

Scheme 2.40

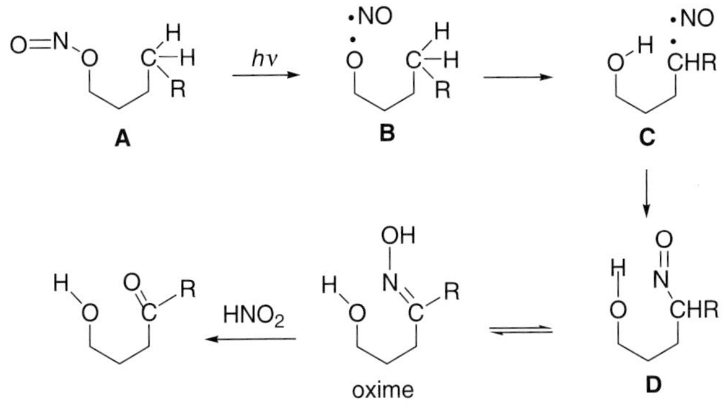

Scheme 2.41

Fragmentation

The radical generated from **2.49** is trapped to give a mixture of products **2.50** and **2.51** in 1:1.8 ratio. But in the presence of magnesium bromide ($MgBr_2$) or $Yb(OTf)_3$ in CH_2Cl_2, a major amount of *S*-isomer **2.50** is obtained via radical **2.52**. The selectivity is further improved in the presence of an additive[44,45] (or chiral ligand).

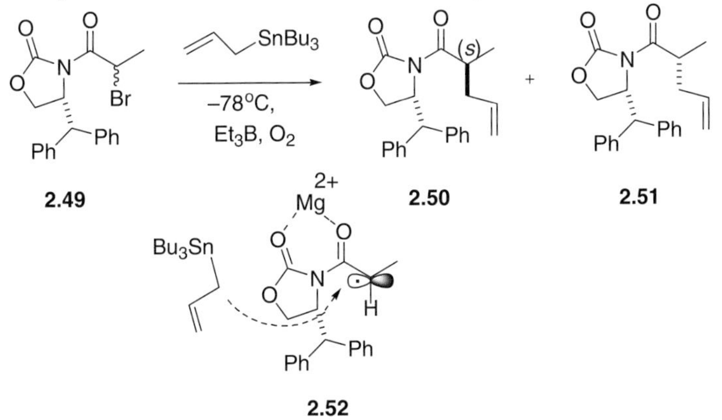

Elimination reactions

The eliminative reduction of vicinal-glycols via a dixanthate to alkenes involves the radicals as illustrated in Scheme 2.42.

Coupling reactions

Radical cage effect and coupling (recombination): Radical coupling reactions do not dominate free radical chemistry as most radicals have very short lifetimes and are present in very low concentrations. Consequently, if short-lived radicals are to contribute to useful synthetic procedures by way of a radical coupling, all the events leading up to the coupling must take place in a solvent cage.

When a pair of radicals is formed by homolysis, they are briefly held in proximity by the surrounding solvent molecules. Rapid decomposition to other radicals may occur, but until

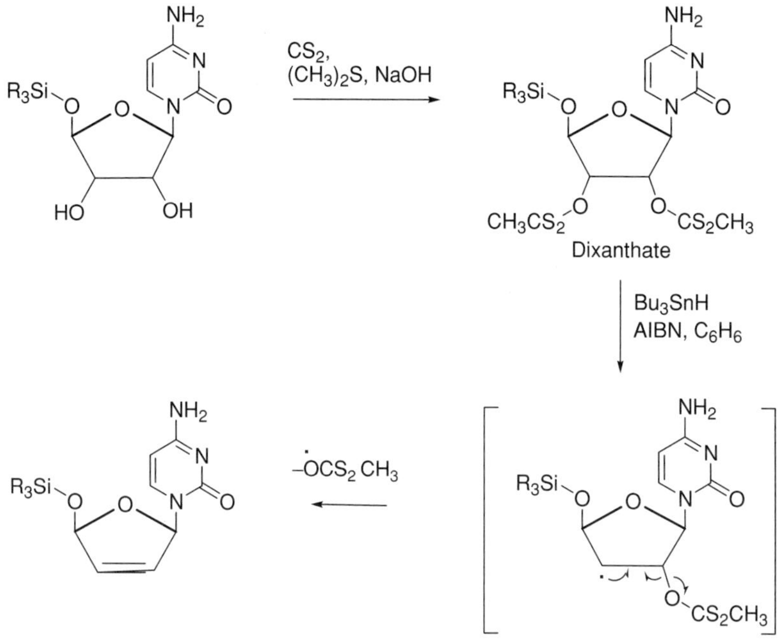

Scheme 2.42

one or both of these radicals escape the solvent cage, a significant degree of coupling may occur which is known as the **radical cage effect**.

Triphenylmethyl radicals couple to the **Gomberg dimer**[46–49] **2.53**, rather than the hexaphenylethane, Ph$_3$C–CPh$_3$ (**2.54**), as Gomberg originally proposed. The reason is that it is energetically more favourable for the dimeric compound to lose aromatic stabilization from one ring than to form the sterically strained **2.54**.

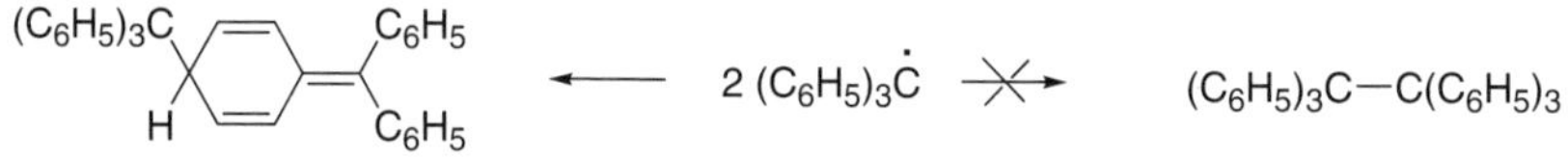

Radical coupling (recombination) reactions having activation energies near zero are very fast, and thus preserve the configuration of the generating species.

Addition to carbon–carbon double bond

It is well known that HBr adds on to propene to give isopropyl bromide (**2.55**) in accordance with the Markovnikov's rule. The reaction proceeds via a 2°-carbocation intermediate (the secondary carbocation being more stable than the primary carbocation) (Scheme 2.43). Normally, when a molecule HX adds to a carbon–carbon double bond, the hydrogen becomes attached to the carbon with the greatest number of hydrogens, and the X group becomes attached to the carbon with the fewest hydrogens. This is known as **Markovnikov's rule** (see section 2.1.2) (also see the addition of HCl to propene; Scheme 2.4).

However, in the presence of peroxide the addition of HBr follows anti-Markovnikov's addition to give *n*-propyl bromide. This is often known as the **peroxide effect**[50,51], and the addition proceeds by the free radical mechanism (Scheme 2.44).

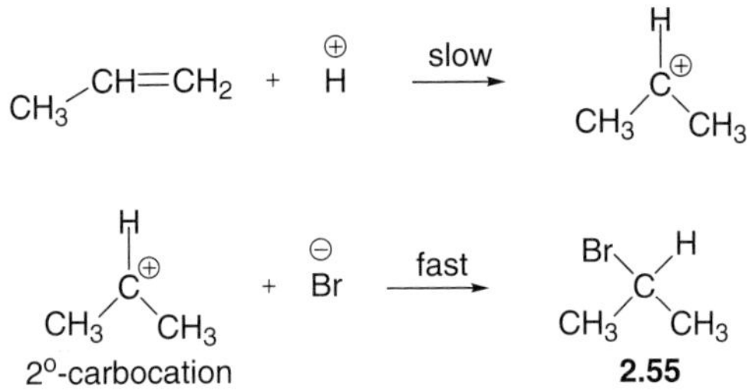

Scheme 2.43

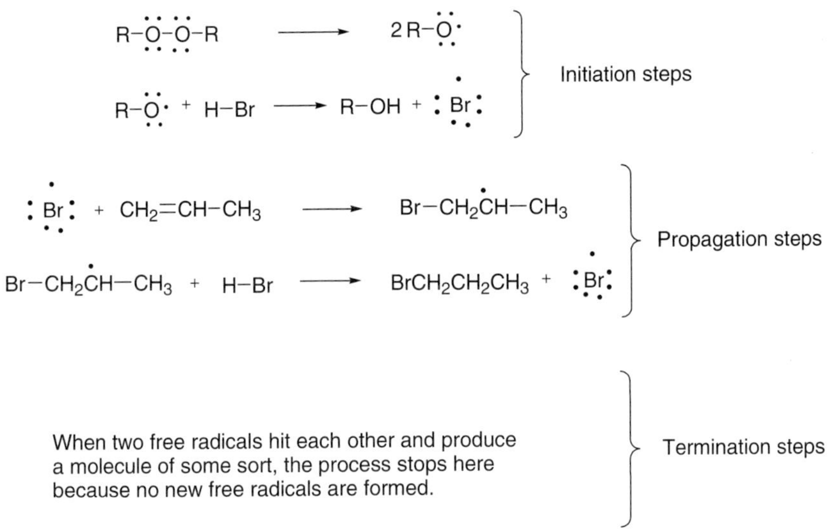

Scheme 2.44

In a similar manner, addition of HBr to 2-methyl-2-butene in the presence of benzoyl peroxide (2.36) gives 2-bromo-3-methyl butane in 55% yield.

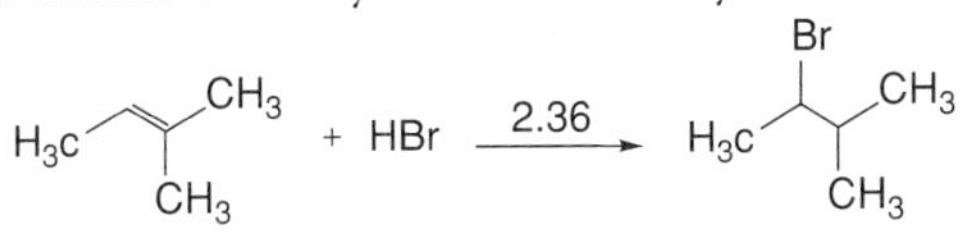

Free radical addition to conjugated dienes involves addition of a radical at position 1 of the diene, generating a resonance-stabilized radical **A** which will attack other molecules, e.g. X–Y via its 2- or 4-position, yielding 1,2- and 1,4-addition products **B** and **C**, respectively (Scheme 2.45).

Free radical addition is less selective than electrophilic addition and the product ratio is more difficult to control (due to high reactivity).

The conjugate addition of the radical generated from *i*-PrI by Bn$_3$SnH (**2.43**) to α,β-unsaturated ketone **2.56** takes place diastereoselectively in the presence of Lewis acid such as Yb(OTf)$_3$ to give addition products[52] **2.57** and **2.58** in CH$_2$Cl$_2$ and THF (tetrahydrofuran; in 4:1 ratio).

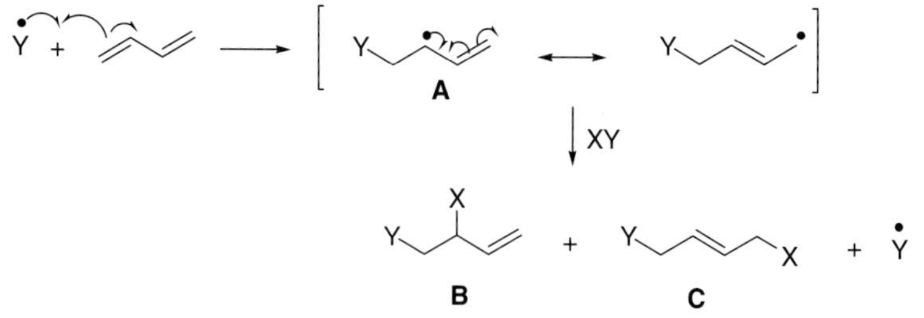

Scheme 2.45

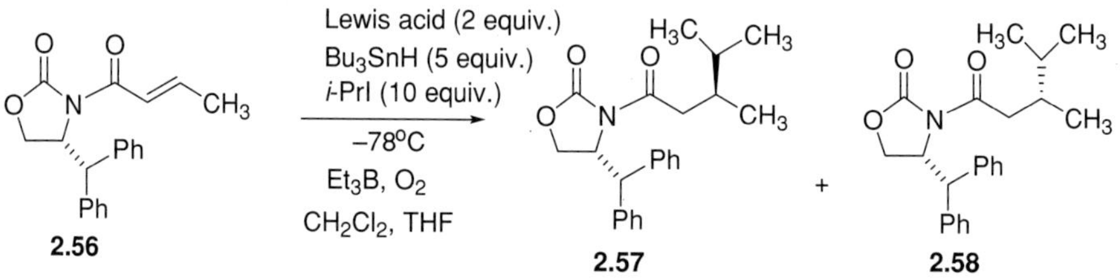

2.57:2.58

	2.57:2.58
Without Lewis acid	1.3:1
Yb(OTf)$_3$	25:1

The mechanism of this addition involves the conjugate addition of the isopropyl radical to **2.56** to give the radical **2.59**. The radical **2.59** then reacts with **2.43** to give the final products **2.57** and **2.58** (Scheme 2.46).

In the presence of stoichiometric or catalytic chiral ligand[53] **2.60**, a stereoselective radical conjugate addition is observed.

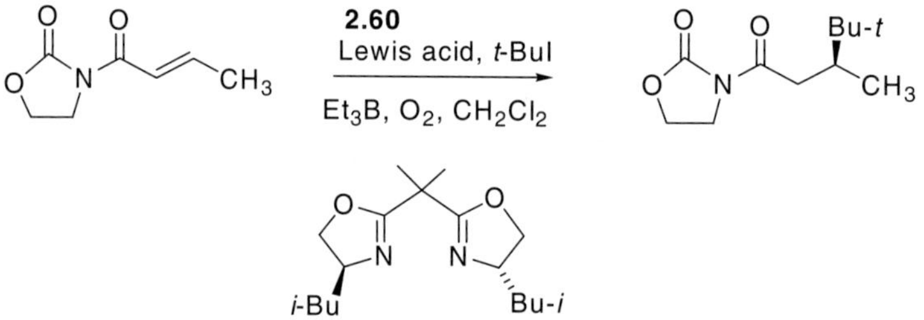

(S,S)-chiral ligand (**2.60**)

	Yield	ee
MgI$_2$ + ligand **2.60** (stoichiometric amounts of both Lewis acid and ligand)	88%	74%
MgI$_2$ + ligand **2.60** (0.2% of Lewis acid)	73%	66%

Cyclization by intramolecular addition reactions

If a radical is joined to a double bond by a chain of three or more carbons, intramolecular addition generates a ring. The regioselectivity of such additions is governed more by stereoelectronic factors than by substituents on the double bond. Thus, five-membered ring formation by way of a 1°-cyclized radical dominates the products, as shown in Scheme 2.47.

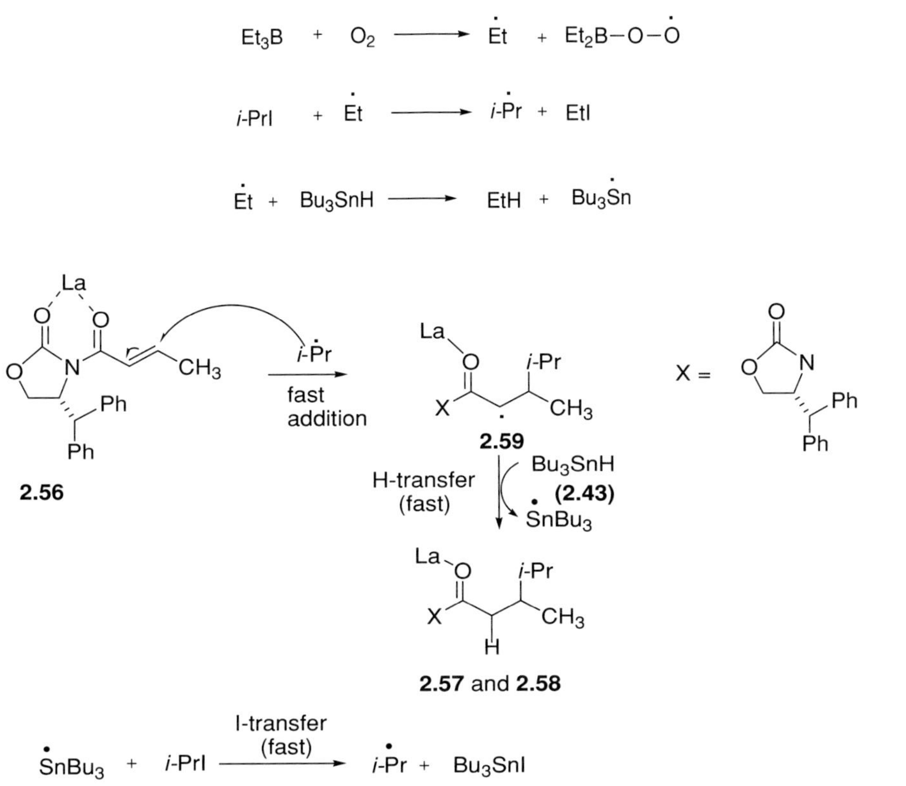

Et$_3$B + O$_2$ $\longrightarrow$ Ėt + Et$_2$B–O–Ȯ

i-PrI + Ėt $\longrightarrow$ i-Ṗr + EtI

Ėt + Bu$_3$SnH $\longrightarrow$ EtH + Bu$_3$Sṅ

Scheme 2.46

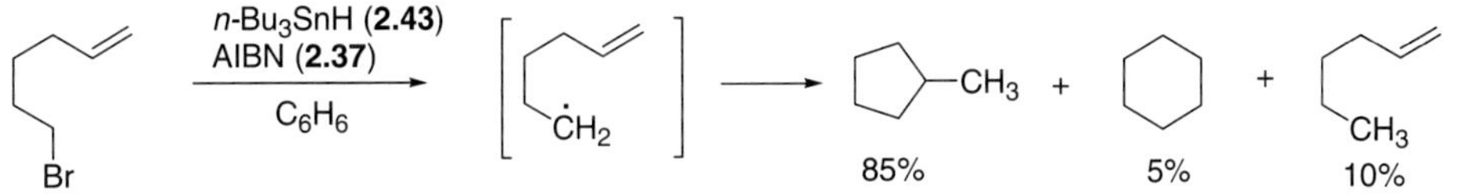

Scheme 2.47

The stereoelectronic factor in this reaction is defined by the preferred mode of approach of a radical as it forms bond to the π-electrons system of an alkene function. As shown below, this is at an angle nearly 20° off the perpendicular to the plane of the double bond. Because of this requirement, many cyclizations to moderately sized rings proceed by radical attack at the nearest carbon of the double bond, regardless of substitution.

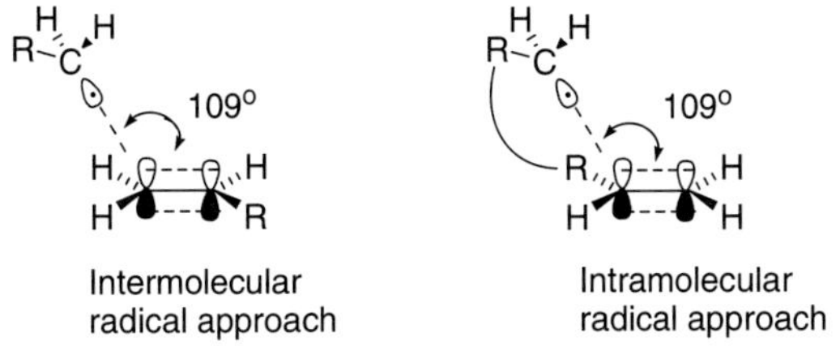

Intermolecular
radical approach

Intramolecular
radical approach

Reactions with oxygen, iodine, NO and metals

The triphenylmethyl radical reacts with a number of reagents including oxygen to give the peroxide, iodine to give the iodide, nitric oxide to give the nitroso compound.

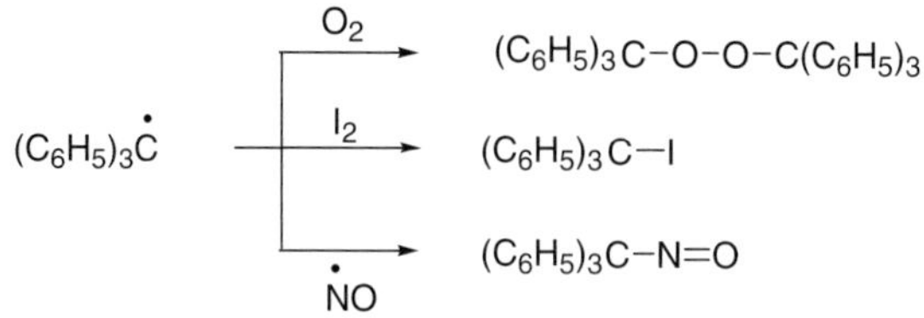

Autoxidation of cumene (**2.61**) is the most important industrial synthesis of cumene hydroperoxide (**2.62**), which gives phenol and acetone.

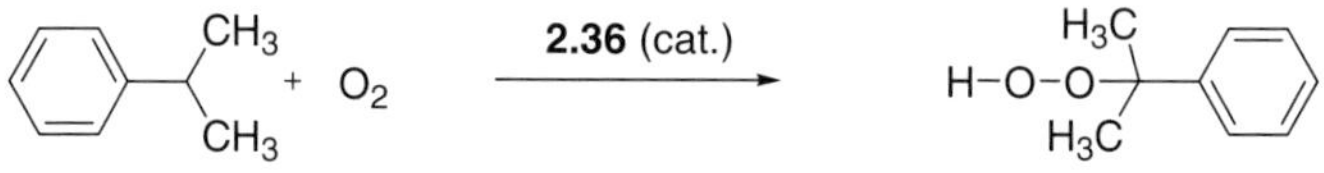

Oxidation of organic molecules with O_2 (flameless) is referred to as **autoxidation**. The synthesis of cumene hydroperoxide from cumene is initiated by catalytic amounts of **2.36** as the radical initiator, which generates the cumyl radical **A**. The cumyl radical **A** reacts with O_2 to give the radical **B** in the first propagation step and regenerated in the second propagation step, in which cumene hydroperoxide (**2.62**) is also formed (Scheme 2.48).

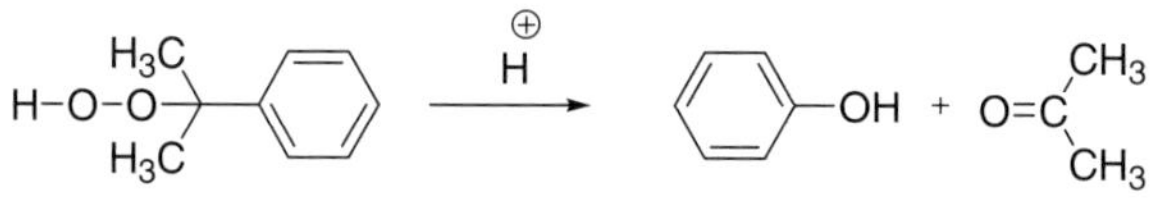

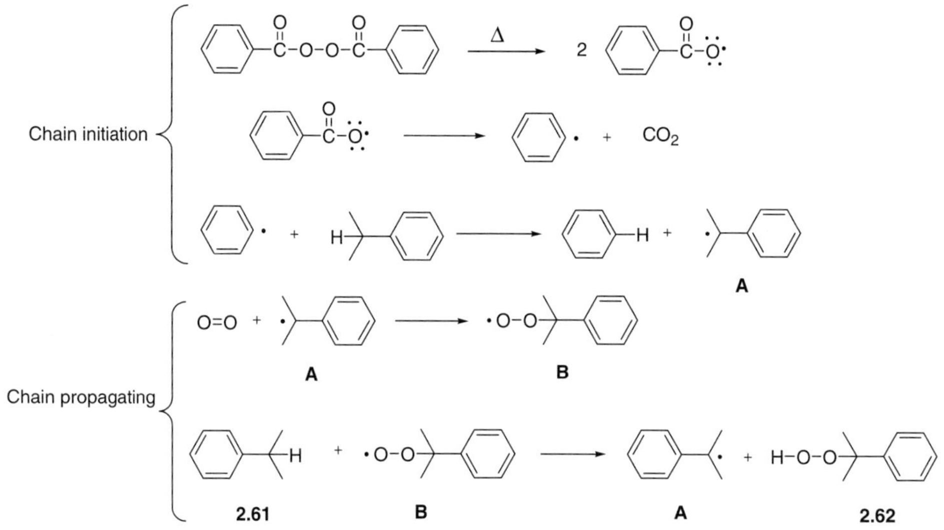

Scheme 2.48

Autoxidation of ethers to ether peroxides also involves the formation of free radicals. THF forms the corresponding hydroperoxide **2.63** at the α-position.

$$CH_3CH_2-O-CH_2CH_3 \xrightarrow[\text{light}]{\text{air}} CH_3-CH_2-O-\overset{\overset{\displaystyle O-O-H}{|}}{C}H-CH_3$$

2.63

Tetralin (**2.64**) is selectively oxidized to the hydroperoxide **2.65**.

2.64   **2.65**

In a similar manner, autoxidation of aldehydes gives the corresponding carboxylic acids.

Rearrangements

Like carbocations and carbanions, the free radicals also undergo rearrangement reactions, although rearrangement reactions of radical intermediates are less common. Few examples of radical rearrangements are given below. The rearrangement of **2.66** to **2.68** involves the migration of Ph (instead of Me) via the bridged intermediate **2.67**.

2.66   **2.67**   **2.68**

The rearrangement of 2-vinyl- and acyl-substituted 2,2-dimethylethyl radicals **2.69** and **2.72** to **2.71** and **2.74** proceeds through cyclopropyl intermediates **2.70** and **2.73**, respectively.

2.69   **2.70**   **2.71**

2.72   **2.73**   **2.74**

The migration of alkynyl and cyano substituents is slower because of the reduced stability of the intermediates **2.75** and **2.76** derived by cyclization of the triply bond substituents.

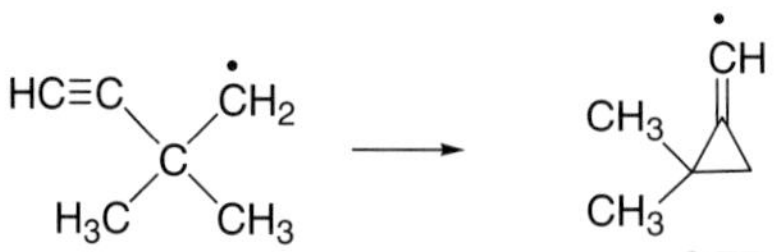

2.75

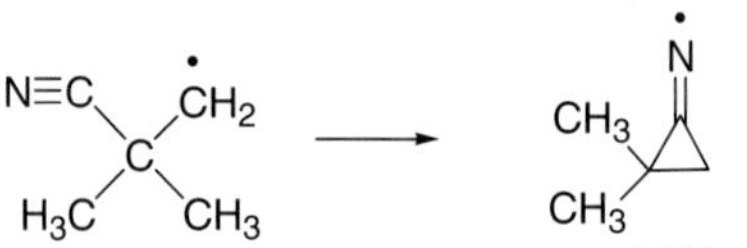

2.76

The Bergmann reaction

The conversion of an ene-diyne **2.77** into a 1,4-benzenediyl diradical **2.78** on heating is known as the **Bergmann reaction**[54,55]. The 1,4-aromatic radical **2.78** may be converted into benzene or may react with CCl_4 to give *p*-dichlorobenzene (**2.79**). Bergmann *et al.* converted deuterium-labelled hexa-3-ene-1,5-diyne **2.80** on heating at 200°C into deuterium-labelled hexa-3-ene-1,5-diyne **2.81** in which both deuterium atoms were shifted from the terminal acylene positions to vinyl positions at the interior of the chain.

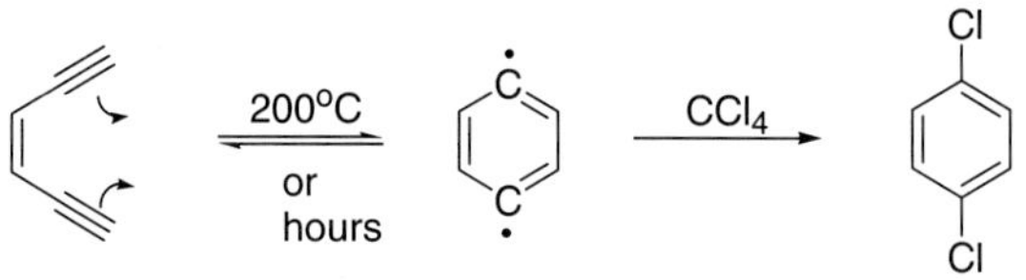

2.77   **2.78**   **2.79**

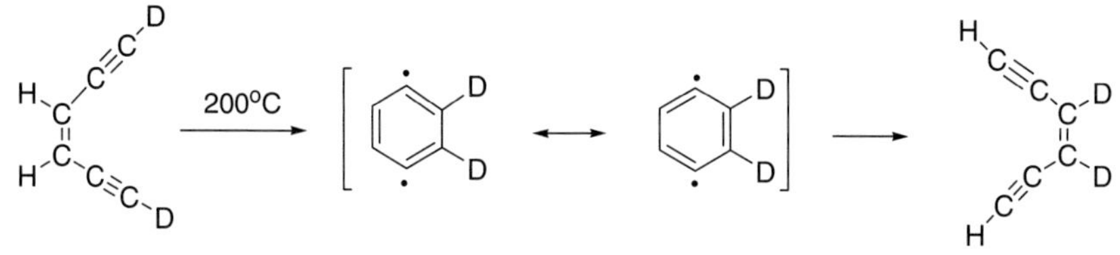

2.80   **2.81**

Gomberg–Bachmann reaction[56,57]

Base-catalyzed reaction between an aryl diazonium salt and an aromatic hydrocarbon (or aromatic heterocycle) to form a diaryl compound (or aryl-substituted heterocycle) involves radical coupling reaction, as shown in Scheme 2.49.

Glaser coupling[58,59]

Homocoupling of terminal alkynes in the presence of cuprous chloride and O_2 involves radical coupling, as shown in Scheme 2.50.

2.4   Carbenes

Carbenes are uncharged, electron-deficient molecular species that contain a divalent carbon atom surrounded by a sextet of electrons. Although the non-bonding electron pair on the

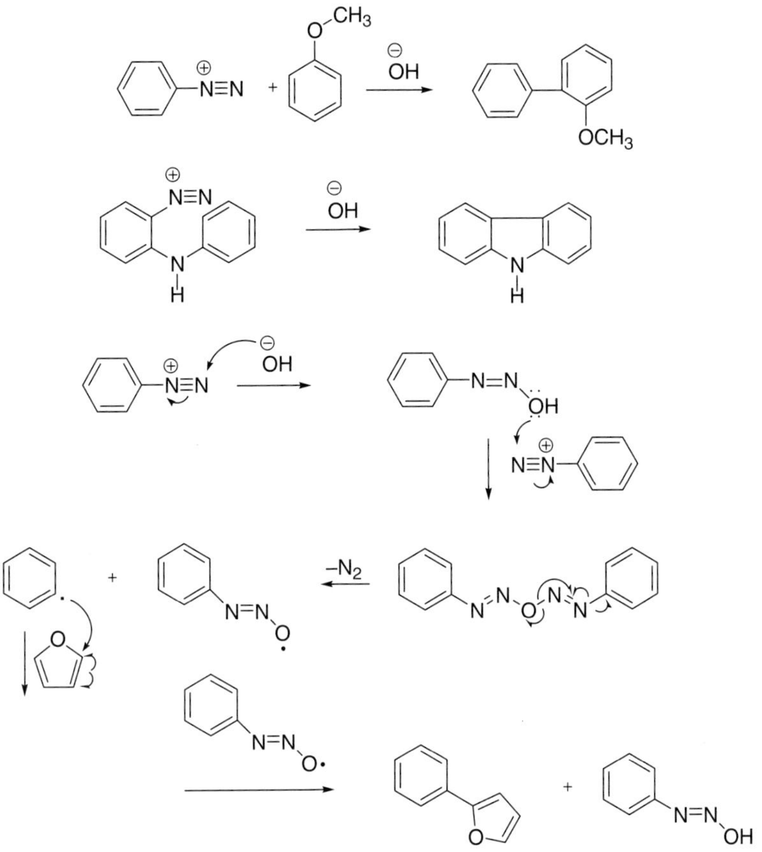

Scheme 2.49

carbon atom gives carbenes the nucleophilic character, as a rule the electrophilic character dominates the carbene reactivity.

The carbene :CH$_2$ derived by removal of two hydrogen atoms from CH$_4$ is known as methylene, :CCl$_2$ is known as dichlorocarbene or dichloromethylene, and :C(C$_6$H$_5$)$_2$ is known as diphenylcarbene or diphenylmethylene.

:CH$_2$	:CCl$_2$	:CPh$_2$
Methylene	Dichlorocarbene	Diphenylcarbene
	or	or
	dichloromethylene	diphenylmethylene

Names for few acyclic and cyclic carbenes are given below:

H$_2$C=C: CH$_2$=CHCH: :   CH$_3$—CH:

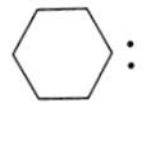

Ethenylidene   Prop-2-en-1-ylidene   Cyclohexylidene   Ethylidene

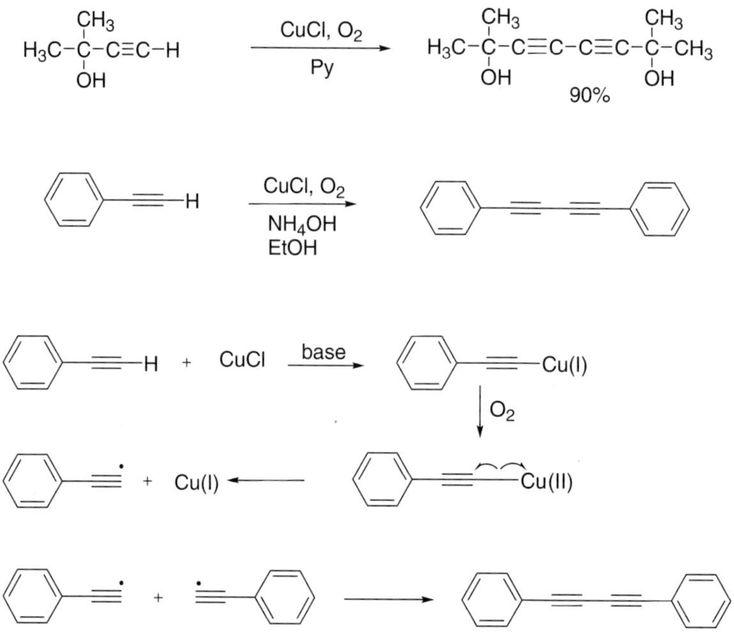

Scheme 2.50

2.4.1   Structure and stability of carbenes

If two non-bonding electrons are spin paired (or have antiparallel spins) then carbene is in **singlet state**, which is the dominant form in solution. But if they are spin non-paired (or have parallel spins) then carbene is in **triplet state**, which is the dominant form in the gas phase. There is no magnetic moment for singlet state. On the other hand, triplet state has magnetic moment. The substituents affect the ground-state multiplicity of a carbene. The singlet and triplet states can interconvert through a process known as **intersystem crossing**.

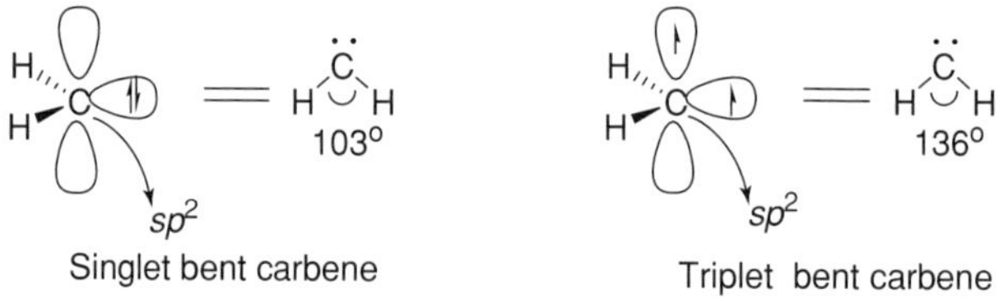

Singlet bent carbene   Triplet bent carbene

Thus, singlet carbene is carbocation-like in nature with trigonal planar geometry. The triplet carbene is a diradical. It has been investigated by esr measurements to be a bent molecule with an angle of about 136°. However, esr measurements cannot be made with singlet carbene. On the basis of electronic spectra of $:CH_2$, formed in the flash photolysis of diazomethane, it was found that the singlet carbene is also a bent molecule with an angle of about 103°. Singlet dichlorocarbene ($:CCl_2$) and dibromocarbene ($:CBr_2$) have also been found to be bent molecules with angles of 100 and 104°, respectively.

When steric bulkiness increases, the bond angle increases; thus triplet is favoured.

$CH_3 \overset{\cdot\cdot}{\underset{}{C}} CH_3$	
111°	152°
Singlet	Triplet

Bent carbenes

Although several carbenes were generated by Staudinger and Kupfer[60] by the decomposition of diazo compounds and ketenes around 1910, the growth in divalent carbenes began in 1950. In 1964, Fischer and Maasbol (see Reference 23, Chapter 3) prepared stable carbene complexes. In 1975, first time a carbene containing the parent methylene group was isolated. The stability of diamino carbenes such as imidazol-2-ylidenes **2.82** was first time demonstrated in 1991 by Arduengo[62–66]. Carbenes in which the carbene carbon is attached to two atoms, each bearing a lone pair of electrons, are more stable due to resonance.

2.82

The silylenes, germylenes and stannylenes are the analogues of carbenes.

2.4.2 Generation of carbenes

Depending on the mode of generation, a carbene may be initially formed in either the singlet or triplet state, irrespective of its stability. Common methods used for the generation of carbenes include photolytic, thermal, or metal catalyzed decomposition of diazocompounds, elimination of halogenfrom gem-dihalides, elimination of Hx from CHX_3, decomposition of ketenes, thermolysis of α-halo-mercury compounds and cycloelimination of shelf stable substrates such as cyclopropanes, epoxides, aziridines and diazirines.

From diazocarbonyl and related diazo compounds

Aliphatic diazocarbonyl and related diazo compounds can be decomposed by either photolysis or pyrolysis (or thermal decomposition), or by transition-metal-catalyzed decomposition generates (Scheme 2.51). For example, pyrolysis of diazoacetic ester at 425°C gives the carbethoxy carbene.

The thermal decomposition of monoanions of tosylhydrazone (Ar = *p*-tolyl) in a protic solvent like diglyme at 130°C or higher temperature generates dialkyl carbenes.[67] Dialkyl carbenes can also be generated photochemically from salts of tosylhydrazones, but it is difficult to perform the reaction on a large scale.

$$R_2C=N-\overset{\ominus}{N}SO_2Ar \xrightarrow{h\nu \text{ or } \Delta} \overset{\ominus}{Ar}SO_2 + R_2C=N_2 \longrightarrow R_2C: + N_2$$

Salts of sulfonylhydrazones

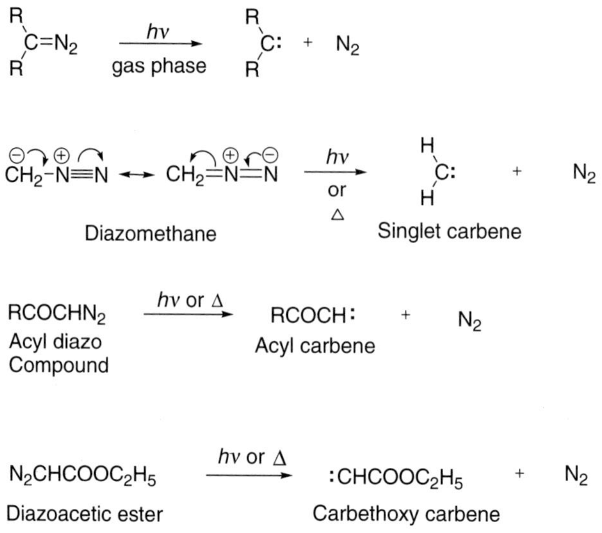

$$\underset{R}{\overset{R}{>}}C=N_2 \xrightarrow[\text{gas phase}]{h\nu} \underset{R}{\overset{R}{>}}C: \; + \; N_2$$

$$\overset{\ominus}{CH_2}-\overset{\oplus}{N}\equiv N \longleftrightarrow CH_2=\overset{\oplus}{N}=\overset{\ominus}{N} \xrightarrow[\Delta]{\overset{h\nu}{\text{or}}} \underset{H}{\overset{H}{>}}C: \; + \; N_2$$

Diazomethane   Singlet carbene

$$RCOCHN_2 \xrightarrow{h\nu \text{ or } \Delta} RCOCH: \; + \; N_2$$

Acyl diazo
Compound   Acyl carbene

$$N_2CHCOOC_2H_5 \xrightarrow{h\nu \text{ or } \Delta} :CHCOOC_2H_5 \; + \; N_2$$

Diazoacetic ester   Carbethoxy carbene

Scheme 2.51

From ketenes
Ketenes can be decomposed thermally or photolytically to generate carbenes.

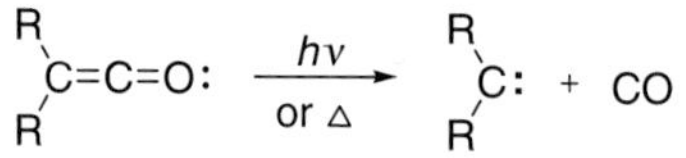

$$\underset{R}{\overset{R}{>}}C=C=O: \xrightarrow[\text{or } \Delta]{h\nu} \underset{R}{\overset{R}{>}}C: \; + \; CO$$

From epoxides
Photolytic decomposition of epoxides generates carbenes.

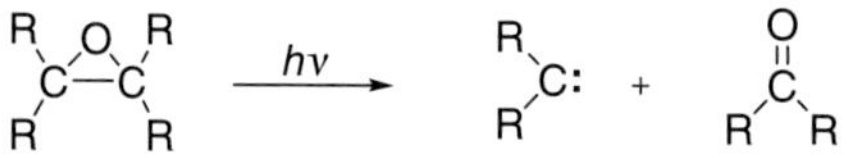

$$\underset{R}{\overset{R}{>}}\underset{R}{\overset{O}{C}}-\underset{R}{\overset{R}{C}} \xrightarrow{h\nu} \underset{R}{\overset{R}{>}}C: \; + \; \underset{R}{\overset{O}{\underset{R}{\overset{\|}{C}}}}$$

From diazirines
Decomposition of diazirines generates carbenes.

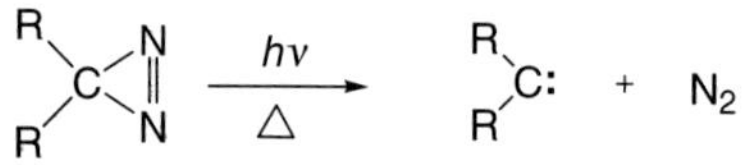

$$\underset{R}{\overset{R}{>}}C\overset{N}{\underset{N}{\|}} \xrightarrow[\Delta]{h\nu} \underset{R}{\overset{R}{>}}C: \; + \; N_2$$

From tetrazoles
Thermal decomposition of tetrazoles generates carbenes.

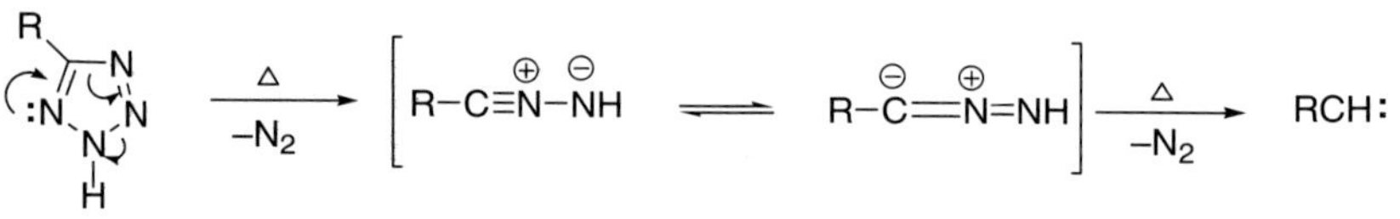

$$\underset{\overset{|}{H}}{\overset{R}{C}}\underset{N}{\overset{N}{\underset{N}{\Big\langle}}} \xrightarrow[-N_2]{\Delta} \left[R-C\overset{\oplus}{\equiv}\overset{\ominus}{N}-NH \rightleftharpoons R-\overset{\ominus}{C}=\overset{\oplus}{N}=NH \right] \xrightarrow[-N_2]{\Delta} RCH:$$

From haloalkanes

α-Dehalogenation (removal of X_2 molecule) and α-dehydrohalogenation (loss of HX) from haloalkanes generate the carbenes. For example, loss of a proton from chloroform by a base followed by expulsion of Cl^- generates dichlorocarbene.

$$CHCl_3 \xrightarrow{\text{KOH}} \overset{\ominus}{C}Cl_3 \longrightarrow :CCl_2 + \overset{\ominus}{Cl}$$

$$CH_2X_2 \xrightarrow{\Delta} :CHX + HX$$

$$CX_4 \xrightarrow{\Delta} :CX_2 + X_2$$

$$CX_2RR' \xrightarrow[\substack{\text{or} \\ hv}]{\Delta} :CRR' + X_2$$

From ylides

Thermal or photolytic decomposition of ylides generates carbenes.

$$\underset{\substack{R \\ \text{Ylide}}}{\overset{R}{\underset{}{C}}}\!\!{-}S(CH_2)_2 \underset{}{\overset{\Delta \text{ or } hv}{\rightleftharpoons}} \underset{R}{\overset{R}{C:}} + S(CH_3)_2$$

Besides the decomposition of diazo compounds, thermal cycloeliminations from carbocycles and heterocycles with odd numbers of ring atoms have been found to be suitable for dialkoxycarbenes[68], which are very difficult to obtain in any other way.

2.4.3 Reactions of carbenes

Carbenes are highly reactive and undergo insertion into σ-bonds, cycloaddition reactions, dimerization, complex formation and intramolecular reactions. The singlet carbene, which often acts as an electrophile, gives different products than the triplet carbene, which behaves as a radical. Despite their very different nature, they manage to produce the same product in some reactions.

Insertion reactions

Carbenes can insert into a σ-bond like C–H, C–X, O–H, N–H, S–H, M–H and M–X (M = metal).

$$\underset{R}{\overset{..}{C}}\!\!{R} + H{-}\overset{\overset{H}{|}}{\underset{\underset{H}{|}}{C}}{-}H \xrightarrow{\text{C–H insertion}} H{-}\overset{\overset{R}{|}}{\underset{\underset{R}{|}}{C}}{-}\overset{\overset{H}{|}}{\underset{\underset{H}{|}}{C}}{-}H$$

$$\underset{R}{\overset{..}{C}}\!\!{R} + H{-}O{-}CH_3 \xrightarrow{\text{O–H insertion}} H{-}\overset{\overset{R}{|}}{\underset{\underset{R}{|}}{C}}{-}O{-}CH_3$$

The reaction of carbene with hydrocarbon is a single-step reaction involving a triangular transition state. This reaction is possible because the singlet is such a strong electrophile that it can attack even the σ-electrons of a C–H bond. Although the triplet carbene gives the insertion product too, the mechanism by which it gets there is entirely different. Since the triplet carbene has a radical nature, it can abstract H atoms just like other organic radicals. The radical pair formed is surrounded by a cage of solvent molecules and the new C–C bond is formed by coupling of two radicals (Scheme 2.52).

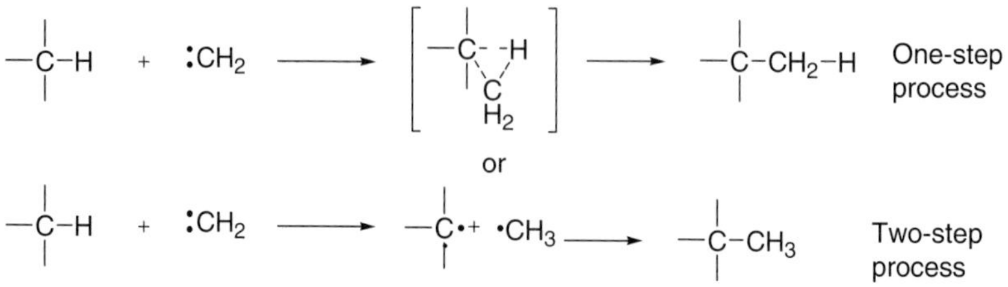

or

Scheme 2.52

Singlet carbenes insert into alkyl C–H bonds randomly, with retention of configuration. Triplet carbenes insert into alkyl C–H bonds selectively, but not stereospecifically.

Cycloadditions

Carbenes add on to multiple bonds like carbon–carbon double bonds, carbon–carbon triple bonds and carbon–nitrogen double bonds.

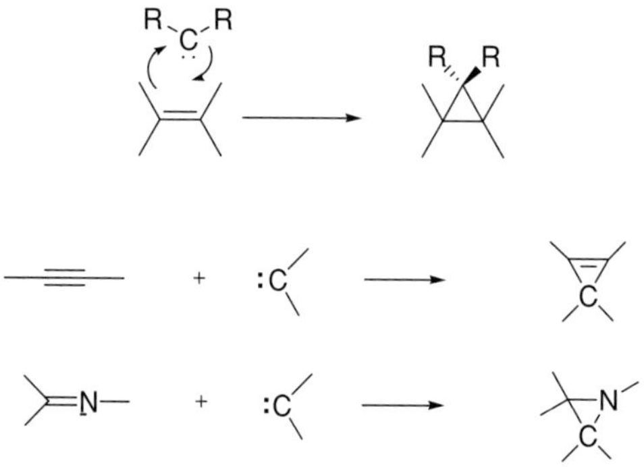

Singlet carbenes add to carbon–carbon double bonds in one step in a stereospecific manner. Triplet carbenes add to carbon–carbon double bonds in two steps in a non-stereospecific manner.

Thus, *cis*-alkene reacts with singlet carbene to give *cis*-cyclopropane, and *trans*-alkene gives *trans*-cyclopropane (Scheme 2.53).

However, triple carbene reacts with *cis*-alkene or *trans*-alkene to give a mixture of *cis*- and *trans*-cyclopropanes (Scheme 2.54).

The stereochemistry of these cycloadditions is so specific that **Skell**[69] used it as a diagnostic test for distinguishing between singlet and triplet carbenes. According to Skell, the addition of singlet carbene to an olefin occurs in a concerted manner and is therefore stereospecific. However, in the case of triplet carbene, both the unpaired electrons cannot form a new covalent bond because of their parallel spins. Therefore, in the latter case the reaction will take place in two steps. In the first step a triplet diradical is formed, which undergoes spin inversion and then ring closure. For this the radical has to wait for the appropriate

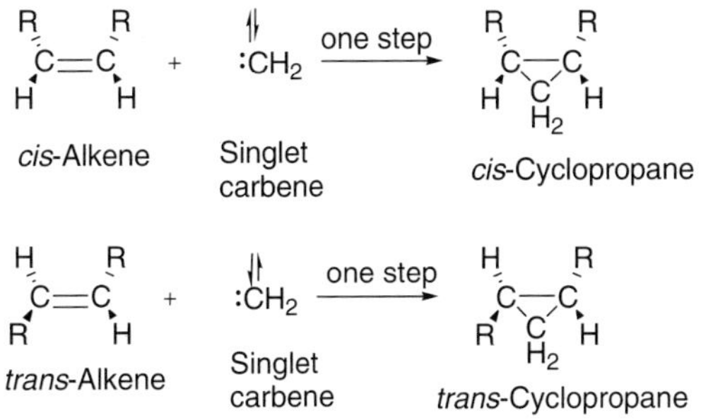

Scheme 2.53

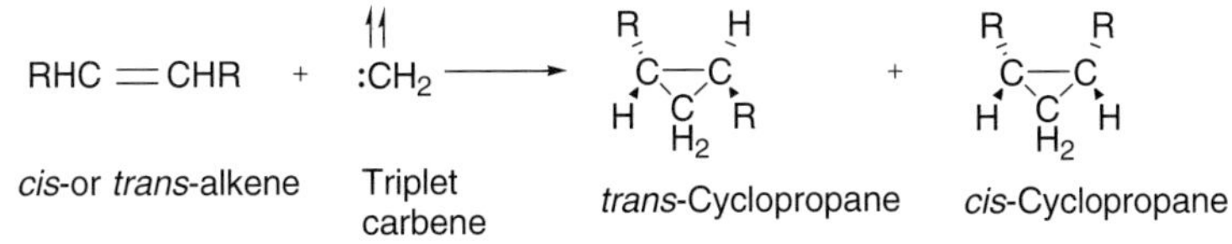

Scheme 2.54

type of collision. During this time, there is free rotation, and a mixture of *cis*- and *trans*-cyclopropanes is obtained (Scheme 2.55).

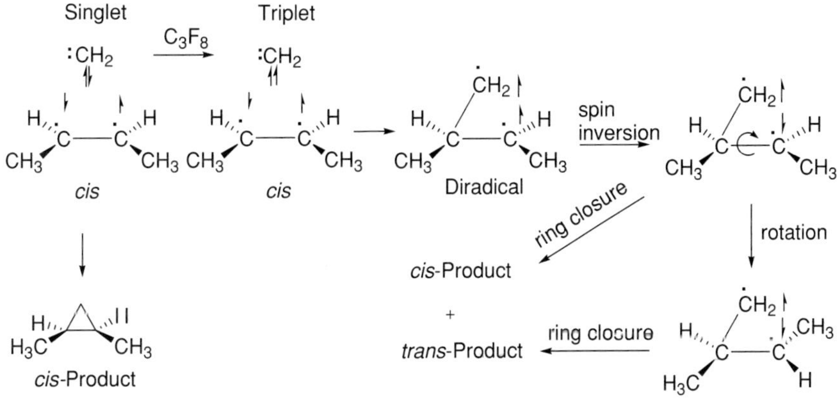

Scheme 2.55

In the above cycloaddition reactions, carbene is generated *in situ*. A more convenient way is to use **Simmons–Smith reagent**[70–74], which transfers methylene from methylene iodide and zinc–copper couple to a carbon–carbon double bond (Scheme 2.56). In the reaction in Scheme 2.56, free carbene is not generated. The intermediate is believed to be ICH_2ZnI, which behaves as an electrophile known as **carbenoid**.

Instead of expensive methylene iodide (diiodomethane), comparatively cheaper methylene bromide (dibromomethane) with zinc dust and cuprous chloride can be used to give better yield of the adduct.

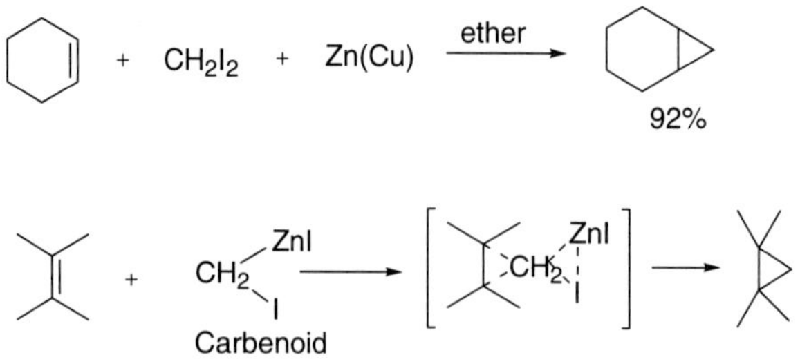

Scheme 2.56

The reaction of alkenes with carbene gives cyclopropane as a major product by cycloaddition reactions but products in which methylene is inserted into the C–H single bond are also obtained.

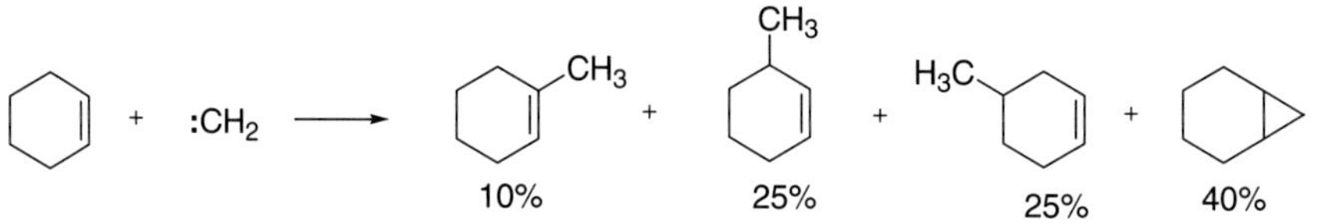

10%   25%   25%   40%

Intramolecular reactions

The intramolecular cycloaddition and insertion reactions of carbene are described in Scheme 2.57.

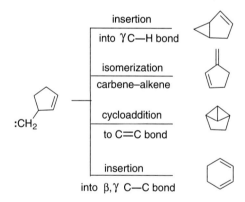

Scheme 2.57

Ring expansion

In certain substrates, addition of carbene involves ring expansion. Thus, the reaction of indene (**2.83**) with dichlorocarbene (:CCl$_2$) gives 2-chloronaphthalene (**2.84**).

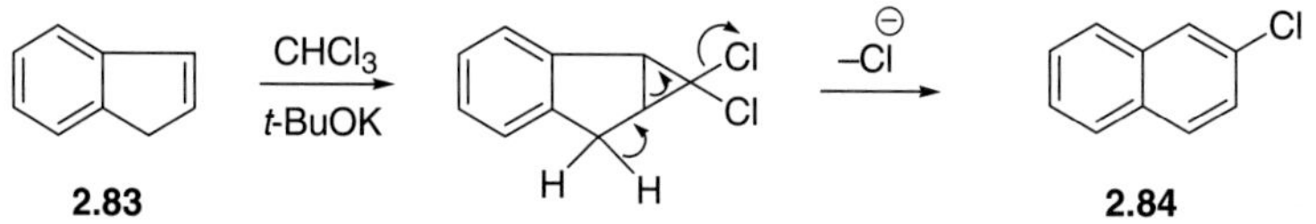

2.83   2.84

In a similar manner, the reaction of pyrrole (**2.85**) with :CCl$_2$ gives 3-chloropyridine (**2.86**).

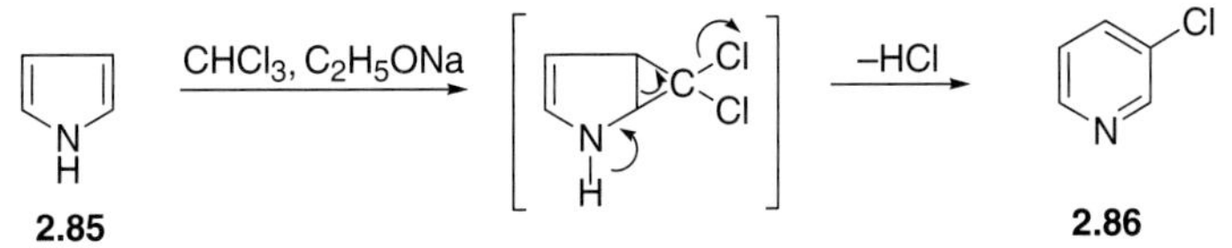

2.85   **2.86**

Wolff rearrangement

A method for converting an acid into its next higher homologue or its derivative is known as the **Arndt–Eistert synthesis**[75-77].

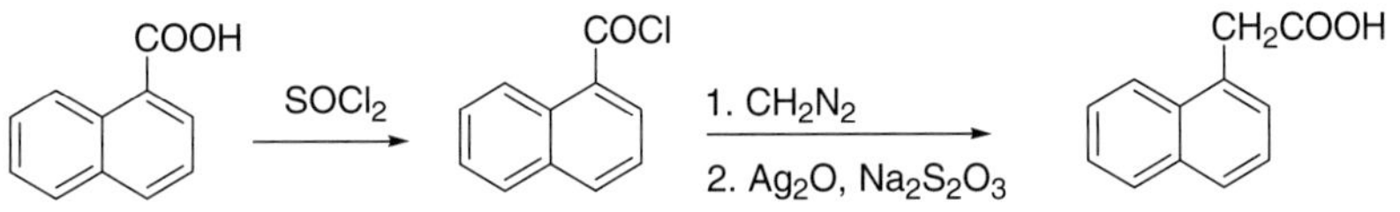

Nitrogen extrusion from α-diazoketone and the 1,2-shift can occur either in a concerted manner or stepwise via a carbene intermediate known as the **Wolff rearrangement**[78] (Scheme 2.58). α-Diazoketones undergo the Wolff rearrangement thermally in the range between room temperature and 750°C in gas-phase pyrolysis. Due to the formation of side products at elevated temperatures, the photochemical or silver-metal-catalyzed variants are often preferred that occur at lower reaction temperature.

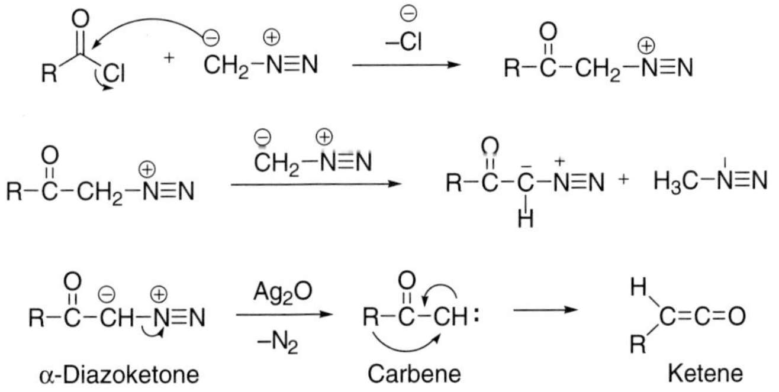

Scheme 2.58

Ketene is converted into acid, ester and amide when treated with water, alcohol and ammonia, respectively. When the rearrangement is carried out in the presence of water or an alcohol, the ketene is directly converted into an acid or ester.

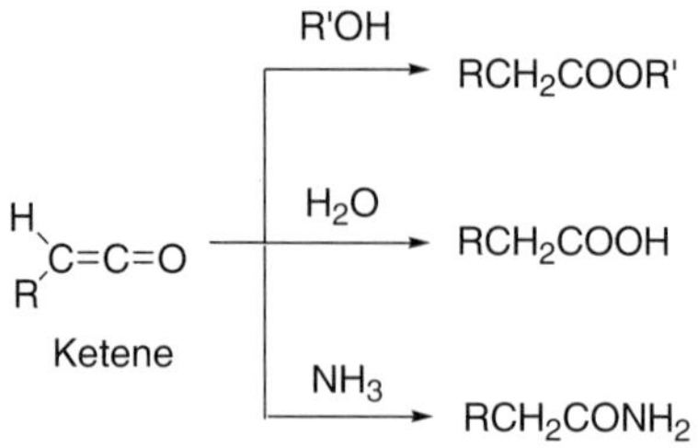

When a cyclic diazoketone is decomposed the rearrangement results in ring contraction (Scheme 2.59).

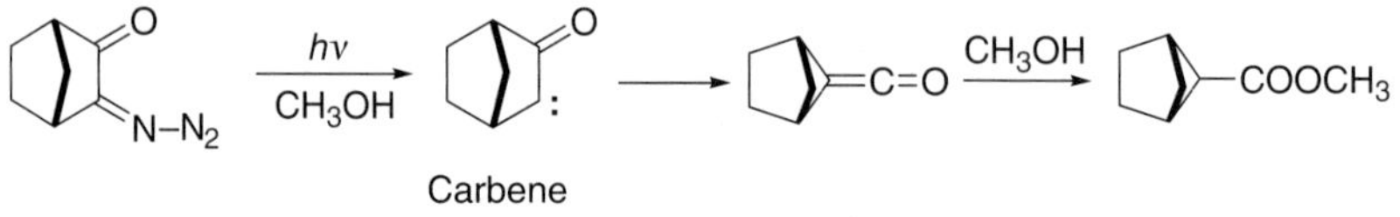

Scheme 2.59

Reaction of stable carbenes

G. Bertrand, in collaboration with W. W. Schoeller, reported recently[79] that certain carbenes, those in which the carbene centre (the divalent carbon atom) is sandwiched between an amine group and an alkyl group, have just enough nucleophilicity to cleave H_2 and NH_3. The resulting fragments (H and/or NH_2) become attached to the carbene centre. With liquid ammonia, the reaction occurs in high yield at temperatures around $-40°C$.

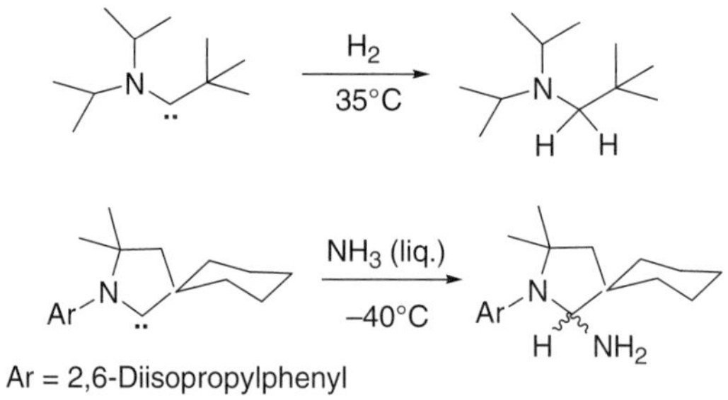

Ar = 2,6-Diisopropylphenyl

Chiral N-heterocyclic carbenes

N-heterocyclic salt such as chiral imidazolium salt **2.87** is the precursor of N-heterocyclic carbene, chiral imidazol-2-ylidene[80–82] **2.88**.

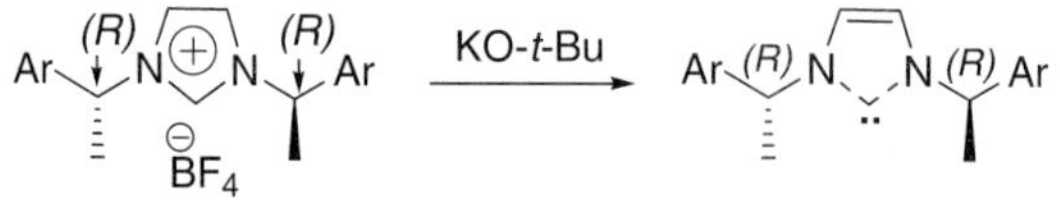

Ar = Ph or α-naphthyl

2.5 Nitrenes

Nitrenes (R–N:) are the nitrogen analogues of carbenes. They are uncharged, electron-deficient molecular species that contain a monovalent nitrogen atom surrounded by a sextet of electrons. High-level ab initio calculation showed that nitrenes are more stable than carbenes. The greater thermodynamic stability of nitrene is attributed to the fact that the N–H bond dissociation energies of aminyl radicals (NR$_2^{\cdot}$)are approximately 20 kcal mol^{-1} lower than the C–H bond dissociation energies of analogous alkyl radicals. This is due to the large amount of $2s$ character of the orbital that is occupied by the lone pair of electrons in nitrenes.

2.5.1 Structure and stability of nitrenes

The nitrogen atom in nitrenes has a sextet of electrons. As in the case of carbenes, the nitrenes exist in singlet and triplet states.

$$R-\overset{..}{\underset{..}{N}}: \quad\quad\quad R-\overset{..}{N}:$$

Singlet nitrene $\quad\quad$ Triplet nitrene

Singlet $\quad\quad\quad\quad$ Triplet

Nitrenes are very reactive species and normally not isolated. A nitrene can be trapped by its reaction with carbon monoxide and alkenes.

$$PhN_3 \xrightarrow[-N_2]{\Delta} Ph\overset{..}{N}: \xrightarrow{CO} PhN=C=O$$

Phenyl isocyanate

$$H\overset{..}{N}: + CH_2=CH_2 \rightleftharpoons \underset{H}{\overset{}{\triangle N}}$$

2.5.2 Generation of nitrenes

From azides
Thermolysis or photolysis of azides gives nitrenes with expulsion of nitrogen. This method is analogous to the formation of carbenes from diazo compounds.

$$RCON_3 \xrightarrow{hv} RCO\overset{..}{N}: + N_2$$

acyl azide $\quad\quad\quad$ Acyl nitrene

$$\left[R-\overset{..}{\underset{}{N}}=\overset{\oplus}{N}=\overset{\ominus}{\underset{..}{N}}: \longleftrightarrow R-\overset{-}{\underset{}{N}}-\overset{+}{N}\equiv N \right] \xrightarrow[\text{or heat}]{hv} R-\overset{..}{N}: + N_2$$

R = alkyl, aryl, H

From isocyanates

Alkyl nitrenes can also be obtained by the photolysis of isocyanates with the expulsion of carbon monoxide, a method analogous to carbene formation from ketenes.

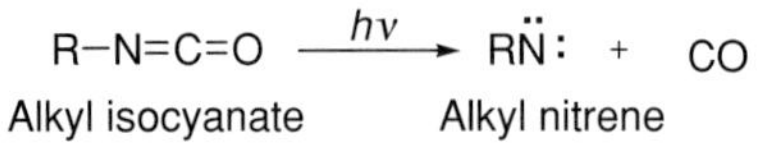

$$R-N=C=O \xrightarrow{h\nu} R\ddot{N}: \ + \ CO$$

Alkyl isocyanate   Alkyl nitrene

From sulfinylamines

Pyrolysis of sulfinylaniline (**2.89**) generates phenylnitrene.

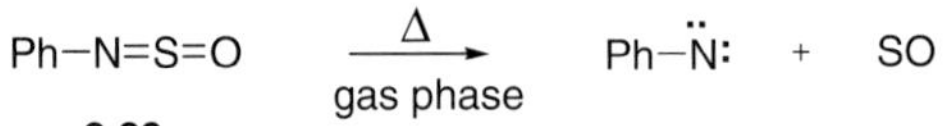

$$Ph-N=S=O \xrightarrow[\text{gas phase}]{\Delta} Ph-\ddot{N}: \ + \ SO$$

2.89

From N-benzenesulfonoxy carbamates

The reaction of *N*-benzenesulfonoxy carbamate (**2.90**) with a base results in the formation of carboalkoxynitrene with the elimination of benenesulfonate anion.

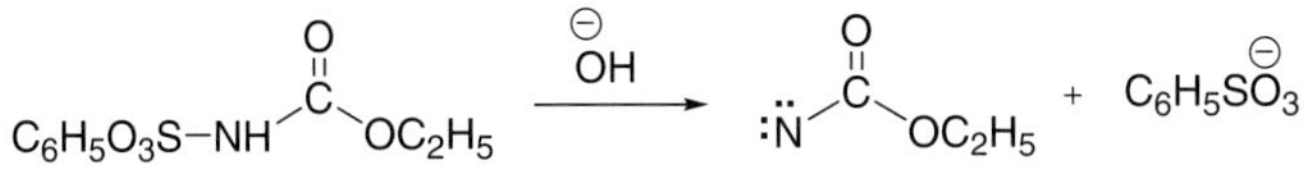

$$C_6H_5O_3S-NH-\overset{\overset{\displaystyle O}{\|}}{C}-OC_2H_5 \xrightarrow{\overset{\ominus}{OH}} :N-\overset{\overset{\displaystyle O}{\|}}{C}-OC_2H_5 \ + \ C_6H_5S\overset{\ominus}{O_3}$$

2.90

2.5.3   Reactions of nitrenes

Cycloaddition

Nitrenes add to carbon–carbon bonds to give aziridine (Scheme 2.60). Like carbenes, the addition of nitrenes to a carbon–carbon bond is stereospecific with singlet and non-stereospecific with triplet nitrenes.

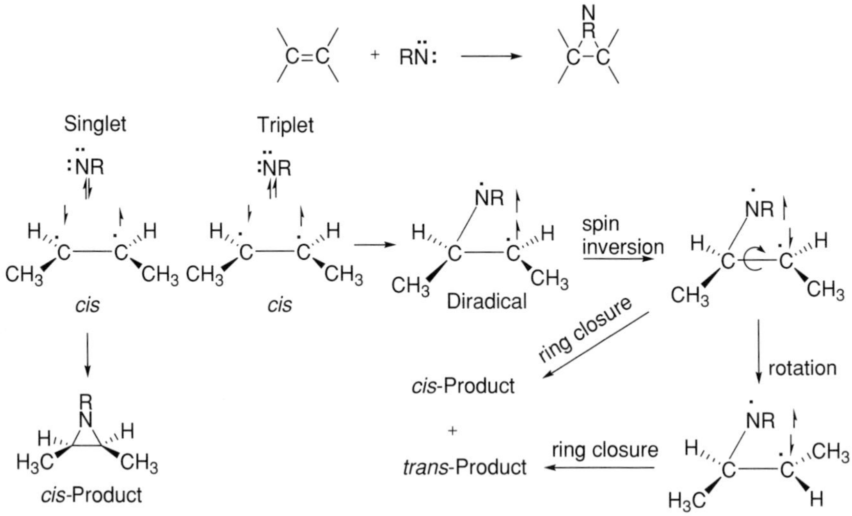

Scheme 2.60

Insertion

Nitrenes insert into C—H bonds (in the order: tertiary C–H > secondary C–H > primary C–H) yielding an amine or amide (Scheme 2.61).

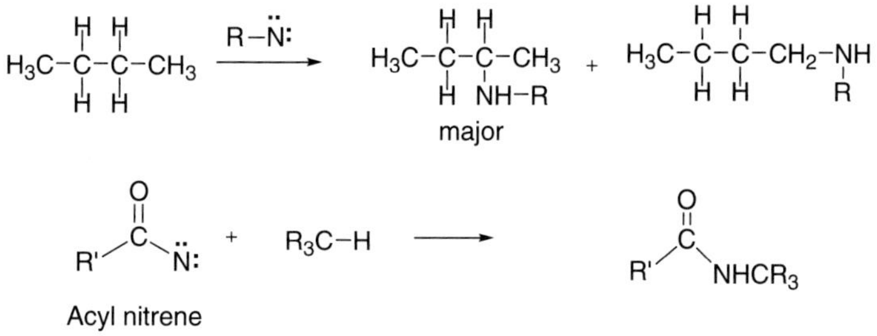

Acyl nitrene

Scheme 2.61

Nitrenes can also undergo insertion into C—H single bonds, leading to ring closure. Thus, vinyl azidothiophenes **2.91** have been found to be useful precursors for annulation of pyrroles and thiophenes.

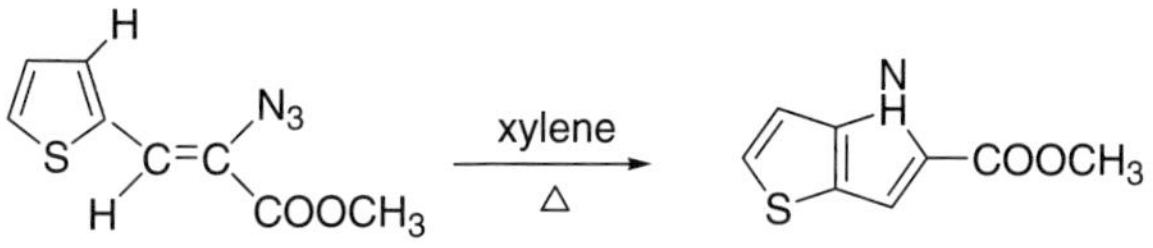

2.91

Singlet nitrenes insert into alkyl C–H bonds selectively, with retention of configuration. Triplet nitrenes do not insert into alkyl C–H bonds.

Hydrogen abstraction

Hydrogen abstraction from the carbon to the nitrogen leads to the formation of imines. This process is of considerable synthetic importance. When hydrogen abstraction takes place from 4- and 5-position which is followed by ring closure, the products formed are pyrrolidines and piperidines, respectively.

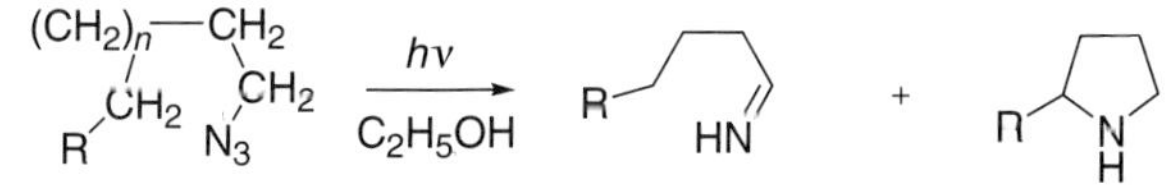

n = 1   pyrrolidines
n = 2   piperidines

Aryl nitrene ring expansion and ring contraction

Aryl nitrenes show ring expansion to seven-membered ring. The mechanism of ring expansion involves the **Wagner–Meerwein rearrangement** (Scheme 2.62).

Nitrenes are also obtained as reaction intermediates in **Hofmann**, **Schmidt** and **Lossen** rearrangements.

The Hofmann rearrangement

The Hofmann rearrangement[83–87] is the conversion of primary amides into primary amines by the action of sodium hypohalite (usually generated *in situ* from halogen and sodium hydroxide). The most important feature of the rearrangement is that the amine formed has

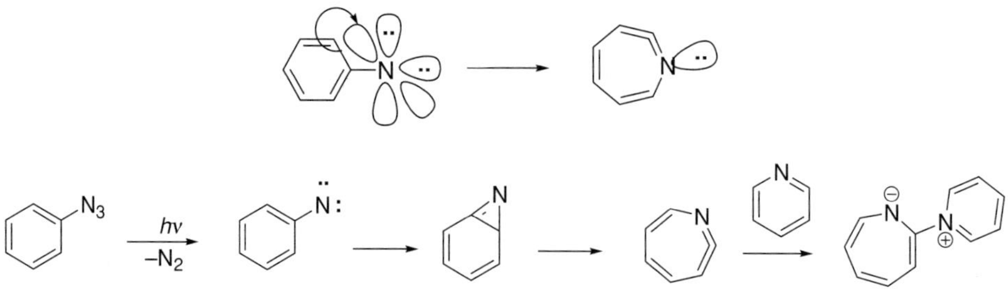

Didehydroazepine

Scheme 2.62

one less carbon atom in the molecule than the original amide. Thus, propanamide gives ethylamine on treatment with sodium hypobromite (or bromine and KOH).

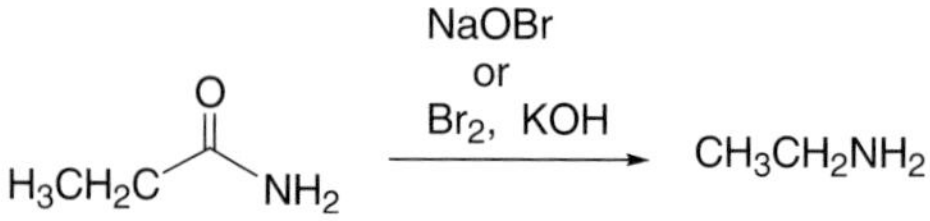

The mechanism of Hofmann rearrangement is shown in Scheme 2.63. In the first step, *N*-bromoamide **A** is formed by the action of alkaline hypobromite on the amide. The *N*-hydrogen atom of *N*-bromoamide **A** becomes acidic because it has electron-withdrawing acyl and electronegative halogen functions. The second step in the reaction involves removal of the acidic hydrogen from the *N*-haloamide **A** by the basic hydroxide ion. Removal of this proton by a base gives the transit nitrogen anionic species **B**, which is unstable and loses bromide ion. Migration of the aryl or alkyl group from the adjacent carbon atom to the nitrogen gives isocyanate **D**. Isocyanate **D** is hydrolyzed under the reaction conditions to N-substituted carbamic acid **E**, which is unstable and finally decarboxylates into the primary amine.

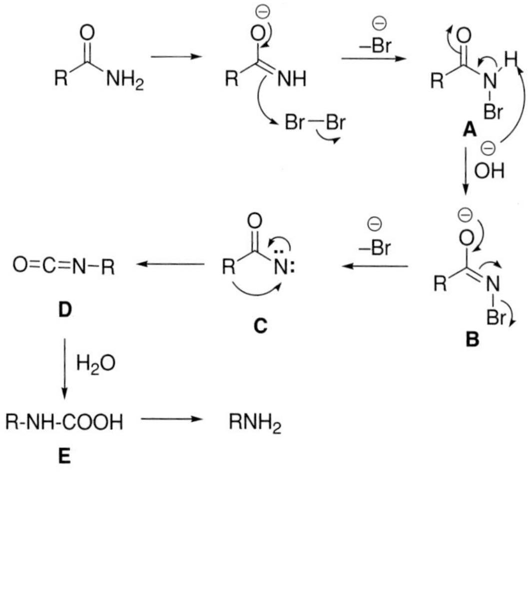

Scheme 2.63

2.6 Benzynes

Benzynes are neutral, highly reactive reaction intermediates in which the aromatic character is not markedly disturbed. Benzynes (or arynes) contain a carbon–carbon triple bond and may be regarded as aromatic counterpart of acetylene. The benzyne bond is not like the triple bond of acetylene where the two carbons form a σ-bond using *sp* orbitals and the remaining *p* orbitals form two π-bonds. Such a structure is not feasible in the case of benzyne due to hexagonal geometry associated with the benzene ring. It is believed that the new bond of benzyne is formed by the overlap of *sp*² orbitals belonging to two neighbouring carbon atoms. As the sideways overlapping is not very effective, the new bond is weak and so the benzyne is strained and is a highly reactive species.

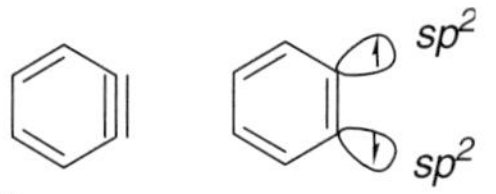

Benzyne

2.6.1 Generation of benzynes

From aryl halides
Aryl halides on treatment with a strong base like KNH_2 and C_6H_5Li generate benzyne.

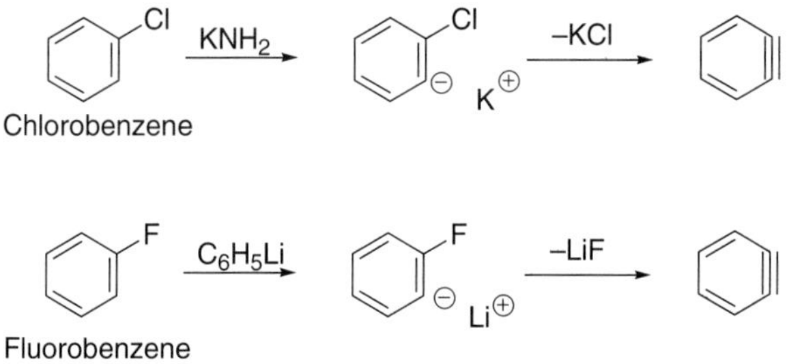

Chlorobenzene

Fluorobenzene

From o-aminobenzoic acid
o-Aminobenzoic acid (**2.92**) on diazotization followed by the decomposition of the diazo compound generates benzyne.

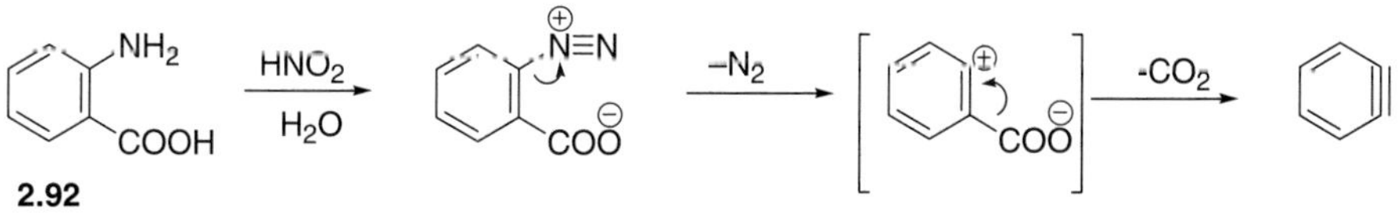

2.92

From phthaloyl peroxide
Phthaloyl peroxide (**2.93**) on photolytic decomposition generates benzyne via lactone intermediate.

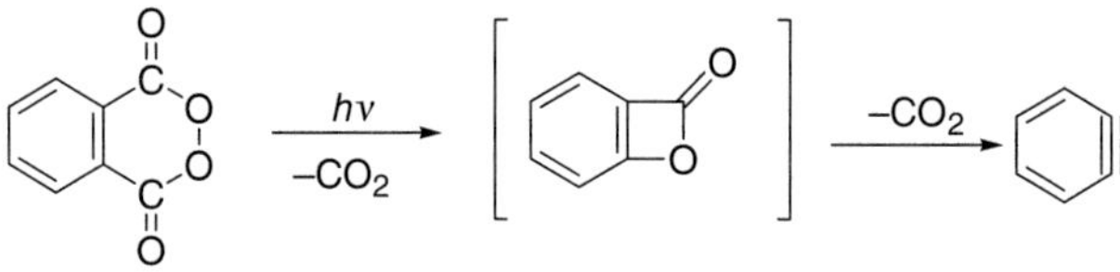

2.93

From benzothiadiazole-1,1-dioxide
Thermal decomposition of benzothiadiazole-1,1-dioxide (**2.94**) generates benzyne.

2.94

From benzenetrifluoromethane sulfonate
Benzenetrifluoromethane sulfonate (**2.95**) on treatment with a strong base generates ben-zyne.

2.95   LDA = Lithium diisopropylamide

From 1,2-dihalobenzene
Benzyne can also be generated from 1,2-bromofluorobenzene via the formation of Grignard reagent or organolithium reagent.

2.6.2   Reactions of benzynes

Benzyne is an extremely reactive species and is generated *in situ* for obtaining various products.

Reaction with nucleophiles
As already stated, aryl halides on treatment with a strong base such as KNH_2 generates benzyne. The benzyne further reacts with nucleophile (NH_2^-) to give aniline.

Bromobenzene   Benzyne   Aniline

When X is I, Br or Cl, the loss of H and X can occur in concerted fashion (E2) (Scheme 2.64). But when X is F, the *ortho*-hydrogen becomes more acidic; thus, the E1cB pathway follows and benzyne is formed in two steps (Scheme 2.65).

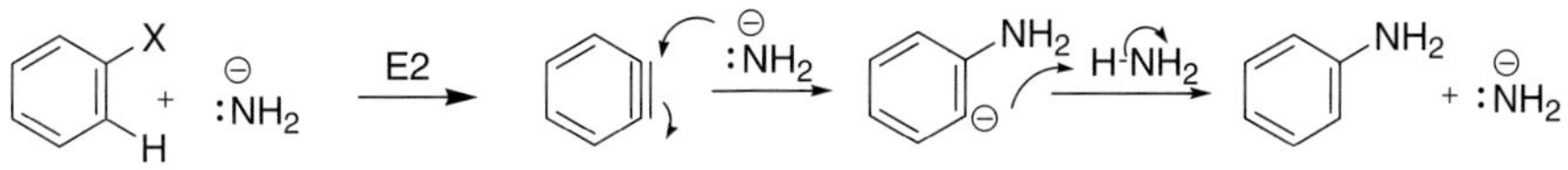

Scheme 2.64

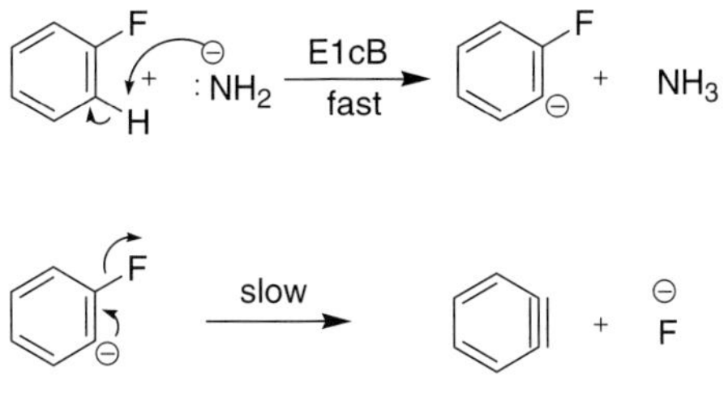

Scheme 2.65

Substituted benzynes give a mixture of products. This is illustrated in Schemes 2.66 and 2.67.

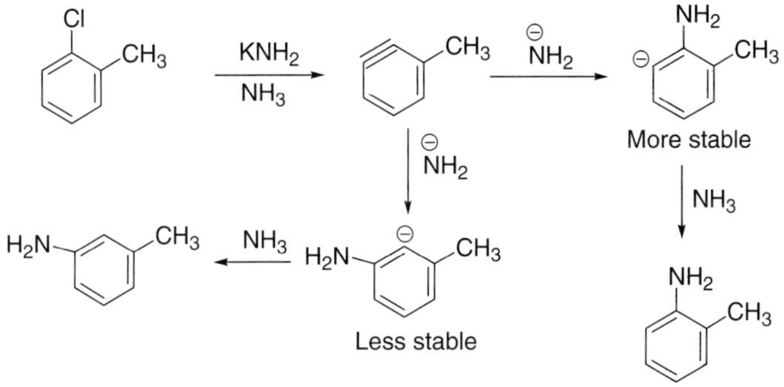

Scheme 2.66

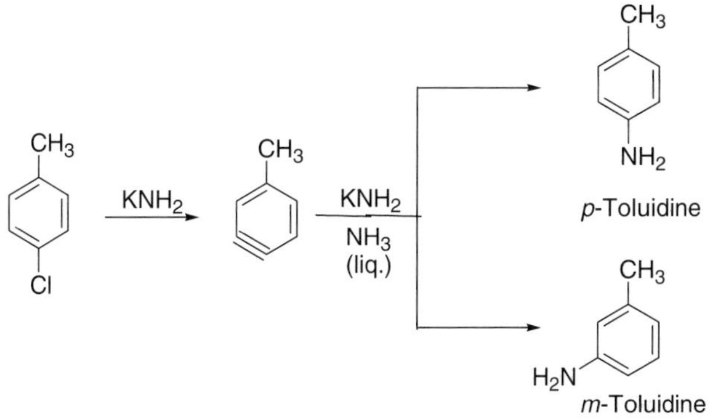

Scheme 2.67

o-Bromoanisole and *m*-bromoanisole on reaction with $NaNH_2$ give only one product, *m*-aminoanisole (Scheme 2.68).

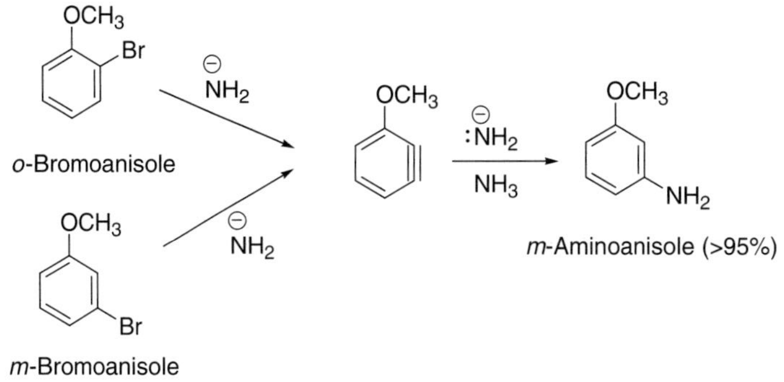

Scheme 2.68

Various products can be synthesized by the reaction of a benzyne with different compounds.

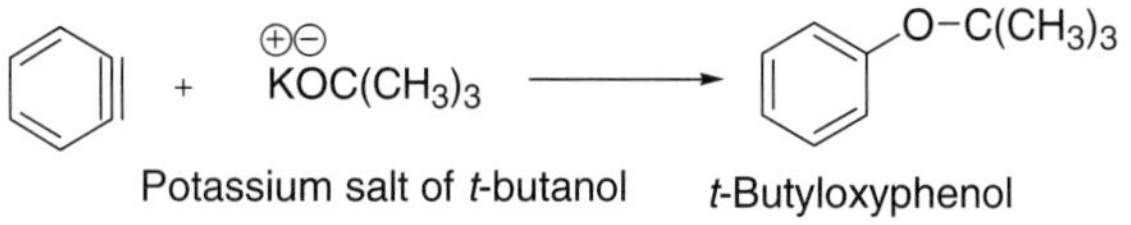

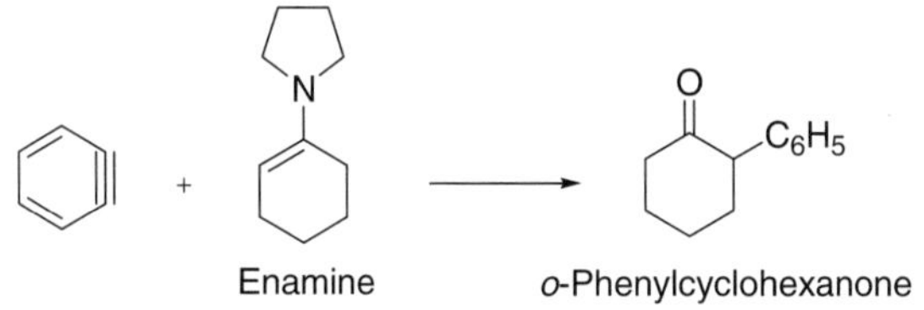

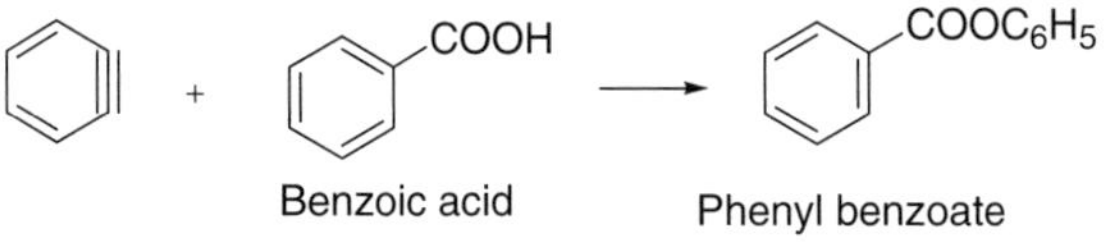

Cycloaddition

Benzyne undergoes [4+2]-cycloaddition reactions in which it acts as a dienophile. The Diels–Alder reaction of benzynes give fused- or bridged-ring systems.

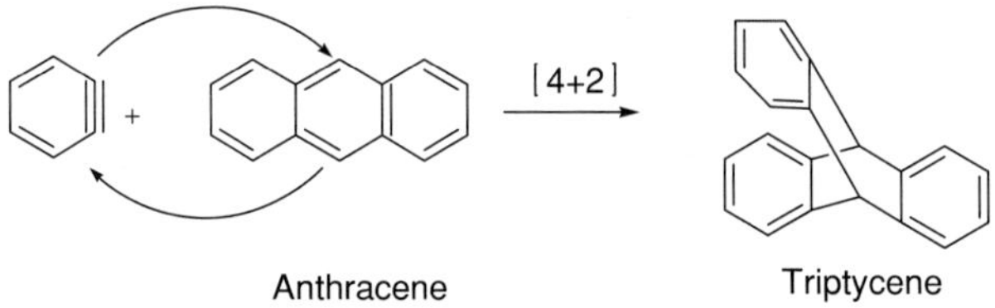

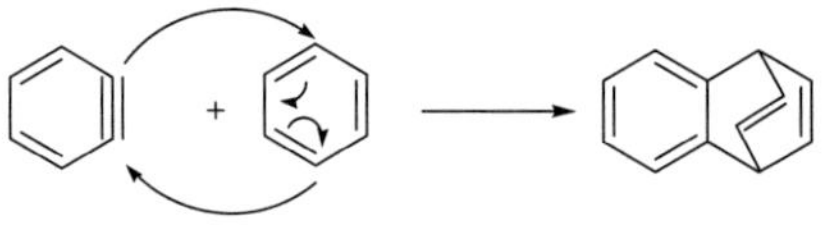

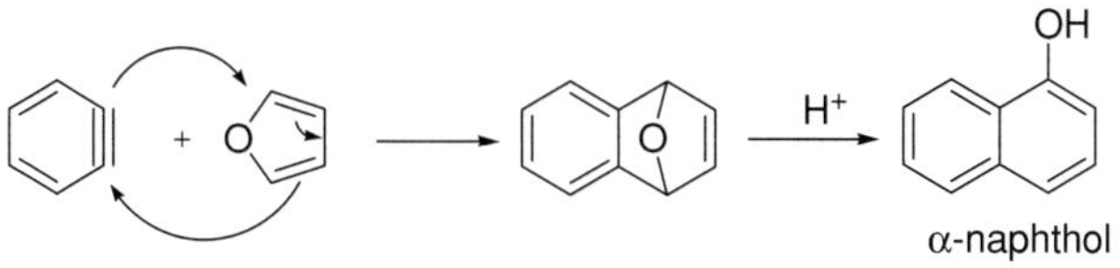

α-naphthol

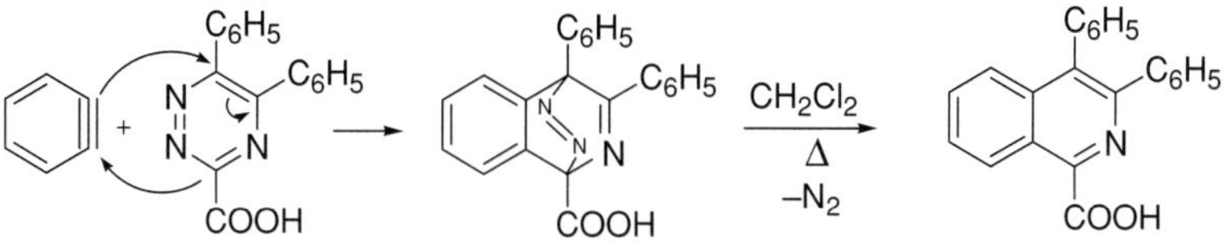

Isoquinoline derivative (70%)

Benzyne undergoes [2+2]-cycloaddition reaction with alkenes to give four-membered rings.

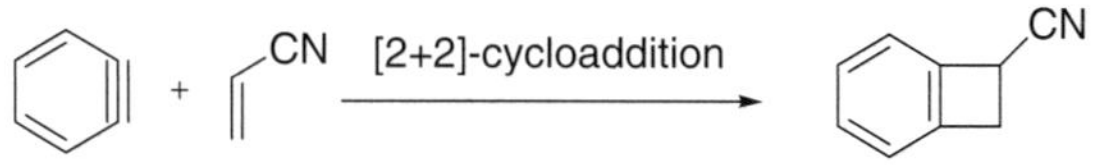

In the absence of any nucleophiles, benzyne undergoes dimerization.

Several heterocyclic compounds are synthesized by the reaction of intramolecular reactions of benzynes.

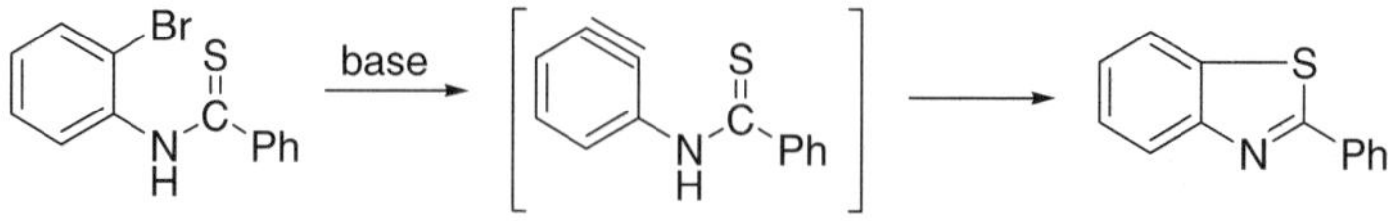

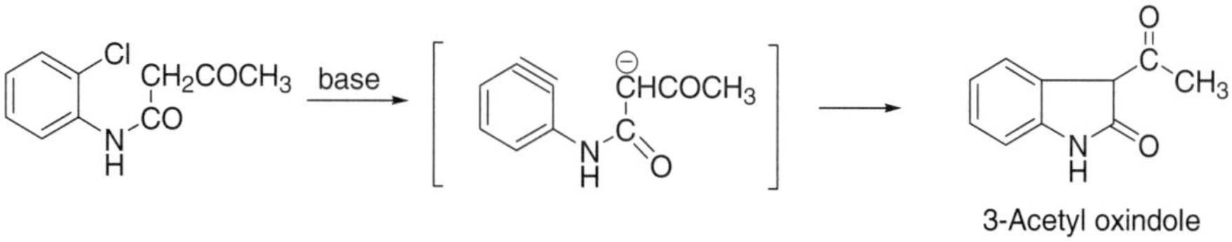

3-Acetyl oxindole

References

1. Platz, M. S., Moss, R. A. and Jones, M., Jr., *Reviews of Reactive Intermediate Chemistry*, Wiley-Inter Sciences April **2007**.
2. Carey, F. A. and Sundberg, R. J., *Advanced Organic Chemistry Part A Structure and Mechanisms,* 2nd edn, Plenum Press, New York, **1984**.
3. March, J., *Advanced Organic Chemistry Reactions, Mechanisms and Structure,* 3rd edn, Wiley, New York, **1885**.
4. Gilchrist, T. C. and Rees, C. W., *Carbenes, Nitrenes and Arynes*, Nelson, London, **1969**.
5. Friedel, C. and Crafts, J. M., *Compt. Rend.,* **1877**, *84*, 1392.
6. Wagner, G., *J. Russ. Phys. Chem. Soc.,* **1899**, *31*, 690.
7. Meerwein, H., *Ann. Chem.,* **1914**, *405*, 129.
8. Starling, S. M., Vonwiller, S. C. and Reek, J. N. H., *J. Org. Chem.,* **1998**, *63, 2262.*
9. Brown, H. C., *The Non-Classical Ion Problem*, Plenum, New York, **1977**.

10. Thomson, V. and Stevens, T. S., *J. Chem. Soc.*, **1932**, 55.
11. Stevens, T. S., Creighton, E. M., Gorden, A. B. and MacNicol, M., *J. Chem. Soc.*, **1928**, 3193.
12. Olsen, R. K. and Currie, J. O., *The Chemistry of the Thiol Group*, Vol. 2 (ed. Patai, S.), Wiley New York, **1974**, p. 561.
13. Hofmann, A. W., *Chem. Ber.*, **1881**, *14*, 659.
14. Cope, A. C. and Trumbull, E. R., *Org. React.*, **1960**, *11*, 320.
15. Cope, A. C., *J. Org. Chem.*, **1965**, *30*, 2163.
16. Favorskii, A. E., *J. Prakt. Chem.*, **1913**, *88*(2), 658.
17. Wallach, O., *Ann. Chem.*, **1918**, *414*, 296.
18. Nace, H. R. and Olsen, B. A., *J. Org. Chem.*, **1967**, *32*(11), 3438.
19. Wagner, R. B., *J. Am. Chem. Soc.*, **1950**, *72*, 972.
20. Schamp, N., Dekimpe, N. and Coppens, W., *Tetrahedron*, **1975**, *31*, 2081.
21. Loftfield, R. B., *J. Am. Chem. Soc.*, **1950**, *72*, 632.
22. Bordwell, F. G. and Scamehorn, R. G., *J. Am. Chem. Soc.*, **1968**, *90*, 6751.
23. Suzuki, N., Azuma, T., Kaneko, Y., Izawa, Y., Tomioka, H. and Nomoto, T., *J. Chem. Soc., Perkin Trans 1*, **1987**, 645.
24. Sandmeyer, T., *Chem. Ber.*, **1884**, *17*, 1633.
25. Sandmeyer, T., *Chem. Ber.*, **1884**, *17*, 2650.
26. Gattermann, L., *Chem. Ber.*, **1890**, *23*, 1218.
27. Fenton, H. J. H., *J. Chem. Soc.*, **1894**, *65*, 899.
28. Kolbe, H., *Ann. Chem. Pharm.*, **1848**, *64*(3), 339.
29. Kolbe, H., *Ann. Chem. Pharm.*, **1849**, *69*(3), 257–372.
30. Brown, A. C. and Walker, J., *Ann. Chem.*, **1891**, *261*, 107.
31. Floreancig, P. E., *Tetrahedron*, **2006**, *62*(27), 6457.
32. Wohl, A., *Chem. Ber.*, **1919**, *52*, 51.
33. Ziegler, K., Spath, A., Schaaf, E., Schumann, W. and Winkelmann, E., *Ann.*, **1942**, *551*, 80.
34. Barton, D. H. R. and McCombie, S. W., *J. Chem. Soc., Perkin Trans 1*, **1975**, 1574.
35. Lopez, R. M., Hays, D. S. and Fu, G. C., *J. Am. Chem. Soc.*, **1997**, *119*(29), 6949.
36. Kirwan, J. N., Roberts, B. P. and Willis, C. R., *Tetrahedron Lett.*, **1990**, *31*(35), 5093.
37. Gimisis, T., Ballestri, M., Ferreri, C., Chatgilialoglu, C., Boukherroub, R. and Manuel, G., *Tetrahedron Lett.*, **1995**, *36*(22), 3897.
38. Zhu, J., Klunder, A. J. H. and Zwanenburg, B., *Tetrahedron*, **1995**, *51*, 5099.
39. Barton, D. H. R., *Tetrahedron Lett.*, **1985**, *26*, 5939.
40. Barton, D. H. R. and Beston, J. M., *J. Am. Chem. Soc*, **1960**, *63*, 2541.
41. Barton, D. H. R., Beaton, J. M., Geller, L. E. and Pechet, M. M., *J. Am. Chem. Soc.*, **1960**, *82*, 2640.
42. Barton, D. H. R. and Beaton, J. M., *J. Am. Chem. Soc.*, **1961**, *83*, 4083.
43. Hobbs, P. D. and Magnus, P. D., *J. Am. Chem. Soc.*, **1976**, *98*, 4594.
44. Sibi, M. P. and Ternes, T. R., Stereoselective radical reactions, In *Modern Carbonyl Chemistry* (ed. Otera, J.), Wiley-VCH, **2000**, pp. 507–538.
45. Sibi, M. P., Manyem, S. and Zimmerman, J., *Chem. Rev.*, **2003**, *103*, 3263.
46. Gomberg, M., *Chem. Ber.*, **1900**, *33*, 3150.
47. Gomberg, M., *J. Am. Chem. Soc.*, **1900**, *22*, 757.
48. Gomberg, M., *Chem. Ber.*, **1897**, *30*, 2043.
49. Gomberg, M., *J. Am. Chem. Soc.*, **1897**, *20*, 773.
50. Kharasch, M. S. and Mayo, F. R., *J. Am. Chem. Soc.*, **1933**, *55*, 2468.
51. Kharasch, M. S., McNab, M. C. and Mayo, F. R., *J. Am. Chem. Soc.*, **1933**, *55*, 2521.
52. Sibi, M. P. and Rheault, T. R., Enantioselective radical reactions, In *Radicals in Organic Synthesis* (eds Renaud, P. and Sibi, M. P.), Wiley-VCH, **2001.**
53. Sibi, M. P., Ji, J. G., Wu, J. H., Gurtler, S. and Porter, N. A., *J. Am. Chem. Soc.*, **1996**, *118*, 9200.
54. Jones, R. G. and Bergmann, R. G., *J. Am. Chem. Soc.*, **1972**, *94*, 660.
55. Bergmann, R. G., *Acc. Chem. Res.*, **1973**, *6*, 25.
56. Lai, Y.-H. and Jiang, J., *J. Org. Chem.*, **1997**, *62*, 4412.
57. Gomberg, M. and Bachmann, W. E., *J. Am. Chem. Soc.*, **1924**, *42*, 2339.
58. Glaser, C., *Chem. Ber.*, **1869**, *2*, 422.
59. Ghose, B. N. and Walton, D. R. M., *Synthesis*, **1974**, 890.
60. Staudinger, H. and Kupfer, O., *Chem. Ber.*, **1912**, *45*, 501.
61. Hine, J., *Divalent Carbon*, The Ronald Press Company, New York, **1964**.
62. Arduengo, A. J., III, Harlow, R. L. and Kline, M., *J. Am. Chem. Soc.*, **1991**, *113*, 361–362.
63. Arduengo, A. J., III, Dias, R. H. V., Harlow, R. L. and Kline, M., *J. Am. Chem. Soc.*, **1992**, *114*, 5530–5534.
64. Kuhn, N. and Kratz, T., *Synthesis*, **1993**, 561–562.
65. Alder, R. W., Allen, D. R. and Williams, S. J., *J. Chem. Soc., Chem. Commun.*, **1995**, 1267.
66. Arduengo, A. J., III, Goerlich, J. E. and Marshall, W. J., *J. Am. Chem. Soc.*, **1995**, *117*, 11027.

67. Chamberlin, A. R. and Bond, F. T., *J. Org. Chem.*, **1978**, *43*(1), 154.
68. Hoffmann, R. W., *Angew. Chem., Int. Ed. Engl.*, **2003**, *8*(10), 529.
69. Dürr, H., Triplet-intermediates from diazo-compounds (carbenes), **1975**, volume *55*, 85–135, Springer, Berlin.
70. Simmons, H. E. and Smith, R. D., *J. Am. Chem. Soc.*, **1959**, *81*, 4256.
71. Smith, R. D. and Simmons, H. E. *Org. Syn. Coll.*, **1973**, *5*, 855.
72. Smith, R. D. and Simmons, H. E. *Org. Syn. Coll.*, **1961**, *41*, 72.
73. Cohen, T. and Kosarych, Z., *J. Org. Chem.*, **1982**, *47*, 4005.
74. Kim, H. Y., Lurain, A. E., Garcia-Carcia, P., Carroll, P. J. and Walsh, P. J., *J. Am. Chem. Soc.*, **2005**, *127*, 13138.
75. Arndt, F. and Eistert, B., *Chem. Ber.*, **1935**, *68*, 200.
76. Smith, A. B., *Chem. Commun.*, **1974**, 695.
77. Walsh, E. J. and Stone, G. B., *Tetrahedron Lett.*, **1986**, *27*(10), 1127.
78. Doz, U., Meier, H. and Zeller, K. P., *Angew. Chem., Int. Ed.*, **1975**, *14*, 32.
79. Frey, G. D., Lavallo, V., Donnadieu, B., Wolfgang, W., Schoeller, W. W. and Bertrand, G., *Science*, **2007**, *316*, 439.
80. Kano, T., Sasaki, K. and Maruoka, K., *Org. Lett.*, **2005**, *7*, 1347.
81. Suzuki, Y., Muramatsu, K., Yamauchi, K., Morie, Y. and Sato, M., *Tetrahedron*, **2006**, *62*, 302.
82. Bourissou, D., Guerret, O., Gabbai, F. P. and Bertrand, G., *Chem. Rev.*, **2000**, *100*, 39.
83. Hofmann, A. W., *Chem. Ber.*, **1881**, *14*, 2725.
84. Hamlin, K. E. and Freifelder, M., *J. Am. Chem. Soc.*, **1953**, *75*, 369.
85. Allen, W., *Org. Syn. Coll.*, **1963**, *4*, 45.
86. Magnieni, E., *J. Org. Chem.*, **1958**, *23*, 2029.
87. Finger, G. C., Starr, L. D., Roe, A. and Link, W. J., *J. Org. Chem.*, **1962**, *27*, 3965.

Chapter 3
Stabilized Carbanions, Enamines and Ylides

A useful assortment of carbon–carbon bond forming reactions such as alkylation of enolates and enamines and carbon–oxygen bond forming reactions such as epoxidation with ylides have been described in this chapter.

3.1 Stabilized carbanions

Aldehydes and ketones exist in equilibrium with just small amounts of enol tautomers:

Keto tautomer   Enol tautomer
>99%

Keto tautomer   Enol tautomer
98.8%   1.2%

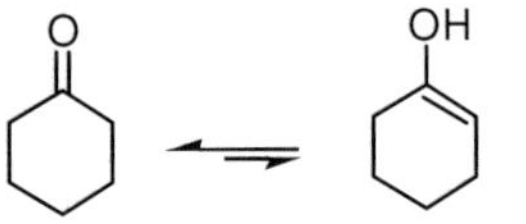

β-Diketones such as 2,4-pentanedione (**3.1**), β-ketoesters such as ethyl acetoacetate (**3.2**) and β-diesters such as diethyl malonate (**3.3**) have much higher enol concentrations than do monocarbonyl aldehydes and ketones. This is due to greater stability of the enol form of β-dicarbonyl compounds, which is attributed to the intramolecular hydrogen bonding.

3.1
2,4-Pentanedione
(Acetylacetone)

Enol form
76%   pure liquid
20%   in water solution
92%   in hexane solution

3.2
Ethyl acetoacetate
(Acetoacetic ester)

Enol form
8%   pure liquid
0.4%   in water solution
46%   in hexane solution

112

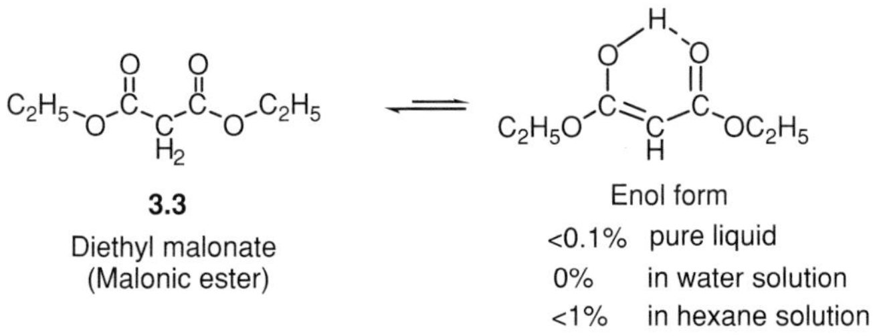

3.3

Diethyl malonate
(Malonic ester)

Enol form

<0.1%   pure liquid

0%   in water solution

<1%   in hexane solution

Enolization may be catalyzed by an acid or a base (Scheme 3.1).

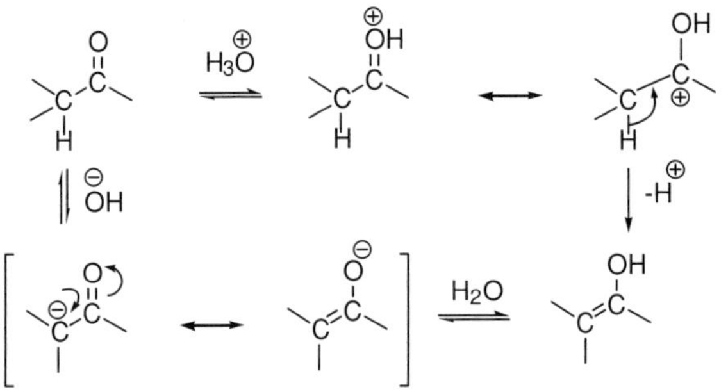

Scheme 3.1   Acid- and base-catalyzed enolization

Enolates are selectively formed by treating ketones with a strong base such as lithium diisopropylamide (LDA) $[LiN(i\text{-}C_3H_7)_2]$ and tetrahydrofuran (THF) at $-78°C$, NaH or alkoxide ion. LDA and THF at $-78°C$ give less stable, i.e. less substituted, enolate (kinetic enolate) **3.5** from unsymmetrical ketone **3.4**, whereas methoxide (CH_3O^-) forms the more stable, i.e. more substituted, enolate (thermodynamic enolate) **3.6** from unsymmetrical ketone **3.4**.

LDA
THF, $-78°C$

3.5
Kinetic enolate
(Less stable)

3.4

$CH_3\overset{\ominus}{O}$

protic solvent

3.6
Thermodynamic enolate
(More stable)

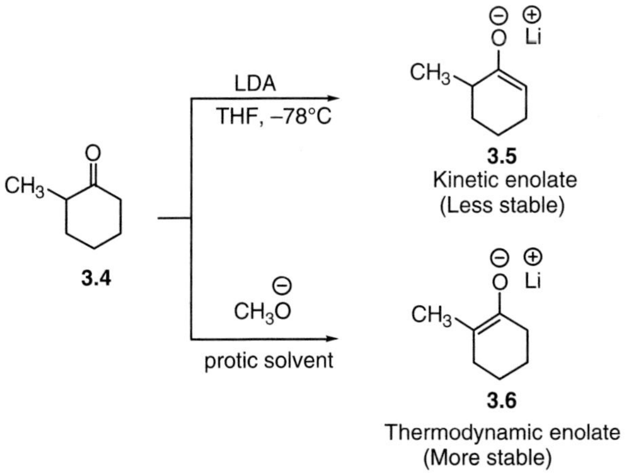

Enolate ions exist as a resonance hybrid in which the negative charge resides primarily on the carbonyl oxygen and the α-carbon.

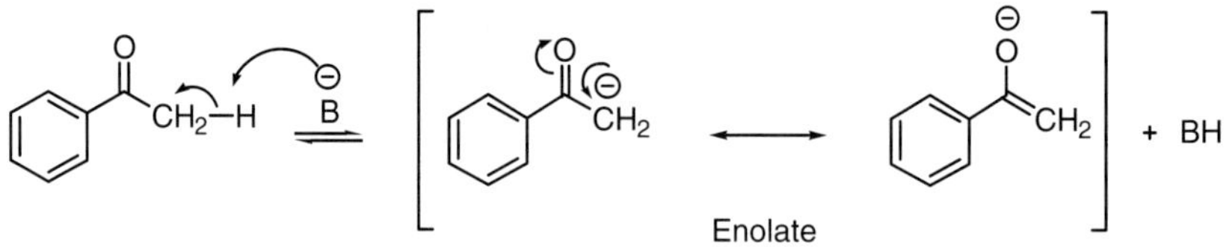

The enols and enolates undergo several types of reactions. The reaction of enolates with electrophiles is referred to as α-substitution of carbonyl compounds.

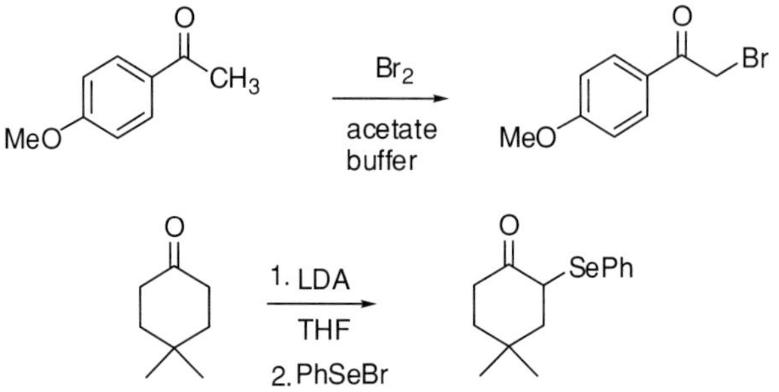

The selenenylation reaction can be viewed as an electrophilic attack on the enolate, or as an S_N2 attack by the enolate on selenium. Selenenylation is useful because the phenyl selenide readily undergoes intramolecular elimination to yield the conjugated enone.

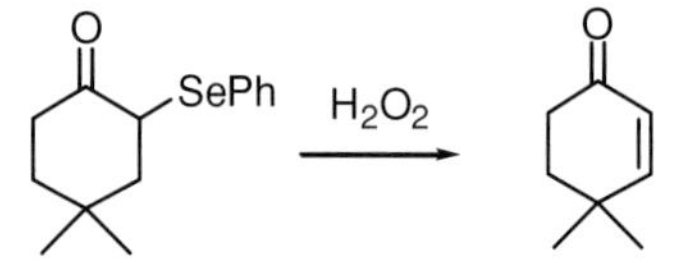

Enolates are useful nucleophilic intermediates which are frequently employed in synthesis particularly for the formation of the carbon–carbon bond.

3.1.1 Reaction of stabilized carbanions (enolates) with alkyl halides (enolate alkylation)

The reaction of enolates with alkyl iodides may give C-alkylated products or O-alkylated products (Scheme 3.2).

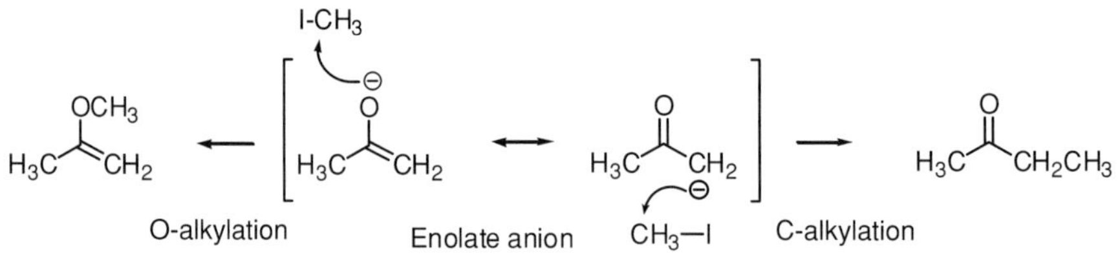

Scheme 3.2

According to the concept of the **hard–soft acid base**, the carbon is considered more nucleophilic than the oxygen, thus more likely to react than the oxygen atom because the leaving

Table 3.1

	ΔH_f (kcal mol^{-1})		
	Reactant	Product	Transition State
O-alkylation	−45.2	−39.5	−24.9
C-alkylation	−44.6	−57.6	−29.0

group, the iodide ion, is also **soft**. Preference for C-alkylation is also observed when the oxygen atom is hydrogen bonded.

The values of heat of formation (ΔH_f) of the reactants, products and transition states for O-alkylation and C-alkylation are reported in Table 3.1.

O-alkylation

C-alkylation

The activation energies, calculated from the change in enthalpy in going from the reactant to the transition state, are +20.3 and +15.6 kcal mol^{-1} for O-alkylation and C-alkylation, respectively. The overall enthalpy change of the reactions, obtained from the differences in the heat of formation between the reactant and the product, is +5.7 and −13.0 kcal mol^{-1} for O-alkylation and C-alkylation, respectively. These results predict that the product obtained for C-alkylation is the preferred product because (i) the activation barrier is smaller (lower in energy) than that in O-alkylation and (ii) the reaction is exothermic while O-alkylation is an endothermic process.

Alkylation of an enolate is an S_N2 reaction.

When ethoxide is used as a base to abstract α-proton, it could react with the alkyl halide (S_N2) to form ether. However, the acid–base equilibrium with the diketone prefers the more stable, less basic enolate over the ethoxide, essentially consuming the ethoxide. Thus, the very stable enolates from 1,3-dicarbonyl compounds are able to undergo S_N2 reactions with alkyl halides without competition from the alkoxide catalyst.

Regioselective monoalkylation of 2-methylcyclohexanone (**3.4**) with PhCH$_2$Br via its manganese enolate **3.7** gives 2-benzyl-6-methylcyclohexanone (**3.8**).[1,2]

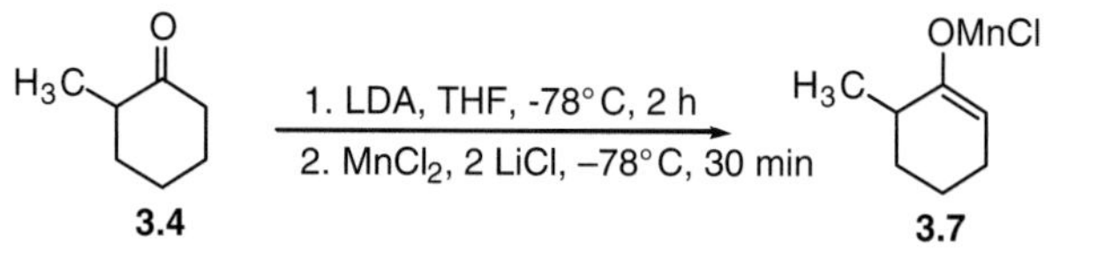
H3C
O
1. LDA, THF, -78°C, 2 h
2. MnCl2, 2 LiCl, –78°C, 30 min
OMnCl
3.4
3.7

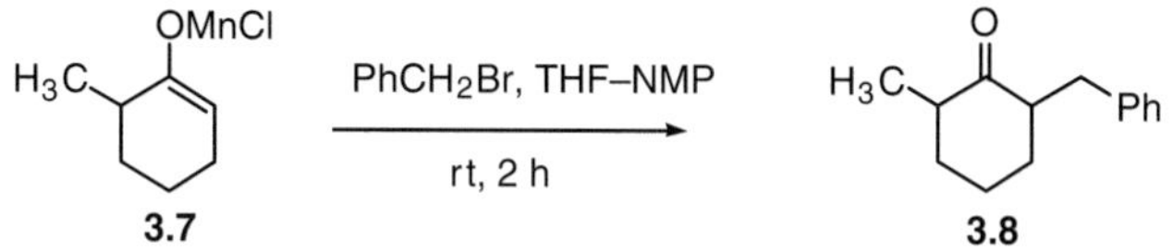
OMnCl
PhCH2Br, THF–NMP
rt, 2 h
Ph
3.7
3.8

N-methyl-2-pyrrolidinone

The other regioisomer, the 2-benzyl-2-methylcyclohexanone (**3.10**), can also be selectively obtained[3] in good yield from the more substituted Mn-enolate **3.9**.

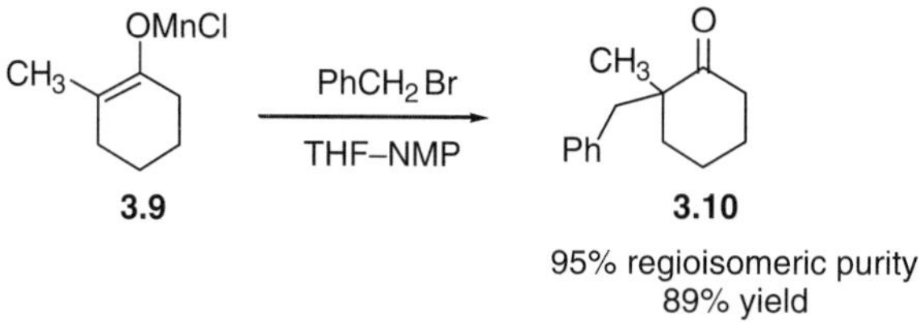
OMnCl
CH3
PhCH2 Br
THF–NMP
Ph
3.9
3.10
95% regioisomeric purity
89% yield

The synthetic utility of alkylation of enolates is utilized in the syntheses of malonic ester (**3.3**) and acetoacetic ester (**3.2**). For example, carbanion generated from malonic ester undergoes an S_N2 reaction with alkyl halide to yield alkyl-substituted malonic ester. The monosubstituted malonic ester still has an active hydrogen atom. The second alkyl group (same or different) can be introduced in a similar manner. Acid-catalyzed hydrolysis or base-catalyzed hydrolysis of mono- or disubstituted derivative of malonic ester followed by acidification gives the corresponding mono- or disubstituted malonic acid, which on decarboxylation yields the corresponding monocarboxylic acid (Scheme 3.3).

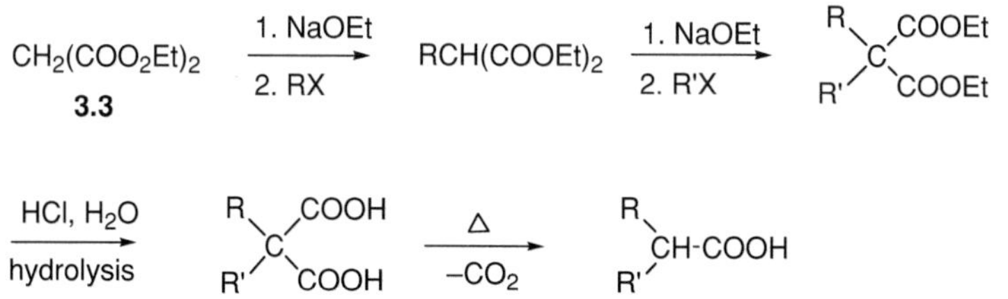
CH2(COO2Et)2
3.3
1. NaOEt
2. RX
RCH(COOEt)2
1. NaOEt
2. R'X
COOEt
HCl, H2O
hydrolysis
COOH
Δ
–CO2
CH-COOH

Scheme 3.3

The base-catalyzed reaction of β-ketoesters such as ethyl acetoacetate (acetoacetic ester) (**3.2**) with the alkyl halide gives monosubstituted β-ketoesters. In a similar manner, on treatment with a base and alkyl halide, a disubstituted ethyl acetoacetate is formed.

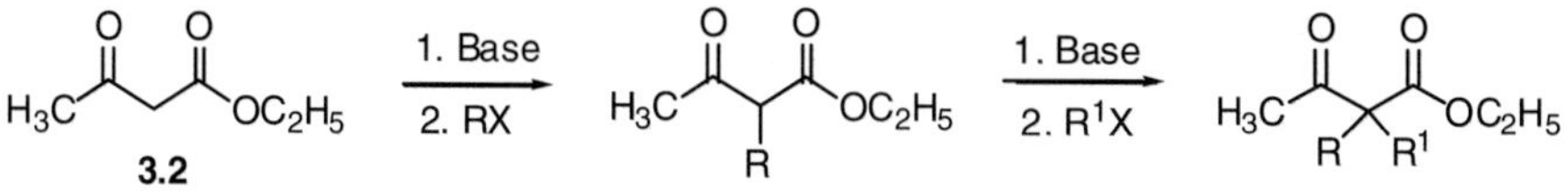
H3C
OC2H5
1. Base
2. RX
1. Base
2. R1X
3.2

Either mono- or disubstituted ethyl acetoacetate on mild hydrolysis with dilute alkali or hydrochloric acid followed by decarboxylation yields substituted acetone and on treatment with a strong alkali gives substituted acetic acids (Scheme 3.4).

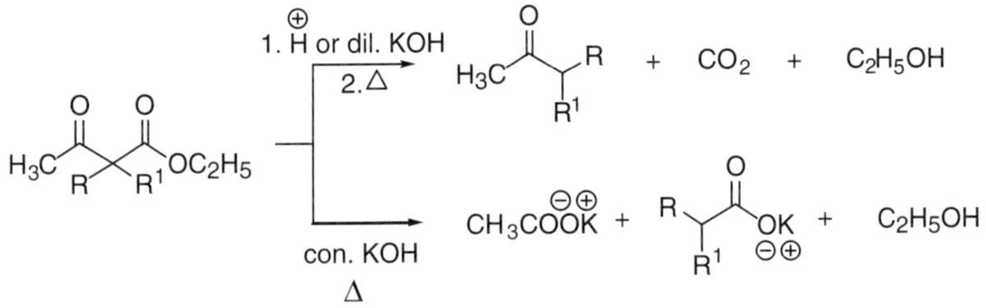

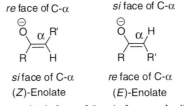

Scheme 3.4

The enolate may be prochiral or chiral and it has two *re* or *si* faces at C-α. The enolate may show *E/Z* isomerism because of the presence of a double bond.

The attack of an electrophile on which face of C-α is favoured will depend on several factors.

The asymmetric glycolate alkylation reaction of *N*-glycolyloxazolidinones[4] **3.11** provides a highly selective method for the synthesis of selectively protected homoallylic alcohols. A variety of protecting groups can be tolerated and the reaction is highly selective even with complex glycolates with other pre-existing stereogenic centres.

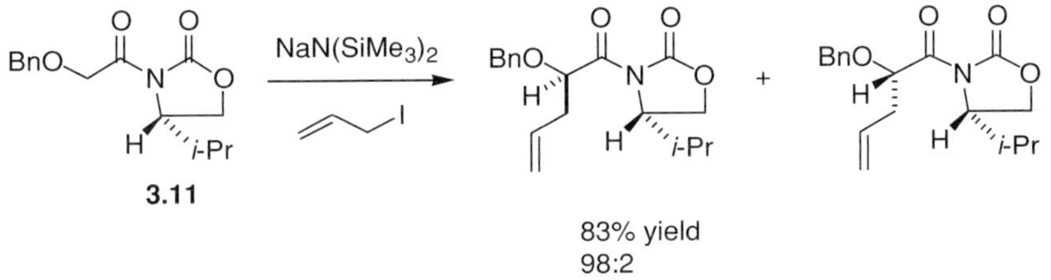

3.11

83% yield
98:2

A highly efficient synthesis of the marine metabolite laurencin[5,6] (**3.12**) involves asymmetric glycolate alkylation (Scheme 3.5).

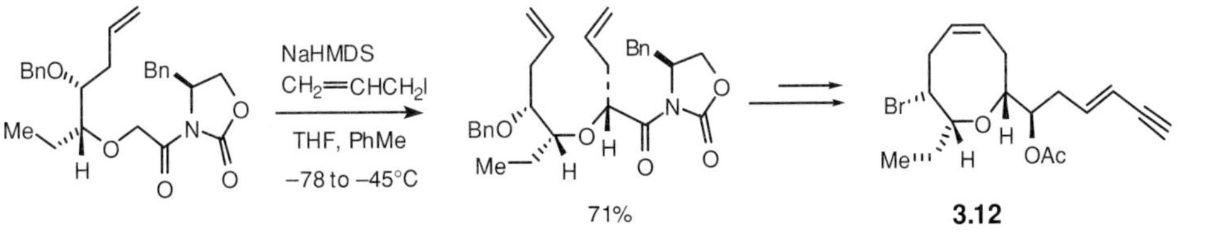

3.12

Scheme 3.5

3.1.2   Reaction of stabilized carbanions with carbonyl compounds

Aldol condensation

The abstraction of α-hydrogen in carbonyl compounds such as acetaldehyde by sodium hydroxide is a reversible reaction and forms an enolate ion that undergoes addition to the carbonyl carbon of another acetaldehyde molecule to give the aldol **3.13**. This is called the aldol condensation[7–9] and its mechanism is shown in Scheme 3.6.

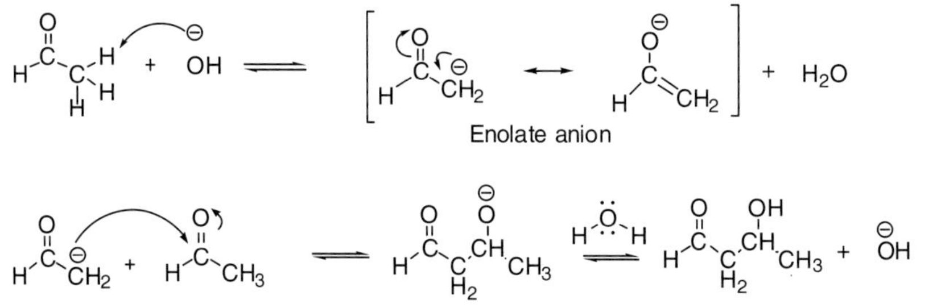

Enolate anion

3.13

Scheme 3.6

The aldol condensation can also be brought about with acid catalysis. There are many aldol-like reactions which involve an aldehyde or ketone. However, the word *aldol* is a common name for the product of these reactions. Generally, the word aldol is used to refer to any β-hydroxyaldehyde or β-hydroxyketone. For example, treatment of acetone with a base results in the aldol reaction. The product, β-hydroxyketone **3.14**, is separated from the reaction mixture as it is formed.

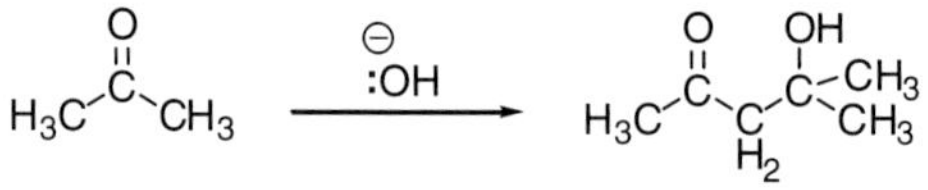

3.14

Aldol condensation reaction is a reversible reaction and equilibrium is favoured towards the product (aldol); only for simple aldehydes, greater substitution favours the starting compounds at equilibrium.

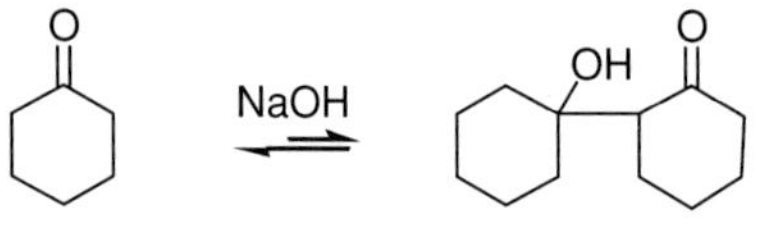

Only 22% at equilibrium

The aldol reaction is often combined with dehydration of the initial aldol product, forming a conjugated double bond (an enone). Thus, β-hydroxyaldehyde and β-hydroxyketone undergo dehydration to produce α, β-**unsaturated aldehyde** and α, β-**unsaturated ketone**, respectively. In fact, it is difficult to isolate β-hydroxyaldehydes and β-hydroxyketones because they are very prone to dehydration.

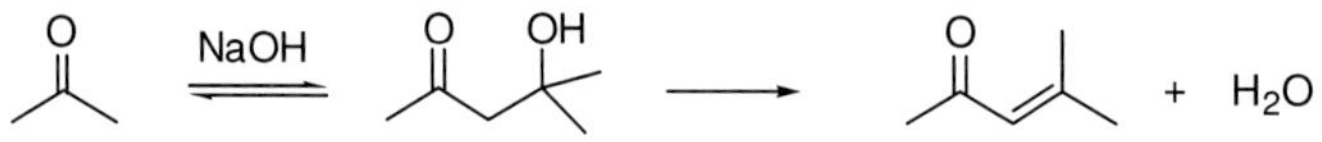

Mixed or crossed aldol condensation: Aldol condensations between different carbonyl reactants are called crossed (or mixed) reactions. Crossed aldol condensation works well if one carbonyl compound has no α-hydrogen(s). For example, acetone reacts with furfural in a crossed-aldol reaction to give the corresponding α,β-unsaturated ketone **3.15**.

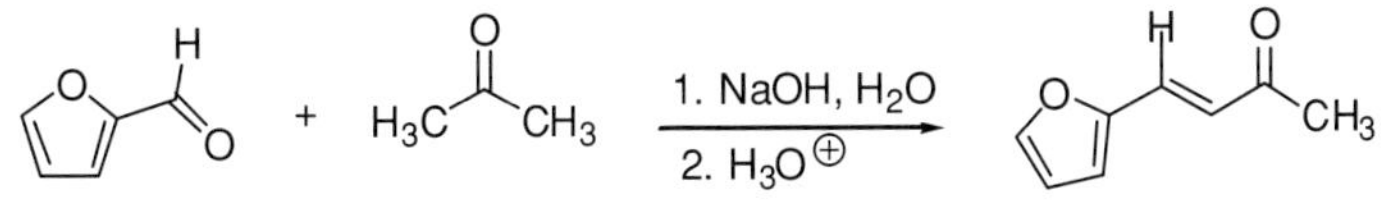

Furfural   **3.15**

The aldol condensation of ketones with aryl aldehydes to form α,β-unsaturated derivatives is called the **Claisen–Schmidt reaction**[10–14].

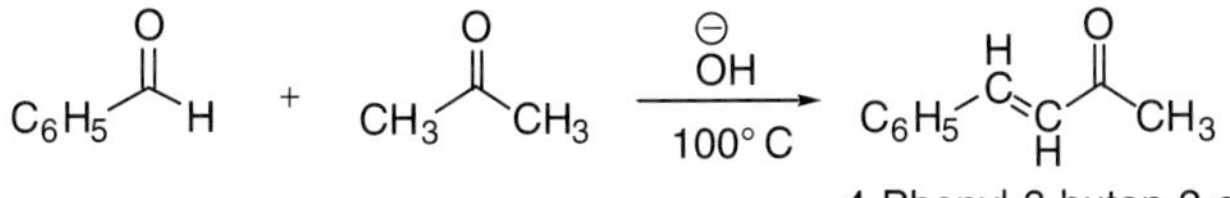

4-Phenyl-3-buten-2-one
(benzalacetone)

Intramolecular aldol condensation: The dicarbonyl compound gives a cyclic product by the intramolecular aldol condensation. Intramolecular aldols are facile when a five- or six-membered ring can be formed as shown below:

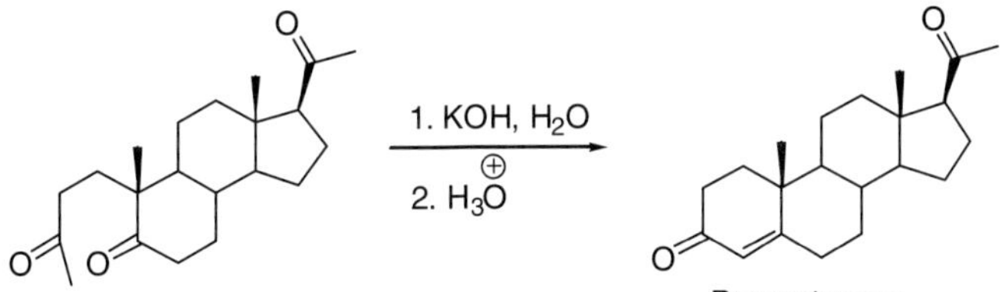

Progesterone

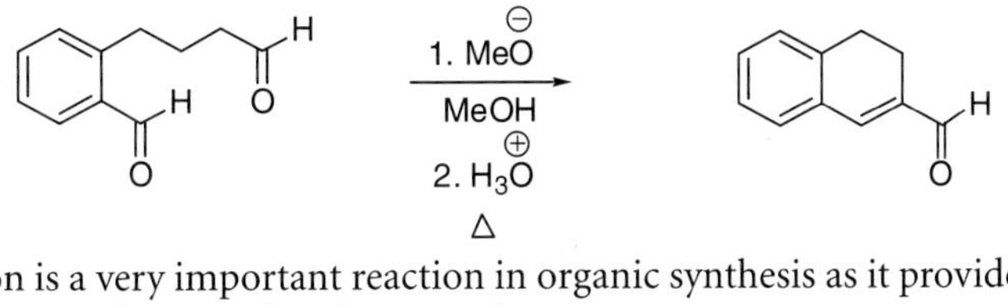

The aldol reaction is a very important reaction in organic synthesis as it provides a method for forming a new C–C bond and at the same time two new stereocentres.

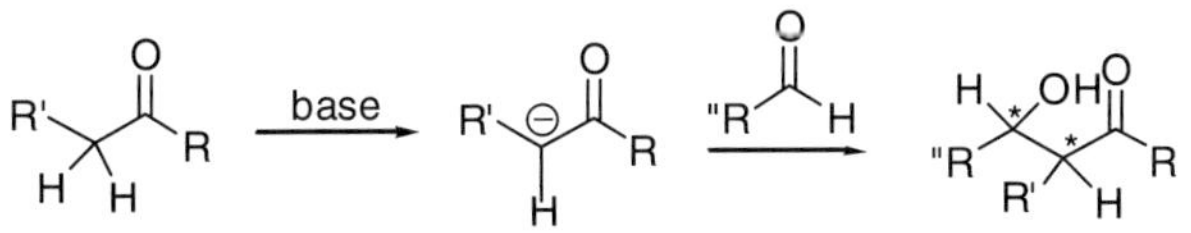

There may be four possible aldol diastereoisomers in an aldol condensation of chiral ketone or aldehyde, as shown below:

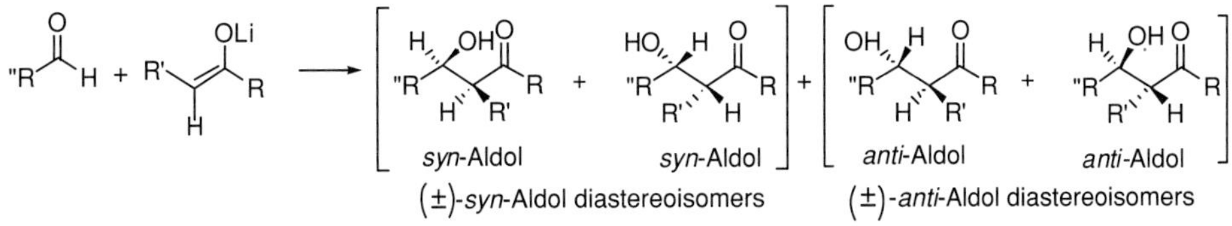

syn-Aldol   syn-Aldol   anti-Aldol   anti-Aldol
(±)-syn-Aldol diastereoisomers   (±)-anti-Aldol diastereoisomers

Mn-enolates can also be hydroxyalkylated by reacting with a vast array of aldehydes to give *syn*-aldol products in good yields. The stereoselectivities obtained from Mn- and Li-enolates are very similar. For example, Mn-enolate obtained from unsymmetrical ketone, *tert*-butyl propyl ketone, with Mn-amides reacts with benzaldehyde to give 81% yield of the *syn/anti* (99:1) hydroxyalkylated ketone[15] **3.16**.

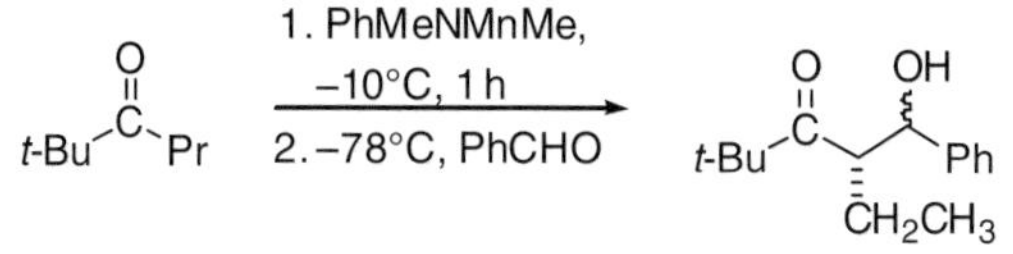

3.16

The chlorotitanium enolates[16] derived from *N*-acyloxazolidinones **3.17**, *N*-acyloxazolidinethiones **3.18** and *N*-acylthiazolidinethiones **3.19** undergo highly selective and efficient aldol reactions with aldehydes. The *N*-acyloxazolidinethione **3.18** and *N*-acylthiazolidinethione **3.19** auxiliaries are also easier to cleave.

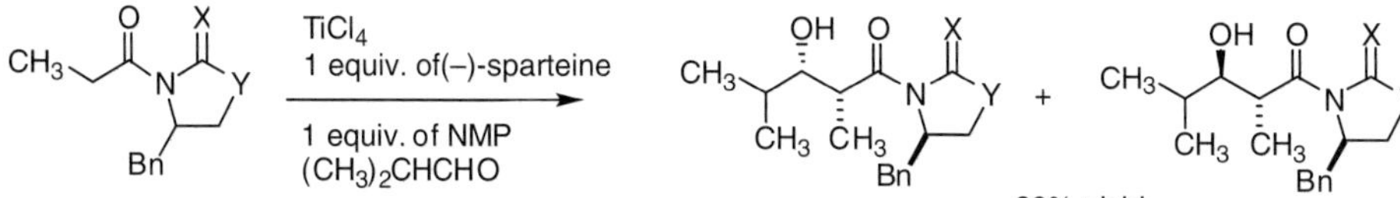

3.17: X = Y = O
3.18: X = S; Y = O
3.19: X = Y = S

N-methyl-2-pyrrolidinone

Although the enantioselective construction of aldol products has long been a workhorse for acyclic stereoselection, there are still new things being done. The use of proline and proline amides[17] **3.20** as a catalyst was reported for the enantioselective addition of ketones to aldehydes.

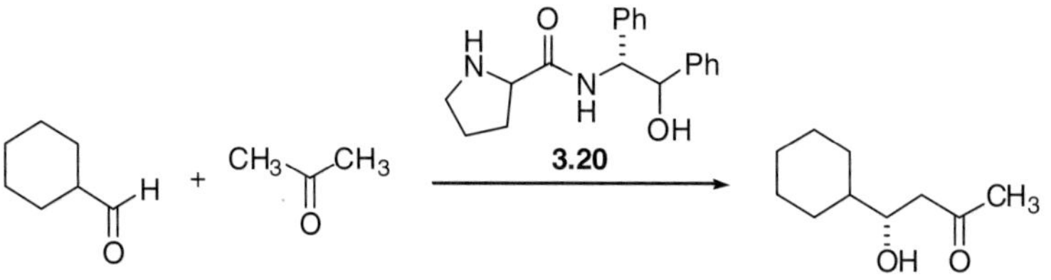

77% yield, 98% ee

Mukaiyama aldol condensation
The aldol condensation of aldehyde and silyl enol ether in the presence of a catalyst such as TiCl$_4$ is called the Mukaiyama aldol condensation[18–22] (Scheme 3.7).

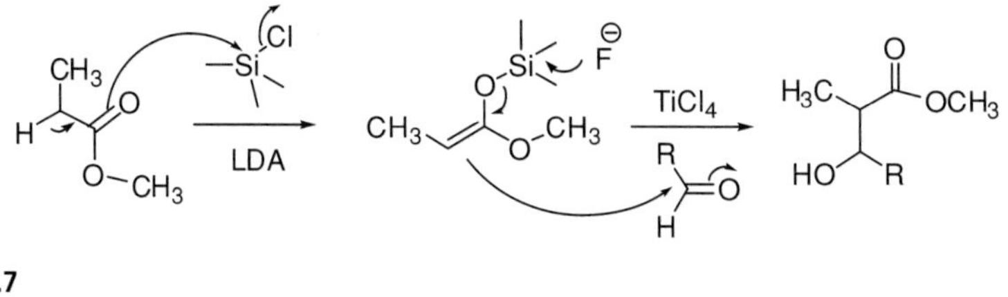

Scheme 3.7

For example, condensation of silyl enol ether (**3.21**) of 3-pentanone with 2-methylbutanal in the presence of $TiCl_4$ gives the Ti-complex **3.22**, which on hydrolysis yields an aldol product, manicone (4,6-dimethyl-4-octen-3-one) (**3.23**), an alarm pheromone. Treatment of 3-pentanone with LDA results in the formation of an enolate, which is trapped with chlorotrimethylsilane to give **3.21**. Other Lewis acids such as tin tetrachloride ($SnCl_4$) and boron trifluoride etherate ($BF_3 \cdot OEt_2$) can also be used.

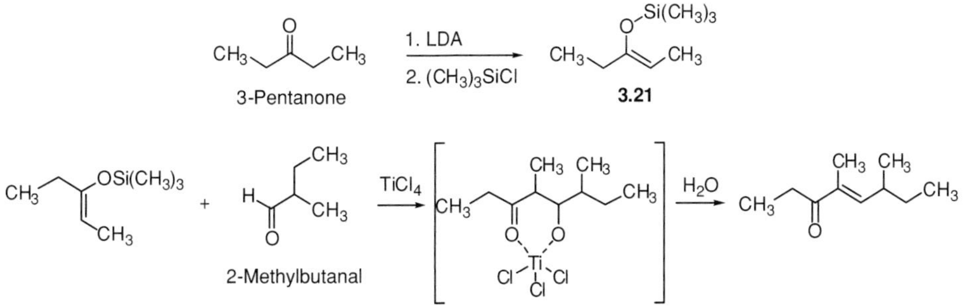

Aldol reaction of Fischer carbene complexes

The pK_a values of Fischer carbene complexes[23,24] such as **3.24** are comparable to those of active methylene compounds such as diethylmalonate. This is due to the resonance stability of the conjugate base of **3.24**.

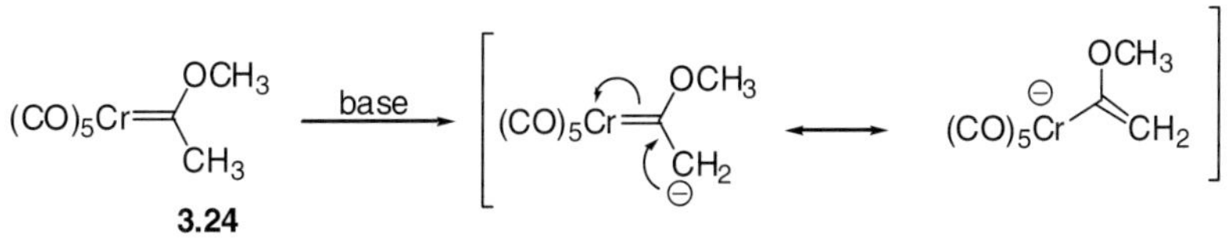

The aldol reactions of these complexes were first observed by Corey in the presence of Lewis acid such as boron trifluoride for ketones and tin chlorides for aldehydes (Scheme 3.8).

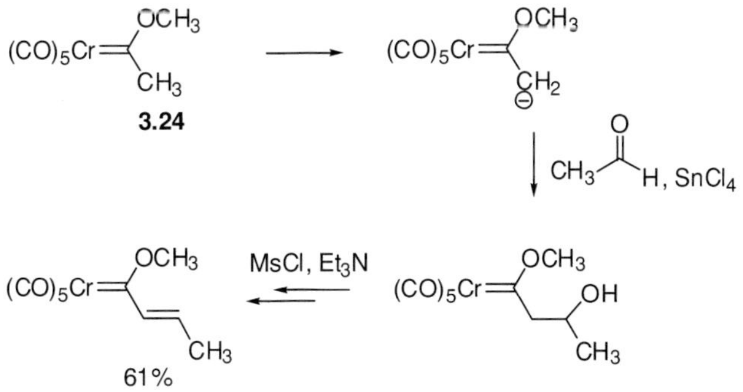

Scheme 3.8

Henry reaction

Nitroalkanes having an α-hydrogen atom undergo aldol-type condensation with aldehydes and ketones in the presence of a base to give β-hydroxy nitro compounds or nitroethylene compounds. The reaction is known as the **Henry reaction**[25–29] or **nitroaldol reaction**.

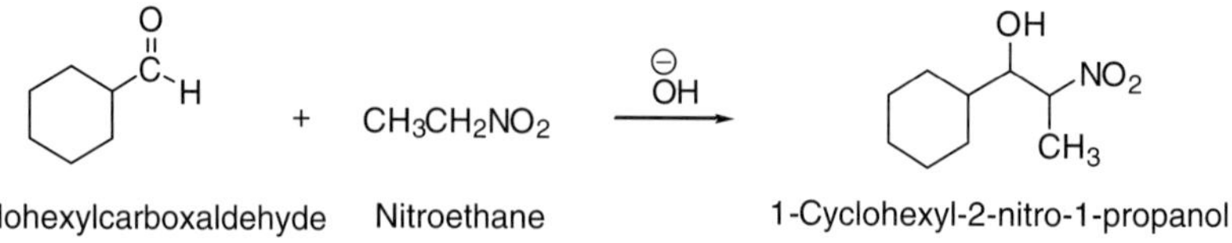

| Cyclohexylcarboxaldehyde | Nitroethane | 1-Cyclohexyl-2-nitro-1-propanol |

The α-hydrogen of nitroalkanes is acidic (less acidic than those of aldehydes and ketones). The anion is also resonance stabilized.

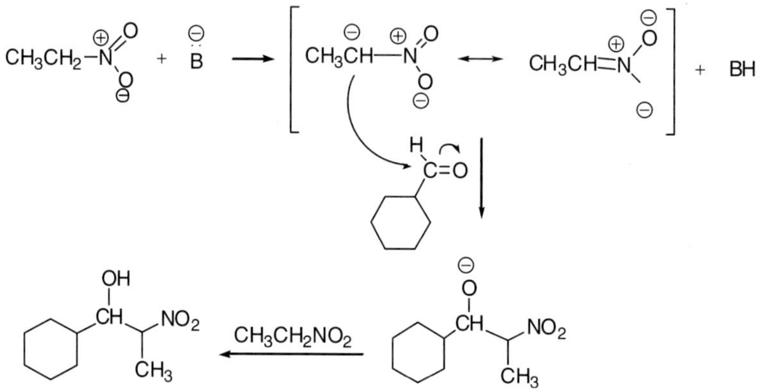

Condensation of sodium salts of acinitroalkanes with the sodium bisulfite addition products of aldehydes in the presence of trace of alkali or weak acid also gives nitro alcohols as reported by Kamlet[30].

When aromatic aldehydes are condensed with nitroalkanes in the presence of a base, α,β-unsaturated nitro compounds are produced directly. For example, benzaldehyde and nitromethane in the presence of a base give α,β-unsaturated compound nitrostyrene (**3.25**). Similarly, the reaction of nitrobutane with furfural in the presence of a base followed by hydrolysis gives 82% yield of 2-substituted furan derivative **3.26**.

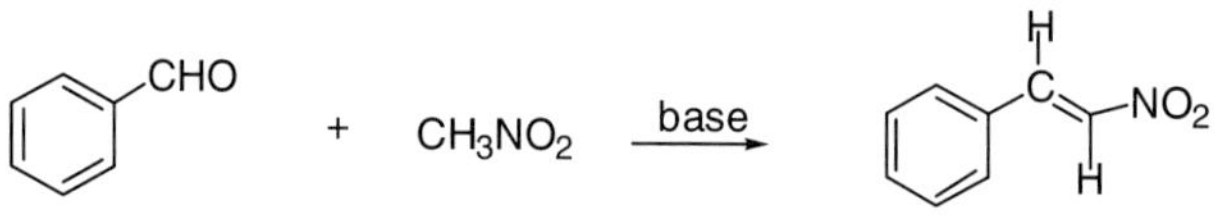

3.25

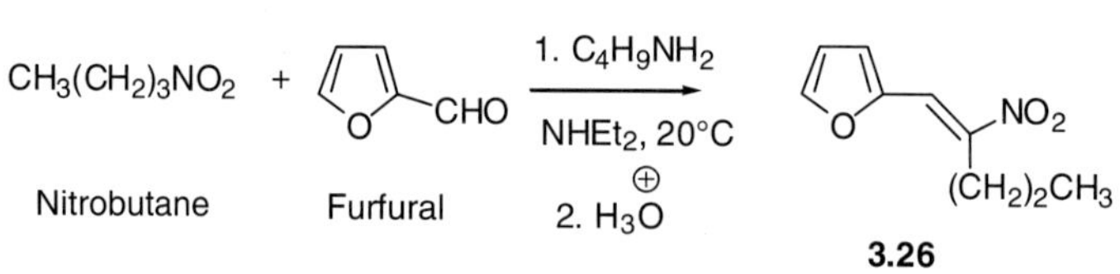

3.26

The enantioselective Henry reaction by using copper acetate-bis(oxazoline) catalyst is reported by Evans[25–29].

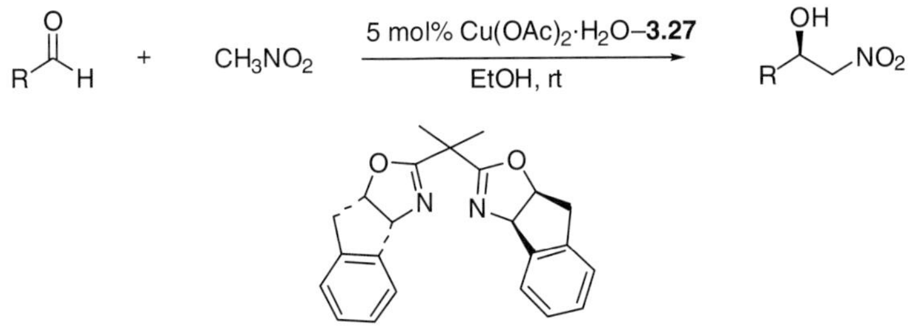

3.27

Claisen condensation

Esters undergo a reaction similar to aldol condensation called the **Claisen condensation**[31]. It is a base-induced reaction of two ester molecules, which usually forms β-ketoester products.

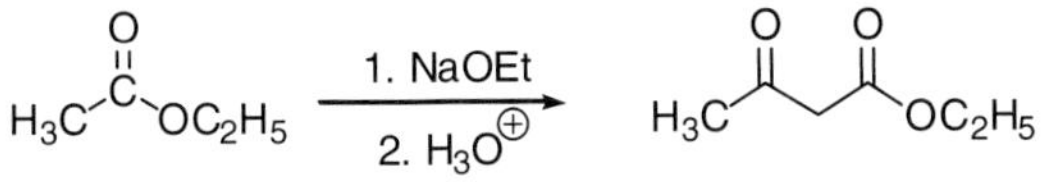

3.2

The first step in the mechanism involves the removal of proton from ester by a strong base to generate an enolate at the α-carbon. The enolate can then add to another ester molecule with loss of the alkoxy group to make the β-ketoester (Scheme 3.9).

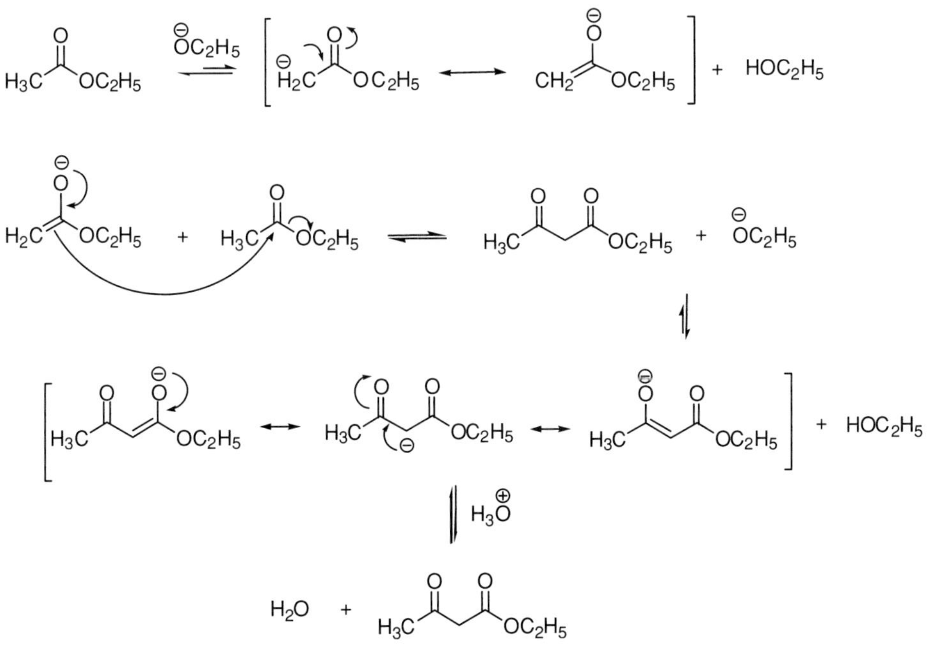

Scheme 3.9

In the case of **mixed Claisen condensations**, a mixture of products is commonly avoided by using an acceptor ester that has no α-hydrogen. Examples of such reactants are ethyl formate

($HCO_2C_2H_5$), diethyl carbonate ($C_2H_5OCO_2C_2H_5$), ethyl benzoate ($C_6H_5CO_2C_2H_5$) and diethyl oxalate ($C_2H_5OCO)_2$. For example, the Claisen condensation between ethyl benzoate and ethyl acetate gives ethyl cinnamate.

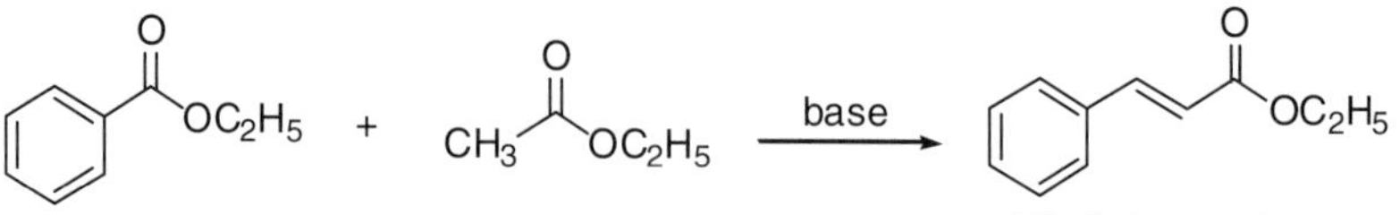

Ethyl cinnamate

The intramolecular Claisen condensation is called the **Dieckmann condensation**[32-35], and can be used to construct five- or six-membered rings.

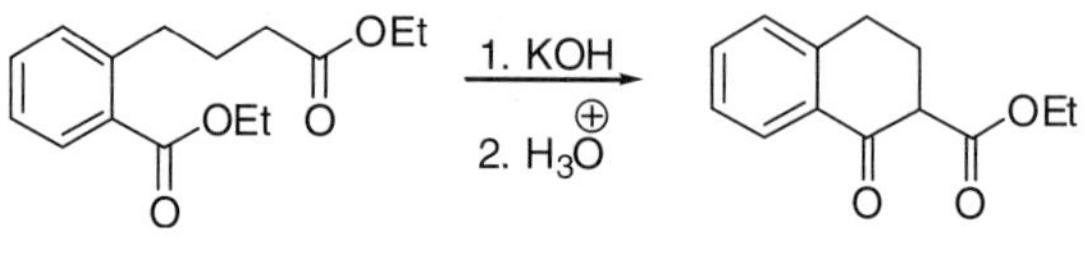

Hiyama aminoacrylate synthesis

The Hiyama aminoacrylate synthesis[36-38] is the synthesis of alkyl ester of 3-aminoacrylic acids or derivatives by aldol-type condensation of nitriles with ester in the presence of a base such as (i-Pr)$_2$NH.

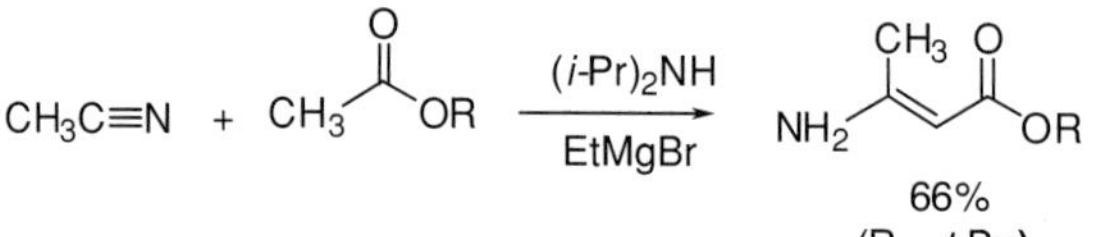

66%
(R = *t*-Bu)

The mechanism of the Hiyama aminoacrylate synthesis reaction is similar to the aldol condensation, as shown in Scheme 3.10.

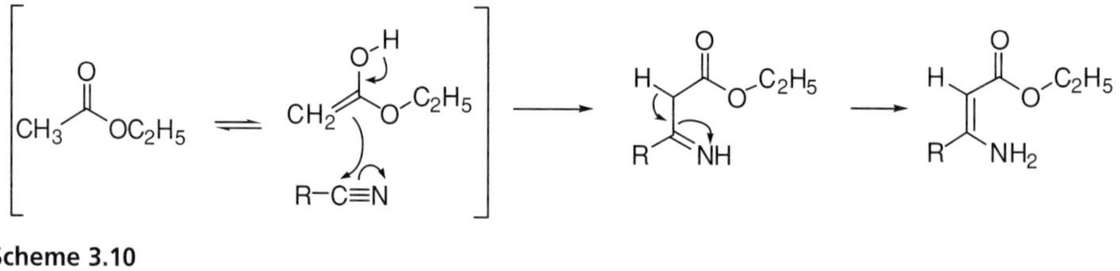

Scheme 3.10

Baylis–Hillmann reaction

The tertiary amine-catalyzed C–C bond forming reaction of aldehydes with α,β-unsaturated compounds is known as the **Baylis–Hillmann reaction**[39-42]. The catalytic amount of tertiary amine such as 1,4-diazabicyclo(2.2.2)octane (**3.28**) is used.

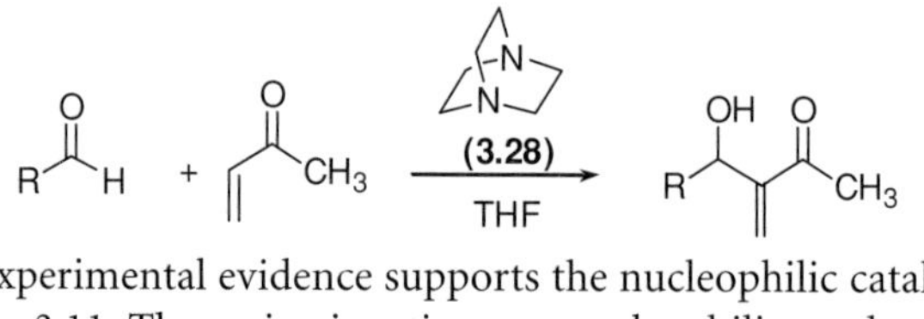

Mechanism: The experimental evidence supports the nucleophilic catalysis mechanism as discussed in Scheme 3.11. The amine is acting as a nucleophilic catalyst rather than a base in the first step. Thus, the conjugate addition of the amine to the enone gives enolate **A**. If the enone is β-substituted, due to steric hindrance the nucleophilic attack of a base does not take place and there is no Baylis–Hillmann reaction. In the next step, the enolate adds to the aldehyde to give **B**. The final product is obtained by the elimination process.

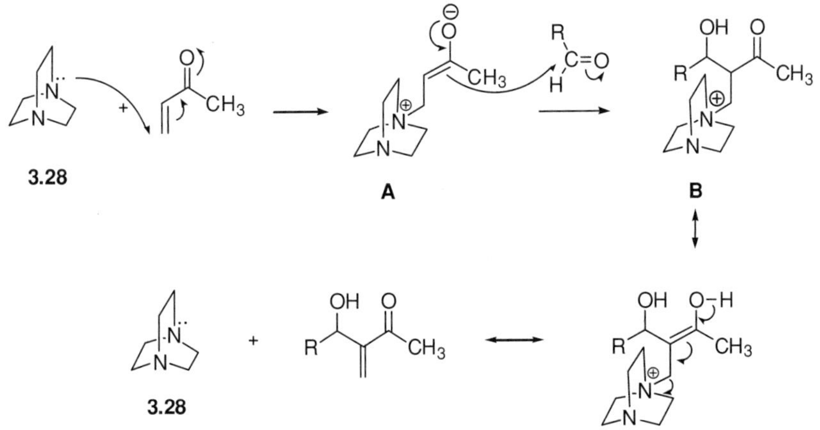

Scheme 3.11

Thus, the catalyst is involved in the reaction as a nucleophile, not as a base. *p*-Dimethylaminopyridine is also used as a catalyst (Scheme 3.12) to avoid the formation of side products, which are sometimes obtained by using **3.28**.

3.1.3 Conjugate addition of enolate to α, β-unsaturated carbonyl compounds

Michael reaction

Conjugate addition of carbon nucleophiles to unsaturated esters, ketones, nitriles, sulfones and other activated double bonds is known as the **Michael reaction**[43–46]. It is a very useful synthetic method for the formation of the carbon–carbon bond.

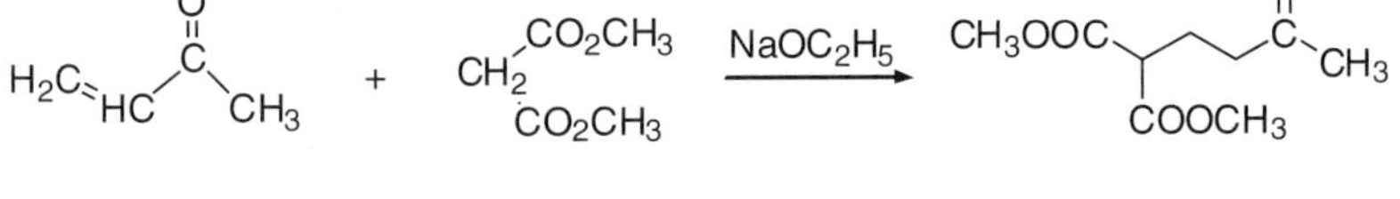

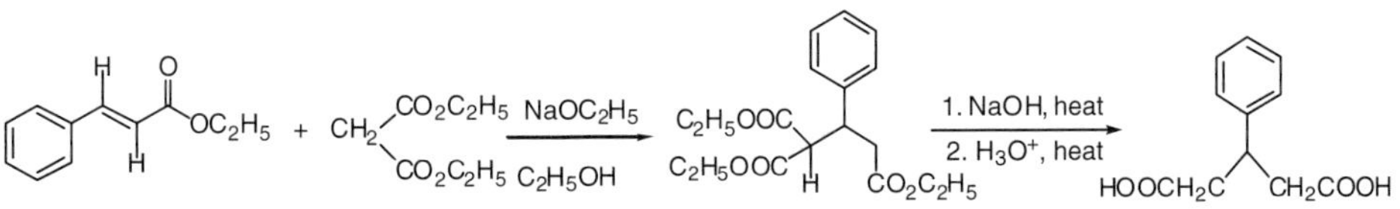

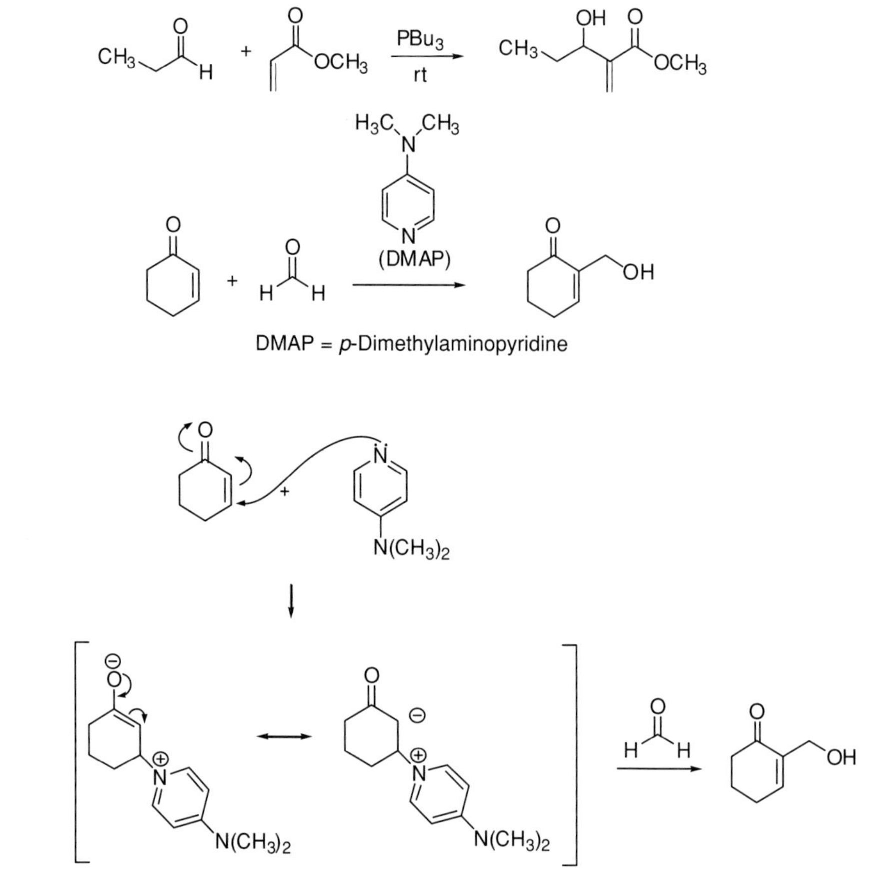

Scheme 3.12

Mn-enolates are useful synthetic reagents for the Michael additions.

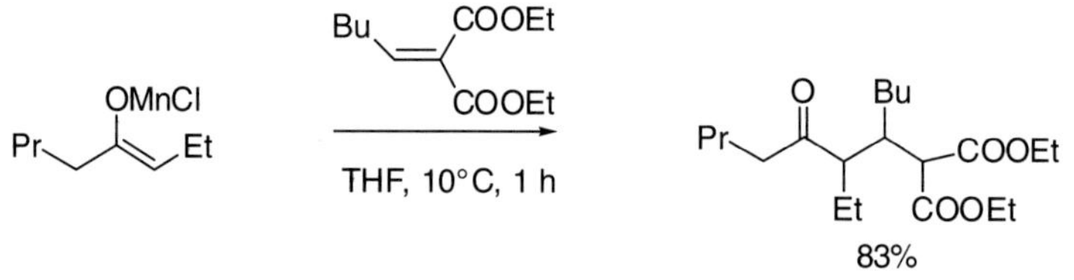

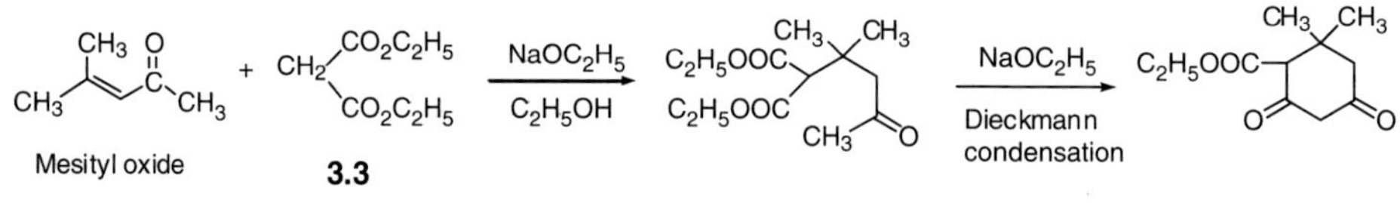

The mechanism of the Michael addition reaction using ester **3.3** and mesityl oxide is shown in Scheme 3.13. The enolate **A** derived from the malonic ester (**3.3**) attacks the β-carbon atom of α,β-unsaturated compounds (conjugate addition). The resulting carbanion **B** is stabilized by the carbonyl group and abstracts a proton from ethanol to give the product **C** and generate the ethoxide ion. The reaction is reversible. In case the Michael addition is followed by an internal **Claisen condensation**, the product obtained is dimedone (**3.29**) (Scheme 3.13).

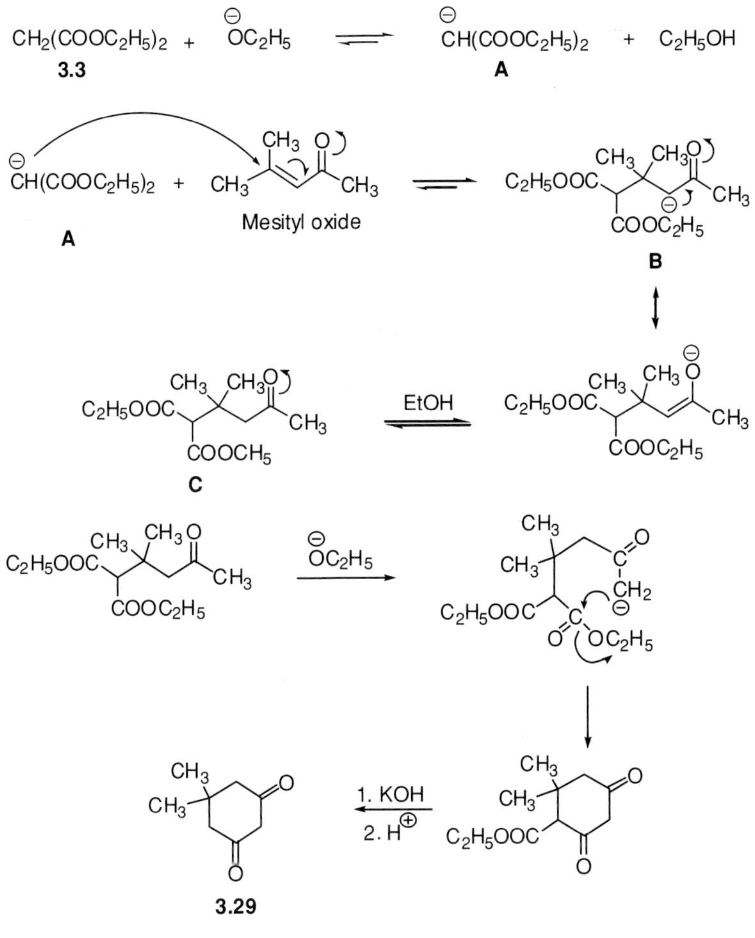

Scheme 3.13

The use of the Michael addition followed by an aldol condensation is an important route for the synthesis of bicyclic ketones and is known as the **Robinson annulation**.

3.1.4   Reaction of enolates with iminium ions or imines

Mannich reaction
Instead of alkyl halides other electrophiles, such as imines or iminium ions, are also used for the nucleophilic attack of enolates. Primary amines react with aldehydes and ketones to produce imines[47–50]. For example, the reaction of acetophenone with methyl amine gives

the initially formed tetrahedral intermediate, which undergoes dehydration to give imine **3.30** (Scheme 3.14).

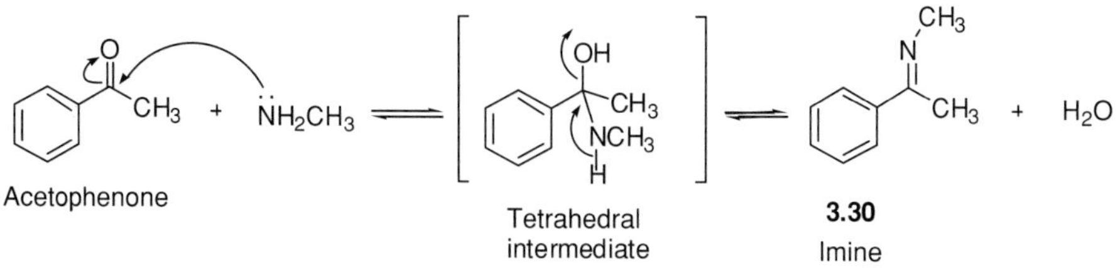

Acetophenone

Tetrahedral intermediate

3.30 Imine

Scheme 3.14

However, the reaction of secondary amine such as dimethyl amine with formaldehyde and acetone gives the corresponding **Mannich base, 3.31**.

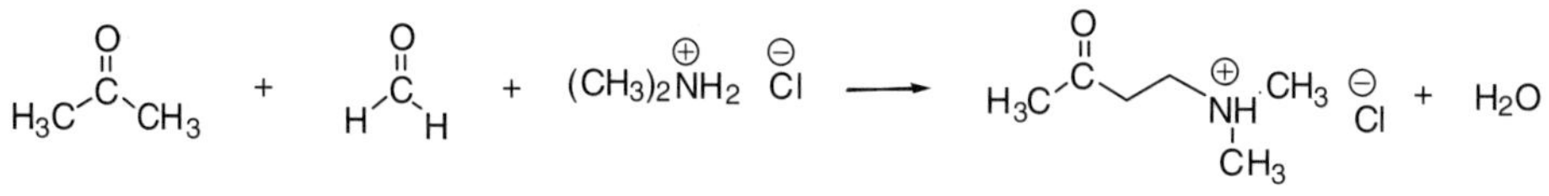

3.31

Mechanism: The reaction is believed to involve the formation of the intermediate methylene-ammonium salt **3.32**, which condenses either with enol form of the ketone (acid catalyzes the conversion of keto form into enol form) or with the carbanion derived from the ketone (small amount of amine acting as a base and abstracts the α-hydrogen) to give the corresponding Mannich base (Scheme 3.15).

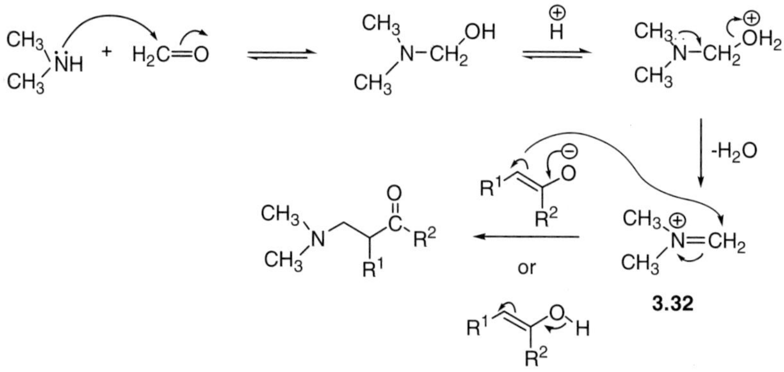

3.32

Scheme 3.15

Although formaldehyde has been most common, other aldehydes have also been used successfully for the formation of iminium ion. The Mannich reaction also proceeds with the other activated hydrogen compounds such as indole, furan, pyrrole and phenols. When primary amine is used, the Mannich base formed is a secondary amine and may undergo further condensation to yield tertiary amine. The Mannich base may eliminate an amine hydrochloride on heating to yield α,β-unsaturated ketones. Another usefulness of Mannich bases is the replacement of the dimethylamino group by the nitrile group, which undergoes hydrolysis to the corresponding acid. For example, the Mannich base **3.33** obtained by the reaction of indole, formaldehyde and dimethylamine on methylation forms the tetraalkyl ammonium salt **3.34**. Treatment of **3.34** with KCN followed by hydrolysis yields β-indole acetic acid (**3.35**).

Catalytic asymmetric alkylation of imines

Two main strategies for the catalytic asymmetric alkylation of imines are (a) chiral Lewis acid approach and (b) chiral nucleophilic approach.

In the chiral Lewis acid approach the transition metal catalyst bounds to a chiral ligand.

R^1 = Ph, 4-cl Ph, 2-furyl
R^4 = OMe; R^2 = R^3 = CH_3

Kobayashi and co-workers,[51] used zirconium-based bromo-BINOL complex for the catalytic enantioselective Mannich-type reaction. The *o*-hydroxyphenyl imine **3.36** chelates the Zr(IV)(BrBINOL)$_2$ to form the activated chiral Lewis acid complex **A**. The ketone acetal **3.37** reacts with the Lewis acid complex **A** to give the complex **B**. The silyl group is then transferred to the β-amino ester to form the product **3.38** and the catalyst Zr(BrBINOL)$_2$ is regenerated, which is ready for binding with another imine molecule (Scheme 3.16).

The second strategy involves the attack of chiral nucleophile to the imine. The mechanism of this reaction involves the formation of chiral nucleophilic complex from nucleophile and chiral ligands **3.39**.

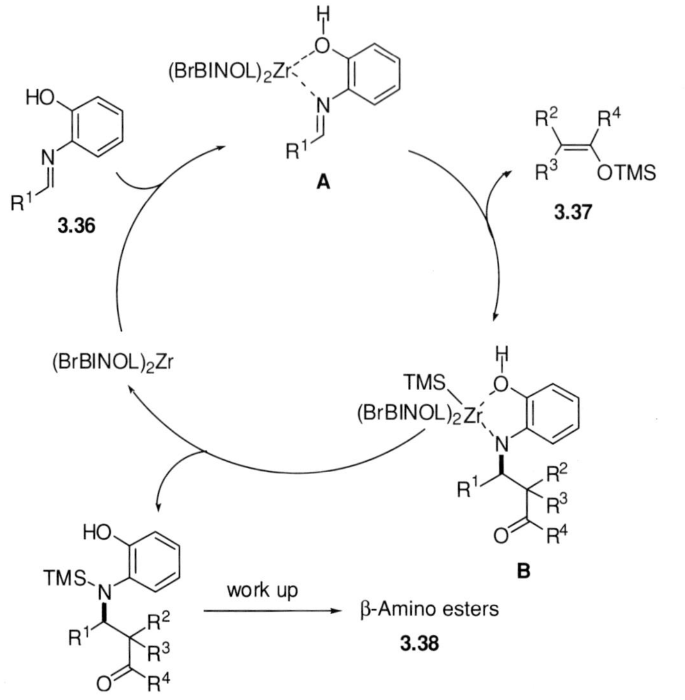

Scheme 3.16

3.2 Enamines

The enamine reaction was introduced by Stork and his co-workers[52–56] and is widely used in organic synthesis. Enamines are α,β-unsaturated amines and are obtained by the reaction of an aldehyde or ketone having α-hydrogen with a secondary amine in the presence of a dehydrating agent (such as catalytic amount of *p*-toluenesulfonic acid).

The reaction of a secondary amine with aldehydes or ketones (containing α-hydrogen) initially follows the same course as that of a primary amine (Scheme 3.14), but now the tetrahedral intermediate cannot form an imine because the nitrogen does not have a second hydrogen atom to lose; consequently, the tetrahedral intermediate undergoes dehydration to form an enamine **3.40**. The reaction is reversible and the removal of water from the equilibrium mixture gives the product in high yield (Scheme 3.17). Imines and enamines may be converted back into aldehydes and ketones by acid-catalyzed hydrolysis, i.e. by reaction with a large excess of water.

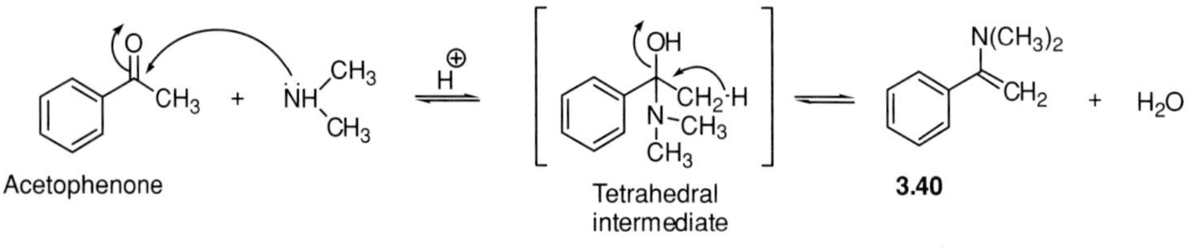

Scheme 3.17

The most common 2° amines used for the preparation of enamines are pyrrolidine, morpholine and piperidine.

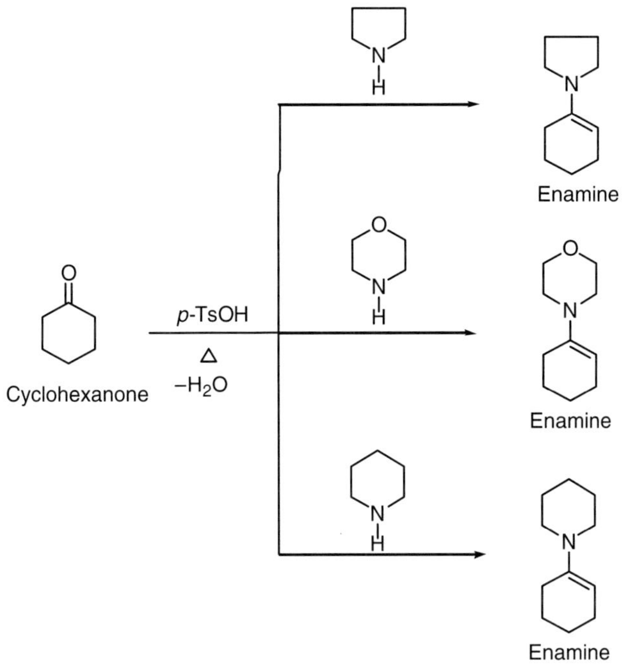

In the case of unsymmetrical ketones, two enamines may be formed. For example, the reaction of pyrrolidine (a secondary amine) with 2-methylcyclohexanone (**3.4**) gives a mixture of enamine isomers **3.41** and **3.42**. The latter, i.e. **3.42**, which is less substituted, predominates (Scheme 3.18).

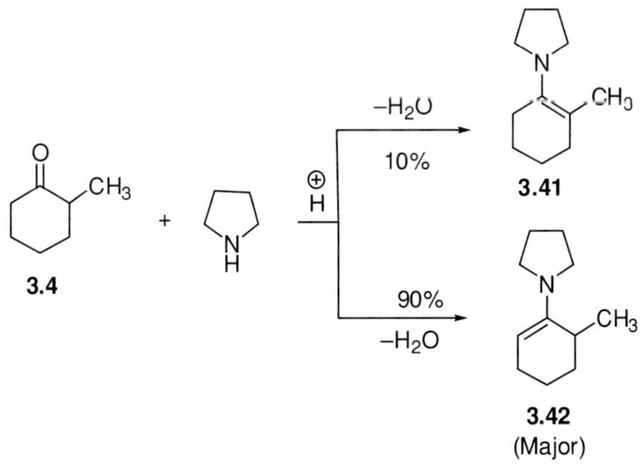

Scheme 3.18

Presumably a steric interaction between the methyl group of the cyclohexyl ring and the methylene group of the pyrrolidine ring reduces the extent of conjugation between the lone pair of electrons on the nitrogen atom and the π system of the double bond in the more substituted case.

Enamines are N analogues of enols, as in both cases similar charge delocalization occurs.

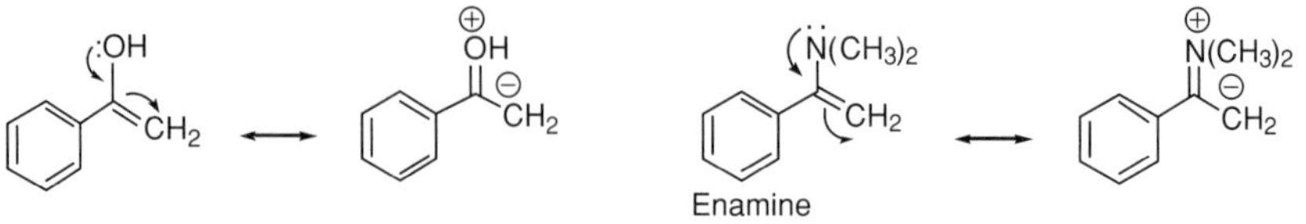

Enamine

Enamines may be regarded as synthetic equivalents of enolate ions and are closely related to the enolates derived from ketones in their reactions with acyl halides, alkyl halides and α,β-unsaturated compounds.

Enamines like enolates are alkylated when treated with reactive alkylating agent. α-Substituted enamines can be converted into aldehydes and ketones by acid-catalyzed hydrolysis. Thus, in the three-step process, alkylation of aldehydes and ketones may be carried out via enamines (**Stork enamine synthesis**) (Scheme 3.19).

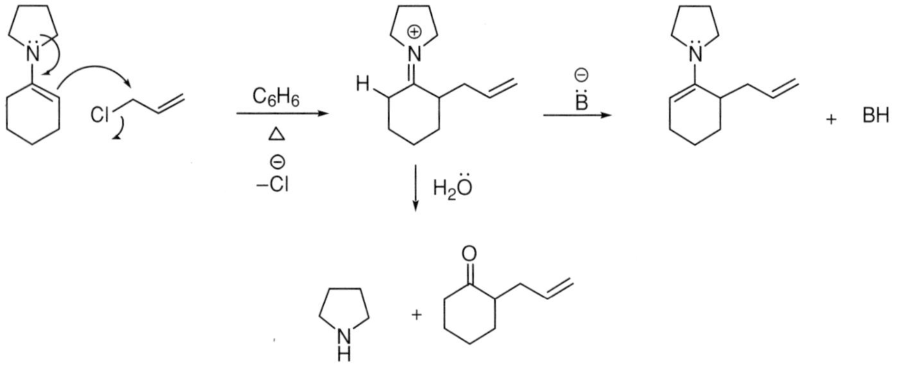

Scheme 3.19

A valuable feature of the enamine reaction is that it is regioselective. Thus, in the alkylation of an unsymmetrical ketone, e.g. **3.4**, the substitution takes place at the less substituted α-carbon atom. However, direct base-catalyzed alkylation of unsymmetrical ketones usually gives a mixture of products. Also in the enamine reactions, since no alkali is used, there is no possibility of side products, which are normally obtained from carbonyl compounds. For example, the reaction of enamine **3.42** with methyl iodide followed by hydrolysis gives almost exclusively 2,6-dimethylcyclohexanone (**3.43**) (Scheme 3.20).

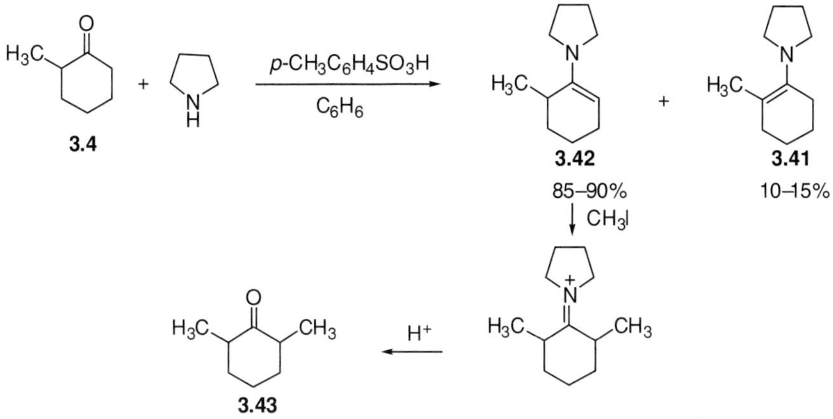

Scheme 3.20

The alkylation procedure has also been used for the synthesis of bicyclic compounds.

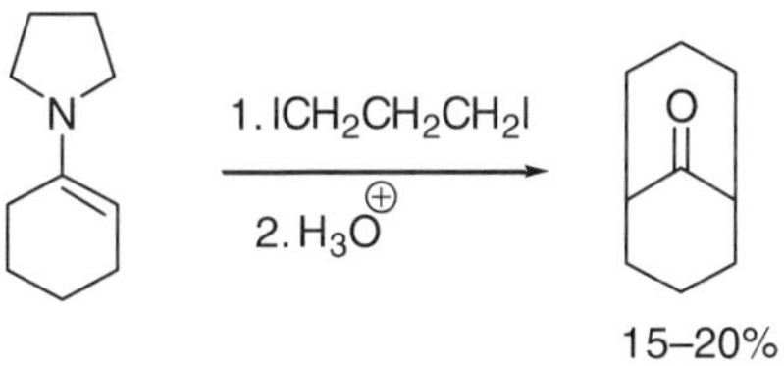

15–20%

Enamines also act as useful intermediates for the acylation of aldehydes or ketones at α-position (Scheme 3.21).

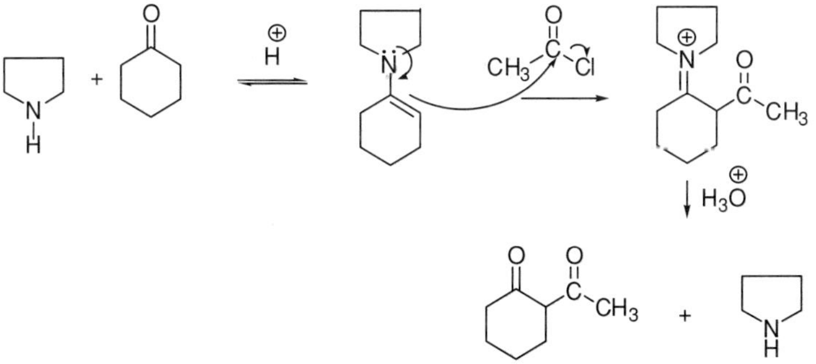

Scheme 3.21

Enamines undergo the Michael addition reaction to give the least substituted product in contrast to the highly substituted product obtained from enolate derived from unsymmetrical ketone by sodium ethoxide (NaOEt) in ethanol (EtOH).

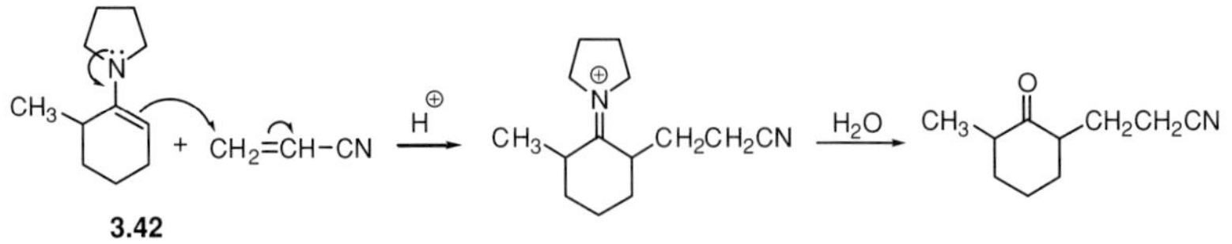

3.42

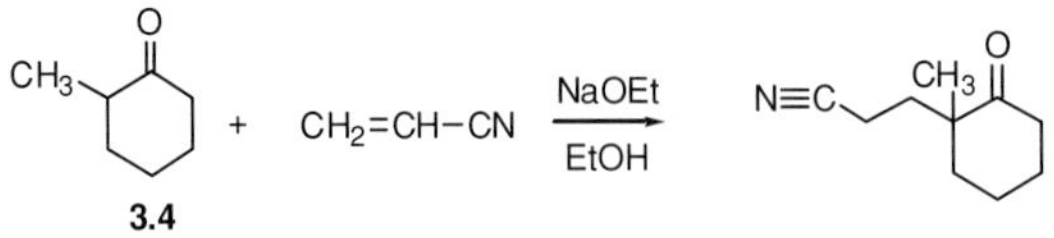

3.4

The Michael addition of enamines to α,β-unsaturated ketones may be coupled with **intramolecular aldol condensations** to produce cyclic ketones. This sequence of reactions is an alternative approach to traditional **Robinson annulations** (Scheme 3.22).

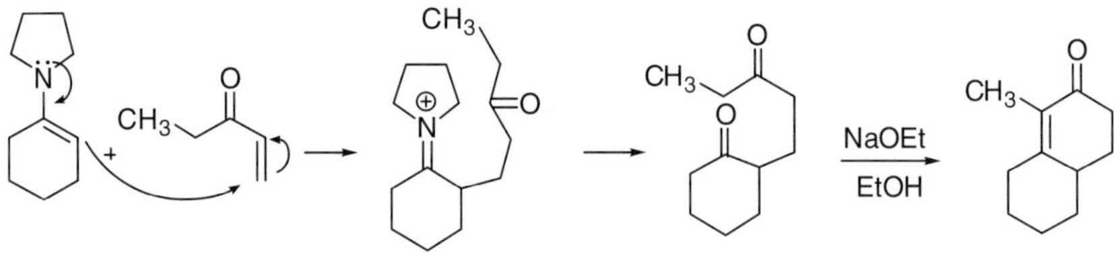

Scheme 3.22

3.3 Ylides

The term ylide was introduced by George Wittig who gave the **Wittig reaction**[57–61] in 1953. Since then, the chemistry of ylides flourished rapidly and has become a powerful and versatile synthetic tool for the synthesis of carbon–carbon bonds.

Ylide can be viewed as a special carbanion in which the negative charge on carbon is stabilized by an adjacent positively charged heteroatom. The most common ylides are phosphonium ylides, sulfur ylides (sulfonium and sulfoxonium ylides) and certain nitrogen-based ylides (ammonium, azomethine, pyridinium and nitrile ylides). In addition to synthetically important phosphorus, sulfur and nitrogen, ylides of tin (Sn) and iodine (I) have been developed in recent years.

Most ylides can be represented by their resonance forms.

$$R_n \overset{\oplus}{Z} - \overset{\ominus}{C}RR' \longleftrightarrow R_n Z = CRR' \qquad \begin{array}{l} Z = P \text{ and } n = 3 \\ \text{or} \\ Z = S \text{ and } n = 2 \end{array}$$

Ylides can be stabilized or non-stabilized. As the electron-donating ability of substituents bonded to the heteroatom increases, the stability of the ylide increases and the reactivity decreases. As the electron-withdrawing ability of substituents bonded to the carbanion carbon increases, the stability of the ylides increases and the reactivity decreases.

3.3.1 Formation of ylides

A number of ylides have been isolated in crystalline state. However, some of the ylides are transient intermediates in some of the reactions.

Nitrogen ylides

The two most important methods for the preparation of ammonium ylides such as **3.44** involve either alkylation of an appropriate amine and subsequent deprotonation by a base or capture of tertiary amines by carbenes.

$$(CH_3)_3\overset{\oplus}{N}CH_3Br \xrightarrow[\text{ether}]{C_6H_5Li} (CH_3)_3\overset{\oplus}{N}-\overset{\ominus}{CH_2} + LiBr$$

Tetramethylammonium
bromide
3.44

$$(CH_3)_3\ddot{N} + :CH_2 \longrightarrow (CH_3)_3\overset{\oplus}{N}-\overset{\ominus}{CH_2}$$

Carbene
3.44

Sulfur ylides

The sulfonium salts on treatment with a base (such as alkyllithium) eliminate an α-hydrogen to form dialkylsulfonium methylide (a sulfur ylide). This is the most commonly used method for the preparation of sulfur ylides. In a similar manner, sulfoxides can be converted into the corresponding sulfoxonium salts. Very strong bases, such as butyllithium, are required for the complete formation of ylides. Sodium hydride (NaH), another powerful base, is insoluble in most solvents, but its reaction with DMSO (dimethylformamide) generates a strong conjugate base, $CH_3S(=O)CH_2^-Na^+$, known as **dimsyl sodium**. This soluble base is widely used for the generation of ylides in DMSO solution. Corey and Chaykovsky[62–65] synthesized dimethylsulfonium methylide ylide (**3.45**) and dimethylsulfoxonium ylide (**3.46**); since formed these ylides have widespread use in synthesis.

$$(CH_3)_2S \xrightarrow{CH_3I} \left[(CH_3)_2\overset{+}{S}-CH_3\right]\overset{\ominus}{I} \xrightarrow[\text{DMSO}]{NaH} \left[(CH_3)_2\overset{+}{S}-\overset{-}{CH_2} \longleftrightarrow (CH_3)_2S=CH_2\right]$$

Trimethylsulfonium
iodide
3.45

$$H_3C-\overset{\overset{\oplus}{C}H_3}{\underset{CH_3}{S}}=O \;\overset{\ominus}{I} \xrightarrow[\text{DMSO}]{RLi \text{ or } NaH} \left[\overset{\ominus}{CH_2}-\overset{\overset{CH_3}{\oplus}}{\underset{CH_3}{S}}=O \longleftrightarrow CH_2=\overset{\overset{CH_3}{}}{\underset{CH_3}{S}}=O\right]$$

Trimethylsulfoxonium
iodide
3.46

Sulfur ylides can also be prepared by the reaction of a carbene with a sulfide.

$$\overset{}{>}C: + :S\overset{R}{\underset{R}{<}} \longrightarrow \overset{}{>}\overset{-}{C}-\overset{+}{S}\overset{R}{\underset{R}{<}}$$

The reaction of arylchlorocarbenes, generated from arylchlorodiazirines, with trimethylenesulfide gives a sulfur ylide as an intermediate[66].

Ar = Ph, p-MeC$_6$H$_4$, p-ClC$_6$H$_4$

The sulfide reacts with a diazo compound in the presence of catalytic amounts of CuSO$_4$ to give a 50:50 mixture of sulfur ylides[67].

However, the generation of only single isomer was achieved by using a shape-selective catalyst[68]. The Cu(I) catalysts coordinated to chiral ligand, **3.47**, **3.48** or **3.49**, are used in the selective synthesis of sulfur ylides.

3.47

3.48

3.49

Phosphorus ylides

Phosphorus ylides are prepared from phosphonium salts by deprotonating them with a strong base. The method consists of the alkylation of triphenylphosphine with alkyl halide. The resulting phosphonium salt is treated with a strong base (phenyllithium or *n*-butyllithium) to give a phosphorus ylide. The simplest ylide is methylenetriphenylphosphorane (**3.50**), which is prepared by the abstraction of a proton from methyltriphenylphosphonium iodide.

Triphenylphosphine   **3.50**

Substituted ylides are also prepared by the alkylation of Ph$_3$P=CH$_2$ with a primary alkyl halide R—CH$_2$—X, followed by deprotonation with a strong base such as C$_4$H$_9$Li.

Carbonyl ylides

Stabilized carbonyl ylides such as **3.52** have been prepared by the treatment of the silyl substituted α-chloro ethers **3.51** with fluoride ions under neutral conditions.

3.51   **3.52**

Non-stabilized carbonyl ylides such as **3.55** can be prepared from 1-iodoalkyl triethylsilyl ethers (**3.53**), probably via bis(1-iodoalkyl) ethers (**3.54**).

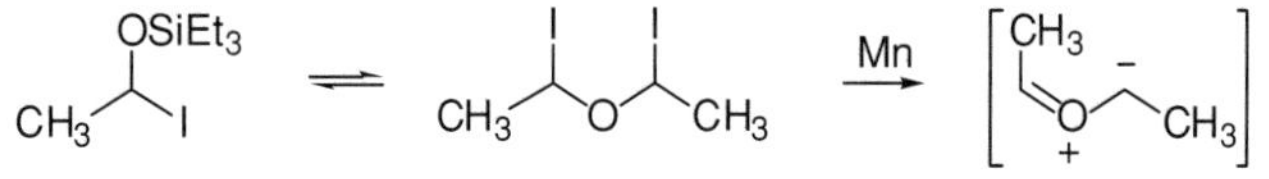

| **3.53** | **3.54** | **3.55** |

3.3.2   Reactions of ylides

Most of the ylides are strong nucleophiles or bases and are very important in organic synthesis. Ylides undergo mainly three types of reactions: alkenation, cyclization to a three-membered ring (epoxidation, cyclopropanation or aziridination) and rearrangement reactions. The negative charge at the carbon makes them reactive and useful for the formation of the carbon–carbon bond. Mainly sulfonium ylides, aminosulfoxonium, arsonium and other related ylides form epoxides on reaction with carbonyl compounds. Epoxides are versatile intermediates in organic chemistry; the electrophilic or nucleophilic ring-opening reactions of them leads to 1,2-difunctionalized systems or to the formation of a new carbon–carbon bond.

The intermediate **A** formed by the reaction of carbonyl compounds, alkenes or imines with ylide may undergo elimination of L_nM to give the cyclization product epoxides, cyclopropanes or aziridines, respectively. However, intermediate **B** may lead to alkenation by the elimination of $L_nM=X$, as shown in Scheme 3.23.

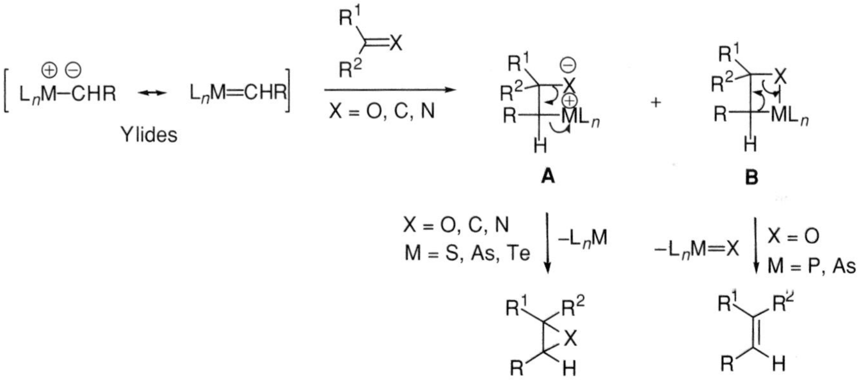

Scheme 3.23   Reaction between carbonyl compounds and an ylide

Carbonyl ylides

The carbonyl ylides are trapped in a [2+3]-cycloaddition with activated alkenes or alkynes. For example, the reaction of the α-chloro ether **3.56** with cesium fluoride (CsF) in the presence of dimethyl fumarate gives only the *trans* cycloaddition product **3.57**, while the reaction with dimethyl maleate gives exclusively the *cis*-isomer **3.58** (Scheme 3.24).

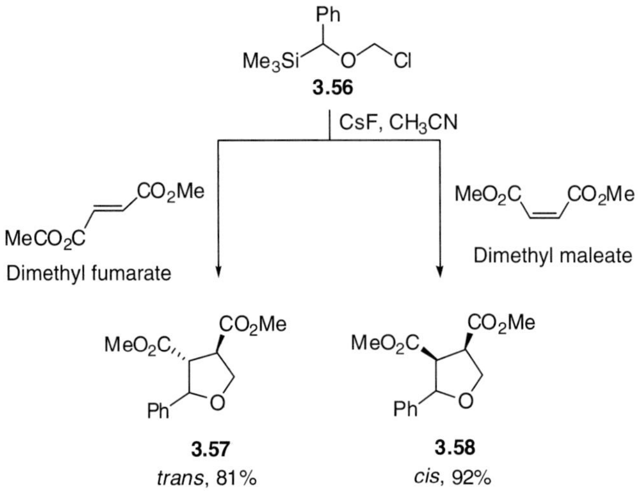

Scheme 3.24

Nitrogen ylides

The nitrogen ylides (**3.60**) are formed as intermediates in the **Sommelet rearrangement**[69–74]. The method is used for the conversion of the tetraalkylammonium halides **3.59** (having hydrogen in an α-position to N) into tertiary aromatic amines **3.61** (Scheme 3.25).

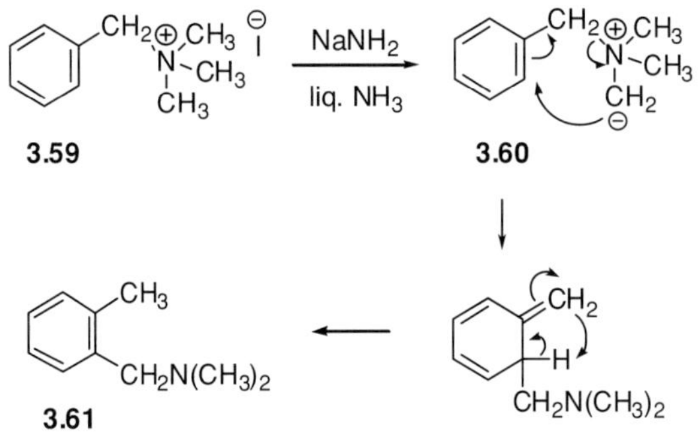

Scheme 3.25

Ammonium ylides such as **3.62** in which none of the alkyl groups possesses a β-hydrogen atom, but one has a β-carbonyl group undergoes the **Stevens rearrangement**[75–79] to give the amino ketone **3.63** (Scheme 3.26).

The migrating group migrates with nearly complete retention of configuration. Thus, the reaction is intramolecular and concerted. However, there is evidence which supports the free radical dissociation–recombination mechanism (Scheme 3.27).

Other heteroatom analogues such as sulfur (S) also undergo the Stevens rearrangement.

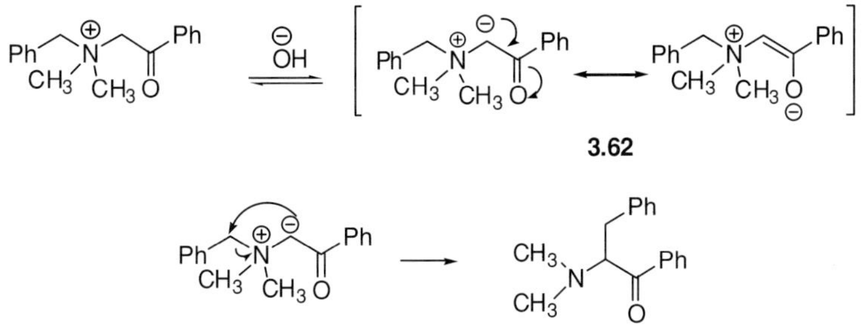

3.62

3.62 3.63

Scheme 3.26

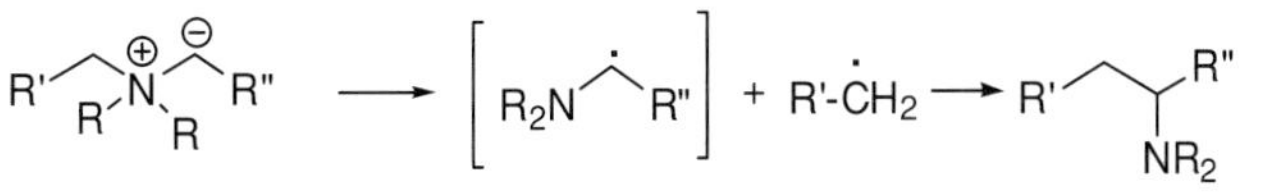

Scheme 3.27

Sulfur ylides

Unstabilized sulfonium ylides such as dimethylsulfonium methylide (**3.45**) and stabilized sulfoxonium ylides such as dimethylsulfoxonium methylide (**3.46**) are the most widely used sulfur ylides.

$$(CH_3)_2\overset{\oplus}{S}-\overset{\ominus}{C}H_2 \qquad (CH_3)_2\underset{\oplus}{\overset{O}{\overset{\|}{S}}}-\underset{\ominus}{C}H_2$$

3.45 **3.46**

The sulfur ylide **3.45** on reaction with cyclohexanone gives oxirane **3.64**, while the phosphorus ylide **3.65** gives the alkene **3.66**.

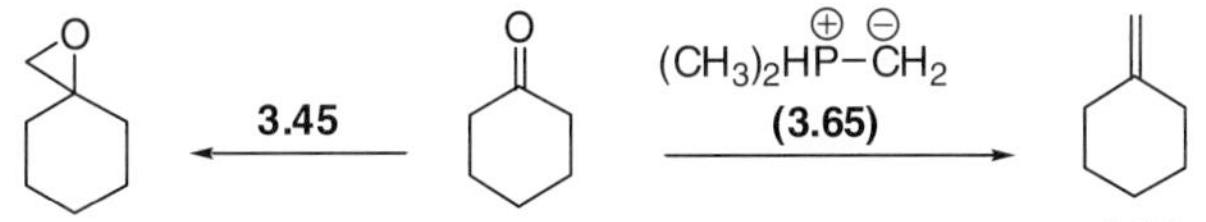

3.64 **3.66** 99%

The reaction of sulfur ylides with aldehydes and ketones was first reported by Johnson[80] in 1961, but is for some reason better known as the **Corey–Chaykovsky reaction**. The reaction between sulfur ylides and an electrophilic carbon atom of a C=O gives a betaine intermediate. The favoured reaction path is therefore an internal S_N2 process which furnishes an epoxide with regeneration of the sulfide (Scheme 3.28).

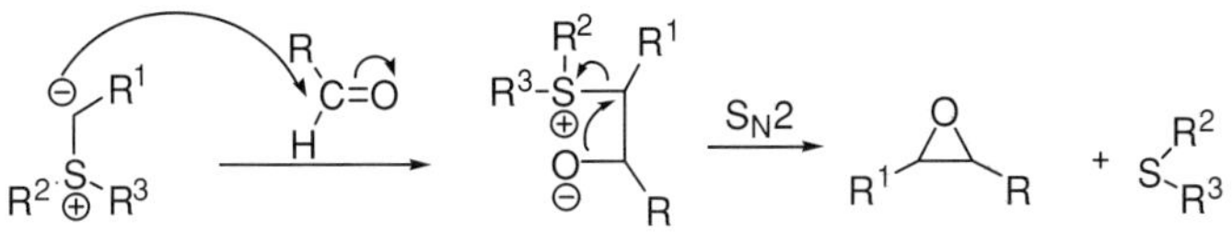

Scheme 3.28

The reaction of **3.45** with aromatic aldehydes is useful for the preparation of epoxides and heterocyclic compounds from appropriate substrates (Scheme 3.29).

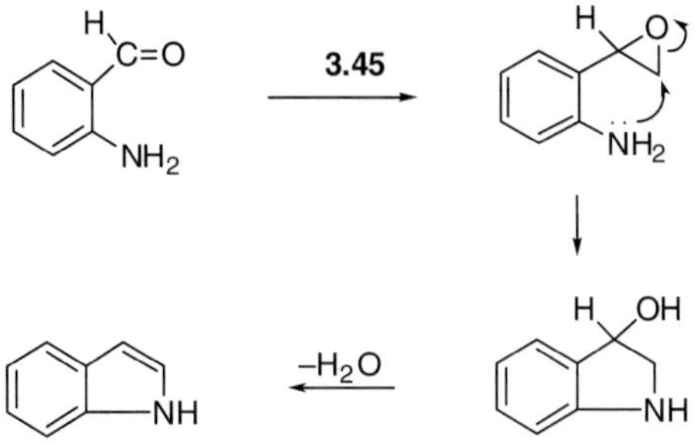

Scheme 3.29

Trost[81] showed that the sulfur ylide **3.68** prepared by the deprotonation of **3.67** with *n*-butyllithium (*n*-BuLi) on reaction with benzaldehyde under the same reaction conditions gave styrene oxide **3.69** as a racemic mixture.

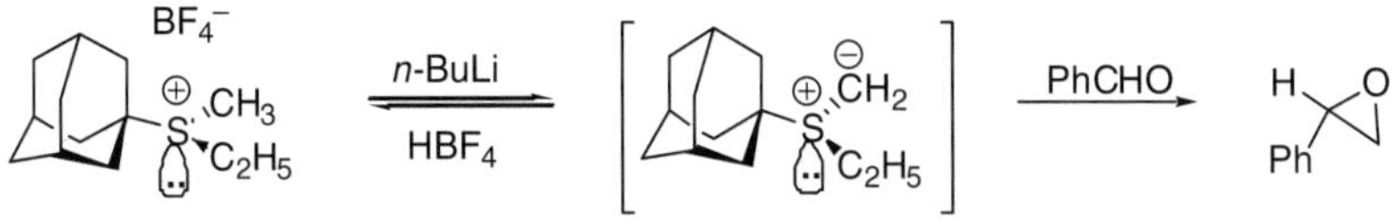

Unstabilized sulfonium ylides and stabilized sulfoxonium ylides show different reactions with α,β-unsaturated carbonyl compounds; the former give epoxides and the latter give cyclopropanes. The epoxide formation (i.e. 1,2-addition) is kinetically favourable while cyclopropane formation (i.e. 1,4-addition, **Michael addition**) is energetically favourable.

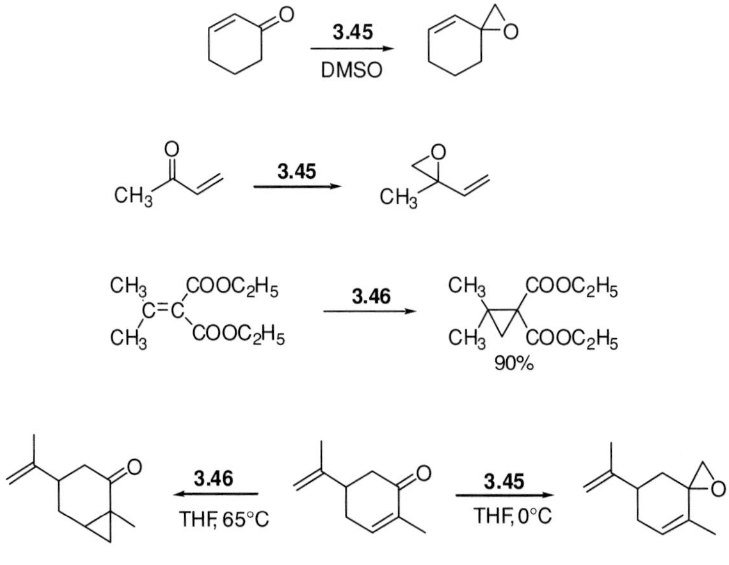

Sulfur ylides undergo a number of rearrangements such as [2,3]-sigmatropic rearrangement[82-85], as shown in Scheme 3.30.

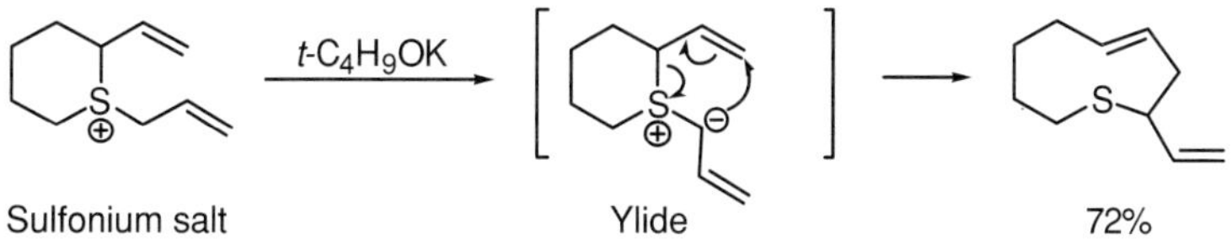

Sulfonium salt   Ylide   72%

Scheme 3.30

Other heteroatom analogues such as oxonium, selenium and iodonium ylides also undergo [2,3]-sigmatropic rearrangements.

Phosphorus ylides

The well-known reaction of phosphorus ylides is their reaction with carbonyl compounds (aldehydes and ketones) to give alkenes. This alkenation methodology is universally known as the **Wittig reaction**[57-61].

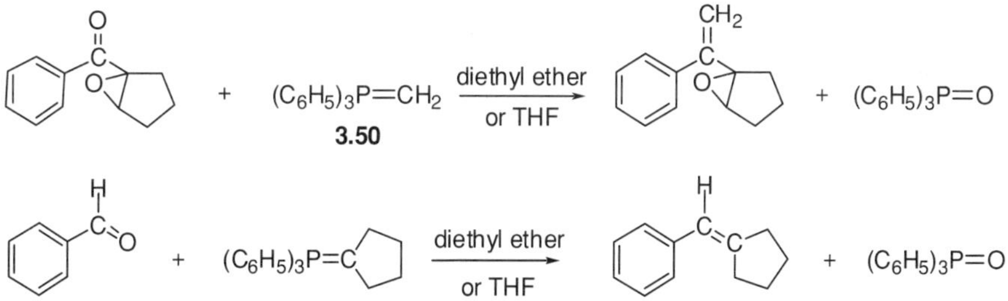

The Wittig reaction is very useful for the introduction of exocyclic double bond. For example, cyclohexanone is converted into methylenecyclohexane (**3.66**) in 99% yield with **3.50**.

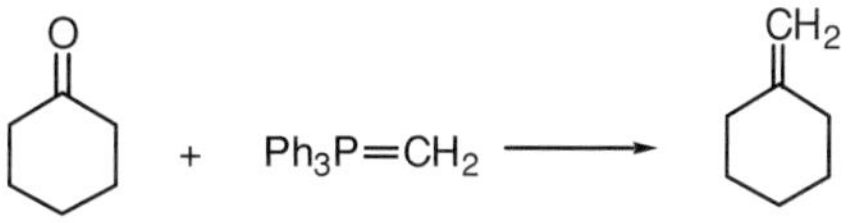

Phosphorus ylide reacts with aldehyde or ketone to form an intermediate betaine **A**, which collapsed into oxaphosphetane **B**. Elimination of triphenylphosphine oxide yields alkene (Scheme 3.31) (also see section 4.3.1).

The Wittig alkenation has found widespread application in synthetic organic chemistry, and numerous papers and reviews have detailed the progress of the Wittig reaction. A principal advantage of alkene synthesis by the Wittig reaction is that the location of the double bond is absolutely fixed in contrast to the mixture often produced by alcohol dehydration. With simple substituted ylides Z-alkenes are favoured.

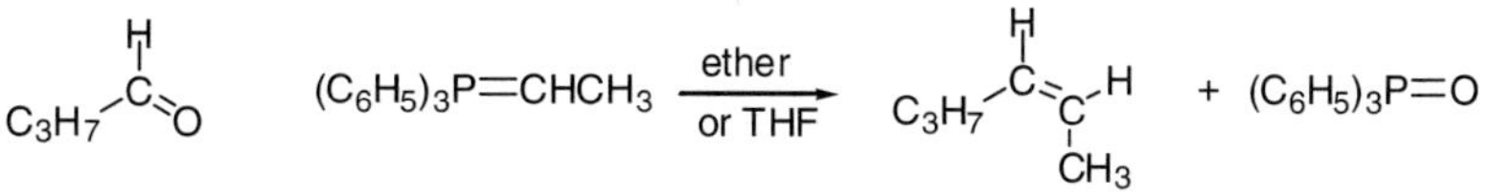

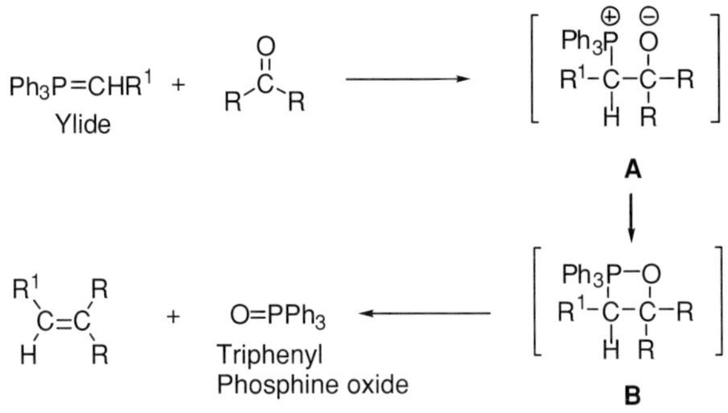

Scheme 3.31

When a phosphonate enolate base is used as the nucleophile the reaction is known as the **Horner–Wadsworth–Emmons reaction**[86–92] (Scheme 3.32) (also see section 4.3.1).

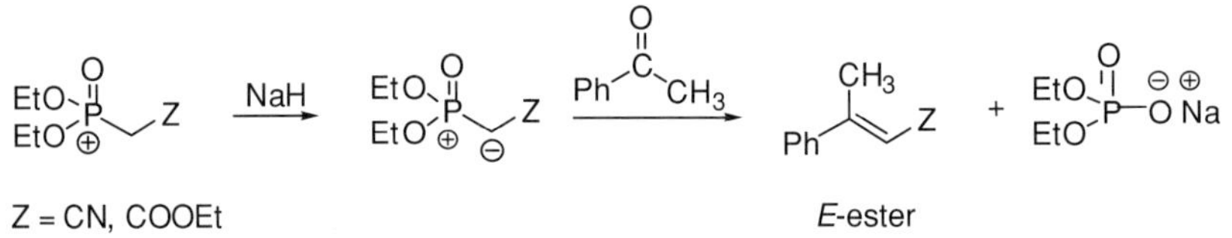

Z = CN, COOEt   *E*-ester

Scheme 3.32

Stabilization of the ylide-like carbanion leads to an *E*-configuration of the product double bond.

Mono-, di- and trisubstituted alkenes can all be prepared in good yield by the Wittig reaction. A large variety of ketones and aldehydes are effective in the reaction, although carboxylic acid derivatives such as esters fail to react usefully. The carbonyl compound can tolerate several groups such as OH, OR, aromatic nitro and even ester groups.

3.3.3 Asymmetric ylide reactions

Asymmetric ylide reactions such as epoxidation, cyclopropanation, aziridination, [2,3]-sigmatropic rearrangement and alkenation can be carried out with **chiral ylide** (reagent-controlled asymmetric induction) or a chiral C=X compound (substrate-controlled asymmetric epoxidations). Non-racemic epoxides are significant intermediates in the synthesis of, for instance, pharmaceuticals and agrochemicals.

In reagent-controlled epoxidation the asymmetric induction has its origin in a chiral ylide. The reaction of an achiral aldehyde or ketone with a chiral ylide gives optically active compounds.

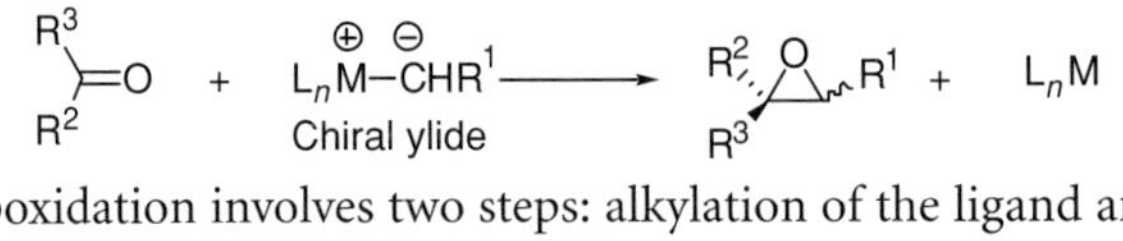

The process of epoxidation involves two steps: alkylation of the ligand and isolation of the resulting salt, followed by treatment with a base in the presence of a carbonyl compound.

Catalytic cycles have also been developed where only a catalytic amount of ligands are required (Scheme 3.33).

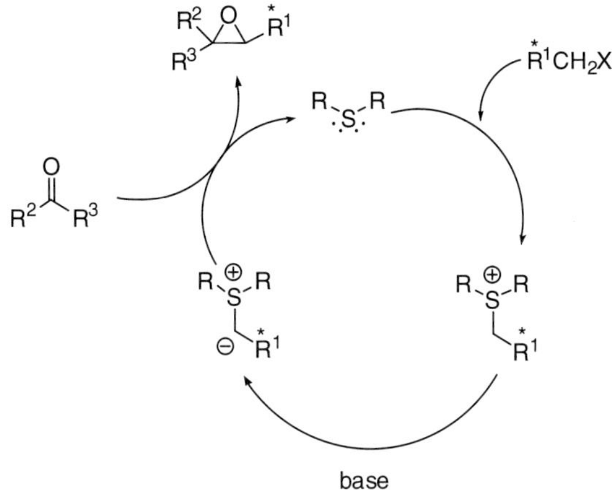

Scheme 3.33

The successful example[93] of catalytic asymmetric chiral sulfonium epoxidation was reported with enantiomeric excess (43%) of *trans*-stilbene oxide (**3.71**) by the reaction of 4-chlorobenzaldehyde with benzyl bromide in acetonitrile at room temperature by using 0.5 equiv. of optically active sulfide **3.70** (with *exo*-OH group). Powdered KOH was used as a base.

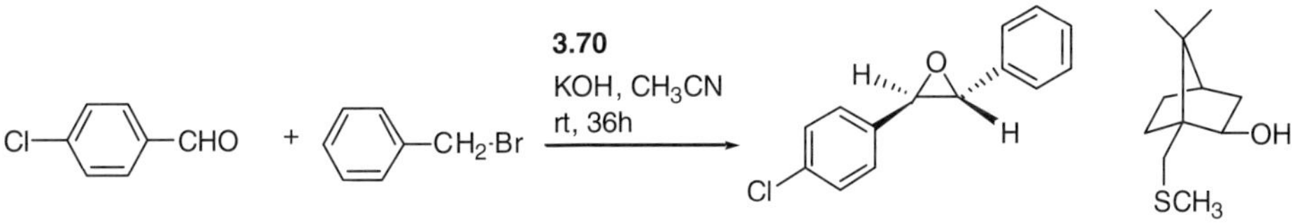

3.71
(*R,R*) (+)-*trans*- 50% yield,
43% ee

Johnson *et al.* were the first to prepare enantiomerically pure aminosulfoxonium ylide[94–97] **3.73** from **3.72**, which on reaction with benzaldehyde gave (*R*)-styrene oxide (**3.69**) in 20% ee. The reaction of aminosulfoxonium ylide **3.74** with heptaldehyde gave the corresponding epoxide **3.75** with opposite enantioselectivity (39% ee, *S*) as expected.

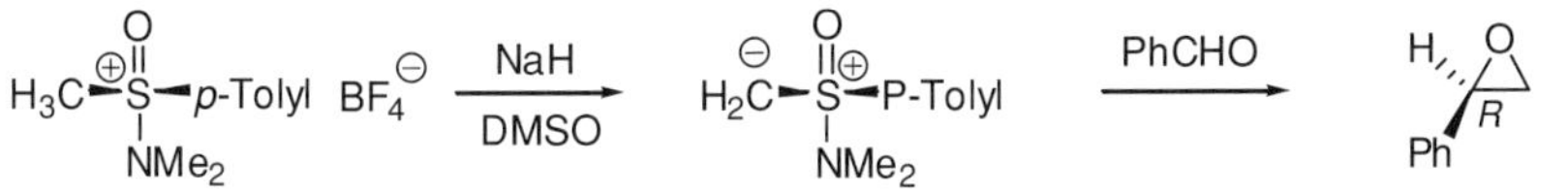

3.72   **3.73**   **3.69**
               (*R*)-(−)-Aminosulfoxonium ylide   (*R*)-Styrene oxide
                                               60% yield, 20% ee

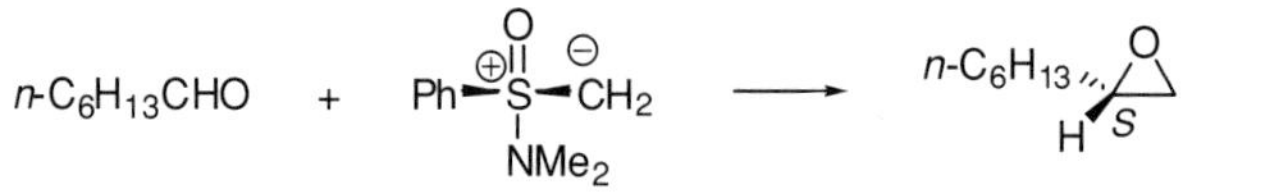

3.74   **3.75**

n-C$_6$H$_{13}$CHO   Ph—S—CH$_2$ n-C$_6$H$_{13}$ ... epoxide

Heptaldelyde   Chiral ylide   (S)-Epoxide
                                                                              39%

A two-step mechanism (Scheme 3.34) for epoxidation was proposed in which intermediate betaine **A** and **B** are obtained from the carbonyl compound and sulfonium ylides irreversibly and from aminosulfoxonium ylide reversibly (step 1). Betaine (**A** or **B**) then undergoes ring closure (step 2) irreversibly.

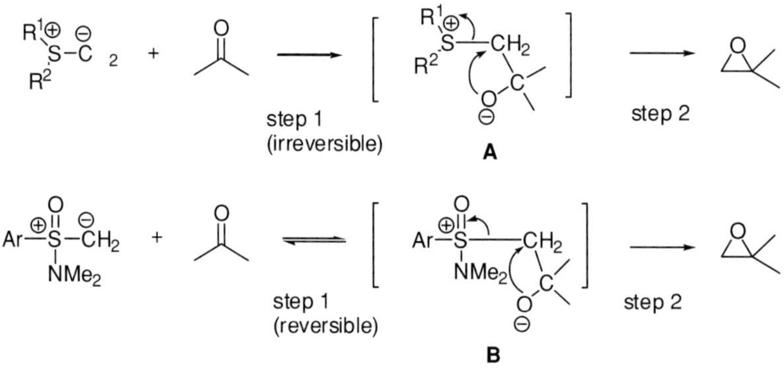

Scheme 3.34

Durst and co-workers.[98] employed C_2-symmetric sulfonium ylides **3.76**, **3.77** and **3.78** obtained from the corresponding sulfonium salts to be used in asymmetric epoxidations (Scheme 3.35) to overcome the problems with low selectivities Furukawa and co-workers[99] had encountered. The epoxidations were conducted under phase transfer conditions.

The reaction of benzaldehyde (1 equiv.) with benzyl bromide (2 equiv.) and thiolane, (2R,5R)-dimethylthiolane (**3.79**), gives *trans*-(2S,3S)-diphenyloxirane (**3.80**).

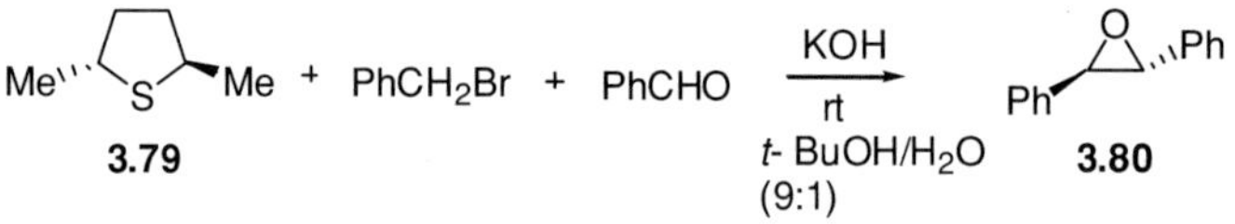

The catalytic cycle (Scheme 3.36) presented by Aggarwal *et al.*[100,101] reduced the conventional two-step sequence for epoxide formation to one step.

In the catalytic cycle itself, ylides **A** are generated *in situ* in the presence of carbonyl compounds in a reaction between a sulfide and a carbenoid **B**, which in turn is generated from a diazo compound **C** and a metal catalyst **D**. At the same time, ylides **A** react with carbonyl compounds to give *trans*-(S,S)-epoxides and are simultaneously regenerated to

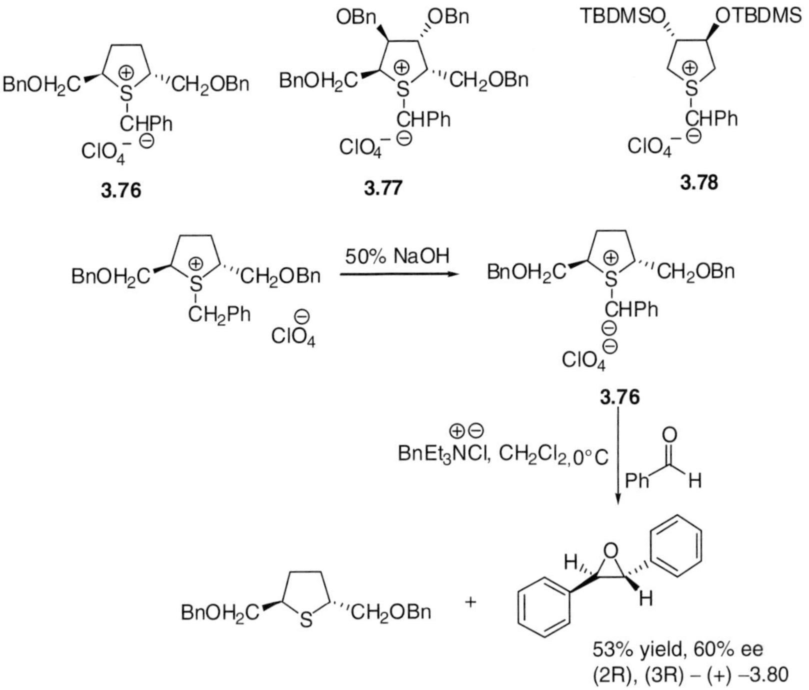

Scheme 3.35

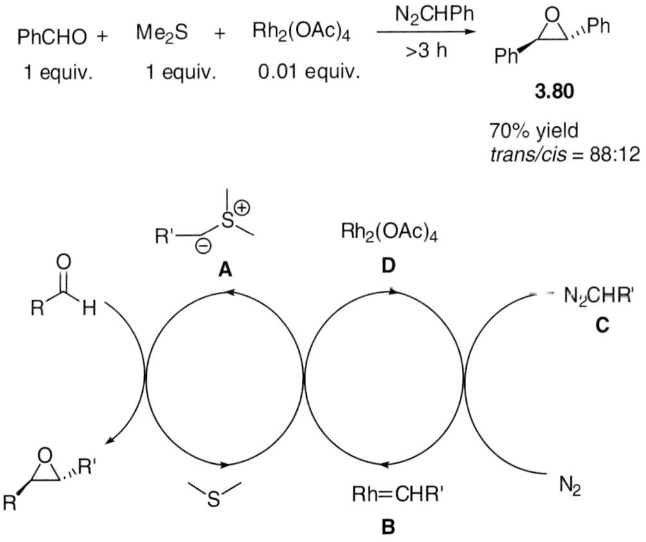

Scheme 3.36

continue the process, as shown in Scheme 3.36. Besides benzaldehyde other aromatic and aliphatic aldehydes worked well, giving good yields of epoxides.

References

1. Cahiez, G., Chau, F. and Blanchot, B., *Org. Synth. Coll.*, **2004**, *10*, 59; **1999**, *76*, 239.
2. Cahiez, G., Chau, F. and Blanchot, B., *Org. Synth. Coll.*, **1999**, *76*, 239.
3. Reetz, M. T. and Haning, H., *Tetrahedron Lett.*, **1993**, *34*, 7395.
4. Crimmins, M. T., Emmitte, K. A. and Katz, J. D., *Org. Lett.*, **2000**, *2*, 2165.
5. Crimmins, M. T. and Emmitte, K. A, *J. Am. Chem. Soc.*, **2001**, *123*, 1533.
6. Crimmins, M. T. and Choy A. L., *J. Am. Chem. Soc.*, **1999**, *121*, 5653.
7. Frost, A. A. and Pearon, R. G., *Kinetics and Mechanism*, Wiley, New York, **1953**, p. 291.
8. Bell, R. P. and Smith, M. J., *J. Chem. Soc.*, **1958**, 1691.
9. Fieser, L. F. and Fieser, M., *Adv. Org. Chem.*, **1961**, 456.
10. Claisen, L. and Claparede, A., *Chem. Ber.*, **1881**, *14*, 2460.
11. Schmidt, J. G., *Chem. Ber.*, **1881**, *14*, 1459.
12. Kohler, E. P. and Chadwell, H. M., *Org. Synth.*, **1941**, *1*, 71.
13. Henecka, H., *Houben-Weyl-Muller*, **1955**, *4*(II), 28.
14. Fieser and Fieser, M. *Adv. Org. Chem.*, **1961**, 467.
15. Cahiez, G., Cléry, P. and Laffitte, J. A., Fr. Demande FR 2/671,085, Fr. Pat. Appl., *1990*, *90/16*, 413; Cahiez, G., Cléry, P. and Laffitte, J. A., *Chem. Abstr.*, **1993**, *118*, 69340b.
16. Crimmins, M. T., King, B. W., Tabet, E. A. and Chaudhary, K., *J. Org. Chem.*, **2001**, *65*, 9084–9085.
17. Tang, Z., Jiang, F., Yu, L.-T., Cui, X., Gong, L.-Z., Mi, A.-Q., Jiang, Y.-Z. and Wu, Y.-D., *J. Am. Chem. Soc.*, **2003**, *125*, 5262.
18. Mukaiyama, T., *Chem. Lett.*, **1982**, 353.
19. Mukaiyama, T., Inomata, K. and Muraki, M., *J. Am. Chem. Soc.*, **1973**, *95*, 967.
20. Mukaiyama, T., Inomata, K. and Muraki, M., *Chem. Lett.*, **1986**, 187.
21. Mukaiyama, T., Inomata, K. and Muraki, M., *Org. React.*, **1982**, *28*, 187.
22. Banno, K. and Mukaiyama, T., *Chem. Lett.*, **1976**, 279.
23. Fischer, E. O. and Maasbol, A., *Angew. Chem., Int. Ed. Engl.*, **1964**, *3*, 580.
24. Bernasconi, C. F., *Adv. Phys. Org. Chem.*, **2002**, *37*, 137.
25. Henry, H., *Compt. Rend.*, **1895**, *120*, 1265.
26. Hass, H. B., Susie, A. G. and Heilder, R. L., *J. Org. Chem.*, **1950**, *15*, 8.
27. Varma, R. S., Dahiya, R. and Kumar, S., *Tetrahedron Lett.*, **1997**, *38*, 5131.
28. Palomo, C., Oiarbide, M. and Laso, A., *Angew. Chem., Int. Ed. Engl.*, **2005**, *44*(25), 3881.
29. Evans, D. A., Seidel, D., Rueping, M., Lam, H. W., Shaw, J. T. and Downey, C. W., *J. Am. Chem. Soc.*, **2003**, *125*, 12692.
30. Kamlet, J., U.S. Pat. 2,151,517, **1939**; Kamlet, J., *Chem. Abstr.*, **1939**, *33*, 5003.
31. Claisen, L., *Chem. Ber.*, **1887**, *20*, 655.
32. Dieckmann, W., *Chem. Ber.*, **1894**, *27*(102), 965.
33. Hauser, C. R. and Hudson, B. E., *Org. React.*, **1942**, *1*, 274.
34. Thyagarajan, B. S., *Chem. Rev.*, **1954**, *54*, 1029.
35. Leonard, N. J. and Schimelpfenig, C. W., *J. Org. Chem.*, **1958**, *23*, 1708.
36. Hiyama, T., *Bull Chem. Soc.*, **1987**, *60*, 2127.
37. Hiyama, T. and Kobayashi, K., *Tetrahedron Lett.*, **1982**, *23*, 1597.
38. Kobayashi, K. and Hiyama, T., *Tetrahedron Lett.*, **1983**, *24*, 3509.
39. Baylis, A. B. and Hillman, M. E. D, Ger. Pat. 1972, 2155113; *Chem. Abstr.*, **1972**, *77*, 34174 q.
40. Basavaiah, D., Gowriswari, V. V. L., Sarma, P. K. S. and Rao, P. D., *Tetrahedron Lett.*, **1990**, *31*, 1621.
41. Rezgui, F. and Gaied, M. M., *Tetrahedron Lett.*, **1998**, *39*, 5965.
42. Shi, M., Li, C.-Q. and Jiang, J.-K., *Chem. Commun.*, **2001**, 833.
43. Michael, A., *J. Prakt. Chem.*, **1887**, *35*(2), 349.
44. Bergmann, E. D., Ginsburg, D. and Pappo, R., *Org. React.*, **1959**, *10*, 179.
45. Kohler, E. P. and Butler, F. R., *J. Am. Chem. Soc.*, **1926**, *48*, 1040.
46. Xu, Y., Ohori, K., Ohshima, T. and Shibasaki, M., *Tetrahedron*, **2002**, *58*(13), 2585.
47. Mannich, C. and Krosche, W., *Arch. Pharm.*, **1912**, *250*, 647.
48. Blicke, F. F., *Org. React.*, **1942**, *1*, 303.
49. Schreiber, J., Maag, H., Hashimoto, N. and Eschenmoser, A., *Angew. Chem., Int. Ed.*, **1971**, *10*, 330.
50. Eftekhari-Sis, B., Abdollahifar, A., Hashemi, M. M. and Zirak, M., *Eur. J. Org. Chem.*, **2006**, *22*, 5152.
51. Ishitani, H., Ueno, M. and Kobayashi, S., *J. Am. Chem. Soc.*, **1977**, *119*, 7153.
52. Stork, G., Terrell, R. and Szmuszkovicz, J., *J. Am. Chem. Soc.*, **1954**, *76*, 2029.
53. Whitesell, J. W. and Whitesell, M. A., *Synthesis*, **1983**, 517.
54. Hickmott, P. W., *Tetrahedron*, **1982**, *38*, 1975.
55. Stork, G., Brizzoland, A., Landesman, H., Szmuszkovich, J. and Terrell, R., *J. Am. Chem. Soc.*, **1963**, *85*, 207.
56. Stork, G. and Landesman, H., *J. Am. Chem. Soc.*, **1956**, *78*, 5128.

57. Maryanoff, B. E. and Reitz, A. B., *Chem. Rev.*, **1989**, *89*, 863, and references therein.
58. Vedejs, E., Marth, C. F. and Ruggeri, R., *J. Am. Chem. Soc.*, **1988**, *110*, 3940.
59. Vedejs, E. and Marth, C. F., *J. Am. Chem. Soc.*, **1988**, *110*, 3948.
60. Reitz, A. B., Mutter, M. S. and Maryanoff, B. E., *J. Am. Chem. Soc.*, **1984**, *106*, 1873.
61. Burton, D. J., Yang, Z. Y. and Qiu, W., *Chem. Rev.*, **1996**, *52*, 1641.
62. Corey, E. J. and Chaykovsky, M., *J. Am. Chem. Soc.*, **1962**, *84*(5), 867.
63. Corey, E. J. and Chaykovsky, M., *J. Am. Chem. Soc.*, **1965**, *87*(6), 1353.
64. Corey, E. J. and Chaykovsky, M., *Org. Synth. Coll.*, **1973**, *5*, 755.
65. Corey, E. J. and Chaykovsky, M., *Org. Synth. Coll.*, **1969**, *49*, 78.
66. Romashin, Y. N., Liu, M. T. H. and Bonneau, R., *Tetrahedron Lett.*, **2001**, *42*(2), 207.
67. Ando, W., Yagihara, T., Tozune, S., Inai, I., Suzuki, J., Toyama, T., Nakaido, S. and Migita, T., *J. Org. Chem.*, **1972**, *37*, 1721.
68. Aggarwal, V. K., Bell, L., Coogan, M. P. and Jubault, P., *J. Chem. Soc., Perkin Trans. 1*, **1998**, 2037.
69. Sommelet, M., *Compt. Rend.*, **1937**, *205*, 56.
70. Wittig, W., Mangold, R. and Felletschin, G., *Ann. Chem.*, **1948**, *560*, 116.
71. Hauser, C. R. and Eenam, D. N. V., *J. Am. Chem. Soc.*, **1956**, *78*, 5698.
72. Jones, G. C. and Hauser, C. R., *J. Org. Chem.*, **1962**, *27*, 3572.
73. Burngardner, C. L., *J. Am. Chem. Soc.*, **1963**, *85*, 73.
74. Jones, G. C., Beard, W. Q. and Hauser, C. R., *J. Org. Chem.*, **1963**, *28*, 199.
75. Stevens, T. S., Creighton, E. M., Gordon, A. B. and MacNicol, M., *J. Chem. Soc.*, **1928**, 3193.
76. Wittig, G. and Felletschin, G., *Ann. Chem.*, **1943**, *555*, 133.
77. Kantor, S. W. and Hauser, C. R., *J. Am. Chem. Soc.*, **1951**, *73*, 4122.
78. Brewster, J. H. and Kline, M. M., *J. Am. Chem. Soc.*, **1953**, *74*, 5179.
79. Jenney, E. F. and Druey, J., *Angew. Chem., Int. Ed.*, **1962**, *1*, 155.
80. Johnson, W. A. and LaCount, R. B., *J. Am. Chem. Soc.*, **1961**, *83*(2), 417.
81. Trost, B. M. and Melvin, L. S., Jr., *Sulphur Ylides*, Academic Press, New York,
82. Ollis, W. D., Plackett, J. D., Smith, C. and Sutherland, I. O., *J. Chem. Soc. Commun.*, **1968**, 186.
83. Doyle, M. P., Griffin, J. H., Chinn, M. S. and Leusen, D., *J. Org. Chem.*, **1984**, *49*, 1917.
84. Gassman, P. G., Miura, T. and Mossman, A., *J. Org. Chem.*, **1982**, *47*, 954.
85. Doyle, M. P., Tamblyn, W. H. and Bagheri, V., *J. Org. Chem.*, **1981**, *46*, 5094.
86. Horner, L., Hoffmann, H. M. R. and Wippel, H. G., *Chem. Ber.*, **1958**, *91*, 61.
87. Horner, L., Hoffmann, H. M. R., Wippel, H. G. and Klahre, G., *Chem. Ber.*, **1959**, *92*, 2499.
88. Wadsworth, W. S., Jr. and Emmons, W. D., *J. Am. Chem. Soc.*, **1961**, *83*, 1733.
89. Wadsworth, W. S., Jr. and Emmons, W. D., *Org. Synth. Coll.*, **1973**, *5*, 547.
90. Wadsworth, W. S., Jr. and Emmons, W. D., *Org. Synth. Coll.*, **1965**, *45*, 44.
91. Wadsworth, W. S., Jr., *Org. React.*, **1977**, *25*, 73–253.
92. Blanchette, M. A., Choy, W., Davis, J. T., Essenfeld, A. P., Masamune, S., Roush, W. R. and Sakai, T., *Tetrahedron Lett.*, **1984**, *25*, 2183.
93. Furukawa, N., Sugihara, Y. and Fujihara, H., *J. Org. Chem.*, **1989**, *54*(17), 4222.
94. Johnson, C. R. and Schroeck, C. W., *J. Am. Chem. Soc.*, **1968**, *90*, 6852.
95. Johnson, C. R. and Schroeck, C. W., *J. Am. Chem. Soc.*, **1973**, *95*, 7418.
96. Johnson, C. R. and Schroeck, C. W., *J. Am. Chem. Soc.*, **1971**, *93*, 5303.
97. Johnson, C. R., Schroeck, C. W. and Shanklin, J. R., *J. Am. Chem. Soc.*, **1973**, *95*, 7424.
98. Breau, L., Ogilvie, W. W. and Durst, T., *Tetrahedron Lett.*, **1990**, *31*, 35.
99. Furukawa, N., Sugihara, Y. and Fujihara, H., *J. Org. Chem.*, **1989**, *54*, 4222.
100. Aggarwal, V. K., Abdel-Rahman, H., Jones, R. V. H., Lee, H. Y. and Reid, B. D., *J. Am. Chem. Soc.*, **1994**, *116*(13), 5973.
101. Aggarwal, V. K., Ford, J. G., Thompson, A., Jones, R. V. H. and Standen, M. C. H., *J. Am. Chem. Soc.*, **1996**, *118*(29), 7004.

Chapter 4
Carbon–Carbon Double Bond Forming Reactions

4.1 Introduction

Alkene synthesis is the most important and most widely used reaction in organic chemistry. Many review articles and research papers describe methods for the positioning and stereospecific introduction of carbon–carbon double bonds[1-3]. The three most important methods for the synthesis of alkenes described in this chapter are (1) elimination reactions, (2) alkenation of carbonyl compounds and (3) reduction of alkynes.

4.2 Elimination reactions

In an elimination reaction the carbon skeleton is pre-assembled. There can be two problems, namely the positioning of the double bond and that of double bond geometry.

4.2.1 β-Eliminations

When 2-bromopropane (**4.1**) is allowed to react with the methoxide ion in methanol, only 40% of the starting material is converted into methyl isopropyl ether (**4.2**) (substitution) and the rest (60%) is transformed into propene (**4.3**) (elimination). Substitution and elimination are often competitive processes. The reaction that produces the alkene involves the loss of an HBr molecule to form a carbon–carbon double bond.

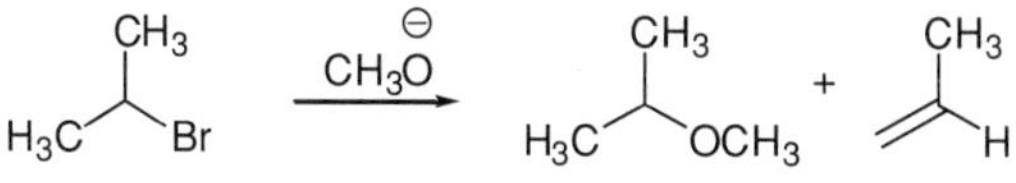

| | **4.1** | **4.2** | **4.3** |
| | | 40% | 60% |

In the E2 elimination reaction of alkyl halides (one-step reaction) a base abstracts a proton adjacent to the leaving group, forcing the removal of the leaving group with the formation of a double bond (Scheme 4.1).

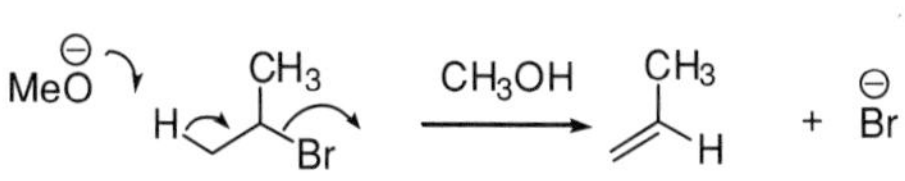

Scheme 4.1

The E2 dehydrohalogenation of alkyl halides is periplanar and occurs most favourably with the proton and halide in an *anti*-configuration. The formation of the *E*- or *Z*-alkene depends on which diastereomer is undergoing elimination. Thus, *erythro*-alkyl halide **4.4** gives *trans*-alkene **4.5**, and *threo*-alkyl halide **4.6** gives *cis*-alkene **4.7** (Scheme 4.2).

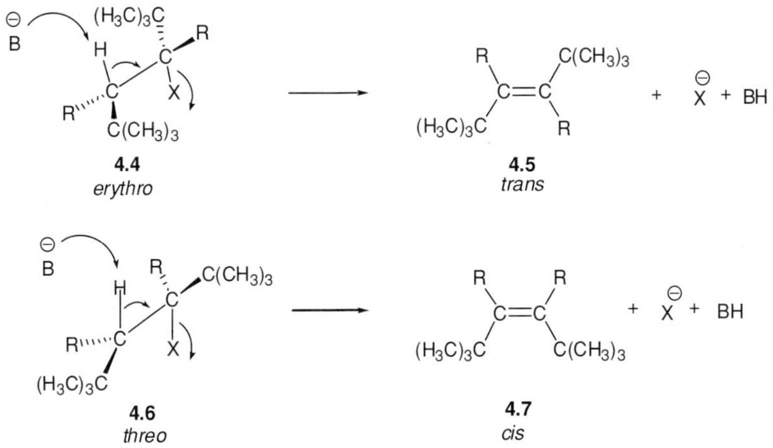

Scheme 4.2

However, a unimolecular E1 reaction undergoes elimination reactions via the formation of the carbocation. It is this step of the reaction that is *rate limiting*. In this case the solvent then acts in the next step of the reaction as the base, removing an H^+ ion from one of the alkyl groups adjacent to the carbocation. The electrons in the C–H bond that is broken are donated to the empty orbital on the carbocation to form a double bond (Scheme 4.3). Thus, alcohols can be dehydrated on heating with 85% H_3PO_4 or 20% H_2SO_4 to alkenes or on heating with $KHSO_4$.

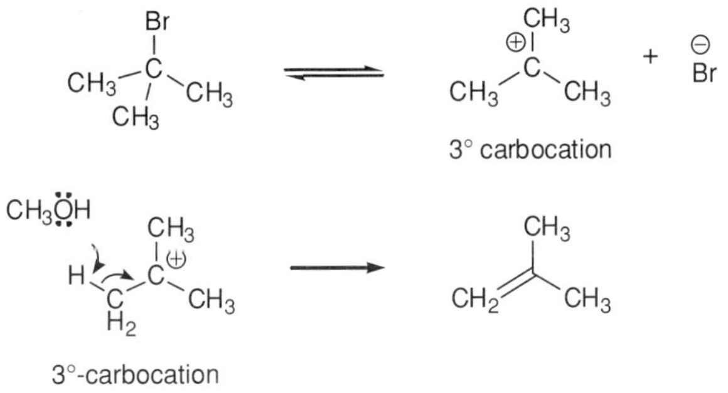

Scheme 4.3

And the most stable alkene (most substituted) is the usual product (**Saytzeff rule**).

The Saytzeff rule (or Zaitsev's rule or Saytsev's rule named after Alexander Mikhailovich Zaitsev) implies that base (sterically unhindered) induced eliminations will lead predominantly to the alkene in which the double bond is more highly substituted. However, if the base, for example, is potassium *t*-butoxide [$(CH_3)_3COK$], the bulkiness prohibits the base from pulling the proton off of the most substituted carbon. In such cases, the less substituted alkene, i.e. **Hofmann elimination**, is preferred.

Cerium chloride[4] ($CeCl_3 \cdot 7H_2O$) in the presence of NaI catalyzes the diastereoselective dehydration of β-hydroxyketones and esters such as **4.8** to the corresponding α,β-unsaturated compound **4.9**.

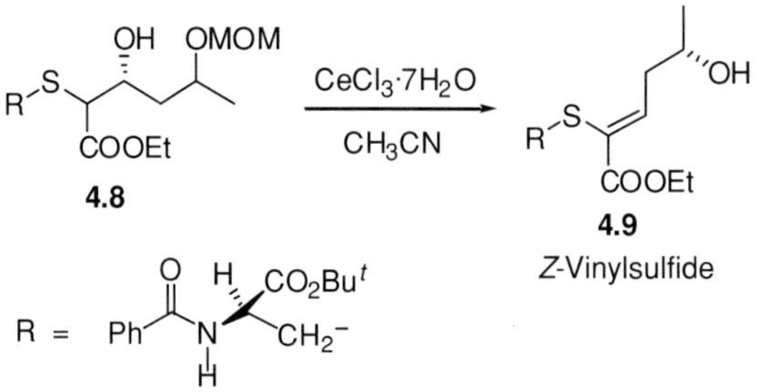

MOM = Methoxymethyl group

The β-eliminations[5] of silicon and halogen can occur from β-haloalkyltrichlorosilanes ($RCHXCH_2SiCl_3$) **4.10** with dilute alkali, Grignard reagent or on heating to give alkenes. These reactions are similar to dehydrohalogenation of alkyl halides as both reactions involve removal of an element more electropositive than carbon (silicon or hydrogen) together with an element more electronegative than carbon (halogen). The most effective reagents used for this elimination are alcoholic bases, aqueous alkali, Grignard reagent, aluminium chloride or potassium acetate in glacial acetic acid. However, the β-chloroethyltriethylsilane (**4.11**) gives alkene only on heating even without the presence of a base.

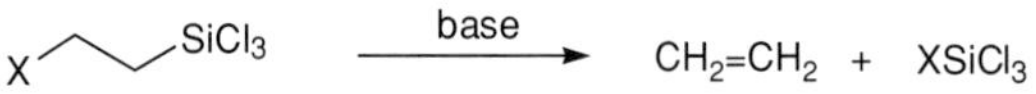

The mechanism of the above elimination reaction is similar to the E2 mechanism of dehydrohalogenation. The reaction proceeds by nucleophilic attack of a base on silicon, simultaneous elimination of halide ion and formation of double bond (Scheme 4.4).

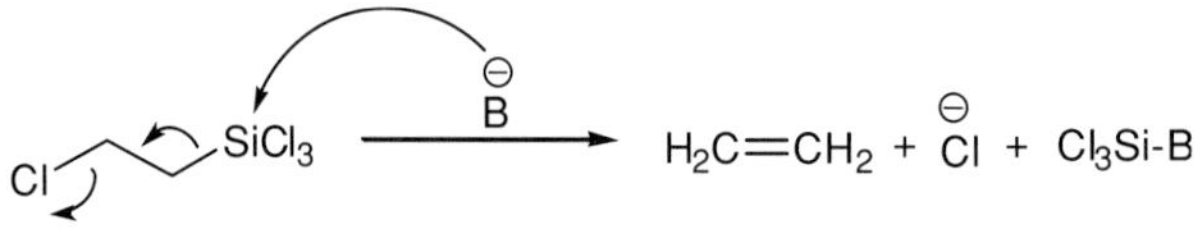

Scheme 4.4

However, in the presence of aluminium chloride the reaction proceeds with the electrophilic attack of the metal halide on halogen followed by the β-elimination of $SiCl_3^+$ from carbocation to give alkene (Scheme 4.5).

The debromination of *vic*-dibromides to alkenes is of some importance in organic synthesis. For example, bromination and debromination of alkenes is a standard method of protecting carbon–carbon double bond.

Debromination of 1,2-dihalo compounds to alkenes can be effected by using a metal such as Na, Sm, In, Mg or Zn in THF (tetrahydrofuran) or MeOH at reflux temperature.

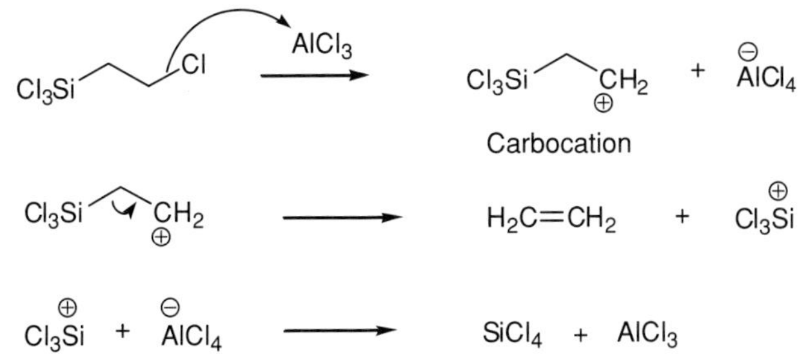

Scheme 4.5

E2 debromination of vicinal *vic*-dibromides is most favoured when the two bromine atoms are periplanar and *anti* (Scheme 4.6).

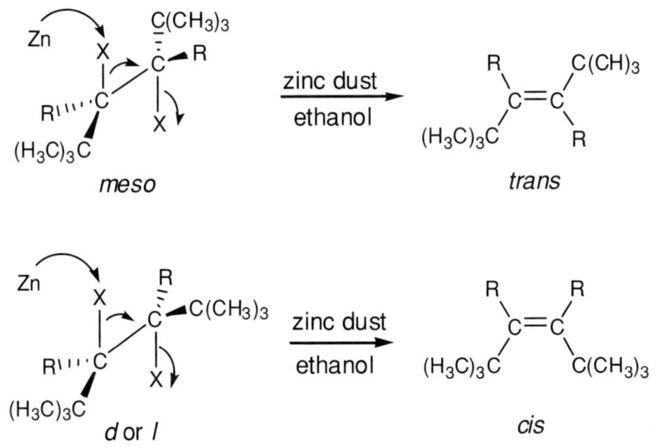

Scheme 4.6

Aryl-substituted *vic*-dibromides undergo debromination to produce the corresponding *E*-alkenes when treated with indium metal in MeOH. Since debromination occurs by the usual *trans*-elimination, *meso/erythro-* and *d,l-/threo-vic*-dibromides would give *trans*- or *cis*-alkenes, respectively, as shown in Scheme 4.6. It is thus suggested that in this case the reaction occurs via a common relatively stable radical or anion intermediate, which directly collapses to *E*-alkene.

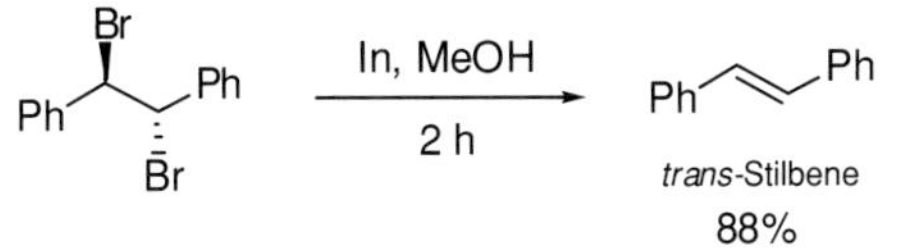

More recently organotellurides have been used for the stereoselective debromination of *vic*-dibromides[6].

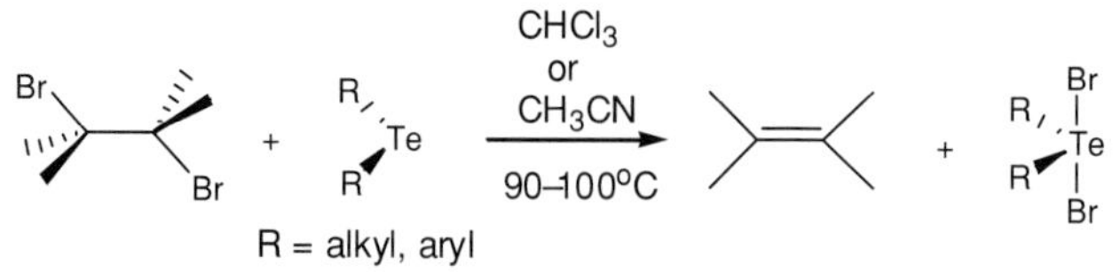

R = alkyl, aryl

Mechanism: The bromonium ion intermediate is formed by the displacement of a Br⁻ by nucleophilic attack of a neighbouring bromine atom. The R_2Te acts as a scavenger of the Br⁺ and alkene is formed as shown in Scheme 4.7.

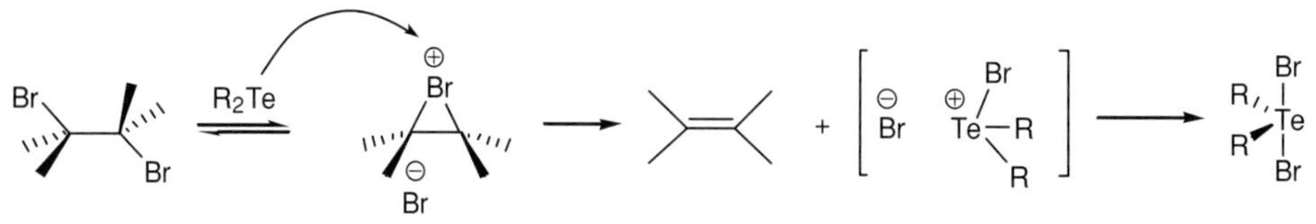

Scheme 4.7

The debromination is stereoselective; thus, *erythro*-dibromides would yield *trans*-alkenes whereas *threo*-dibromides would yield *cis*-alkenes. The *erythro*-dibromides are also more reactive than *threo*-dibromides. This is due to the greater stability of intermediate bromonium ion **A** formed from *erythro*-dibromides than of **B** formed from *threo*-dibromides. The mechanism given in Scheme 4.8 explains the formation of *trans*-alkenes from *erythro*-1,2-dibromo-1,2-diphenylethane; however, *threo*-1,2-dibromo-1,2-diphenylethane yields a mixture of *cis*- and *trans*-alkenes. This is due to the steric repulsion between the eclipsing phenyl groups in intermediate **B** formed from the *threo*-isomer. The ring opening followed by free rotation leads to the formation of **A**. Thus, from the *threo*-isomer, both *cis*- and *trans*-alkenes are formed.

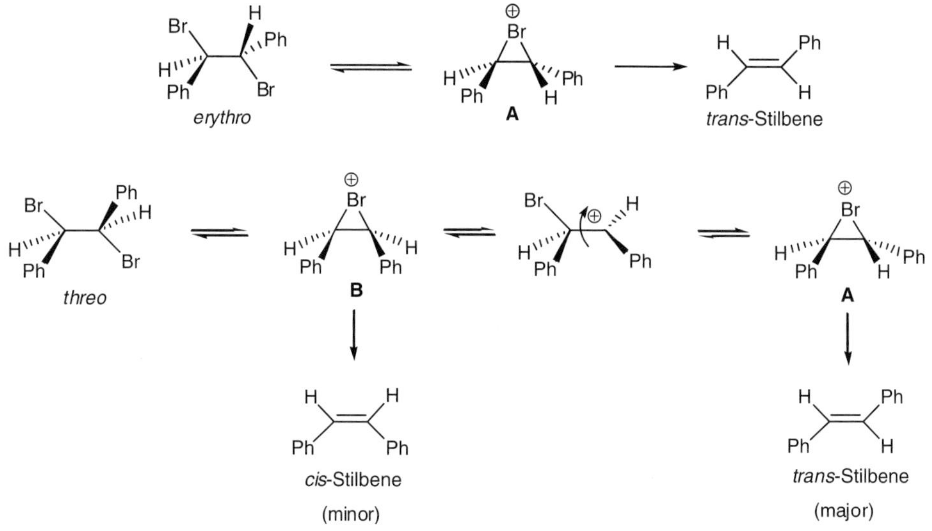

Scheme 4.8

Debromination has also been achieved in the presence of an **ionic liquid** as catalyst and solvent. 1-Methyl-3-pentylimidazolium fluoroborate, [pmIm]BF$_4$, an ionic liquid, is a suitable solvent for the stereoselective debromination under microwave irradiation[7].

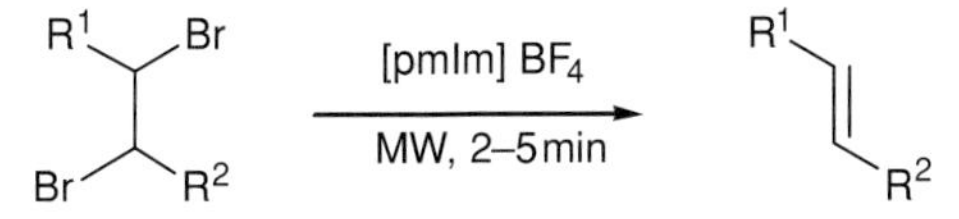

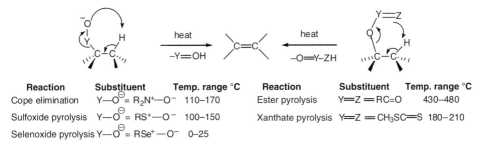

R^1, R^2 = alkyl, aryl, CN, CO$_2$Me, CO$_2$Et, COPh, NO$_2$

A plausible mechanism[7] for the stereoselective debromination using this ionic liquid is shown in Scheme 4.9.

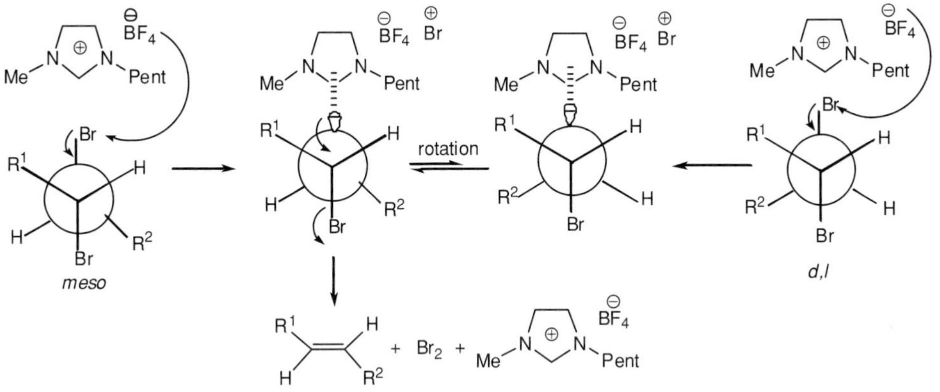

Scheme 4.9 Mechanism of debromination using the ionic liquid [pmIm]BF$_4$

4.2.2 Unimolecular syn-eliminations

E2 elimination reactions occur preferentially when the leaving groups are in an *anti* coplanar arrangement in the transition state. However, there are a few thermal, unimolecular *syn*-eliminations that produce alkenes. For example, pyrolysis of several closely related amine oxides, sulfoxides, selenoxides, acetates, benzoates, carbonates, carbamates and thiocarbamates gives alkenes on heating (Scheme 4.10). The *syn* character of these eliminations is enforced by a five- or six-membered cyclic transition states by which they take place.

Reaction	Substituent	Temp. range °C	Reaction	Substituent	Temp. range °C
Cope elimination	Y—O$^{\ominus}$ = R$_2$N$^+$—O$^-$	110–170	Ester pyrolysis	Y=Z = RC=O	430–480
Sulfoxide pyrolysis	Y—O$^{\ominus}$ = RS$^+$—O$^-$	100–150	Xanthate pyrolysis	Y=Z = CH$_3$SC=S	180–210
Selenoxide pyrolysis	Y—O$^{\ominus}$ = RSe$^+$—O$^-$	0–25			

Scheme 4.10 Pyrolytic *syn*-eliminations

Cope elimination,[8,9] i.e. elimination of amine oxide that proceeds well at temperatures near or slightly above 100°C (Scheme 4.11), is a modification of **Hofmann elimination**[10].

Amine oxides can be prepared by the oxidation of the corresponding amine with an oxidant such as H_2O_2, Ag_2O or *m*-CPBA (*meta*-chloroperoxybenzoic acid).

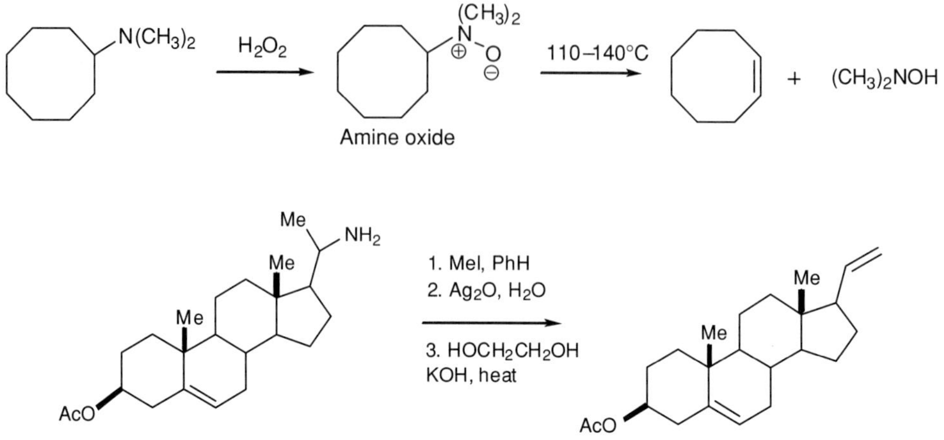

Scheme 4.11

Sulfoxides, prepared by the oxidation of sulfides with $NaIO_4$ or peracid, undergo elimination to give the alkenes at temperatures similar to the amine oxides (Scheme 4.12).

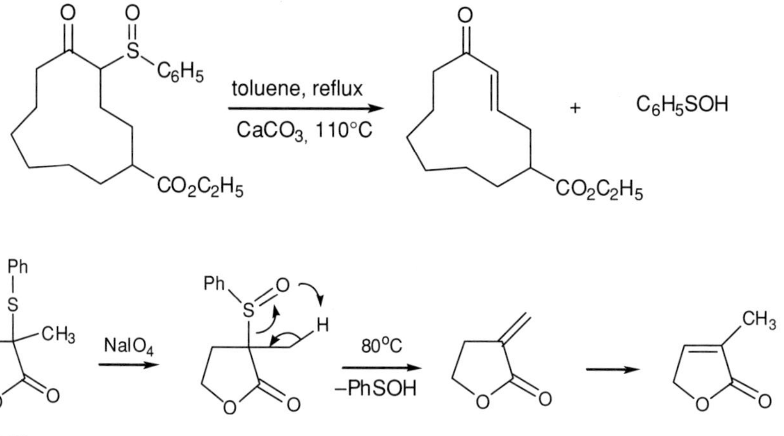

Scheme 4.12

Selenoxides undergo elimination at temperatures even lower than room temperature. The **Grieco elimination**[11] is an organic reaction describing the elimination reaction of an aliphatic primary alcohol through a selenide to a terminal alkene.

Reaction of alcohol **4.12** with *o*-nitrophenyl selenocyanate (**4.13**) and tributylphosphine gives **4.14**. Oxidation of **4.14** with hydrogen peroxide gives selenoxide **4.15**. Aryl selenoxide **4.15** bearing a β-hydrogen atom is unstable and undergoes thermal *syn*-elimination to give styrene (**4.16**) with the expulsion of a selenol **4.17** in a fashion similar to that of the Cope elimination (Scheme 4.13).

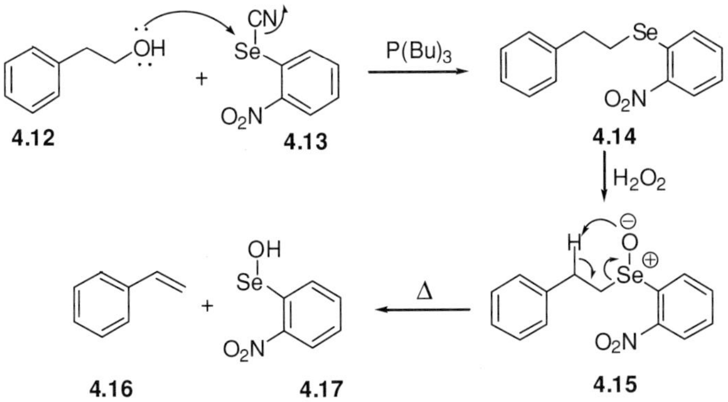

Scheme 4.13

Another example of alkene synthesis by the pyrolysis of selenoxide is given in Scheme 4.14. The enolate derived from **4.18** reacts with either PhSeBr or PhSeSePh to form selenide **4.19**. Oxidation of **4.19** gives selenoxide **4.20**, which undergoes *syn*-elimination to give α,β-unsaturated carbonyl compound **4.21**.

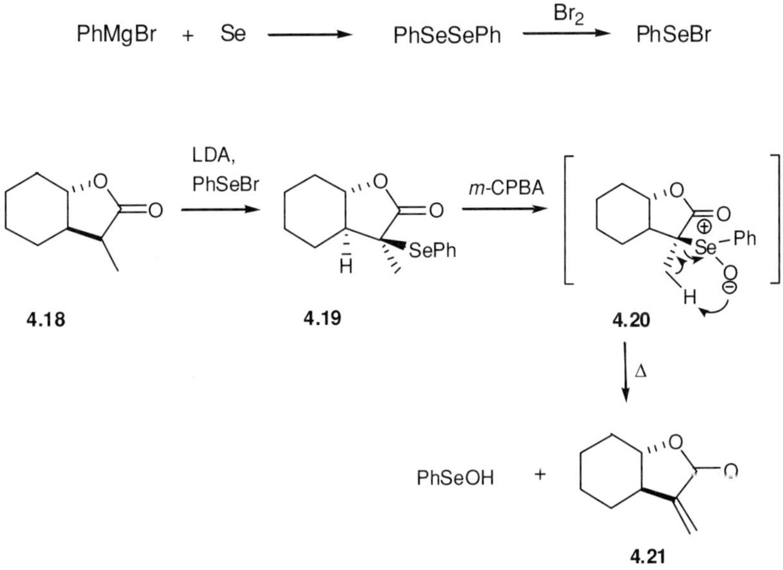

$$PhMgBr \ + \ Se \ \longrightarrow \ PhSeSePh \ \xrightarrow{Br_2} \ PhSeBr$$

Scheme 4.14

Esters and thioesters of alcohols require higher temperatures for elimination (Schemes 4.15 and 4.16). This is expected because of the stronger C–O bond and the lower polarity of C=Z bond.

The thioester function of xanthate derivatives of alcohols undergoes elimination at much lower temperatures than do carboxylic esters, reflecting the favourable bond energy change in going from O–C=S in the xanthate to S–C=O. Thus, xanthate is a better leaving group than carboxylate. Pyrolysis of xanthate esters to alkenes is known as the **Chugaev (or Tschugaev) reaction**[12,13] (Scheme 4.16).

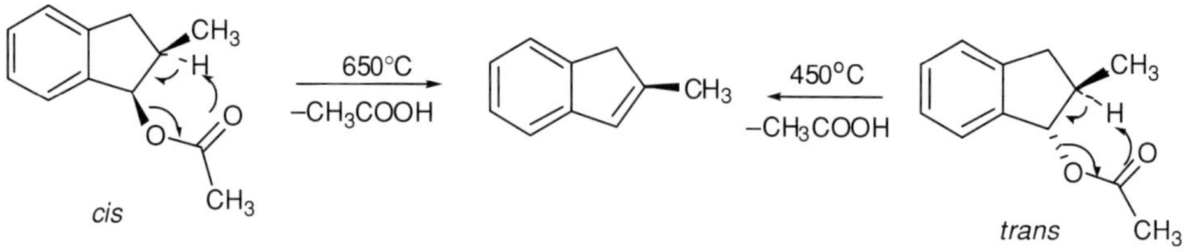

Scheme 4.15

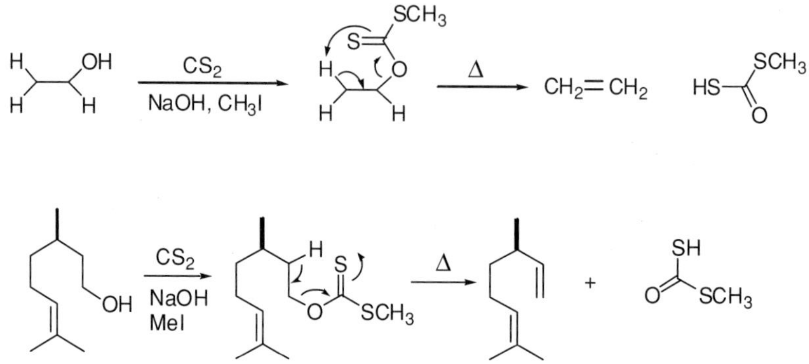

Scheme 4.16

The potassium xanthate salts of some tertiary alcohols on pyrolysis yield alkenes[14]. The overall yield in this case is often markedly better than those obtained from the corresponding S-methyl xanthates (the classical Chugaev procedure).

4.2.3   Reactions from epoxides, thionocarbonates and episulfides

The affinity of trivalent phosphorus for oxygen (and sulfur) has been put to use for the deoxygenation of epoxides and desulfurization of episulfides, which leads to alkenes.

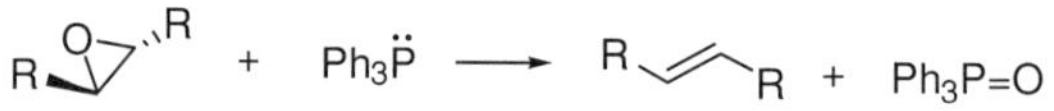

Corey and Winter[15] converted 1,2-diol into a cyclic thionocarbonate **4.23** on heating with thiocarbonyldiimidazole (**4.22**) in toluene or xylene. Thionocarbonates **4.23** can also be prepared by the reaction of diol with n-butyllithium, followed by the reaction with carbon disulfide and methyl iodide. Desulfurization–decarboxylation is carried out by heating the cyclic thionocarbonate **4.23** with trimethylphosphite, and alkene is produced (Scheme 4.17).

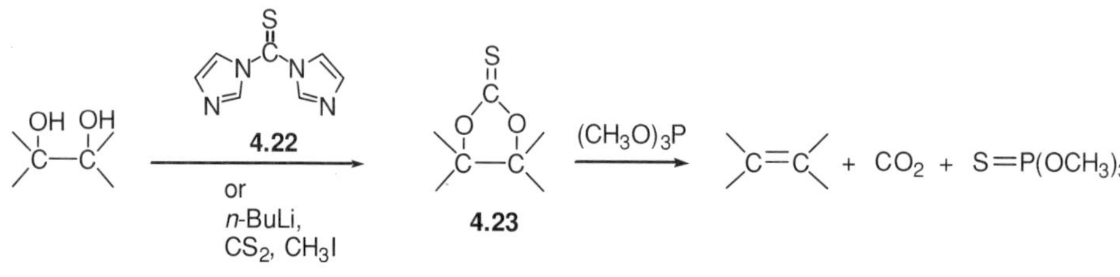

Scheme 4.17

Episulfides, generated from diazo compounds and a thioketone, can also undergo desulfurization with phosphines to give alkenes (see section 4.3.7).

4.3 Alkenation (alkylidenation) of carbonyl compounds

Alkenation of carbonyl compounds aldehydes, ketones, acids, amides, esters and lactones is one of the most important processes for the construction of organic molecules. In these reactions two smaller carbon units can be conjoined to make alkene with unambiguous positioning of the double bond. To understand the stereoselectivity of each of these processes, it is necessary to understand the mechanism of these reactions.

A variety of reagents are used for the construction of carbon–carbon double bonds from carbonyl compounds. For example, ylides such as **Wittig**-type reagents or carbanions stabilized by an α-heteroatom react with carbonyls to give the corresponding alkenes.

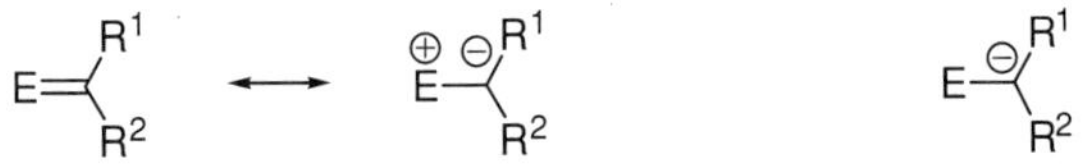

Ylides

E = PR$_3$ (**Wittig reagent**)

E = Heteroatom S or Si
(Carbonion stabilized by α-heteroatom)

The reaction can proceed by way of a four-membered intermediate, and in a final step removal of the oxygen and heteroatom (E) converts the carbonyl group into the carbon–carbon double bond (Scheme 4.18).

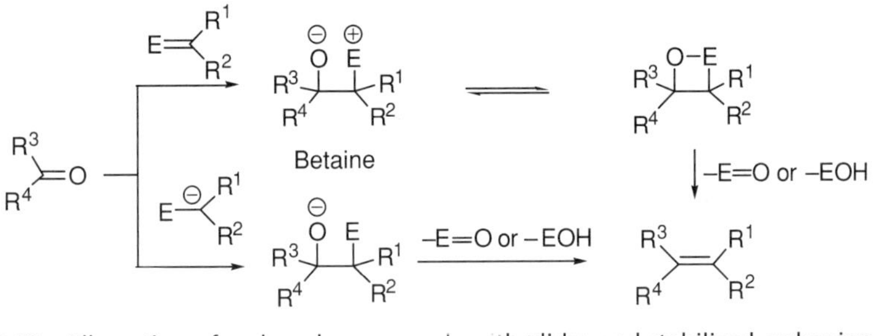

Scheme 4.18 Alkenation of carbonyl compounds with ylides and stabilized carbanion by an α-heteroatom

Geminal dimetallic reagents ($L_nM^2CH_2M^1L_n$) can also add to a carbonyl carbon to give a β-oxymetal-substituted organometallic compound which undergoes elimination to give an alkene (Scheme 4.19). Metal carbene complexes (metallocarbenes, $L_nM=CH_2$) also react with carbonyl compounds to give alkenes, as shown in Scheme 4.20. The condensation reactions of active methylene compounds with carbonyl compounds are also important methods of alkenation of the carbonyl group, which are discussed in Chapter 3, section 3.1.2.

The following methods for the alkenation of carbonyl compounds are discussed in this chapter: Wittig reaction using phosphorus ylides, Peterson reaction using organosilicon reagents, Julia alkenation using organosulfur reagents, use of titanium-based reagents

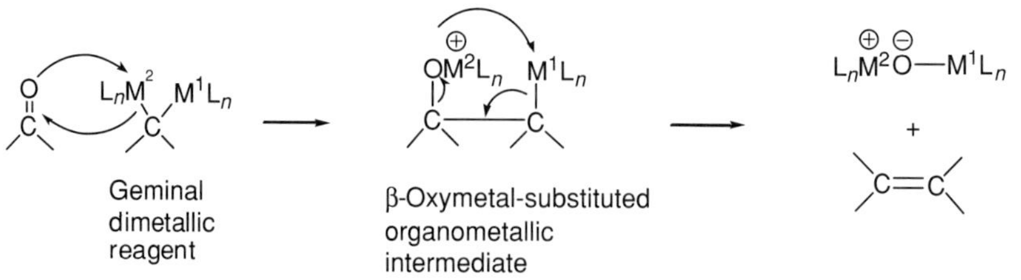

Scheme 4.19

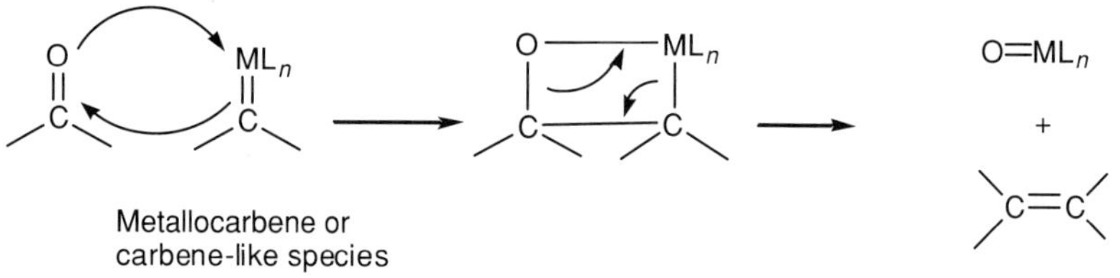

Scheme 4.20

(McMurry coupling, Tebbe reagent, Petasis reagent and Takeda reagent) and catalytic alkenation of aldehydes and ketones plus the related reactions in each case.

4.3.1 Wittig reactions

The reaction of Wittig reagent (phosphonium ylide form or phosphorane form) with an aldehyde or ketone to yield an alkene is known as the **Wittig reaction**[16,17] (Scheme 4.21). The Wittig reaction has become a popular method for alkene synthesis precisely because of its simplicity, effectiveness and generality, mostly because of the specificity of the location of the double bond. Unlike elimination reactions (such as dehydrohalogenation of alkyl halides and dehydration of alcohols), which produce a mixture of alkene regioisomers (more substituted and less substituted structural isomers), in the Wittig reaction there is no ambiguity regarding the site of the double bond.

The Wittig reagent is prepared by the treatment of phosphonium salt with a strong base (such as PhLi, n-BuLi or lithium diisopropylamide). The phosphonium salts are prepared by the reaction of phosphines with alkyl halides.

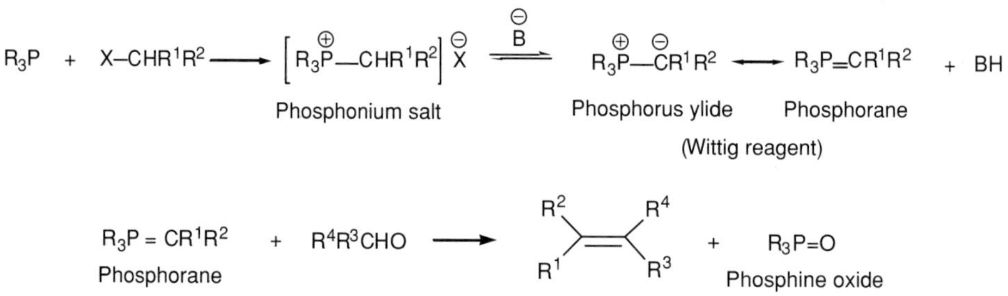

Scheme 4.21

Phosphoranes are reactive towards all types of aldehydes and ketones (aldehydes are much reactive than ketones). Generally, no reaction occurs with nitriles, esters, amides, nitro, amino, hydroxy, halogen and alkoxy groups.

The intramolecular Wittig reaction is very useful for the synthesis of cycloalkenes and heterocyclic rings. Thus, optically active phosphorane **4.24** on heating in toluene gave *p*-nitrobenzyl 2-methyl-(5*R*)-penem-3-carboxylate (**4.25**) in 89% yield[18].

4.24   **4.25**

High selectivity for *Z*- or *E*-alkenes is obtained depending on the particular circumstances, such as the type of ylide, type of carbonyl compound or reaction conditions. The **non-stabilized ylides** (R^2, $R^1 = H$, alkyl) are very reactive and provide the thermodynamically less stable *Z*-alkenes. On the other hand, **stabilized ylides**, in which the electron-withdrawing group (such as CN, CHO, COR or COOR) is present in the α-position of the ylides, are more stable and less reactive.

Non-stabilized ylides: $R^1 \neq R^2 = H$ or alkyl
Stabilized ylides: $R^1 \neq R^2 = CN, CHO, COR, COOR$

Stabilized ylide

Stabilized reagents give thermodynamically more stable *E*-isomers preferentially. For example, amino aldehyde **4.26** was alkenated with the phosphorane **4.27** to afford mainly the *E*-enoate **4.28** in 95% yield[19].

4.26   **4.27**   **4.28**
                                                         95% *E*

Mechanism: In early papers, Wittig described that the reaction proceeds with the formation of betaine, which collapsed to four-membered cyclic oxaphosphetane. Either of these two intermediates decomposes to form an alkene. The decomposition of oxaphosphetane to alkene is a stereospecific process; i.e. *cis*-oxaphosphetane gives *Z*-alkene whereas *trans*-oxaphosphetane gives *E*-alkene. For example, addition of non-stabilized ylide **4.29** to 2-butanone gives a mixture of betaines **A** and **B**, which forms the corresponding oxaphosphetanes **C** and **D**. The *anti*-oxaphosphetane **C** undergoes *syn*-elimination to form the *E*-alkene **4.30**, whereas the *syn*-oxaphosphetane leads to the *Z*-alkene **4.31** (Scheme 4.22). Under salt-free conditions, the reaction of non-stabilized ylide appears to be under kinetic control, and the formation of *erythro*-betaine **B** is irreversible, which gives the *Z*-alkene **4.31**.

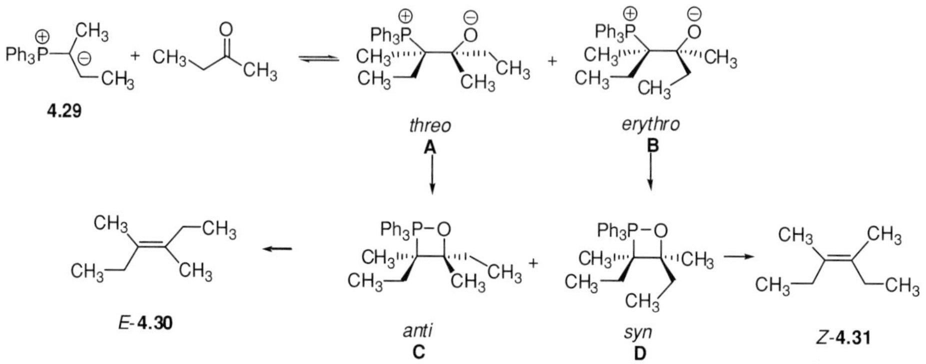

Scheme 4.22   Mechanism of the Wittig reaction of non-stabilized ylides under salt-free conditions

However, in the case of stabilized ylides the formation of **E** and **F** is reversible and equilibrium favours formation of the thermodynamically more stable *threo*-betaine **E**, which leads to the *E*-alkene (Scheme 4.23).

Currently accepted mechanism of the Wittig reaction of aldehydes with non-stabilized ylides involves the formation of oxaphosphetanes through a [2+2]-cycloaddition-like reaction[20]. The oxaphosphetanes are thermally unstable and collapse to alkene and phosphine oxide below room temperature. Under salt-free conditions there is no formation of betaine intermediates. The salt-free ylides can be prepared by the reaction of phosphines with carbenes generated *in situ*. Vedejs *et al.*[20] proposed a puckered 4-centre cyclic transition state **I** for *syn*-oxaphosphetane and planar structure **J** for *anti*-oxaphosphetane. In general, the *anti*-oxaphosphetane **J** is more stable than the *syn*-oxaphosphetane **I**, and under equilibrium conditions (when stabilized ylides are used) the *E*-alkene product is favoured (Scheme 4.24). However, kinetic control conditions, which appear to dominate when non-stabilized ylides are used, would lead to *Z*-alkene.

A variety of factors[21] affect the stereoselectivity of the cycloaddition and the degree of equilibration. These include cation, anion, solvent, temperature, concentration, carbonyl compound and substituents on the phosphine.

Schlosser modification[22] **of Wittig reaction:** The presence of soluble metal salts such as lithium salts decreases the *cis/trans*-selectivity. The normal Wittig reaction of non-stabilized ylides with aldehydes gives *Z*-alkenes. The **Schlosser modification** of the Wittig reaction of non-stabilized ylides furnishes *E*-alkenes. In the presence of lithium halides oxaphosphetanes can often be observed, but betaine–lithium halide adducts are also formed. If lithium salts are added to the equilibrium, oxaphosphetane formation and elimination of

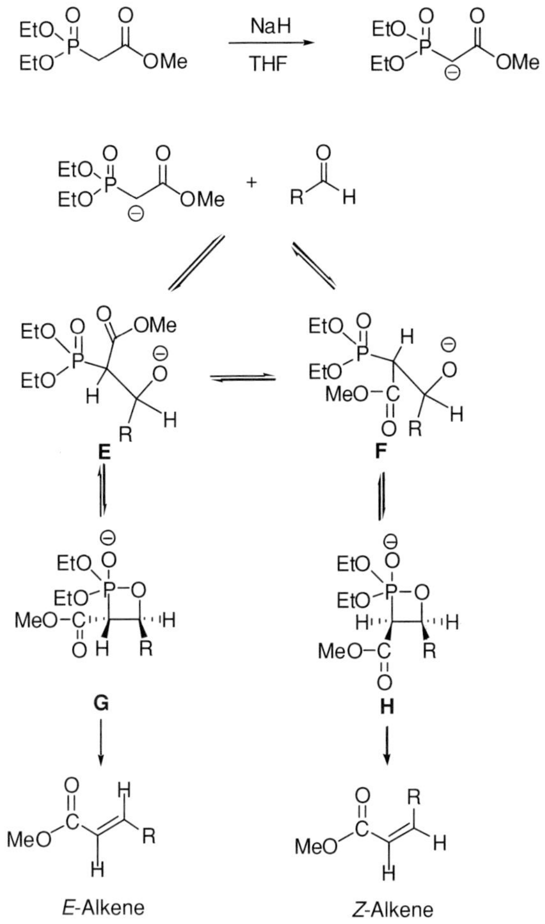

Scheme 4.23

the kinetic *erythro*-betaine are inhibited. This equilibrium favours the *trans*-product, which collapsed to the *anti*-alkene (Scheme 4.25).

Horner–Wittig modification: Alternatively, phosphine oxide[23] reacts with aldehydes in the presence of a base (sodium amide, sodium hydride or potassium *t*-butoxide) to give an alkene. The phosphine oxide can be prepared by the thermal decomposition of alkyl-triphenylphosphonium hydroxide. Deprotonation of phosphine oxide with a base followed by addition to aldehyde yields salt of β-hydroxy phosphineoxide, which undergoes further *syn*-elimination of the anion $Ph_2PO_2^-$. The lithium salt of β-hydroxy phosphineoxide can be isolated, but Na and K salt of β-hydroxy phosphine oxide undergoes *in situ* elimination to give alkene (Scheme 4.26).

The elimination step is stereospecific; thus, *erythro*-β-hydroxy phosphineoxide, obtained by the reaction of benzaldehyde with phosphine oxide, gives Z-isomer and *threo*-isomer gives E-isomer (Scheme 4.27).

Horner–Wadsworth–Emmons (HWE) reaction: There is a close relationship between the original Wittig alkenation and its phosphonate ester modification which is known as

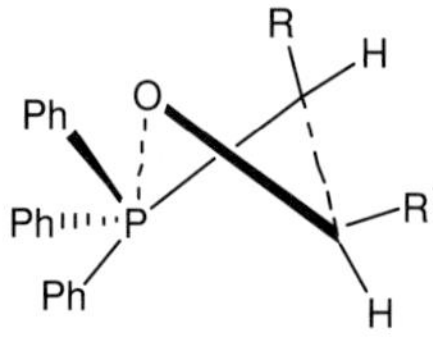

I
Puckered

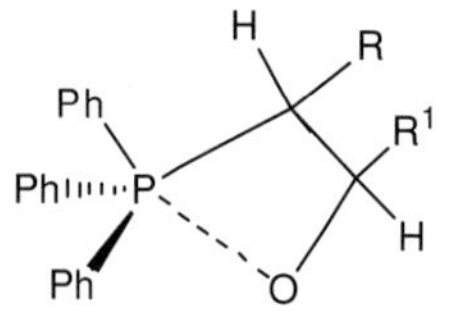

J
Planar (more stable)

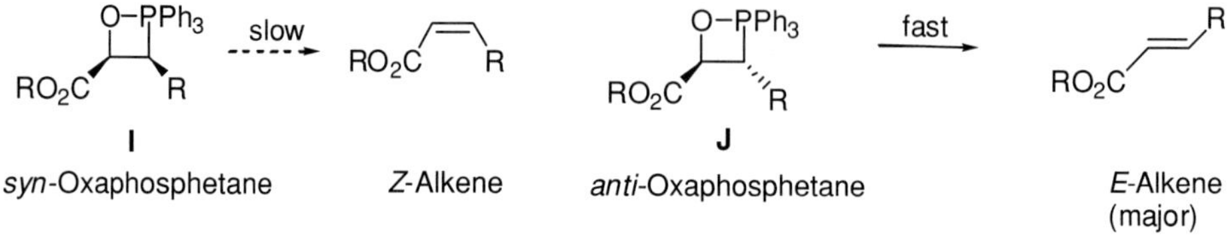

I
syn-Oxaphosphetane

Z-Alkene

J
anti-Oxaphosphetane

E-Alkene
(major)

Scheme 4.24

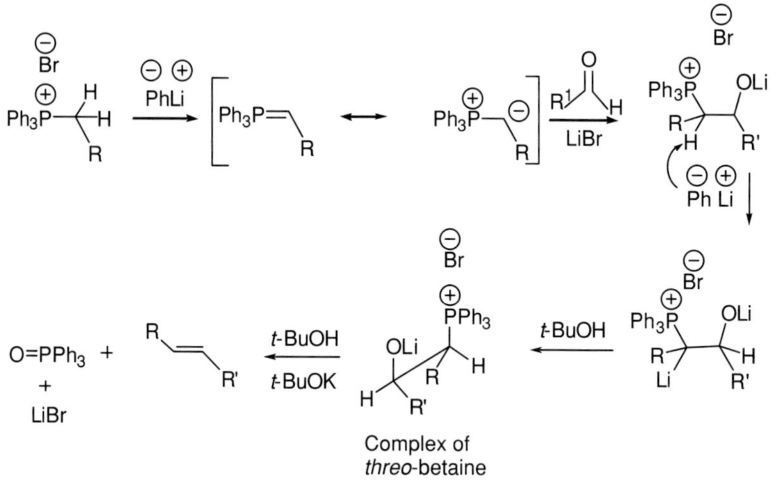

Complex of
threo-betaine

Scheme 4.25

the **Horner–Wadsworth–Emmons (HWE) reaction**[24]. In contrast to phosphonium ylides used in the Witting reaction in the HWE reaction, phosphonate-stabilized carbanion **4.32** is used.

Phosphonate esters can be prepared either by the **Arbuzov reaction (rearrangement)** or by the alkylation of esters with DMMP (dimethyl methylphosphonate) (Scheme 4.28). Variation of the reaction conditions and reagents allows one to synthesize *E*- and *Z*-alkenes with a high degree of stereoselectivity[25], as shown in Scheme 4.29. For example, 6-heptenal (**4.33**) on reaction with **4.34** in the presence of NaH gives *E*-nonadienoate (**4.35**); on the other hand, in the presence of a base *n*-BuLi, *Z*-nonadienoate (**4.36**) is the major product.

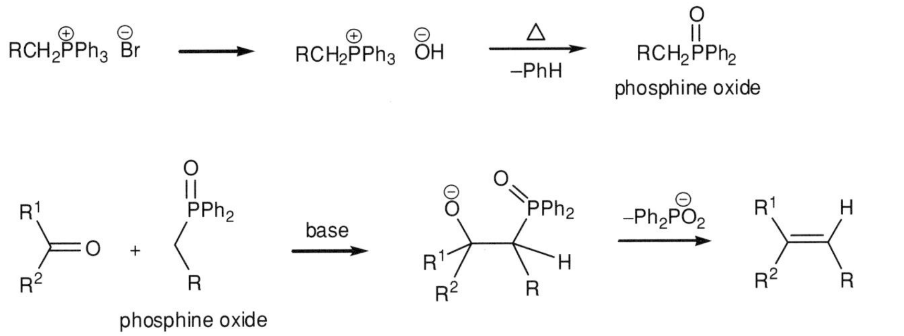

RCH$_2$$\overset{\oplus}{P}Ph_3$ $\overset{\ominus}{B}$r $\longrightarrow$ RCH$_2$$\overset{\oplus}{P}Ph_3$ $\overset{\ominus}{O}$H $\xrightarrow[-PhH]{\Delta}$ RCH$_2$PPh$_2$ (O)

phosphine oxide

R^{1}R^{2}C=O + (phosphine oxide, PPh$_2$, R) $\xrightarrow{\text{base}}$ [R^{1}R^{2}C(O$^\ominus$)–CH(PPh$_2$O)(R)(H)] $\xrightarrow{-Ph_2PO_2^{-}}$ alkene (R^1, R^2, H, R)

phosphine oxide

Scheme 4.26

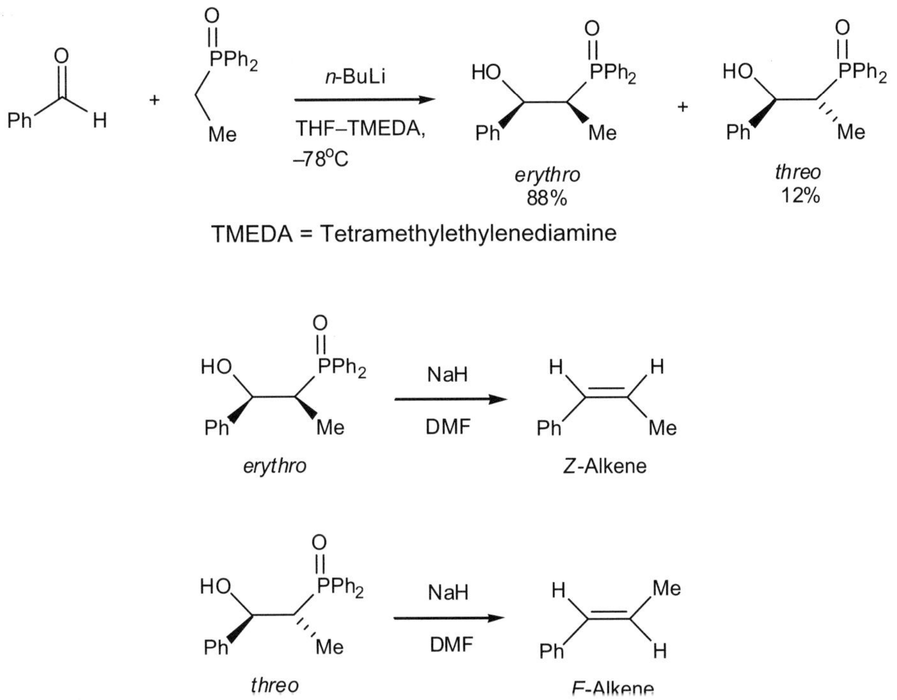

Ph–CHO + (PPh$_2$O, Me) $\xrightarrow[\text{THF–TMEDA, }-78°C]{n\text{-BuLi}}$ erythro 88% + threo 12%

TMEDA = Tetramethylethylenediamine

erythro $\xrightarrow[\text{DMF}]{\text{NaH}}$ Z-Alkene

threo $\xrightarrow[\text{DMF}]{\text{NaH}}$ E-Alkene

Scheme 4.27

Low reactivity of α,β-unsaturated ketones and sterically hindered ketones as well as separation of reaction products from phosphine oxide are a few disadvantages of the classical Wittig reaction. Phosphonate anions are stronger nucleophiles than related phosphoranes; thus, alkenate hindered ketones under mild conditions. The dialkylphosphate (R$_2$PO$_4$), the other product of the HWE reaction, can easily be removed from the reaction mixture because it is soluble in water.

The mechanism of the HWE reaction is similar to that of the classic Wittig reaction.

Still–Gennari modification[26] **of the HWE reaction:** Z-isomers are predominantly formed under conditions of kinetic control, e.g. at low temperatures, in weak polar solvents and under the action of strong bases. The use of phosphonates with an electron-withdrawing

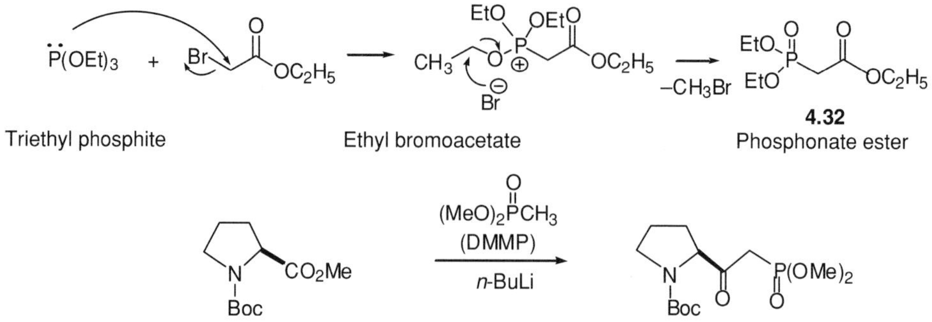

Triethyl phosphite   Ethyl bromoacetate

4.32
Phosphonate ester

Scheme 4.28

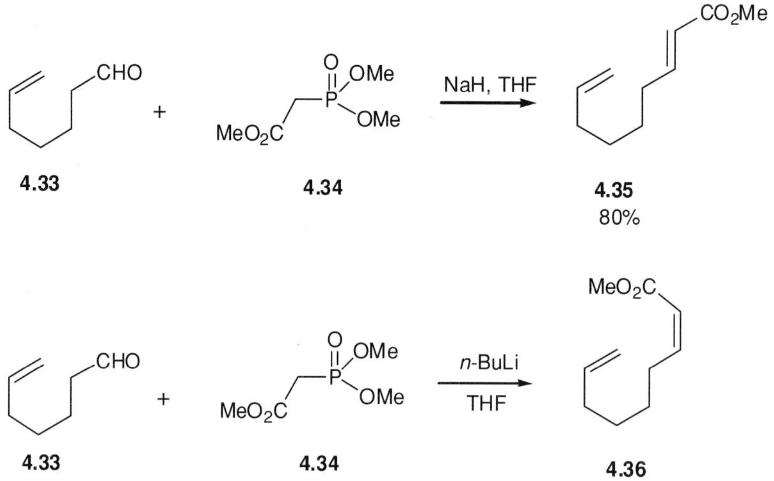

4.33   **4.34**   **4.35**
80%

4.33   **4.34**   **4.36**
39%

Scheme 4.29

group such as trifluoroethyl together with strongly dissociating conditions such as potassium hexamethyldisilazide (KHMDS) and 18-crown-6 in THF also favours the formation of alkenes having Z-configuration. For example, the reaction of amino aldehyde, Boc-D-alaninal (**4.37**), with phosphonate **4.38** in the presence of 18-crown-6 gave the corresponding Z-enoate **4.39**.

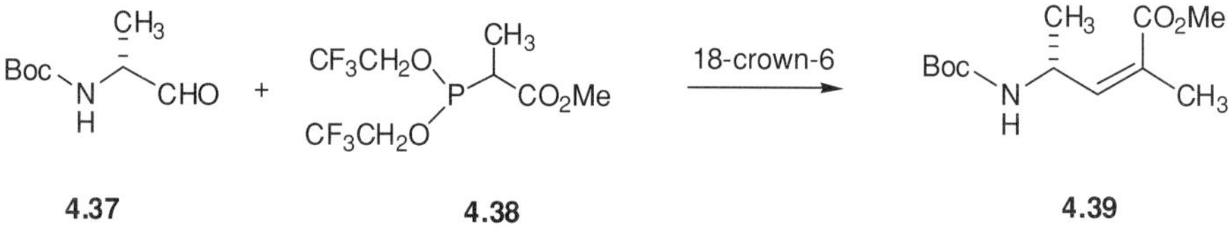

4.37   **4.38**   **4.39**
Z-Enoate
96%

Asymmetric Wittig-type reaction: The possibility of carrying out asymmetric version of the Wittig-type reaction is not readily apparent as there is no sp^3 stereocentre formed. However, asymmetric Wittig-type alkenation reactions[27] have been carried out by placing stereogenic centres adjacent to or near the reaction centre. Three types of asymmetric Wittig-type reactions have been reported: (i) by kinetic resolution of racemic carbonyl compounds, (ii) desymmetrization of ketones and (iii) preparation of optically active allenes.

The reaction of chiral phosphonium ylide or related reagent with a 4-substituted cyclohexanone gives an axially dissymmetrical alkene. For example, alkenation of 4-methylcyclohexanone (**4.40**) with chiral ylide **4.41** containing stereogenic centre on phosphorus gives optically active alkene (*S*)-(+)-**4.42** in 43% yield[28].

(*S*) -(+)-Alkene, 43% ee

The *meso*-dialdehydes **4.43** and **4.46** undergo the HWE reaction with chiral phosphonate **4.44** in the presence of the base KHMDS and 18-crown-6 followed by the treatment with NaBH$_4$ to give major amounts of monoaddition products **4.45** and **4.47**, respectively, with good diastereoselectivities and high *E*-selectivities[29]. Similarly, *meso*-diketone **4.48** reacts with chiral phosphonate **4.49** to give α,β-unsaturated ester **4.50** in 95% yield with high *E*-selectivity (98% ee)[29].

Desymmetrization of ketones also achieved by the reaction of an achiral ketone **4.51** with an achiral phosphonium ylide in the presence of chiral ligand **4.52** give the α,β-unsaturated ester **4.53** in 58% yield with good *E*-selectivity (57% ee)[30].

4.52

4.51

4.53

58% yield, 57% ee

4.3.2 Julia alkenation and modified Julia alkenation (Julia–Kocienski alkenation)

The **Julia alkenation**, also known as the **Julia–Lythgoe alkenation**[31], is a reaction of phenyl sulfones **4.54** with aldehydes or ketones followed by reductive elimination with sodium amalgam to give alkenes.

X = Ac or Bz

The Julia–Lythgoe alkenation procedure gives predominantly *E*-alkene[32], depending on the reaction conditions. For example, oxidation of sulfide **4.55** with *m*-CPBA gave sulfone **4.56**, which on treatment with a base and aldehyde **4.57** followed by reduction with sodium–mercury (Na–Hg) and disodium hydrogen phosphate in MeOH afforded the *E*-alkene **4.58** in 68% yield[32].

4.55   **4.56**   **4.58**

Mechanism: The *n*-BuLi abstracts a proton from sulfone and produces the phenyl sulfonyl carbanion **A**, which reacts with an aldehyde to form an alkoxide **B**. The alkoxide **B** is esterified *in situ* to yield the corresponding ester **C**. The sodium amalgam reduction of ester

C may proceed through a vinylic radical species to give the thermodynamically more stable *trans*-alkene (Scheme 4.30).

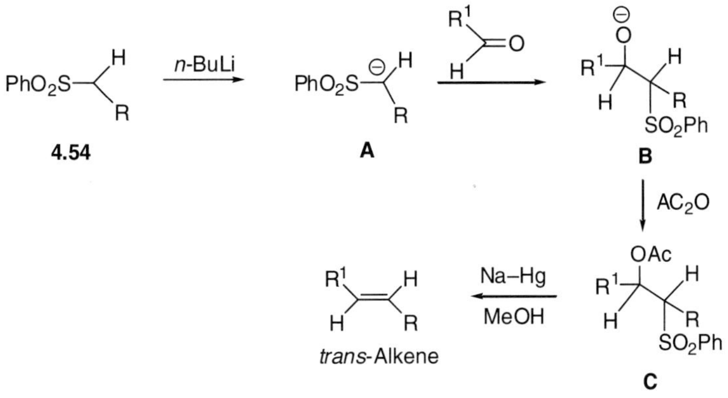

Scheme 4.30

The originally proposed mechanism of Na–Hg amalgam reductive elimination of acetoxy sulfones **C** (Julia–Lythgoe alkenation) is shown in Scheme 4.31.

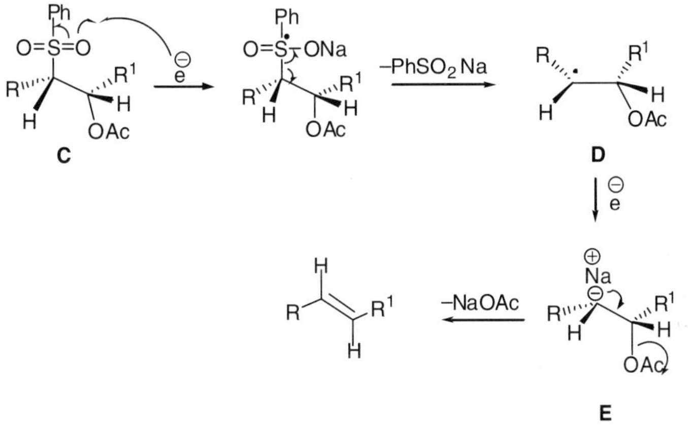

Scheme 4.31 Mechanism of Na–Hg reduction of acetoxy sulfones

Although the reaction of carbanion with aldehyde followed by esterification produces acyloxy sulfone **C** diastereomers, both the diastereomeric acyloxy sulfones **C** are converted into the same *trans*-alkene (Scheme 4.32).

This is because the mechanism of reduction proceeds through a planar radical **D** that can rotate freely about the carbon–carbon bond (Scheme 4.33). Thus, both diastereomers **C** would pass through the same radical intermediate, which can be used to explain the *E*-selectivity. Even though the carbanion **E** is not configurationally or conformationally stable, it will prefer an arrangement with the R groups further apart that will later lead to the *E*-alkene.

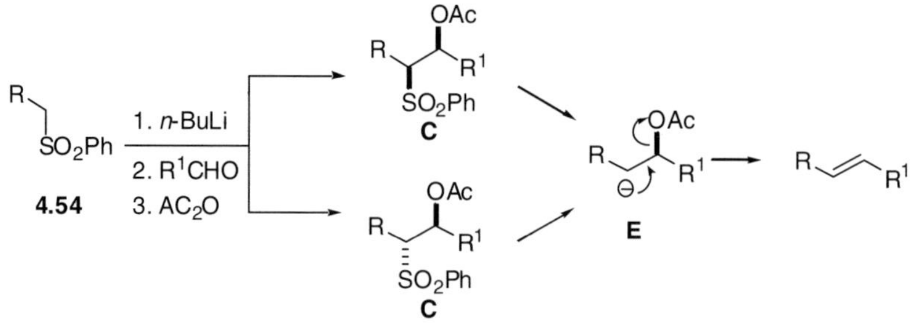

Scheme 4.32   Diastereomeric acetoxy sulfones

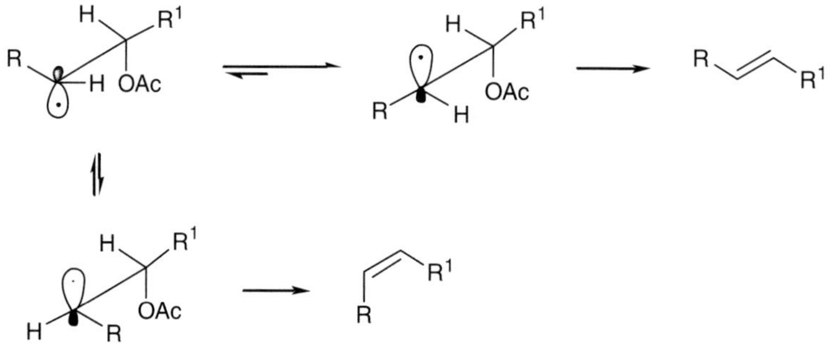

Scheme 4.33

Alternatively, the classical Julia alkenation of acetoxy sulfones **C** with sodium amalgam in MeOH might possibly proceed via an initial formation of an alkenyl sulfone **F**, which would then undergo homolytic cleavage involving single electron transfer (Scheme 4.34). This mechanism is proposed on the basis of deuterium incorporation studies performed by Keck et al.[33].

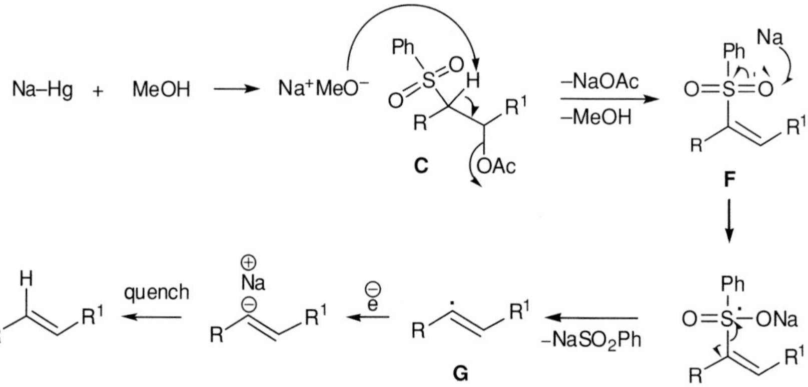

Scheme 4.34

The *cis-* and *trans*-vinyl radicals **G** can equilibrate at this stage, and the *trans*-radical is the more stable of the two. Therefore, both diastereomeric acetoxy sulfones **C** would still lead selectively to the same product (Scheme 4.35).

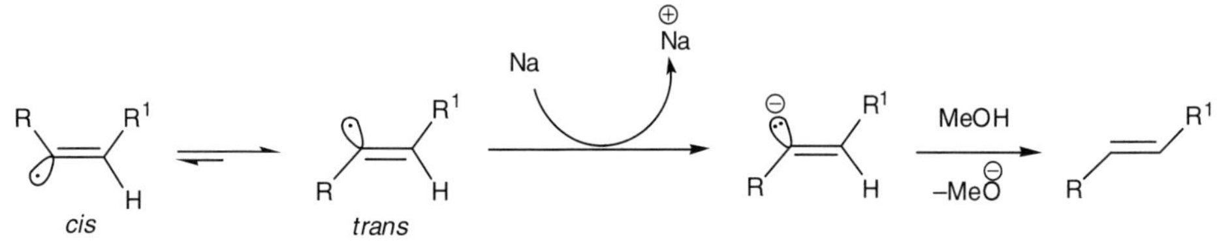

Scheme 4.35

The less toxic and more selective reducing agents such as magnesium or samarium diiodide (SmI_2) can also be used as an alternative to Na–Hg amalgam in the Julia–Lythgoe alkenation. For example, vinyl sulfone **F** prepared by the elimination of the corresponding acetoxy sulfone **C** with 1,8-diazaobicyclo[5.4.0]undec-7-ene (DBU) on reductive elimination with SmI_2 in the presence of 1,3-dimethyl-3,4,5,6-tertahydro-2(1 *H*)-pyrimidinone (DMPU) and MeOH gives high selectivity for the *E*-alkenes (Scheme 4.36).

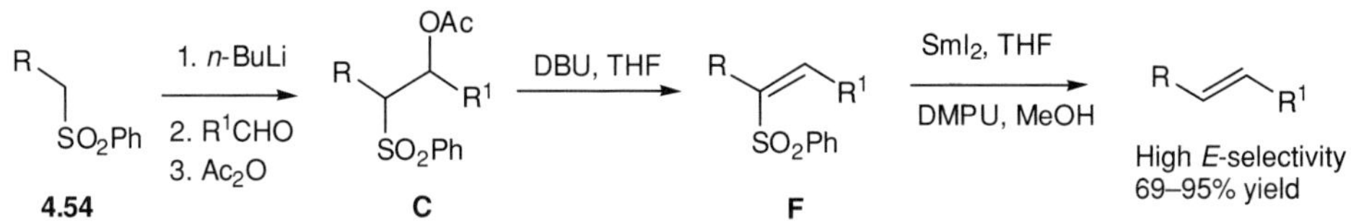

Scheme 4.36

The SmI_2-promoted mechanism follows a pathway originally proposed for the Na–Hg reductions (Scheme 4.31). The exact role of DMPU is not clear, but rate enhancement may be due to lowering of the oxidation potential of SmI_2.

A disadvantage of the Julia–Lythgoe alkenation is its low tolerance for reducible functional groups. The *E*-selectivity is generally good to very good for alkenes with a low degree of substitution, while the selectivity improves as a function of increased branching in the substitutents.

The modified Julia–Lythgoe alkenation known as **Julia–Kocienski alkenation** is one-step synthesis of alkenes from benzothiazol-2-yl sulfones (RCH_2SO_2BT) **4.59** and aldehydes, which is an alternative procedure that leads to the alkene in one step and offers very good *E*-selectivity.

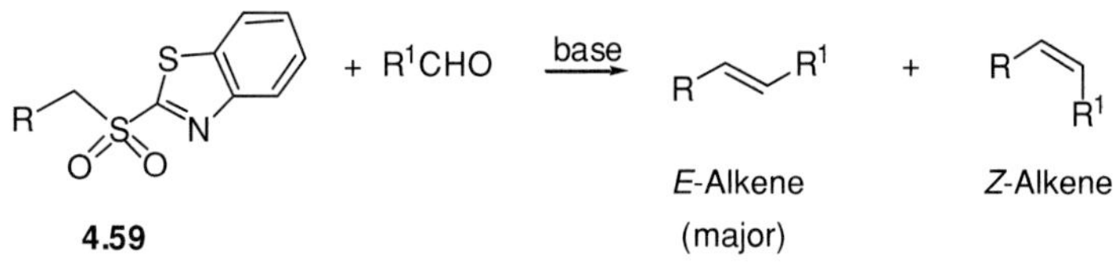

The reaction pathway greatly alters when benzothiazole sulfone **4.59** is used instead of phenyl sulfone. The initial addition of the sulfonyl anion **4.60** to the aldehyde is reversible and gives alkoxide intermediates **4.61** and **4.62** (Scheme 4.37). Whether the *anti*- or *syn*-alkoxide intermediate (**4.61** or **4.62**) is generated can be influenced to some extent by the choice of reaction conditions.

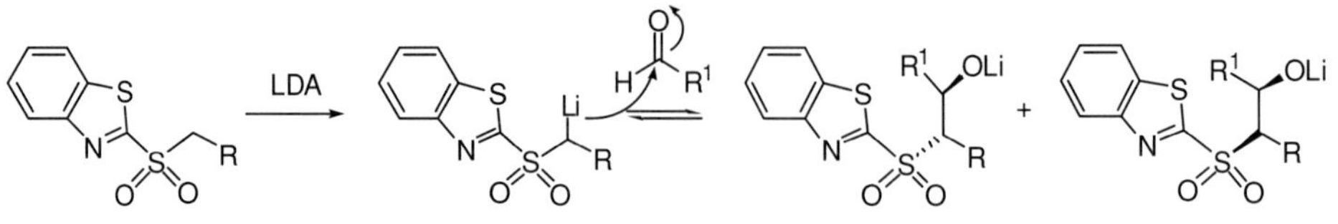

4.59   **4.60**   **4.61**   **4.62**
                                   anti   *syn*

Scheme 4.37

The alkoxide intermediate **4.61** or **4.62** formed in this case is unstable and forms first the adduct **4.63** or **4.65** and then the sulfinate salt **4.64** or **4.66**. The sulfinate salt **4.64** or **4.66** spontaneously eliminates sulfur dioxide and lithium benzothiazole to give the corresponding alkene (Scheme 4.38).

Like the benzothiazolyl (BT) group, other heterocyclic groups such as pyridyl (py), phenyltetrazolyl (PT) and *tert*-butyltetrazolyl (TBT) can also assume these roles and offer somewhat different selectivity.

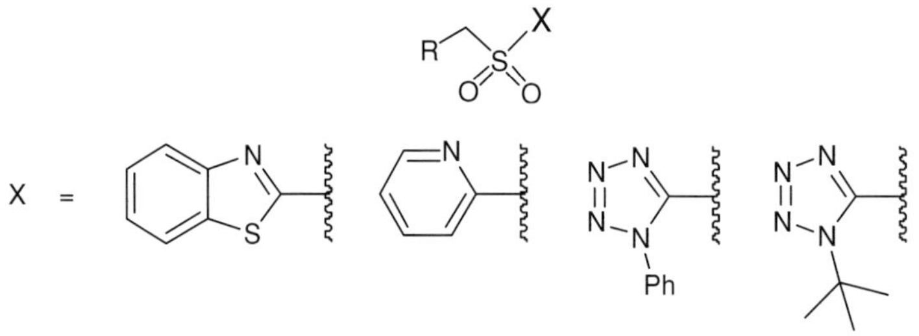

BT   Py   PT   TBT

For example, a stereoselective synthesis of *trans*-alkene **4.68** by the condensation of cyclohexanecarboxaldehyde with sulfone **4.67** containing a phenyltetrazole system was described by Kocienski and co-workers.[34].

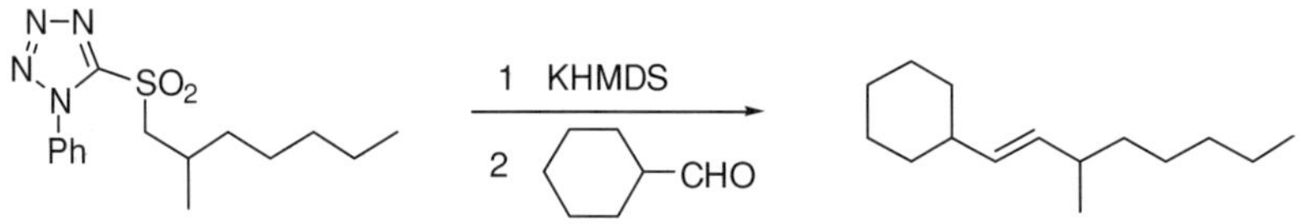

4.67   **4.68**

                                       59%, *E/Z* > 99:1

The stereoselective synthesis of alkenyl halides[35] is also possible via Julia–Lythgoe alkenation. The other two methods for the preparation of alkenyl halides – Wittig reaction (using

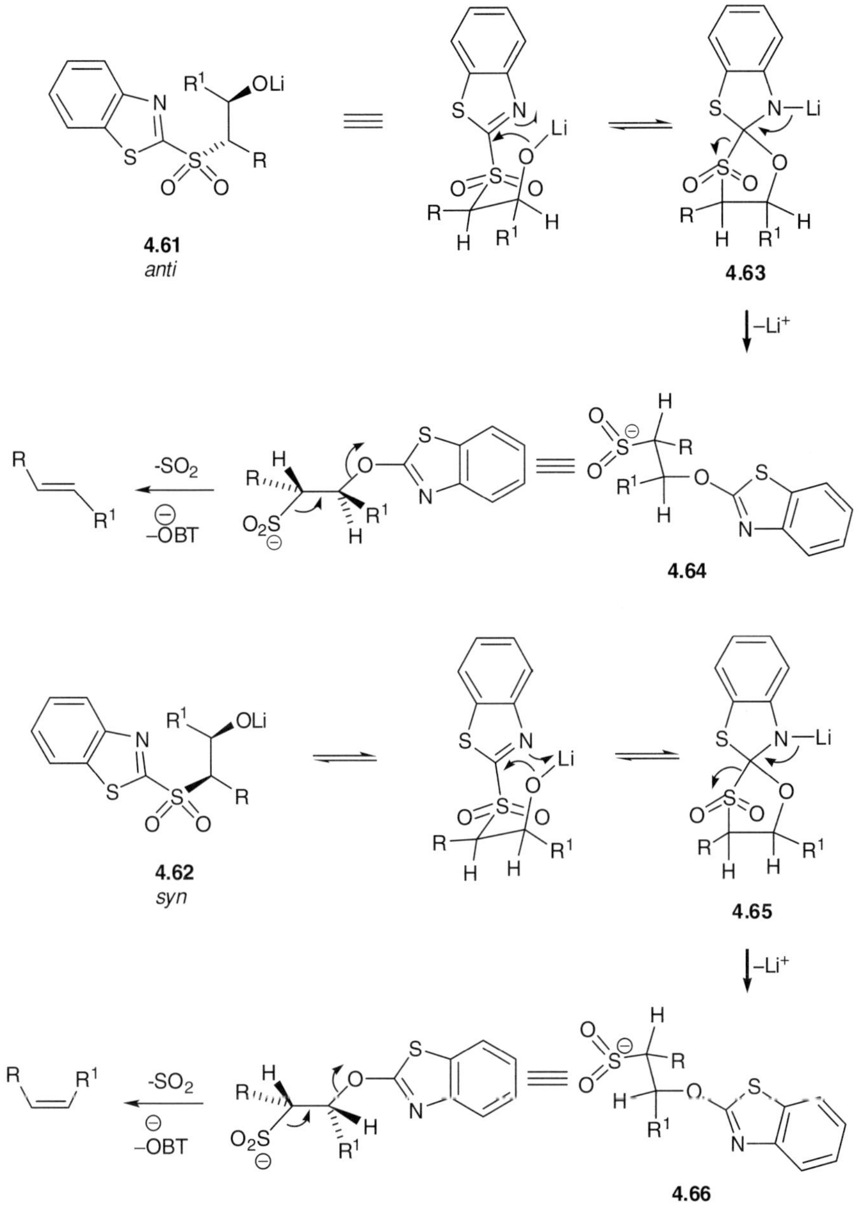

Scheme 4.38 Mechanism of Julia–Kocienski alkenation

α-halomethyltriphenylphosphorane) and CrCl$_2$ reduction of trichloroalkanes – give high *Z*-stereoselectivity. An optimized Julia alkenation between readily available α-halomethyl sulfones and a variety of aldehydes afforded alkenyl halides in good to excellent yields with high *E/Z*-stereoselectivities[35].

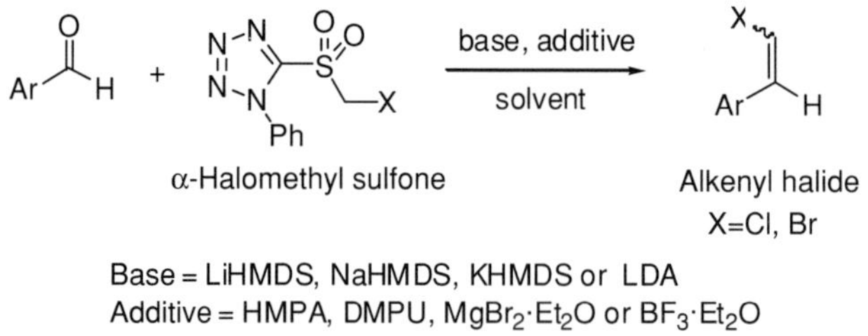

α-Halomethyl sulfone

Alkenyl halide
X=Cl, Br

Base = LiHMDS, NaHMDS, KHMDS or LDA
Additive = HMPA, DMPU, $MgBr_2 \cdot Et_2O$ or $BF_3 \cdot Et_2O$

In contrast to the classical Julia alkenation, the modified Julia alkenation offers the possibility of saving one or two steps. The *E/Z*-selectivity can be controlled by varying the sulfonyl group, solvent and base.

4.3.3 Peterson reaction

The **Peterson reaction**[36] allows the preparation of alkenes from α-silylcarbanions and carbonyl compounds (ketones and aldehydes) via the intermediate β-hydroxysilanes. Addition of the silylcarbanion to a carbonyl compound and subsequent aqueous work up produces diastereomeric β-hydroxysilanes, which may be isolated. It is possible to prepare either *cis*- or *trans*-alkenes from the same β-hydroxysilanes intermediate (Scheme 4.39).

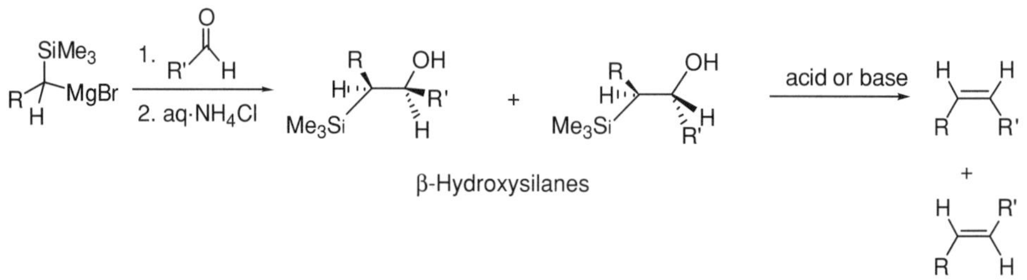

β-Hydroxysilanes

Scheme 4.39

For example, *threo*-5-trimethylsilyloctan-4-ol (**4.69**) under acidic conditions (H_2SO_4 or BF_3) undergoes *anti*-elimination to give *cis*-oct-4-ene (**4.71**) and under basic conditions (NaH or KH) undergoes *syn*-elimination to give *trans*-oct-4-ene (**4.72**). The *erythro*-5-trimethylsilyl octan-4-ol (**4.70**) gives opposite result (Scheme 4.40).

Mechanism: The Peterson alkenation offers the synthesis of desired alkene stereoisomer by careful separation of the two diastereomeric intermediate β-hydroxysilanes and subsequently performing elimination under two different conditions.

The base-catalyzed elimination requires the trialkylsilyl and hydroxyl groups to be in an antiperiplanar relationship. Under basic conditions both groups adopt a synperiplanar relationship. The action of a base on a β-hydroxysilane results in a *syn*-elimination (Scheme 4.41).

The base-catalyzed elimination may proceed via a 1,3-shift of the silyl group after deprotonation or with the formation of a pentacoordinate 1,2-oxasiletanide that subsequently undergoes cycloreversion.

Treatment of the β-hydroxysilane with acid results in protonation followed by an *anti*-elimination to form the alkene (Scheme 4.42).

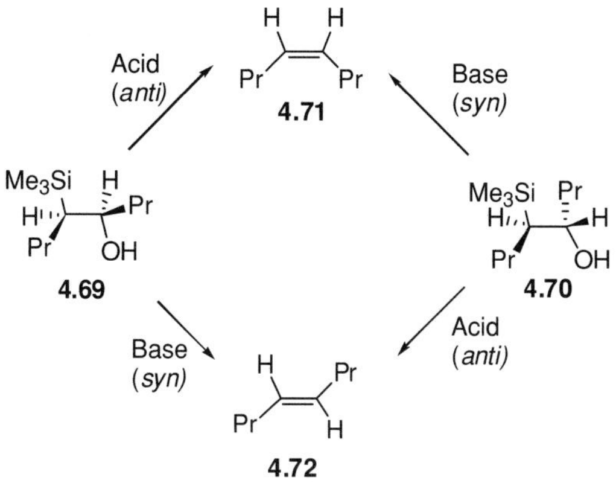

Scheme 4.40   Peterson elimination of β-hydroxysilanes

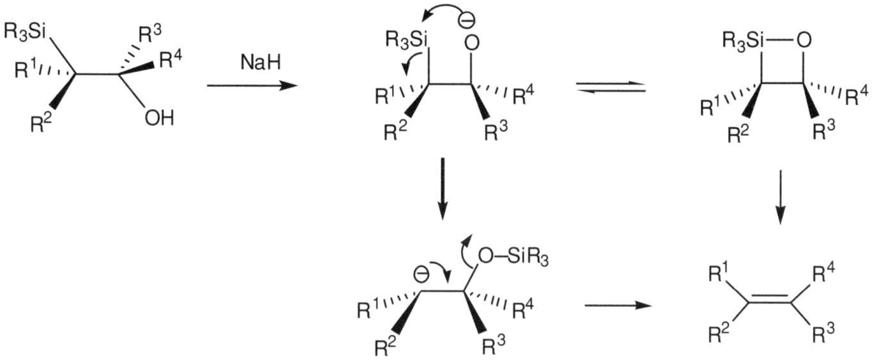

Scheme 4.41

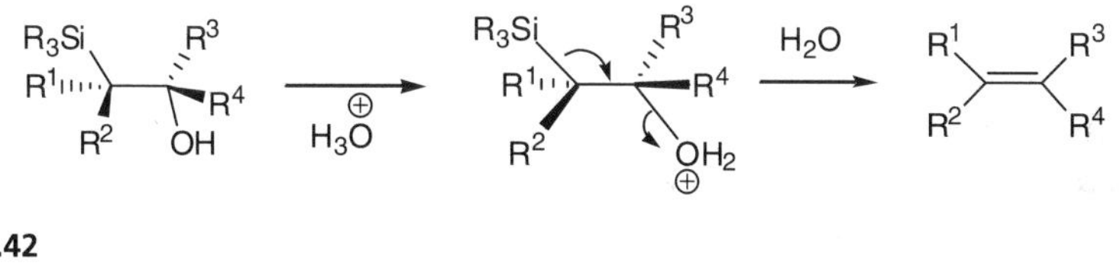

Scheme 4.42

The order of reactivity of alkoxides – K > Na ≫ Mg – is consistent with the higher electron density on oxygen, hence increasing the alkoxide nucleophilicity. When the α-silylcarbanion contains electron-withdrawing substituents, the Peterson alkenation directly forms the alkene, without the isolation of β-hydroxysilane.

Due to low diastereoselectivity in the addition step, the non-stabilized reagents are usually not selective. However, some of these reactions are **stereoselective** and may be rationalized with simple models (Scheme 4.43). For example, the reaction of benzaldehyde and a silyl-carbanion gives the *threo*-product if the silyl group is small, i.e. the trimethylsilyl group. This implies that in the transition state, the two sterically demanding groups are *anti*. As the silyl group becomes more sterically demanding than trimethylsilyl, the selectivity shifts towards the *erythro*-isomer.

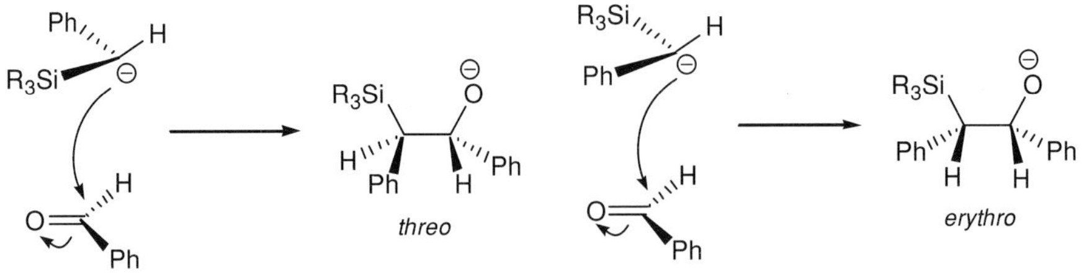

Scheme 4.43

The reaction of α-*tert*-butyldiphenylsilyl carbonyl compounds **4.73** with organometallics occurred with a high diastereoselectivity to give *erythro*-β-hydroxysilanes **4.74**, which under acidic and basic elimination conditions gave *E*- and *Z*-alkenes, respectively (Scheme 4.44)[37].

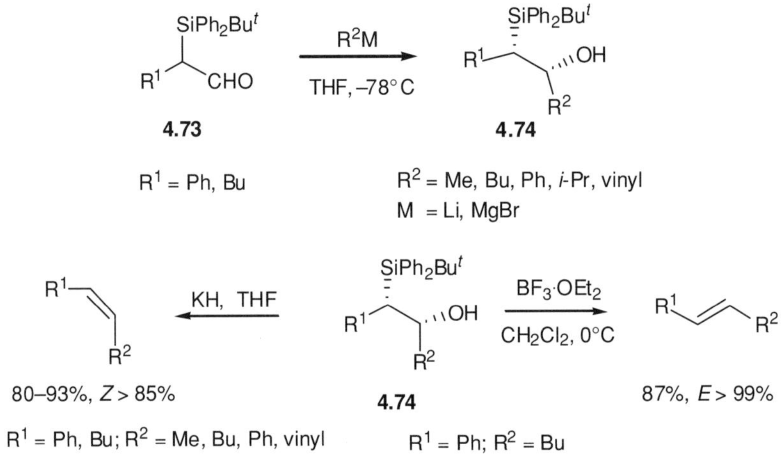

Scheme 4.44

Recently, germyl, stannyl and plumbyl Peterson reactions have also been reported which demonstrate the vibrant development of this methodology.

4.3.4 Use of titanium-based reagents

Tebbe reagent, **Grubbs reagent**, **Petasis reagent**, **Takeda reagent** and **Takai reagent** are the most common titanium reagents used for the alkenation of carbonyls. Tebbe, Grubbs and Petasis reagents react with carbonyl via the same species, titanocarbene ($Cp_2Ti=CH_2$) (**4.75**).

McMurry alkenation

McMurry[38,39] developed a reduction procedure that is used for alkenation of carbonyl compounds in the presence of low-valent titanium (LVT) reagent. The reagent (thought to be a mixture of Ti(0) and Ti(II) species) is formed by the reduction of $TiCl_4$ or $TiCl_3$ with a suitable reducing agent (Zn–Cu alloy, $LiAlH_4$ or alkali metal are the most commonly used).

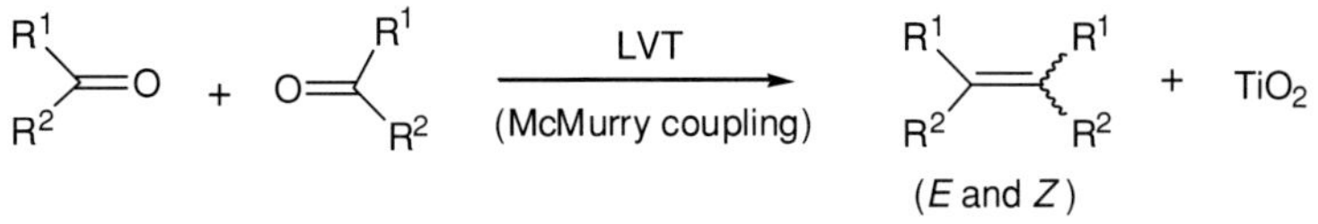

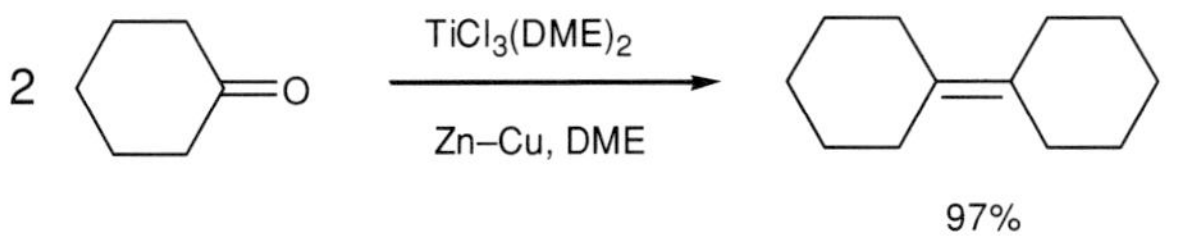

$$2 \quad \bigcirc\!\!=\!O \quad \xrightarrow[\text{Zn–Cu, DME}]{\text{TiCl}_3(\text{DME})_2} \quad \bigcirc\!\!=\!\!\bigcirc$$

97%

DME = 1,2-Dimethoxyethane

Mechanism: The mechanism of the McMurry reaction is still not well understood, but mechanistic studies have suggested that this reaction occurs on the surface of the reduced titanium metal. It is generally accepted[40] that the reaction takes place in two successive steps: (i) formation of carbon–carbon bond by reductive dimerization of the ketone or aldehyde and (ii) deoxygenation of the 1,2-diolate intermediate (pinacolate) to the corresponding alkene (Scheme 4.45). The TiO_2 is formed as a by-product. In many cases the pinacol can be isolated. The cleavage of carbon–oxygen bond is more facile in aromatic carbonyls as compared to aliphatic ones. This is due to the relatively weak benzyl–oxygen bond in comparison to the alkyl–oxygen bond.

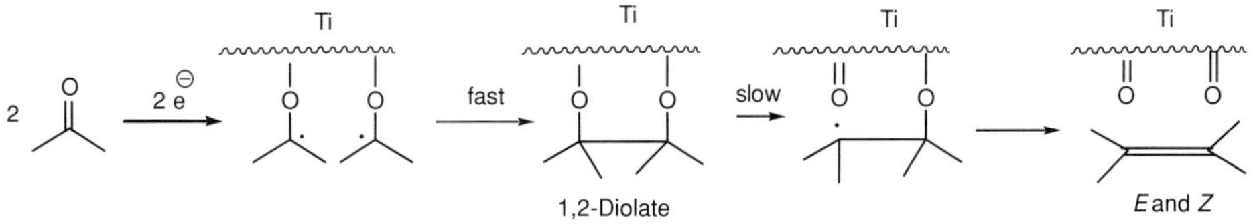

1,2-Diolate   *E* and *Z*

Scheme 4.45 Mechanism of the McMurry reductive deoxygenation of aldehydes and ketones

The dimerization of carbonyl compounds gives symmetrical alkenes[41,42]. For example, 2-adamantanone (**4.76**) and retinal (**4.78**) on reaction with $LiAlH_4$–$TiCl_3$ were converted into adamantylideneadamantane (**4.77**) and β-carotene (**4.79**), respectively.

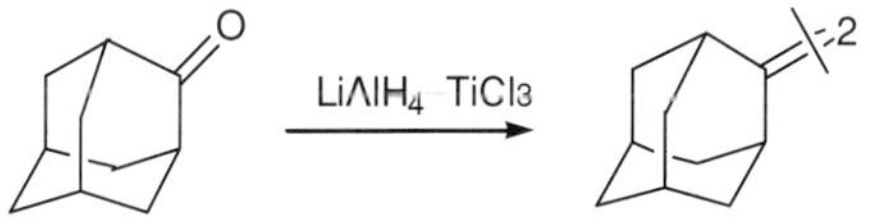

4.76   4.77

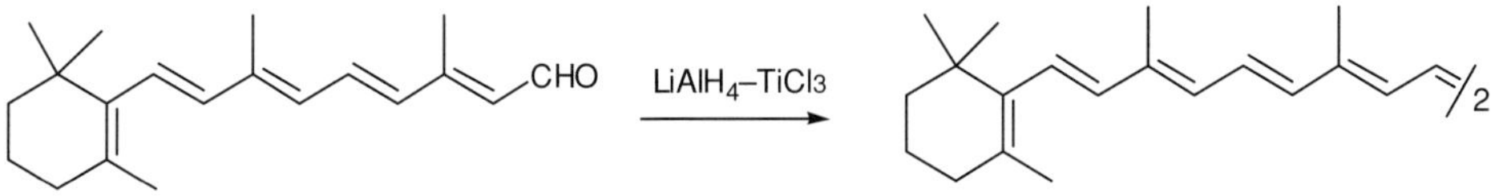

4.78   4.79

The dimerization of aromatic aldehydes is used for the synthesis of substituted stilbenes[43]. The intermolecular unsymmetrical McMurry reaction is not very important in organic syntheses because of the formation of a mixture of products. However, the reaction of two aryl ketones **4.80** and **4.81** functionalized by a sulfonyl and a hydroxyl group, respectively, gave the cross-coupling product **4.82** in a complete stereocontrolled manner (*Z*-isomer > 99%).

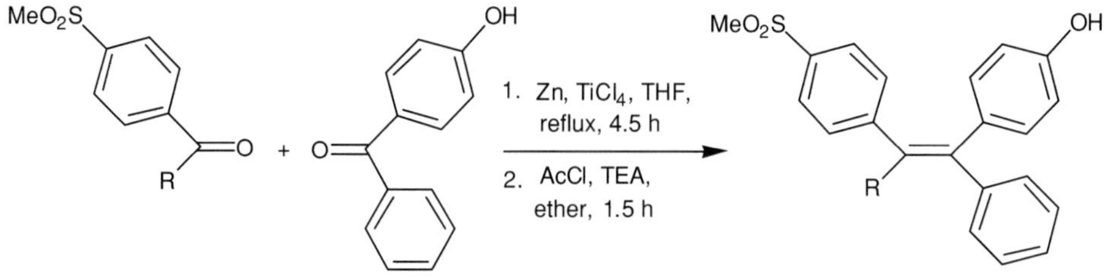

| 4.80 | 4.81 | | 4.82 |

The *Z*-isomer arises from a consecutive induction of active metallic titanium surface to the polydented pinacolic intermediate formed by homolytic coupling of a radical anion species generated from reduction of two carbonyl compounds that is followed by subsequent demetallation and deoxygenation reactions. In this regard, the phenoxy-Ti-sulfone induction plays the key role for *Z*-stereoselection by forcing the phenoxy and sulfone moieties to be positioned on the same side.

Stuhr-Hansen[44] has reported on a microwave-induced high yielding synthesis of alkenes by McMurry coupling of aldehydes and ketones. With microwave heating, the corresponding alkenes have been produced much faster with higher yields (>80%) as compared to conventional heating.

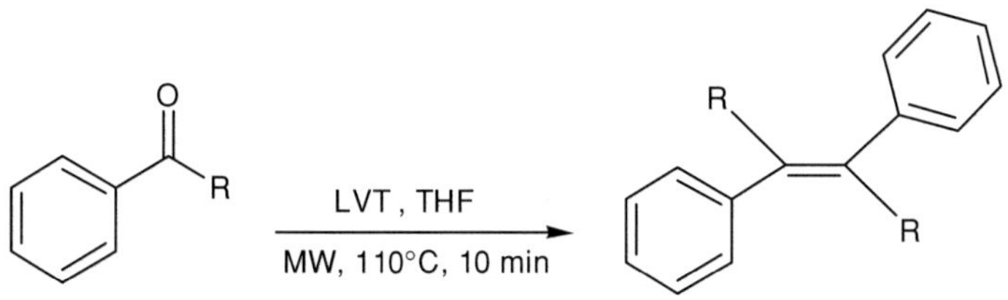

R = H, alkyl, aryl

82–93%

The McMurry coupling is most useful for the intramolecular reduction[45] of aldehydes and ketones to form cycloalkenes with ring size varying from 3 to 36. The medium and large rings are usually formed with the *E*-configuration[46].

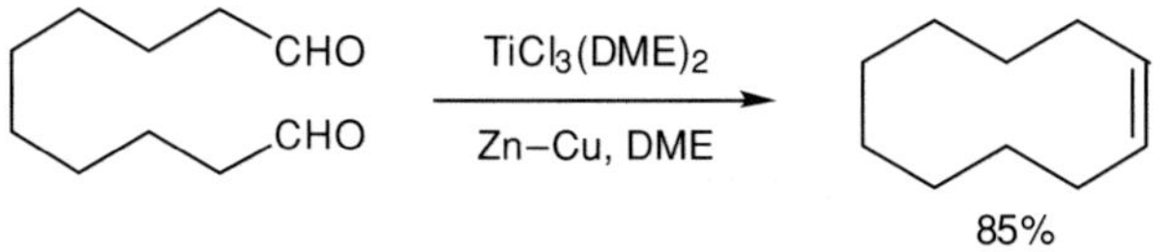

85%

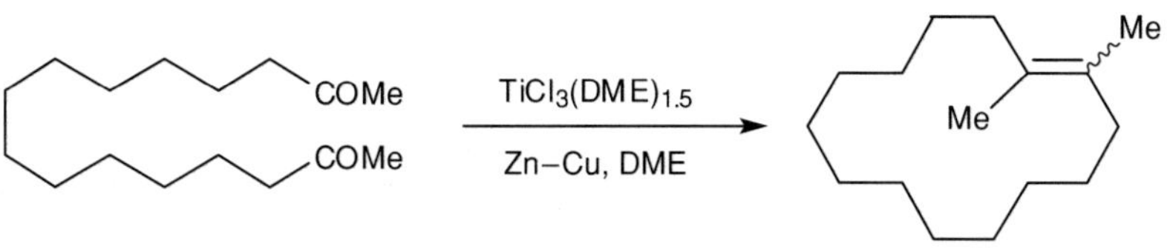

82%, *E/Z*= 92:8

Intramolecular McMurry condensation of bisbenzaldehydes **4.83** in the presence of $TiCl_3(DME)_3$ and Zn–Cu yielded the corresponding phenanthrenes **4.84** in 45–57% yields[47a].

4.83   **4.84**

$R^{ij} = H; 4',5'-O-CH_2-O-; 3'-OCH_3; 4'-OBn$

Balu et al.[47b] used **Tyrlik's reagent** ($TiCl_3$, Mg, THF) for the coupling reaction of aldehydes and ketones. However, in the presence of pyridine or hydroxy auxiliary (such as catechol, ethylene glycol or mannitol) the main product found is the diol and not the alkene.

E and Z   50:10   d,l and meso

Tebbe alkenation

The Tebbe reagent[48] (**4.85**) is used for methylenation of carbonyl compounds[49].

R = alkyl or aryl; R' = H, NR_2, alkyl   **4.85**

The Tebbe reagent (**4.85**) is dimetallomethylene, containing two cyclopentadienyl (Cp) rings bonded to titanium. The titanium and aluminium atoms are bridged by both CH_2 and Cl groups. Aluminium is also bonded to two methyl groups. The most convenient synthesis of the Tebbe reagent (**4.85**) is by the reaction of trimethyl aluminium with titanocene dichloride[50,51].

$$Cp_2TiCl_2 \ + \ 2\ Al(CH_3)_3 \ \rightarrow \ CH_4 \ + \ Cp_2TiCH_2ClAl(CH_3)_2 \ + \ Al(CH_3)_2Cl$$

4.85

Mechanism: In the presence of a donor ligand such as pyridine or THF, the Tebbe reagent forms the highly reactive methylene–titanocene complex **4.75**. It reacts with carbonyl compounds to give a oxatitanacyclobutane **4.86** intermediate, which breaks down immediately to produce the corresponding alkene[52] (Scheme 4.46). The high reactivity of the Tebbe reagent (**4.85**) compared to the Wittig reagent (see section 4.3.1) appears to be due to the high oxophilicity of Ti(IV); further, the Tebbe reagent is less basic than the Wittig reagent and therefore does not give the β-elimination product.

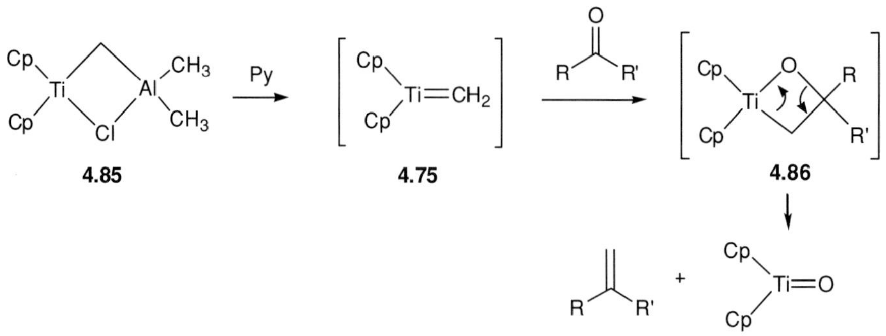

Scheme 4.46

However, in the absence of a base, the mechanism[52] shown in Scheme 4.47 is proposed.

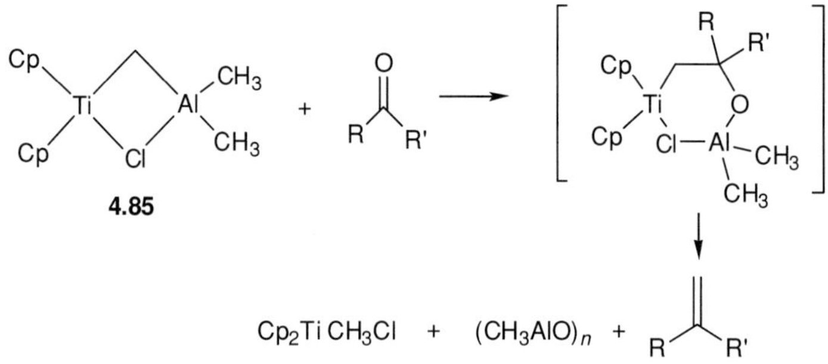

Scheme 4.47

The Tebbe reagent (**4.85**) converts aldehydes and ketones to alkenes[53]. The reaction of the Tebbe reagent (**4.85**) with cyclohexanones in toluene produces the corresponding methylenecyclohexanes in >65% yields. For example (4-methylenecyclohex-1-yl)benzene (**4.88**) was prepared from 4-phenylcyclohexanone (**4.87**) in 96% crude yield on reaction with the Tebbe reagent (**4.85**)[54].

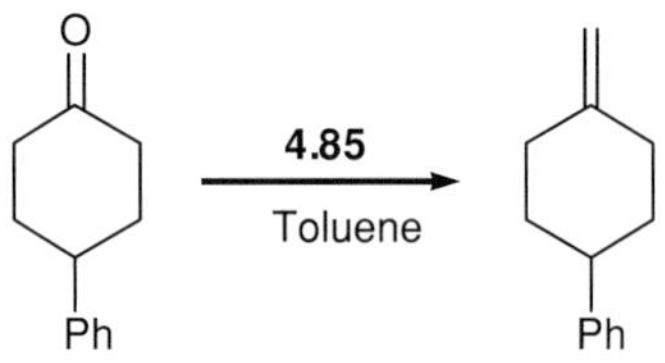

The Tebbe reagent (**4.85**) gives better yield of alkenes with sterically hindered ketones than the Wittig reagent[55].

Tebbe reagent (**4.85**): 77%
Wittig reagent: 4%

The Tebbe reagent (**4.85**) converts esters[56] to enol (vinyl) ethers.

Lactones[57] are also alkenated with **4.85**. For example, dihydrocoumarin (**4.89**) on treatment with a toluene solution of **4.85**, prepared *in situ* in THF, afforded the 3,4-dihydro-2-methylene-2 *H*-1-benzopyran (**4.90**) in 76% yield.

Amides are converted into enamines[58] with **4.85**.

The Tebbe reagent reacts with acid chlorides to give titanium enolate.

In compounds containing both ketone and ester groups, the ketone selectively reacts in the presence of 1 equiv. of the Tebbe reagent (**4.85**), but with excess amount of the Tebbe reagent (**4.85**) both carbonyl groups are alkenated.

The Tebbe reagent (**4.85**) has found applications in reactions of sugars because it methylenates carbonyls without racemizing a chiral α-carbon. Thus, methylenation[59] of the formyl functions of 1-O-formylglycoside **4.91** with **4.85** produced 1-O-vinyl glycosides **4.92**.

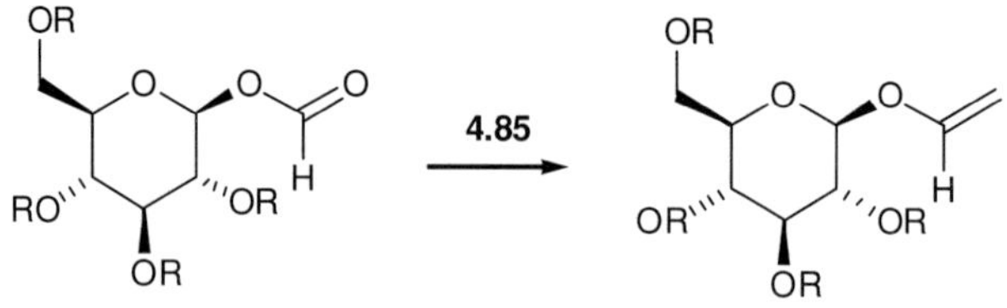

4.91   **4.92**

The reagent **4.85** can also be used for the alkenation of thioesters and carbonates[60].

Petasis reagent

The **Petasis reagent**,[61] dimethyl titanocene (**4.93**) can also be used for the methylenation of carbonyl compounds. The **Petasis reagent** (**4.93**) is prepared by the reaction of methyl magnesium chloride or methyllithium with titanocene dichloride (Cp_2TiCl_2). Carbonyl compounds on heating with **4.93** at 60–65°C in a toluene solution give the corresponding alkenes or enol ethers.

$$Cp_2TiCl_2 \ + \ 2 \ MeLi \ \longrightarrow \ Cp_2TiMe_2 \ + \ 2 \ LiCl$$
4.93

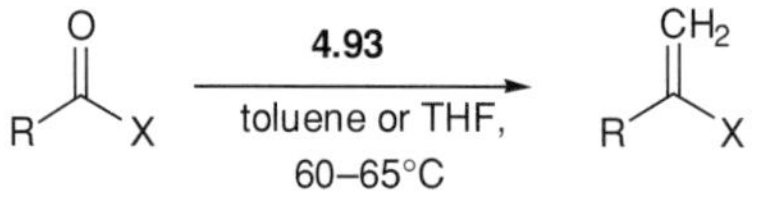

X = H, alkyl, aryl, vinyl, OR; R = alkyl or aryl

This aluminium-free Petasis reagent (**4.93**) is an alternative to the Tebbe (**4.85**) and Grubbs reagents, which is more stable and can be isolated as a pure solid, or used directly as a solution in toluene or THF. Unlike the Wittig reaction, the Petasis reagent (**4.93**) can react with a wide range of aldehydes and ketones in toluene to give the corresponding alkene in 40–90% yields. A few examples are shown below:

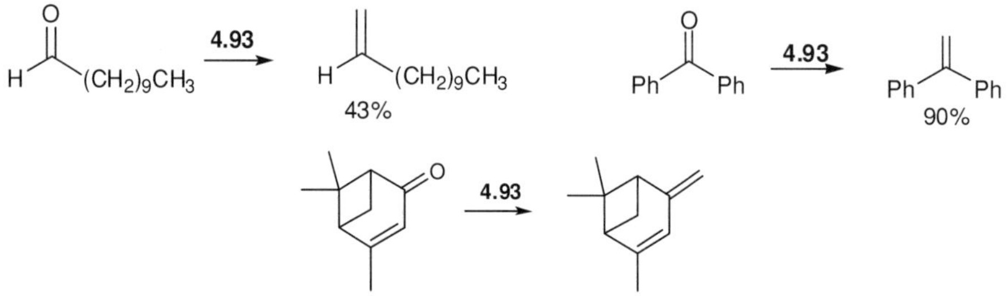

Esters and lactones are converted into the corresponding enol ethers in 40–80% yields with **4.93**, although these reactions are slow.

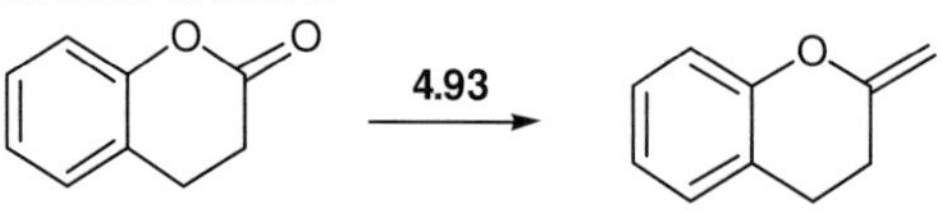

4.89   **4.90**
                                            80%

Lactams, imides, acid anhydrides, carbonates, acylsilanes, thioesters and selenoesters can also be converted to the corresponding alkenes.

Mechanism: The mechanism of alkenation appears to involve an active alkenating reagent, Cp_2TiCH_2 (**4.75**), which is formed by heating the Petasis reagent (**4.93**) in toluene or THF to 60°C with the elimination of methane. However, on the basis of some deuterium incorporation experiments an alternate mechanism is also proposed. This involves carbonyl complexation to Cp_2TiMe_2 (**4.93**) followed by methyl transfer to the carbonyl. The resulting adduct may then undergo loss of methane and titanocene oxide to form the alkenes.

Takeda reaction

The low valent titanium LVT reagent **4.94** is prepared by the reaction of Cp_2TiCl_2 with magnesium turnings and $P(OEt)_3$ in THF under dry conditions.

The reaction of reagent **4.94** with thioacetal and ketones or aldehydes produces the corresponding alkene[62–66].

Alkenation of esters by this method yields enol ethers, and lactones give cyclic enol ethers

Intramolecular alkenation[67] of esters **4.95** with **4.94** produces enol ethers **4.97** via titanium–carbene complexes **4.96** (Scheme 4.48).

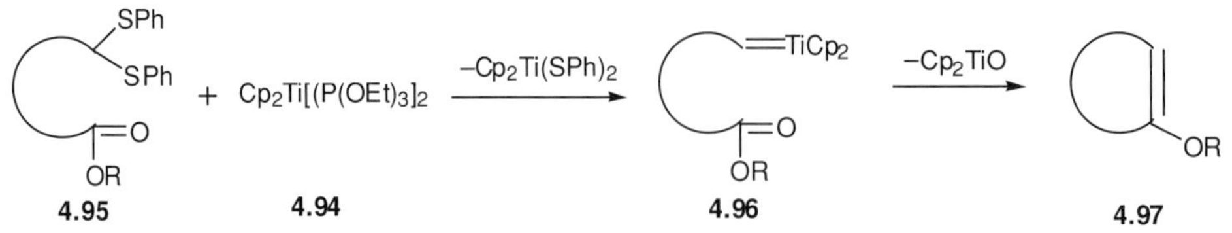

Scheme 4.48

For example, methyl 2-phenyl-7,7-bis(phenylthio)heptanoate (**4.98**) on treatment with **4.94** at room temperature produces the cyclic vinyl ether **4.99** in 68% yield.

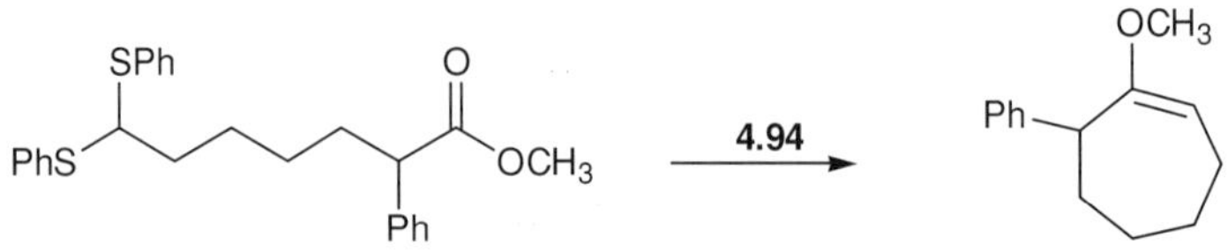

Mechanism: Two pathways are suggested for this reaction (Scheme 4.49). The titanium–carbene complex **A** is formed as a key intermediate, which reacts with carbonyl compound to form an alkene via the oxatitanacyclobutane **B** (Path A). Alternatively, the addition of gem-dimetallic species **C** to a carbonyl compound gives the adduct **D**, which eliminates (TiCp$_2$ RS)$_2$O to give an alkene (Path B).

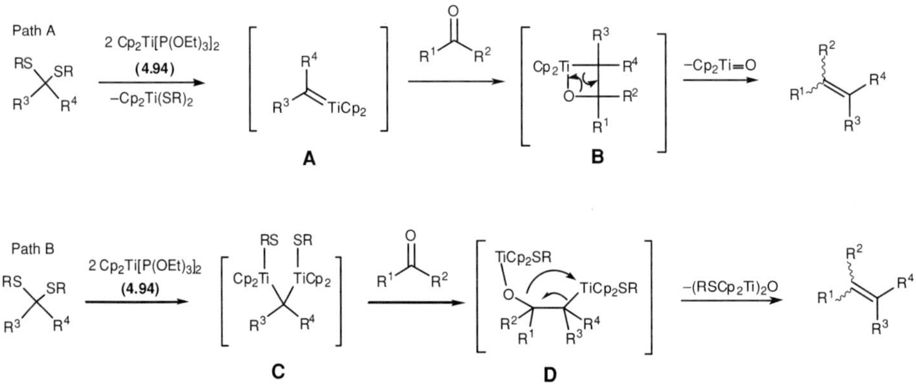

Scheme 4.49

4.3.5   Use of Zinc (Zn) and zirconium (Zr) reagents for the alkenation of ketones and aldehydes

Although transition metal alkylidene complexes are successfully used for the alkenation of carbonyl compounds, various 1,1-bimetalloalkanes, often prepared by the hydrometallation of alkenyl organometallics, are also useful reagents for the alkenation of carbonyl compounds.

Nysted used a Zn reagent (**Nysted reagent, 4.100**) for the methylenation of ketones and aldehydes[68,69].

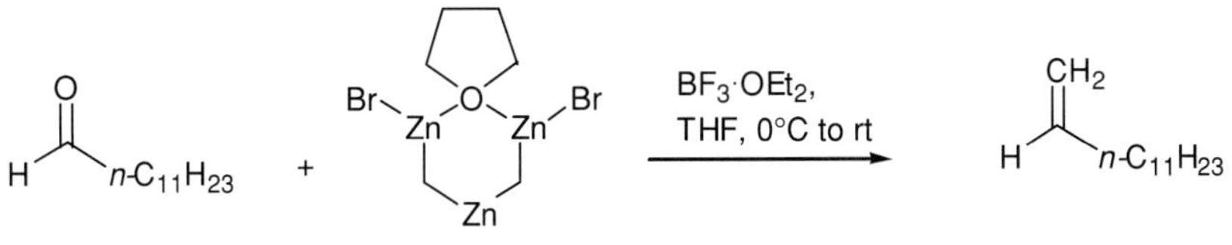

4.100   85%

Reductive coupling of polyhaloalkanes with carbonyl compounds in the presence of Zn is used in the synthesis of various halogen-substituted alkenes. For example, the reaction of benzaldehyde with the organozinc reagent **4.101** gives the trifluoromethyl-substituted alkene **4.102** (Scheme 4.50).

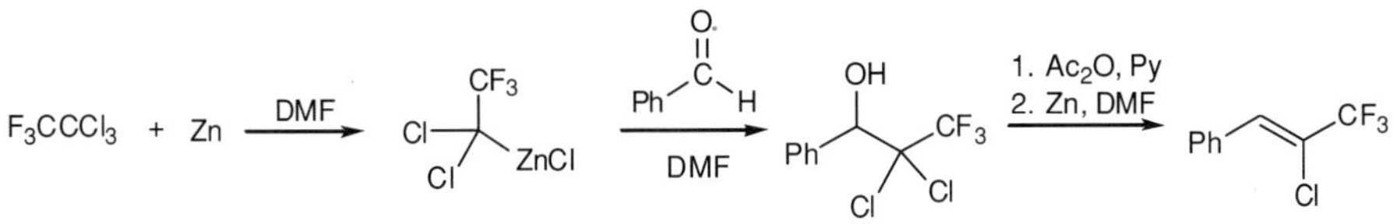

Scheme 4.50

Zinc and zirconium 1,1-bimetallic reagents **4.104**, prepared by the hydrozirconation of alkenylzinc halides **4.103**, react with carbonyl compounds to produce alkenes with high *E*-stereoselectivity (Scheme 4.51). Ketones give an *E/Z* mixture of stereoisomers[70].

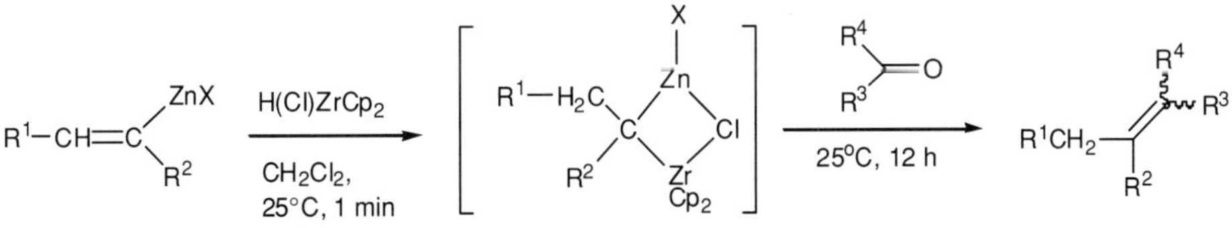

4.103   **4.104**

Scheme 4.51

Zirconocene dichloride reacts with dibromomethane in the presence of Zn to yield a methylene–zirconium complex **4.105**, which is used for the methylenation of carbon compounds to produce the terminal alkenes[71].

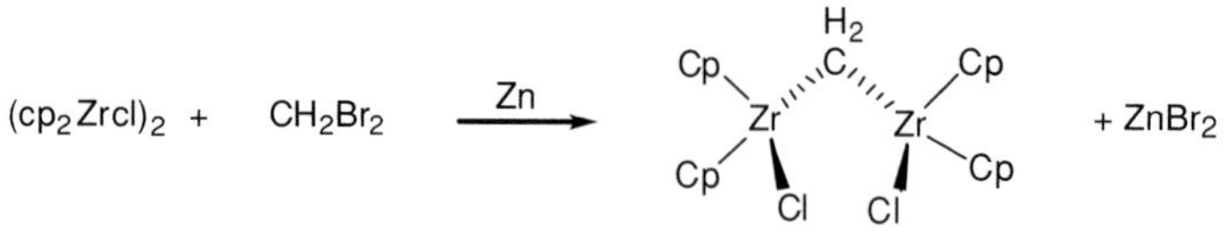

4.105

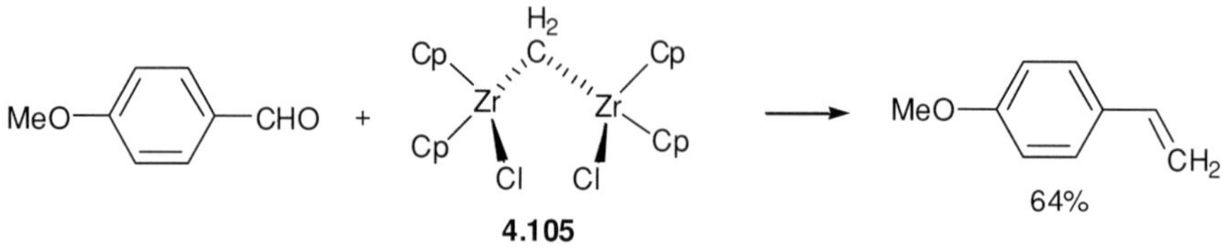

64%

4.105

The mechanism of the above reaction includes coordination of the zirconium atom to the carbonyl oxygen atom and an attack of the nucleophilic carbon atom on carbonyl carbon to yield metallaoxacyclobutane; subsequent elimination of zirconocene gives the alkene (Scheme 4.52).

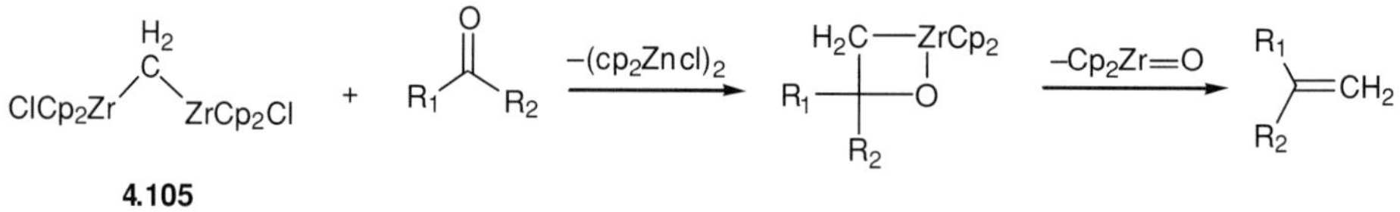

4.105

Scheme 4.52

4.3.6   Bamford–Stevens reaction and Shapiro reaction

The **Bamford–Stevens reaction**[72] is a tosylhydrazones, **4.106**, decomposition with a base to form alkenes.

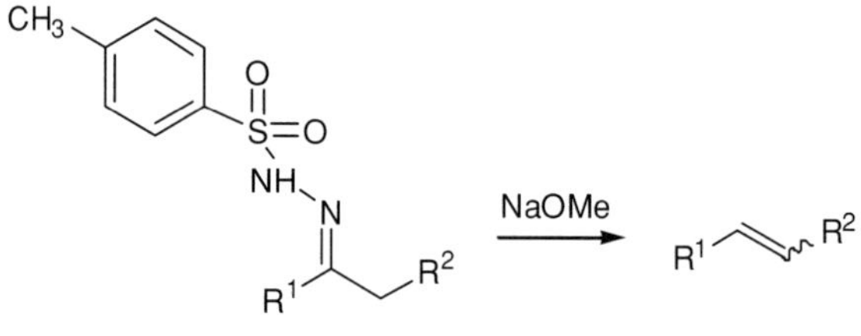

4.106

Mechanism: The first step of the Bamford–Stevens reaction is the formation of the diazo compound **A** by the treatment of tosylhydrazone with a base. The reaction mechanism involves a carbene **B** in an aprotic solvent (Path A) and carbocation **C** in a protic solvent (Path B) (Scheme 4.53). When an aprotic solvent is used, predominantly Z-alkenes are obtained, while a protic solvent gives a mixture of E- and Z-alkenes. If there is a choice of product, the more substituted alkene is produced predominantly.

Ketones and aldehydes are converted to less substituted alkenes through an intermediate tosylhydrazone in the presence of 2 equiv. of a strong base typically alkyllithium

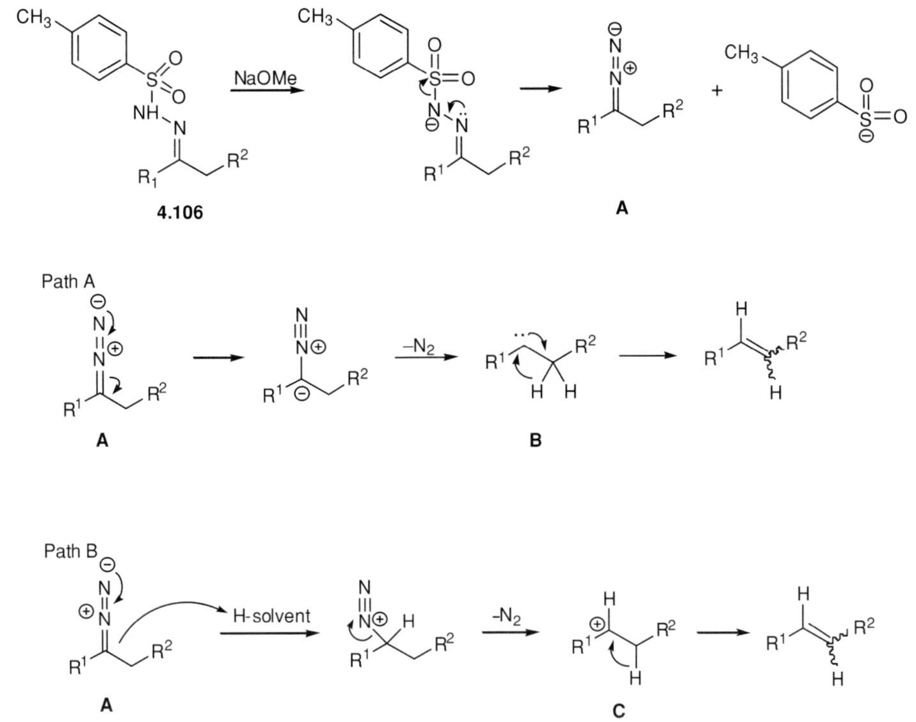

Path A

Path B

Scheme 4.53   Mechanism of the Bamford–Stevens reaction

(*n*-butyllithium) or lithium dialkylamide. This reaction is known as the **Shapiro reaction**[73,74]. For example, camphor (**4.107**) gives 2-bornene[75] (**4.109**) through intermediate tosylhydrazone **4.108** on treatment with a strong base.

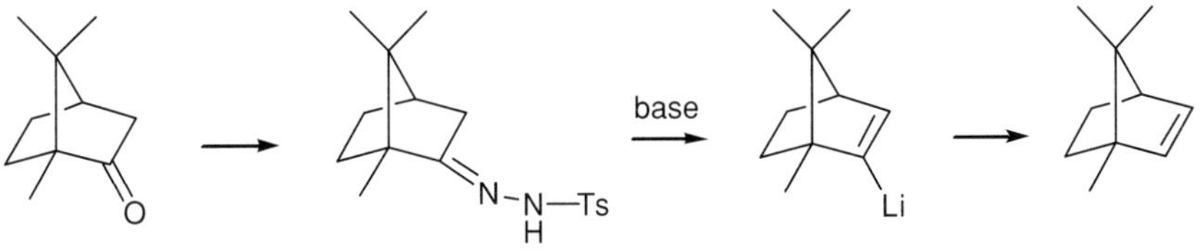

Two equivalents of a strong base such as *n*-butyllithium abstract first the tosylhydrazone proton and then the less acidic α-carbonyl proton, leaving a dianion **A**. The **A** eliminates first *p*-toluenesulfinate and then N_2. The resulting vinyllithium compound **B** on protonation gives alkene (Scheme 4.54).

4.3.7   Barton–Kellogg reaction

The **Barton–Kellogg reaction**[76–78] is a coupling reaction between a diazo compound and a thioketone to give an episulfide. Desulfurization of the episulfide to alkene can be accomplished by phosphine or by copper powder. This reaction has been pioneered by Hermann Staudinger[79] and therefore the reaction also goes by the name **Staudinger-type diazothioketone coupling**.

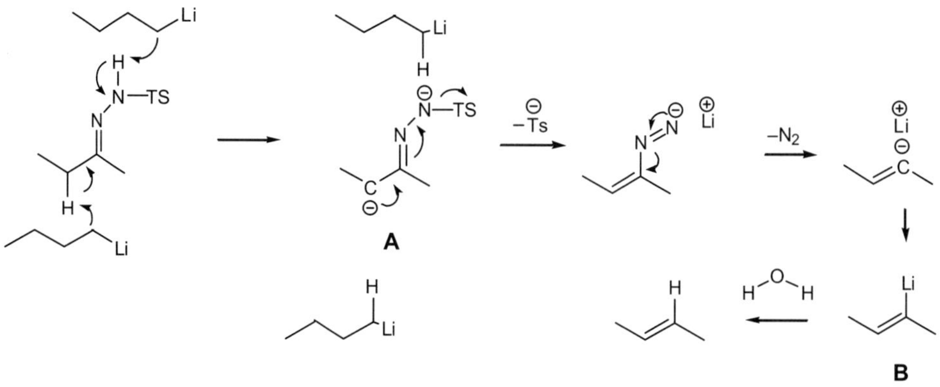

Scheme 4.54 Mechanism of the Shapiro reaction

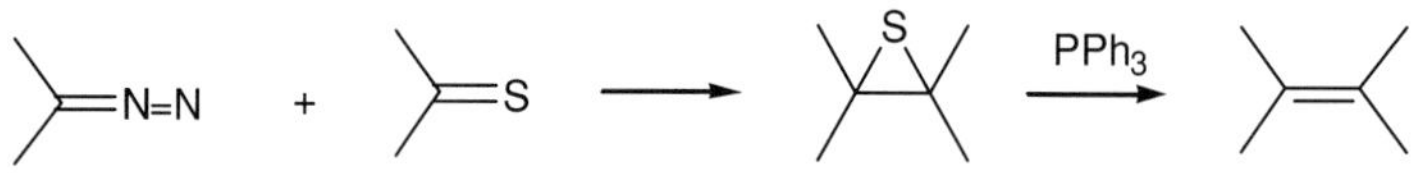

Diazo compound Thioketone Episulfide

Mechanism: The diazo compound can be obtained from a ketone by the reaction with hydrazine followed by oxidation with silver(I) oxide or [bis(trifluoroacetoxy)iodo]benzene. The thioketone required for this reaction can also be obtained from a ketone by the reaction with phosphorus pentasulfide. The 1,3-dipolar cycloaddition of diazo compound with the thioketone gives thiadiazoline **A**. This intermediate is unstable and through nitrogen gas expulsion and formation of an intermediate thiocarbonyl ylide **B** it forms a stable episulfide. Triphenylphosphine opens the three-membered ring and then a sulfaphosphatane **C** is formed in a manner similar to the Wittig reaction (section 4.3.1). In the final step, a sulfaphosphatane **C** is decomposed to form alkene and triphenyl phosphinesulfoxide (Scheme 4.55).

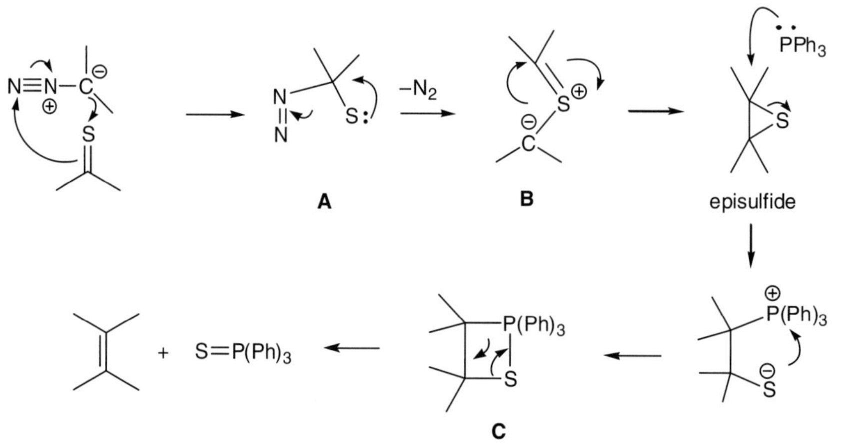

Scheme 4.55 Mechanism of the Barton–Kellogg reaction

The main advantage of this reaction over the **McMurry reaction** (see section on 'McMurry alkenation') is the notion that the reaction can take place with two different ketones. In this regard the diazo-thioketone coupling is a cross-coupling rather than a homocoupling.

4.3.8   Catalytic aldehyde and ketone alkenation[80]

Although the Wittig reaction (section 4.3.1) and its modified versions provide highly effective and general methods for the aldehydes and ketones alkenation, there are several drawbacks in their use. To overcome these drawbacks, new methods which employ the transition metal complex reagents[80–91] as catalysts have been developed. A variety of catalytic aldehyde alkenation reactions have been reported with Mo catalyst; other metals such as Re, Ru, Rh and Fe are also established as useful catalysts. Very mild conditions are required for catalytic alkenation; short reaction times and high selectivities are generally observed.

Several aldehydes were converted to alkenes by the use of 10% of the catalyst $MoO_2(S_2CNEt_2)$ in the presence of triphenylphosphine and diazoacetate[92]. Alkenes yields of up to 83% were reached at 80°C and the *E*-isomers were the main products. In contrast to the reactivity pattern of the Wittig reaction, the aliphatic aldehydes are more reactive than the aromatic aldehydes and the presence of electron-donating group enhances the yields.

The reaction of *p*-chlorobenzaldehyde with phenyldiazomethane in the presence of $(MeO)_3P$ and catalytic amounts of *meso*-tetraphenylporphyrin iron chloride (ClFeTPP) resulted in the formation of the corresponding alkenes with an *E/Z*-selectivity of 86:14, but the yield was low (30%). When phenyldiazomethane is generated *in situ* from the corresponding potassium tosylhydrazone salt, the olefin yield increases to 92% with *E/Z*-selectivity of 97:3. Thus, high levels of *E*-selectivity are obtained with semistabilized ylides by this method[93]. This process is applied to a wide range of aldehydes and is practical as compared to standard Wittig reaction, and therefore finds applications in industry.

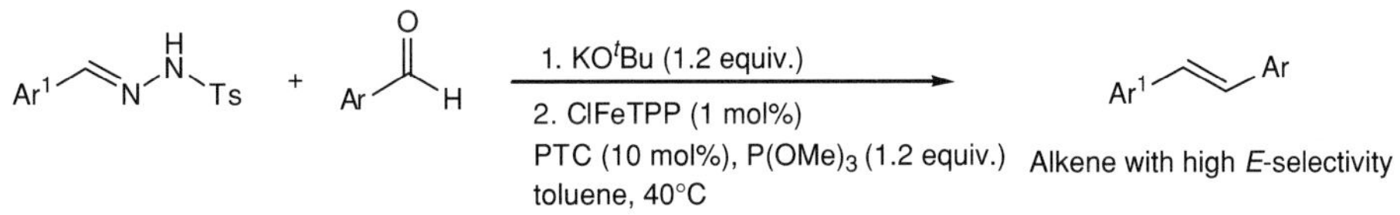

These reactions are presumed to proceed with the formation of the phosphorus ylide by carbene transfer (Scheme 4.56).

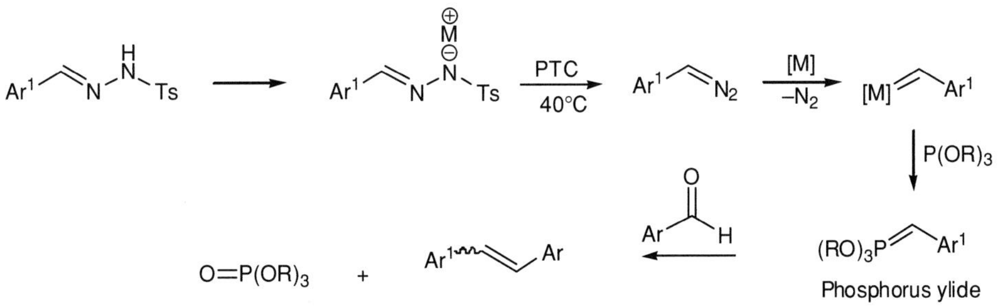

Scheme 4.56 Mechanism of alkenation of aldehydes and ketones with tosylhydrazones in the presence of a catalyst (ClFeTPP)

The efficiency of this new method has been demonstrated in the synthesis of the anticancer compound **4.110**. The standard Wittig reaction gives low selectivity; however, this method gave the desired compound with 97:3 *E/Z*-selectivity.

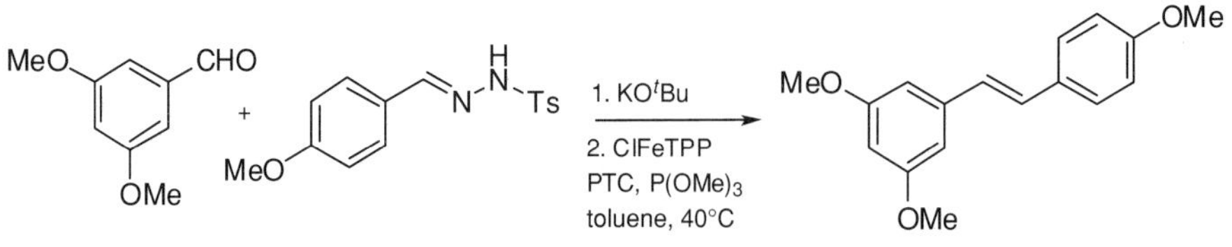

4.110

Another example of catalytic alkenation of carbonyl compounds such as acetophenone with ethyl diazoacetate catalyzed by Fe(TPP)Cl in the presence of an additive benzoic acid yielded a mixture of *Z/E*-alkenes **4.111** in 59:41 ratio in 84% yield.

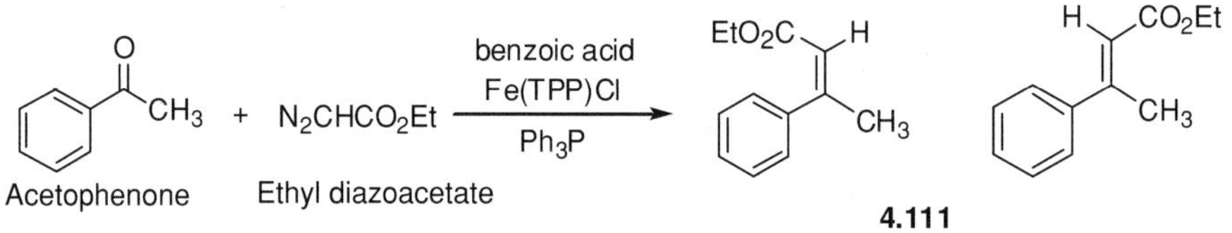

4.111
Alkene
Z/E 59:41 (84%)

Mechanism: The mechanism of this reaction is assumed to proceed via a metallocarbene-phosphorane. The Fe(III) porphyrin is reduced *in situ* by diazo reagents to the catalytically active Fe(II) porphyrin. The activation of ketones by acids operates possibly through protonation of the carbonyl oxygen, which would render the carbonyl group a stronger electrophile towards reaction with phosphorane (Scheme 4.57).

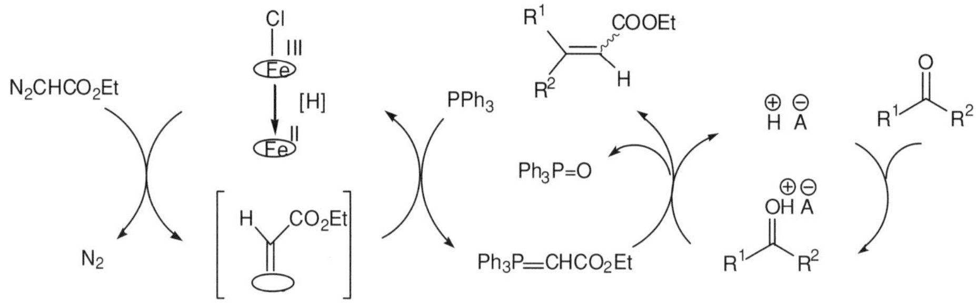

Scheme 4.57 Mechanism of alkenation of aldehydes and ketones in the presence of a catalyst (ClFeTPP)

4.4   Reduction of alkynes

Alkynes are reduced to *trans*-alkenes under dissolving-metal conditions.

There are several ways to perform the *cis* reduction of alkynes. For example, catalytic hydrogenation using a poisoned catalyst and hydroboration followed by protonation yields *cis*-alkenes. For examples and mechanism, see Chapter 6, section 6.2.

References

1. Li, A.-H., Dia, L.-X. and Aggarwal, V. K., *Chem. Rev.*, **1997**, *97*, 2341.
2. Aube, J., *Chemtracts – Org. Chem.*, **1988**, *1*, 461.
3. Yamaguchi, M. and Hirama, M., *Chemtracts – Org. Chem.*, **1994**, *7*, 401.
4. Marcantoni, E., Massaccesi, M. and Petrini, M., *J. Org. Chem.*, **2000**, *65*, 4553.
5. Sommer, L. H., Bailey, D. L. and Whitmore, F. C., *J. Am. Chem. Soc.*, **1948**, *70*, 2869.
6. Butcher, T. S., Zhou, F. and Detty, M. R., *J. Org. Chem.*, **1998**, *63*(1), 169.
7. Ranu, B. C. and Jana, R., *J. Org. Chem.*, **2005**, *70*(21), 8621.
8. Cope, A. C., *Tetrahedron Lett.*, **1949**, *71*, 3929.
9. Bluth, M., *Tetrahedron Lett.*, **1984**, *25*, 2873.
10. Cope, A. C. and LeBel, N. A., *J. Am. Chem. Soc.*, **1960**, *82*(17), 4656.
11. Grieco, P. A., Gilman, S. and Nishizawa, M., *J. Org. Chem.*, **1976**, *41*(8), 1485.
12. Mori, K.,*Tetrahedron Lett.*, **1978**, *19*(37), 3447.
13. Chugaev, L., *Chem. Ber.*, **1899**, *32*, 3332.
14. Rutherford, K. G., Ottenbrite, R. M. and Tang, B. K., *J. Chem. Soc.*, **1971**, 582.
15. Corey, E. J. and Winter, R. A. E., *J. Am. Chem. Soc.*, **1963**, *85*, 2677.
16. Horton, D., *Tetrahedron Lett.*, **1964**, *5*(36), 2531.
17. Corey, E. J. and Hopkins, P. B., *Tetrahedron Lett.*, **1982**, *23*, 1979.
18. Maryanoff, B. E. and Reitz, A. B., *Chem. Rev.*, **1989**, *89*, 863.
19. Pihko, P. M. and Koskinen, A. M. P., *J. Org. Chem.*, **1998**, *63*, 92.; Narita, M., Otsuka, M., Kobayayashi, S., Ohno, M., Umezawa, Y., Morishima, H., Saito, S., Takita, T, Umezawa, H. *Tetrahedron, Lett.*, **1982**, 23, 523.; Itaya, T., Fujii, T., Evidente, A., Randazzo, G., Surico, G., Iacobellis, N. S., *Tetrahedron Lett.*, **1986**, 27, 6349.
20. Vedejs, E., Meier, G. P. and Snoble, K. A. J., *J. Am. Chem. Soc.*, **1981**, *103*, 2823.
21. Schlosser, M., Shaub, B., Oliveira-Nato, J. D. and Jegananthan, S., *Chimia*, **1986**, *40*, 244.
22. Schlosser, M., *Angew. Chem., Int. Ed. Engl.*, **1966**, *5*, 126.
23. Horner, L., Hoffman, H., Wippel, H. G. and Klahre, G., *Chem. Ber.*, **1959**, *92*, 2499.
24. Wadsworth, W. S. and Emmons, W. D., *J. Am. Chem. Soc.*, **1961**, *83*, 1733.
25. Denmark, S. E. and Middleton, D. S., *J. Org. Chem.*, **1998**, *63*, 1604.
26. Still, W. C. and Gennari, C., *Tetrahedron Lett.*, **1983**, *24*, 4405.
27. Rein, T. and Reiser, O. *Acta Chem. Scand.*, **1996**, *50*, 369.
28. Bestmann, H. J. and Lienert, J., *Angew. Chem., Int. Ed. Engl.*, **1969**, *8*, 763.
29. Kann, N. and Rein, T., *J. Org. Chem.*, **1993**, *58*, 3802.
30. Toda, F. and Akai, H., *J. Org. Chem.*, **1990**, *55*, 3446.
31. Julia, M. J. and Paris, M., *Tetrahedron Lett.*, **1973**, *14*, 4833.
32. Hart, D., Li, J., Wu, W.-L. and Kozikowski, A. P., *J. Org. Chem.*, **1997**, *62*, 5023.
33. Keck, G. E., Savin, K. A. and Weglarz, M. A., *J. Org. Chem.*, **1995**, *60*, 3194.
34. Blakemore, P. R., Cole, W. J., Kocienski, P. J. and Morley, A., *Synlett*, **1998**, 26.
35. Lebrun, M.-E., Marquand, P. L. and Berthelette, C., *J. Org. Chem.*, **2006**, *71*, 2009.
36. Peterson, D. J., *J. Org. Chem.*, **1968**, *33*(2), 780.
37. Barbero, A., Blanco, Y., Garcia, C. and Pulido, F. J., *Synthesis*, **2000**, 1223.
38. McMurry, J. E., *Chem. Rev.*, **1989**, *89*, 1513.
39. Blaszczak, L. C. and McMurry, J. E., *J. Org. Chem.*, **1974**, *39*, 258.
40. Balu, N., Nayak, S. K. and Banerji, A., *J. Am. Chem. Soc.*, **1996**, *118*(25), 5932.
41. Lenour, D., Malwitz, D. and Meyer, E., *Tetrahedron Lett.*, **1984**, *25*, 2965.
42. McMurry, J. E. and Fleming, M. P., *J. Am. Chem. Soc.*, **1974**, *96*, 4708.
43. Backer, K. B., *Synthesis*, **1983**, 341.
44. Stuhr-Hansen, N., *Tetrahedron Lett.*, **2005**, *46*, 5491.
45. McMurry, J. E. and Kees, K. L., *J. Org. Chem.*, **1977**, *42*, 2655.
46. McMurry, J. E. and Rico J. G., *J. Org. Chem.*, **1989**, *54*, 3748.
47a Gies, A. E. and Pfeffer, M., *J. Org. Chem.*, **1999**, *64*(10), 3650.
47b Balu, N., Nayak, S. K. and Banerji, A., *J. Am. Chem. Soc.*, **1996**, *118*(25), 5932.
48. Tebbe, F. N., Parashall, G. W. and Reddy, G. S., *J. Am. Chem. Soc.*, **1978**, *100*, 3611.
49. Beadham, I. and Micklefield, J., *Curr. Org. Synth.*, **2005**, *2*, 231.
50. Herrmann, W. A., *Adv. Organomet. Chem.*, **1982**, *20*, 195.
51. Straus, D. A., *Encyclopedia of Reagents for Organic Synthesis*, Wiley, New York, **2000**.
52. Hartley, R. C., Li, J., Main, C. A. and McKiernan, G. A., *Tetrahedron*, **2007**, *63*(23), 4825.
53. Stork, G. and Hagedorn, A. A., *J. Am. Chem. Soc.*, **1978**, *100*, 3609.
54. Cannizzo, L. F. and Grubbs, R. H., *J. Org. Chem.*, **1985**, *50*, 2316.
55. Pine, S. H., Shen, G. S. and Hoang, H., *Synthesis*, **1991**, 165.
56. Pine, S. H., Rzahler, R., Evans, D. A. and Grubbs, R. H., *J. Am. Chem. Soc.*, **1980**, *102*, 3270.

57. Burton, J. W., Clark, J. S., Derrer, S., Stork, T. C., Bendall, J. G. and Holmes, A. B., *J. Am. Chem. Soc.*, **1997**, *119*, 7483.
58. Pine, S. H., Pettit, R. J., Geib, G. D., Cruz, S. G., Gallego, C. H., Tijerina, T. and Pine, R. D., *J. Org. Chem.*, **1985**, *50*, 1212.
59. Yuan, J., Lindner, K. and Frauenrath, H., *J. Org. Chem.*, **2006**, *71*(15), 5457.
60. Hartley, R. C. and McKiernan, G. J., *J. Chem. Soc., Perkin Trans. 1*, **2002**, 2763.
61. Petasis, N. A. and Bzowej, E. I., *J. Am. Chem. Soc.*, **1990**, *112*(17), 6392.
62. Horikawa, Y., Watanbe, M., Fujiwara, T. and Takeda, T., *J. Am. Chem. Soc.*, **1997**, *119*, 1127.
63. Takeda, T., Watanabe, M., Nozaki, N. and Fujiwara, T., *Chem. Lett.*, **1998**, 115.
64. Rahim, M. A., Taguchi, H., Watanabe, M., Fujiwara T. and Takeda, T., *Tetrahedron Lett.*, **1998**, *39*, 2153.
65. Takeda, T., Watanabe, M., Rahim, M. A. and Fujiwara, T., *Tetrahedron Lett.*, **1998**, *39*, 3753.
66. Fujiwara, T., Iwasaki, N. and Takeda, T., *Chem. Lett.*, **1998**, 741.
67. Rahim, M. A., Sasaki, H., Saito, J., Fujiwara, T. and Takeda, T., *Chem. Commun.*, **2001**, 625.
68. Nysted, L. N., *US Patent 3 865 848*, **1975**; *Chem. Abstr. 83*, **1975**, 10406q.
69. Matsubara, S., Sugihara, M. and Utimoto, K., *Synlett*, **1998**, 313.
70. Tucker, C. E. and Knochel, P., *J. Am. Chem. Soc.*, **1991**, *113*, 9888.
71. Tour, J. M., Bedworth, P. V. and Wu, R., *Tetrahedron Lett.*, **1989**, *30*, 3927.
72. Bamford, W. R. and Stevens, T. S., *J. Chem. Soc.*, **1952**, 4735.
73. Shapiro, R. H., *Org. React.*, **1976**, *23*, 405.
74. Shapiro, R. H. and Kolonko, K. J., *J. Org. Chem.*, **1978**, *43*, 404.
75. Shapiro, R. H. and Coll. J. H. D., *Org. Synth.*, **1988**, 6, 172.
76. Barton, D. H. R. and Willis, B. J., *Chem. Commun.*, **1970**, 1225.
77. Kellogg, R. M. and Wassenaar, S., *Tetrahedron Lett.*, **1970**, *11*(23), 1987.
78. Kellogg, R. M., *Tetrahedron*, **1976**, *32*, 2165.
79. Staudinger, H. and Siegwart, J., *Helv. Chim. Acta*, **1920**, *3*, 833.
80. Kuhn, F. E. and Santos, A. M., *Org. Chem.*, **2004**, *1*, 55.
81. Santos, A. M., Romão, C. C. and Kühn, F. E., *J. Am. Chem. Soc.*, **2003**, *125*, 2414.
82. Zhang, X. and Chen, P., *Eur. J. Org. Chem.*, **2003**, *9*, 1852.
83. Cheng, G., Mirafzal, G. A. and Woo, L. K., *Organometallics*, **2003**, *22*, 1468.
84. Mirafzal, G. A., Cheng, G. and Woo, L. K., *J. Am. Chem. Soc.*, **2002**, *124*, 176.
85. Grasa, G. A., Moore, Z., Martin, K. L., Stevens, E. D., Nolan, S. P., Paquet, V. and Lebel, H., *J. Organomet. Chem.*, **2002**, *658*, 126.
86. Lebel, H. and Paquet, V., *Org. Lett.*, **2002**, *4*, 1671.
87. Lebel, H., Paquet, V. and Proulx, C., *Angew. Chem., Int. Ed. Engl.*, **2001**, *40*, 2887.
88. Fujimura, O. and Honma, T., *Tetrahedron Lett.*, **1998**, *39*, 625.
89. Ledford, B. E. and Carreira, E. M., *Tetrahedron Lett.*, **1997**, *38*, 8125.
90. Herrmann, W. A., Roesky, P. W., Wang, M. and Scherer, W., *Organometallics*, **1994**, *13*, 4531.
91. Herrmann, W. A. and Wang, M., *Angew. Chem., Int. Ed. Engl.*, **1991**, *30*, 1641.
92. Lu, X., Fang, H. and Ni, Z., *J. Organomet. Chem.*, **1989**, *373*, 77.
93. Aggarwal, V. K., Fulton, J. R., Sheldon, C. G. and deVicente, J., *J. Am. Chem. Soc.*, **2003**, *125*, 6034.

Chapter 5
Transition Metal-Mediated Carbon–Carbon Bond Forming Reactions

The organometallic compounds are the compounds that contain a carbon and metal atom bond. Organic derivatives of **metalloids** (boron, silicon, germanium, arsenic and tellurium) are also included in this definition. The **organometallic compounds** have played indispensable roles in organic chemistry. Grignard reagent was discovered more than a century ago and every chemist has carried out the Grignard reaction at least once in its life time[1].

Transition metals open up new opportunities for synthesis, because their means of bonding and their reaction mechanisms differ from those of the elements of the s and p blocks. The empty and partially filled d-orbitals that characterize most of these metals enable them to bond reversibly to many functional groups. Thus, transition metals activate many difficult or previously unobserved reactions which are not readily achieved by using conventional reagents. The organometallic chemistry of transition metals has grown explosively in the last decade.

Catalysts are substances that increase or decrease the rate of the chemical reaction without themselves undergoing any permanent chemical change. The **platinum group of metals** form a special category of catalysts. For example, palladium complexes act as catalysts for many key C–C, C–O and C–N bond forming reactions (Scheme 5.1). An example of Pd complex, $Pd[P(o\text{-}CH_3C_6H_4)_3]_2$, catalyzed C–N bond forming reaction is the Hartwig–Buchward coupling reaction (Scheme 5.2).

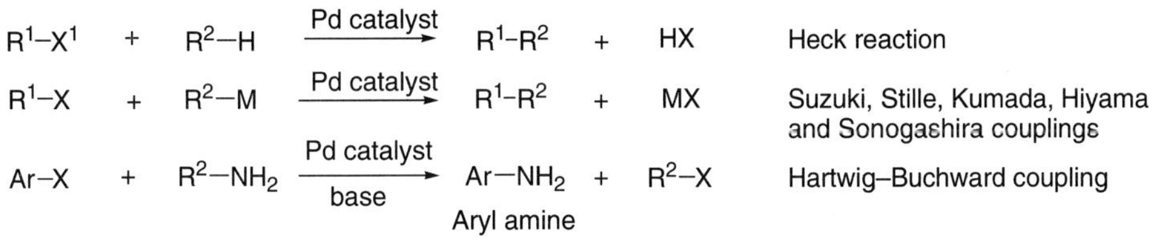

R^1–X^1	+	R^2–H	$\xrightarrow{\text{Pd catalyst}}$	R^1–R^2 + HX		Heck reaction
R^1–X	+	R^2–M	$\xrightarrow{\text{Pd catalyst}}$	R^1–R^2 + MX		Suzuki, Stille, Kumada, Hiyama and Sonogashira couplings
Ar–X	+	R^2–NH_2	$\xrightarrow[\text{base}]{\text{Pd catalyst}}$	Ar–NH_2 + R^2–X		Hartwig–Buchward coupling
				Aryl amine		

Scheme 5.1

Homogeneous catalysis (in which catalysts are soluble in the reaction mixture) has several drawbacks, in particular the problem of recycling of the catalyst. Thus, expensive catalysts and ligands are lost. To solve these problems, **heterogeneous Pd catalysis** (in which catalysts are not soluble in the reaction mixture) is a promising option. In **heterogeneous Pd catalysis**[2,3], Pd is fixed to a solid support, such as activated carbon (charcoal), zeolites, molecular sieves, metal oxides, clays, alkali and alkaline earth salts, organic polymers or porous glass. A number of solid-supported Pd catalysts are commercially available. In

191

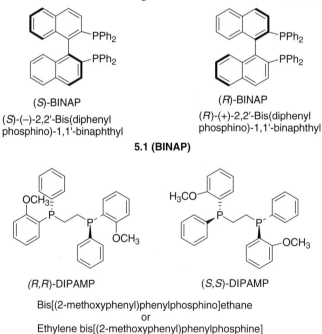

P(o-C₆H₄Me)₃ + Pd(dba)₂ $\xrightarrow{\text{benzene}}$ Pd[P(o-C₆H₄Me)₃]₂

dba = Dibenzylideneacetone (PhCH=CHCOCH=CHPh)

Scheme 5.2

general, carbon–carbon coupling reactions of organometallics with organohalides catalyzed by solid-supported Pd follow the usual reaction mechanism.

An important aim of organic synthesis is to synthesize single enantiomers of chiral compounds. The design and synthesis of new chiral ligands remains an important area of research with respect to developing highly enantioselective transition metal-catalyzed reactions. A successful ligand should therefore be readily accessible, stable and highly tunable since modification of the steric and electronic properties is often necessary for achieving high asymmetric induction. Many excellent chiral phosphine ligands such as BINAP[4–6] (**5.1**), DIPAMP[7] (**5.2**), DIOP[8] (**5.3**) and CHIRAPHOS[9] (**5.4**) have been developed for a variety of catalytic asymmetric reactions. **Chiral transition metal complexes** allow asymmetric transformations and therefore have an important and increasing role in organic synthesis. For example, Ru and Rh complexes of chiral diphosphines, such as BINAP, have been found to be highly efficient for the enantioselective reductions of various functional groups (see Chapter 1, section 1.5, and Chapter 6). The chiral transition metal complexes-catalyzed asymmetric carbon–carbon bond forming reactions are described in this chapter.

(S)-BINAP
(S)-(−)-2,2'-Bis(diphenyl phosphino)-1,1'-binaphthyl

(R)-BINAP
(R)-(+)-2,2'-Bis(diphenyl phosphino)-1,1'-binaphthyl

5.1 (BINAP)

(R,R)-DIPAMP

(S,S)-DIPAMP

Bis[(2-methoxyphenyl)phenylphosphino]ethane
or
Ethylene bis[(2-methoxyphenyl)phenylphosphine]

5.2 (DIPAMP)

(R,R)-(–)-DIOP   (S,S)-(–)-DIOP

trans-4,5-Bis(diphenylphosphinomethyl)-2,2-dimethyl-1,3-dioxolane

5.3 (DIOP)

(R,R)-CHIRAPHOS   (S,S)-CHIRAPHOS

trans-2,3-Bis(diphenylphosphino)butane

5.4 (CHIRAPHOS)

Although this chapter includes the applications of transition metals as catalysts and reagents for C–C bond forming reactions, it is planned from the point of view of the reaction mechanism rather than on the basis of transformation.

5.1 Carbon–carbon bond forming reactions catalyzed by transition metals

The transition metal salts and complexes catalyze several useful transformations; however, very useful applications of these catalysts to reactions in which carbon–carbon bonds are formed are discussed in this chapter. Over the last 50 years many new carbon–carbon bond forming reactions have been discovered. Within that broad genus lies the art of coupling reactions, reactions that catalytically bring together two neutral organic precursors. Transition metal-catalyzed carbon–carbon coupling reactions are one of the most powerful synthetic tools in organic chemistry. One important participant in this particular field is the Heck reaction, which is catalyzed by palladium.

5.1.1 Heck reaction

The Heck reaction was discovered independently in the end of 1960s by T. Mizoroki[10] and R. F. Heck[11,12]. But Heck developed it into a synthetically useful reaction. Since then it has become one of the most important reactions for the synthesis[13–20] of aromatic compounds substituted with alkenes.

The classical Heck reaction involves the Pd(0)-mediated coupling of alkenes with aryl and alkenyl halides at much more convenient laboratory conditions (Scheme 5.3). Hindered amines such as tri-*n*-butyl amine and triethyl amine are used as a base to neutralize HX produced as a by-product of the catalytic cycle. The Heck reaction has trans-selectivity.

The Heck reaction is typically performed in the presence of $Pd(OAc)_2$ or $Pd_2(dba)_3 \cdot CHCl_3$ and a stoichiometric amount of an inorganic or organic base[21]. Although Heck did not initially use phosphine ligands, the use of triphenylphosphine is the standard in his reaction. $Pd(PPh_3)_4$ can be used as such or generated *in situ* from $Pd(OAc)_2$ and PPh_3. Functional group tolerance and the ready availability and low cost of simple alkenes contribute to

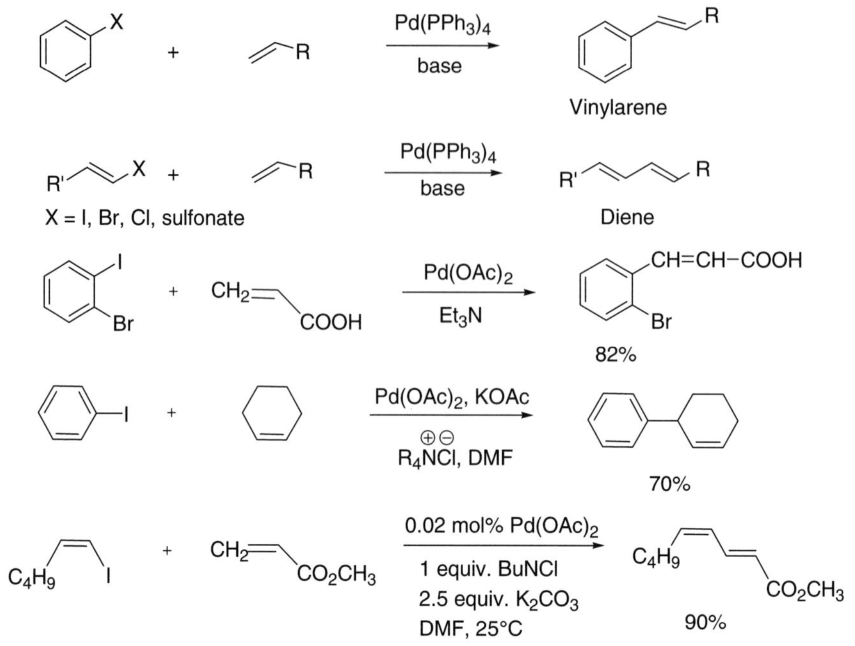

Scheme 5.3 The Heck reaction

the exceptional utility of the Heck reaction. Since the discovery of this reaction, scientists have discovered ways to alter the original procedure to bend the outcome towards more favourable results for their individual needs.

The order of reactivity of various organic halides is Ar–I > Ar–Br > Ar–Cl. Aryl chlorides, the most attractive class of aryl halides, due to their lower price and greater availability as compared with the corresponding bromides and iodides, are very slow in their reactivity in the Heck reaction. Pd[P(t-Bu)$_3$]$_2$ (bis(tri-test-butylphosphine)palladium) in the presence of Cy$_2$NMe (dicyclohexylmethylamine) is an unusually mild and versatile catalyst for the Heck reactions of aryl chlorides. For example, the coupling of chlorobenzene (**5.5**) with butyl methylacrylate (**5.6**) leads to stereoselective synthesis of a trisubstituted alkene, *n*-butyl α-methylcinnamate (**5.7**). Similarly, the coupling of 4-chlorobenzonitrile (**5.8**) with styrene (**5.9**) to give **5.10** demonstrates that Pd[P(t-Bu)$_3$]$_2$ can catalyze the Heck reaction of activated aryl chlorides at room temperature.

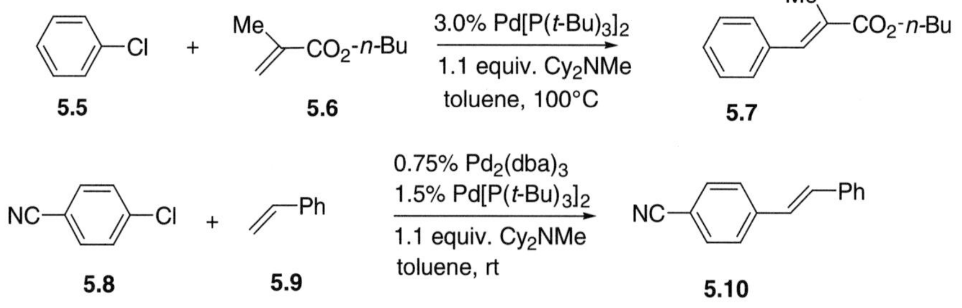

Although intermolecular Heck reactions are common, intramolecular Heck reactions can also be used for carbon–carbon coupling reactions.

The intramolecular Heck reaction[22] has been well established as a powerful tool for the construction of complex polycyclic ring systems in the context of natural product synthesis.

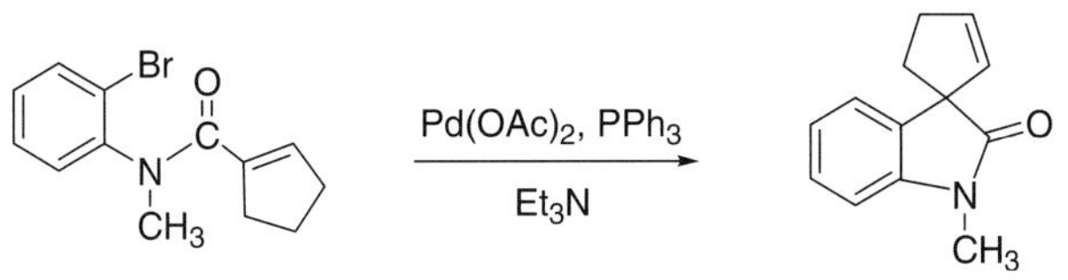

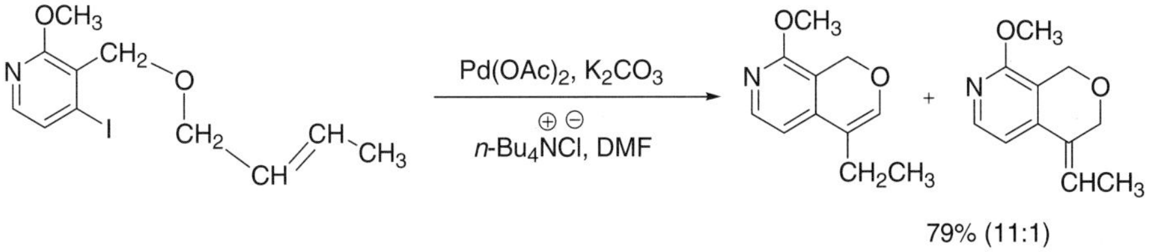

79% (11:1)

Mechanism[23]: The Heck reaction has become a powerful tool in organic synthesis, but the mechanism of this reaction has remained a topic of debate since the reaction's discovery. The general mechanism for the Heck reaction which is widely accepted is outlined in Scheme 5.4.

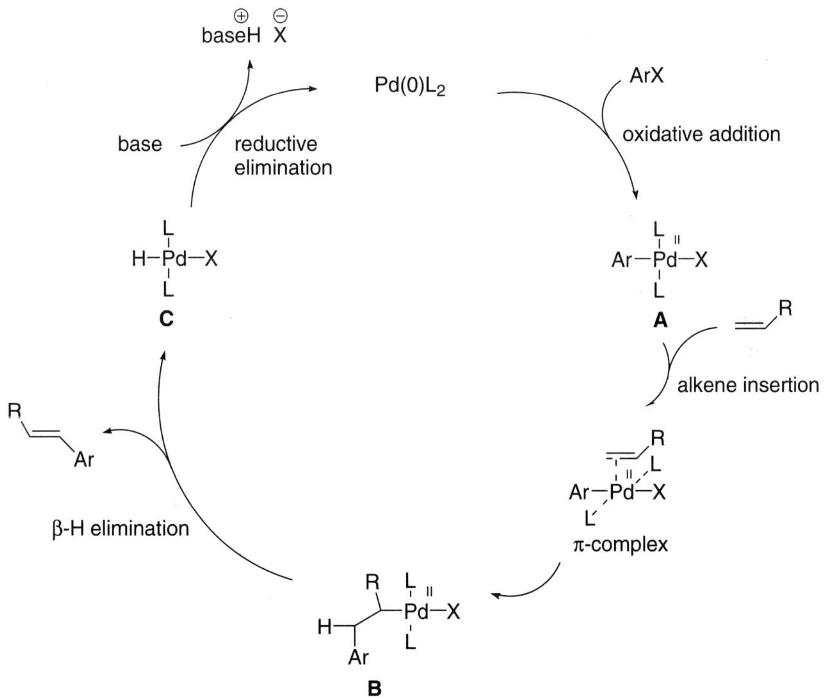

Scheme 5.4 Outline of the catalytic cycle for the Heck reaction

If a Pd(II) source is used in the Heck reaction, it must be reduced to Pd(0) before entering the catalytic cycle. The initial oxidative addition of aryl halide to a Pd(0) catalyst affords an arylpalladium(II) complex **A**. Insertion of an alkene into the Pd–Ar bond provides an alkylpalladium(II) intermediate **B**, which readily undergoes β-hydride elimination to release the alkene. A base is required for conversion of the hydridopalladium(II) complex **C** into the active Pd(0) catalyst to complete the catalytic cycle (Scheme 5.4).

An anionic mechanism is also proposed in which the oxidative addition of Ar–X to $[PdL_2OAc]^-$ gives **A**, which undergoes alkene insertion to give **B**. This is followed by β-H elimination step to form **C** and coupled product. Reductive elimination of **C** regenerates $[PdL_2OAc]^-$ to complete the catalytic cycle (Scheme 5.5).

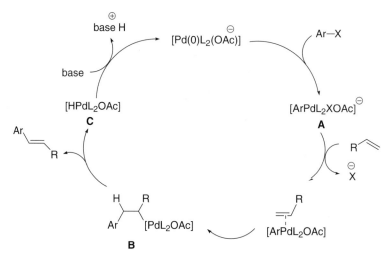

Scheme 5.5 Outline of the anionic mechanism for the Heck reaction

Asymmetric Heck reaction: Asymmetric Heck reactions have been employed as central strategic steps in a wide variety of natural product total syntheses. The chiral ligands at palladium allow control of absolute stereochemistry. A large variety of chiral ligands[24–26] are available for the asymmetric Heck reaction, but Noyori's[27] BINAP ligand is most widely used as a chiral ligand. The Pd–BINAP catalyst derived from Pd(OAc)₂ is far superior to that formed from $Pd_2(dba_3)\cdot CHCl_3$. Shibasaki *et al.* and Overman *et al.* first time reported the enantioselective intramolecular Heck reaction.

Shibasaki and co-workers[28] used (*R*)-BINAP as a chiral ligand for the enantioselective Heck cyclization of prochiral vinyl iodide **5.11** to *cis*-decalin **5.12**. Silver carbonate was used as a base and N-methyl-2-pyrrolidinone (NMP) as a solvent.

CO₂Me	Pd(OAc)₂(3 mol%)– (*R*)-BINAP (9 mol%) Ag₂CO₃ (2 equiv.) 60°C N-methyl-2-pyrrolidinone (NMP)	CO₂Me
5.11		**5.12** 74% yield, 46% ee

During the same year, Overman and co-workers[29] synthesized spirocycle **5.14** in 90% yield with 45% ee by two sequential Heck cyclization of trienyl triflate **5.13** at room temperature in the presence of Pd(OAc)₂–(*R,R*)-DIOP and Et₃N in benzene.

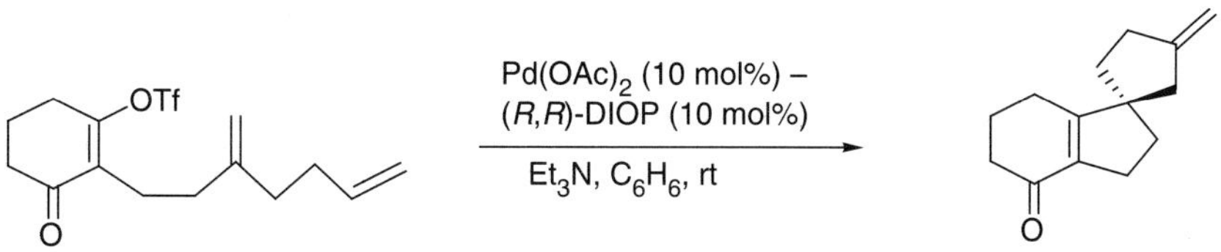

5.13

Pd(OAc)$_2$ (10 mol%) –
(*R,R*)-DIOP (10 mol%)

Et$_3$N, C$_6$H$_6$, rt

5.14

90% yield, 45% ee

Control over regioselectivity and stereoselectivity in the formation of new C–C σ-bond is required to utilize the Heck reaction in complex molecule synthesis. For the intramolecular Heck reaction, the size of the ring formed in the insertion step controls the regiochemistry, with 5-*exo* and 6-*exo* cyclization favoured. A mixture of regioisomers is formed from Heck insertions of acyclic alkenes, whereas cyclic alkenes such as cycloalkenes as a Heck substrate produce a σ-alkylpalladium(II) intermediate **A**, which has only one *syn*-β-hydrogen. Syn-elimination of the hydrogen provides only product **B** (Scheme 5.6).

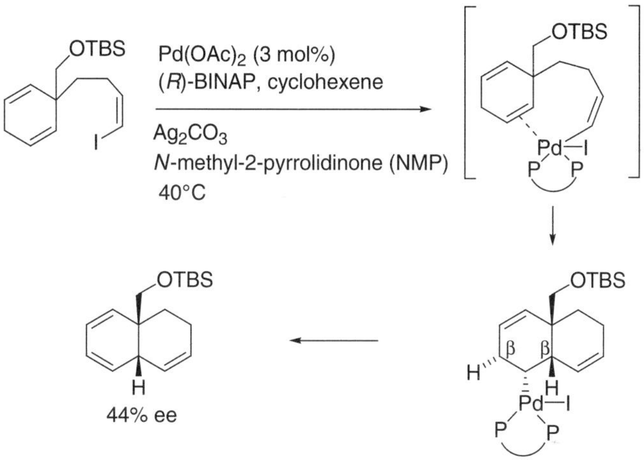

Pd(OAc)2 (3 mol%)
(*R*)-BINAP, cyclohexene

Ag$_2$CO$_3$
N-methyl-2-pyrrolidinone (NMP)
40°C

44% ee

B

A

Scheme 5.6

The asymmetric Heck cyclization of **5.15** takes place with high enantioselectivity (71–98%) with the Pd–BINAP catalyst to form the corresponding 3-alkyl-3-aryloxindole[30] **5.16**. A wide variety of aryl and heteroaryl substituents can be introduced into an oxindole by this method. This asymmetric synthesis of 3-aryl- or 3-heteroaryl oxindoles is very useful for the enantioselective synthesis of a range of indole alkaloids[31].

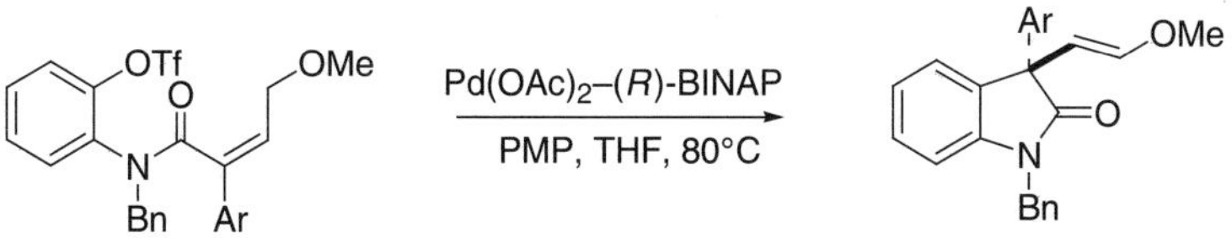

5.15

Pd(OAc)$_2$–(*R*)-BINAP

PMP, THF, 80°C

5.16

Ar = Ph: 86% yield, 84% ee
Ar = 1-naphthyl: 92% yield, 92% ee

Although BINAP (**5.1**) and modified BINAPs have given moderate to acceptable enantiomeric excesses, a better ligand class for intermolecular Heck reactions is the phosphinoxaolines **5.17**.

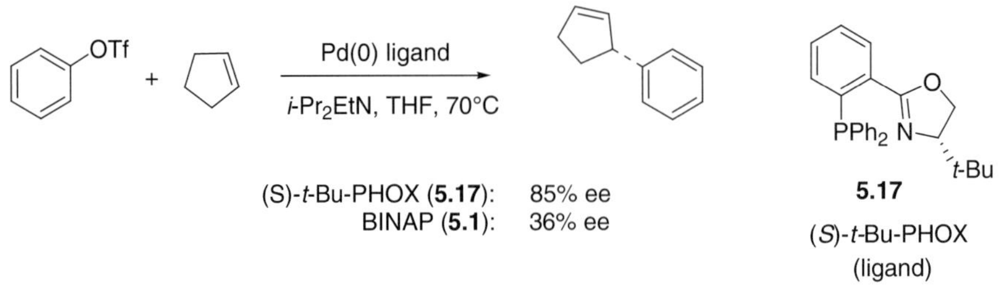

(S)-*t*-Bu-PHOX (**5.17**):   85% ee
BINAP (**5.1**):   36% ee

5.17

(S)-*t*-Bu-PHOX
(ligand)

5.1.2   Allylic substitutions

Transition metal (such as Pd, Ir, Mo and W)-catalyzed asymmetric allylic substitutions with various nucleophiles are widely employed in organic synthesis and played an important role in the area of asymmetric C–C bond formation. Trost, Helmchen, Pfaltz and others have focused primarily on the direct allylation of malonates by prochiral electrophiles[32,33].

In general, reactions involving prochiral nucleophiles are rare; however, examples using activated dicarbonyl compounds and carbonyl compounds that can form only a single enolate under the base-mediated reaction conditions have recently been reported.

The Pd-catalyzed allylic substitution reactions proceed via π-allyl complex, so that branched and linear substrates yield the same products (Scheme 5.7).

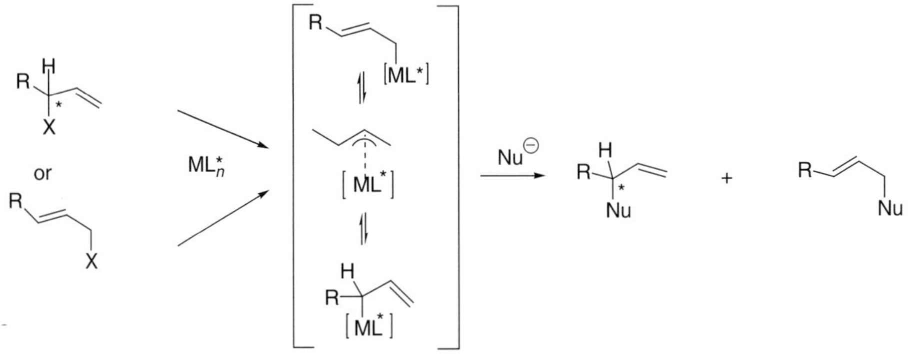

Scheme 5.7

Stereochemistry: A Pd(0) complex displaces allylic leaving groups (with inversion of configuration) to generate cationic π-allyl palladium species **A**. This complex is electronically deficient and undergoes attack of a suitable soft nucleophile (with inversion) to give a product with overall retention (Scheme 5.8).

Regiochemistry: With Pd complexes as catalysts, allylic substitution generally occurs at the less hindered end. If both ends are similar (not identical) then a mixture results (Scheme 5.9). However, recently ligands have been developed that give rise to products in which substitution occurs at the hindered end.

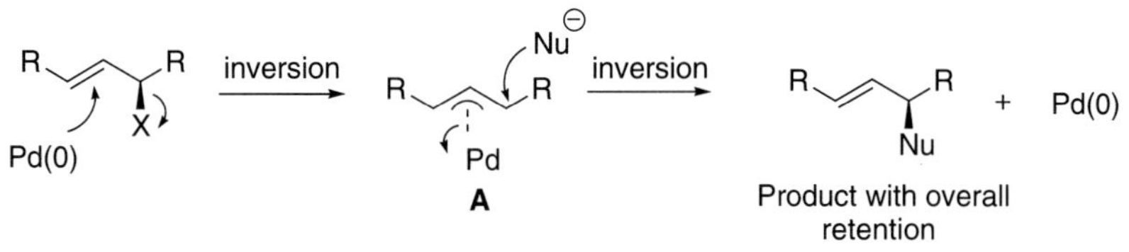

X = halogen, OCOR (carboxylate), O$_2$COR (carbonate)

$\overset{\ominus}{N}u = \overset{\ominus}{R}C(EWG)_2$; EWG = COR, CO$_2R, SO_2$Ph

Scheme 5.8

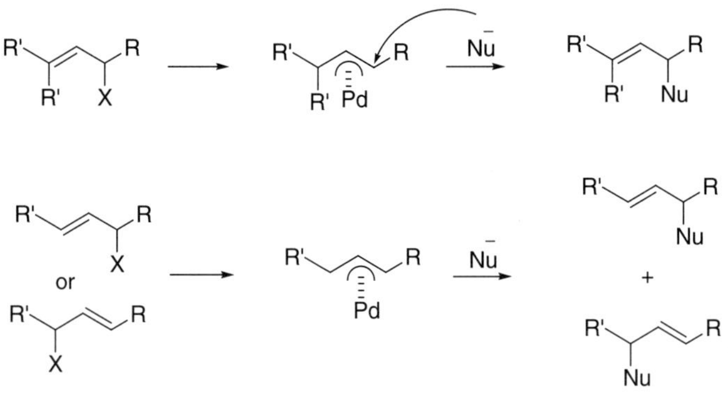

Scheme 5.9

Trost's enantioselective allylic substitutions

Chiral ligand **5.20** constructed from a chiral diamine **5.18** and 2-(diphenylphosphino) benzoic acid (**5.19**) was used by Trost and co-workers[34–36] for the enantioselective allylic substitutions.

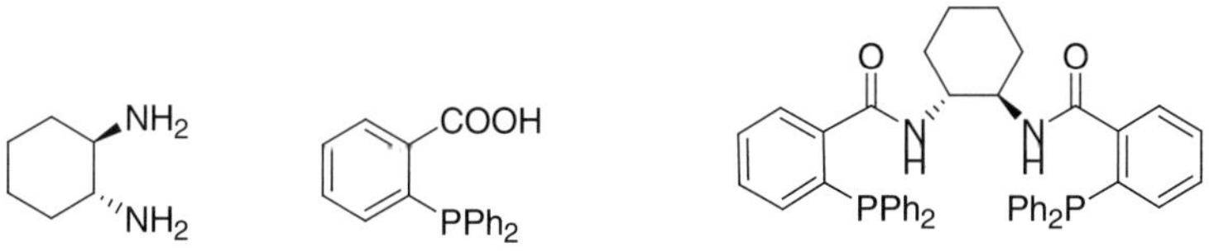

5.18   5.19   5.20

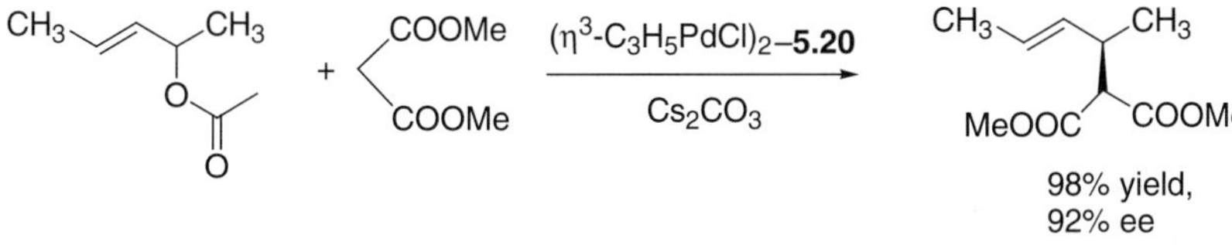

98% yield,
92% ee

A new phosphine ligand **5.21** has been synthesized and employed by Zhang and co-workers.[37] in the Pd-catalyzed enantioselective alkylation of 2-cyclohexenyl ester. The ligand can differentiate quite effectively between the *R*- and *S*-enantiomers of acetates.

(R)- and (S)-Acetates   (R)-Acetate gives product

5.21

The enantioselective Tsuji allylation

The Pd-catalyzed enantioselective Tsuji allylation[38,39] of enol carbonate to allyl ketone reaction is examined using a variety of chiral ligands by Behenna Stoltz[40]. For example, decarboxylative allylation of allyl enol carbonates in the presence of $Pd_2(dba)_3$ and a chiral ligand gives the corresponding α-allylcyclohexanone derivatives with asymmetric α-quaternary cycloalkanones. It was found that the mixed P- and N-type ligands were more effective in inducing asymmetry. For example, the use of the (S)-t-Bu-PHOX ligand **5.17** provided cyclohexanone **5.23** in 96% yield and 88% ee from **5.22**.

5.22   **5.23**   **5.17**

Iridium catalysts generally favour chiral branched products in contrast to palladium catalysts, which typically give rise to linear achiral products. The catalyst system $[Ir(COD)Cl]_2$–$P(OPh)_3$ is effective in allylic alkylation and amination[41].

R = H: 91% yield, 85% ee
R = CH3: 83% yield, 71% ee

The Ir-catalyzed enantioselective allylic substitution reaction can be used for the synthesis of β-substituted α-amino acids[42,43]. For example, the enantioselective Ir-catalyzed allylic substitution of 3-arylallyl diethyl phosphates **5.24** with pronucleophile diphenylimino glycinate **5.25** is achieved up to 98% ee by using bidentate chiral phosphite ligand **5.26**. By changing the base, both diastereoisomeric substitution products **5.27a and 5.27b** could be formed selectively.

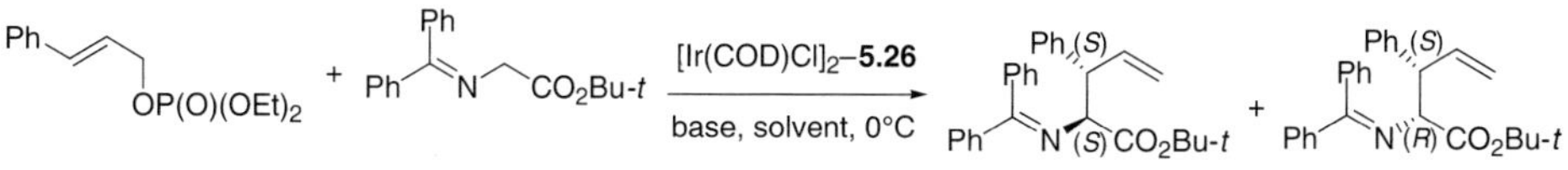

5.24   **5.25**   **5.27a**   **5.27b**

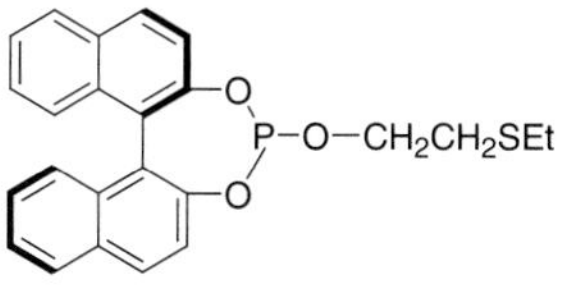

5.26

5.1.3   Cu- and Ni-catalyzed couplings

The Ullmann-type reaction that is homocoupling of aryl or vinyl halides is conveniently mediated by copper at high temperature. The copper powder serves as a zerovalent metal. The classical Ullmann reaction reported in 1901 has long been employed by chemists to generate a carbon–carbon bond between two aromatic nuclei.

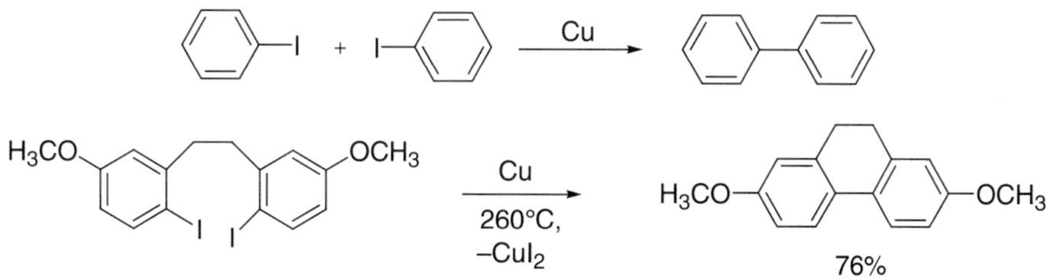

76%

Nickel in its zerovalent state can also be used for the coupling of aryl halides to biphenyls and vinyl halides to 1,3-dienes. Bis(1,5-cyclooctadiene)nickel, Ni(COD)$_2$, is the best source of zerovalent nickel for such couplings.

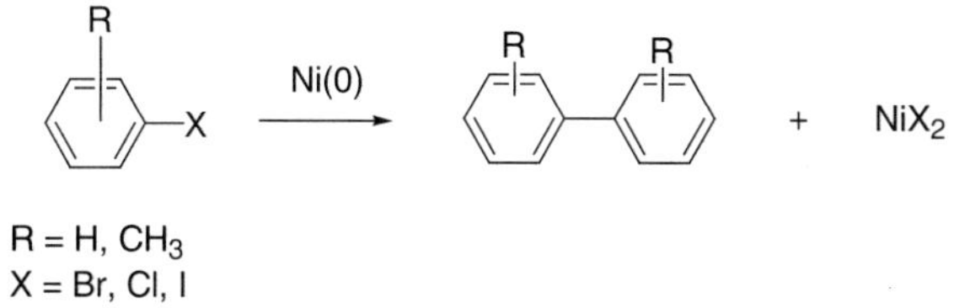

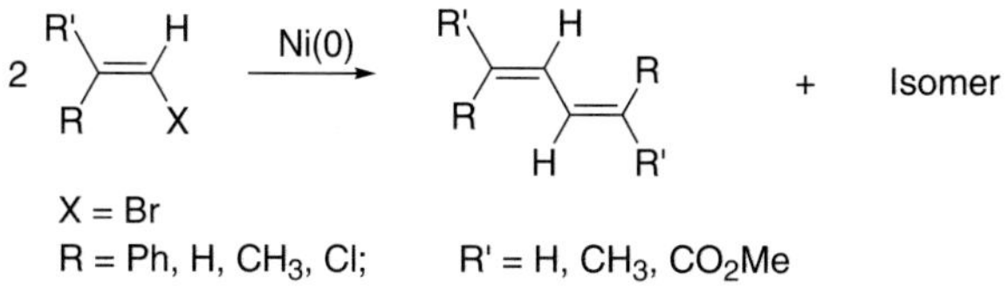

R = H, CH$_3$
X = Br, Cl, I

X = Br
R = Ph, H, CH$_3$, Cl;   R' = H, CH$_3$, CO$_2$Me

The mechanism of the above reactions was studied by Tsou and Kochi[44] (Scheme 5.10). The aryl nickel halides **A**, formed by the oxidative addition of Ar–Br to NiL$_4$, on reaction with aryl halides give biaryls.

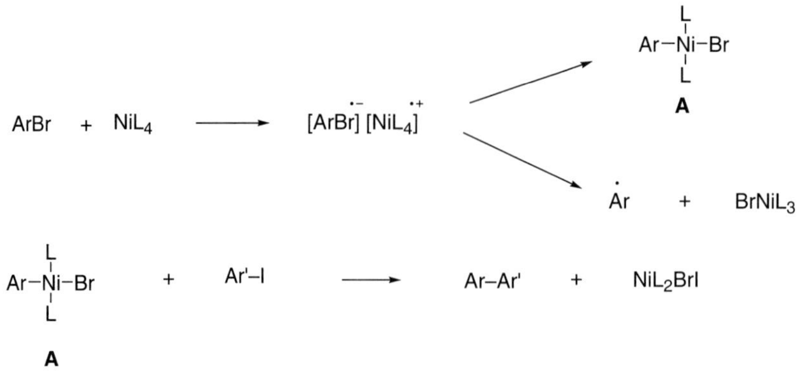

Scheme 5.10

Although copper-catalyzed homocoupling of alkynes and Ni(0)-catalyzed homocoupling of aryl halides were well known, **Cadiot and Chodkiewicz** gave first-time synthetically more useful copper-catalyzed cross-coupling of alkynes.

The copper(I)-catalyzed cross-coupling of a terminal alkyne and an alkynyl halide to yield diyne is known as the **Cadiot–Chodkiewicz coupling**[45].

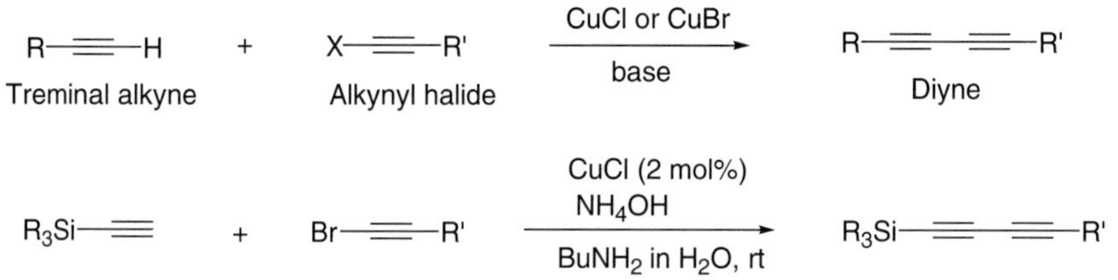

Mechanism: The reaction mechanism involves deprotonation by a base of the acetylenic proton followed by the formation of a copper(I) acetylide **A**. A cycle of oxidative addition and reductive elimination of **B** then forms diynes **C** (Scheme 5.11).

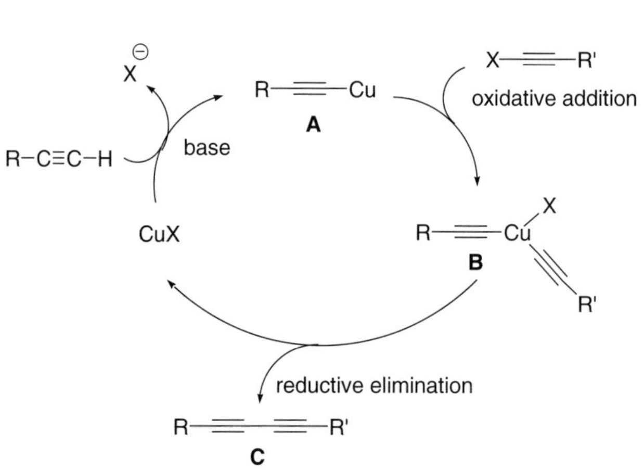

Scheme 5.11

The **Castro–Stephens coupling**[46,47] also involves the reaction of copper(I) acetylide with aryl halides to form diarylacetylenes.

5.2 Transition metal-catalyzed coupling of organometallic reagents with organic halides and related electrophiles

The carbon–carbon bond formation by the reaction of organometallic compounds with organic halides is straightforward, but often poor yields of coupled products are obtained because of several side reactions. The formation of carbon–carbon bonds via the transition metal-catalyzed coupling of organometallic reagents (R^2–M^1) with organic halides (R^1-X) or related reagents has become the most important approach and gives high yields of coupled product[48–54].

$$R^2{-}M^1 \ + \ R^1{-}X \ \xrightarrow{\text{metal catalyst}} \ R^2{-}R^1 \ + \ M^1{-}X$$

The Ni(II) complexes-catalyzed reaction of organomagnesium with alkenyl or aryl halides was reported independently by Kumada and Tamao and Corriu in 1972. The Pd-catalyzed reaction of Grignard reagents was first reported by Murahashi. Negishi *et al.* demonstrated the Pd- and Ni-catalyzed reactions of organoaluminium, zinc and zirconium reagents. After these discoveries, many other organometallic reagents were used successfully as nucleophiles for the transition metal-catalyzed cross-coupling reaction, e.g. organolithiums by Murahashi, organostannanes by Migita and Stille, 1-alkenylcopper(I) by Normant and organosilicon by Hiyama. Although organic groups on boron are weakly nucleophilic, cross-coupling reactions of organoboron reagents catalyzed by Pd(II) halides activated by suitable bases have proven to be a quite general technique for a wide range of selective carbon–carbon bond formation.

X = I, Br, Cl or OTf
M^1 = SnR$_3$ (Stille coupling)
    BR$_2$ or B(OR)$_2$ (Suzuki–Miyaura coupling)
    SiR$_{(3-n)}$F$_n$ (Hiyama coupling)
    Si(OR)$_3$ (Tamao-Ito)

General mechanism: Although nickel and palladium complexes are the most extensively studied metals in cross-coupling reactions, a vast majority of mechanistic studies concern palladium chemistry. Palladium-catalyzed cross-coupling reactions of organohalides (or organotriflates) with organometallic reagents follow a general mechanistic cycle as shown in Scheme 5.12.

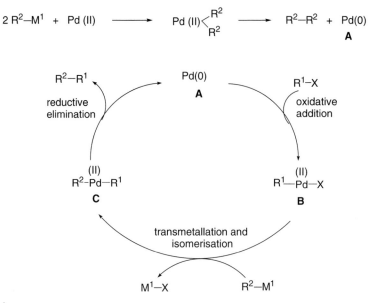

Scheme 5.12   Catalytic cycle of palladium-catalyzed cross-coupling of organometallic reagents

The Pd(0) **A** is sometimes reduced from a Pd(II) species by an organometallic reagent R^2–M^1. The transmetallation product then undergoes a reductive elimination step, giving rise to the Pd(0) species **A**, along with the homocoupling product R^2–R^2. This is one of the reasons why the organometallic coupling partners are often used in a slight excess relative to the electrophilic partners.

When Pd(0) catalyst **A** is generated, the catalytic cycle goes through a three-step sequence: (i) Electrophile R^1–X undergoes an **oxidative addition** to Pd(0) to afford Pd(II) intermediate **B**; (ii) subsequently, **B** undergoes a **transmetallation** with the organometallic reagent R^2–M^1, followed by (iii) **isomerization** step to produce intermediate **C**. Finally, with appropriate *syn* geometry, intermediate **C** undergoes a **reductive elimination** to produce the coupling product R^2–R^1 and regenerate the Pd(0) catalyst **A** to continue the catalytic cycle (Scheme 5.12). Oxidative addition is the rate-determining step in a catalytic cycle.

The Pd catalysts Pd(PPh$_3$)$_4$, Cl$_2$Pd(PPh$_3$)$_2$ and Pd(OAc)$_2$ plus PPh$_3$ or other phosphine ligands have been most frequently used. The use of other Pd catalysts is desirable in some highly demanding cases. The relative reactivity of R^1–X decreases in the order of I > OTf > Br ≫ Cl.

Various electrophiles (R^1–X) used in cross-coupling reactions can be classified by the hybridization of the carbon atom bonded to a halogen or a related leaving group, Csp^2–X: aryl, alkenyl and acyl; Csp–X: alkynyl; Csp^3: allyl, benzyl, propargyl and alkyl. But electrophiles of class Csp^2, i.e. aryl, heteroaryl, alkenyl, allenyl and acyl electrophiles, have proved to be generally satisfactory. The reactivity of sp^3 substrates other than methyl iodide is slow, and the transition metal-catalyzed formation of Csp^3–Csp^3 bonds has been much less successful. This is due to slow oxidative addition of the alkyl halide or sulfonate to the metal centre and fast thermodynamically favoured β-hydride elimination of the resulting alkyl–metal complex **A** (Scheme 5.13). Much work has been carried out in developing suitable reaction conditions for cross-coupling reactions[55] of alkyl halides by facilitating the oxidative addition and reductive elimination steps and by preventing the competing β-hydride elimination.

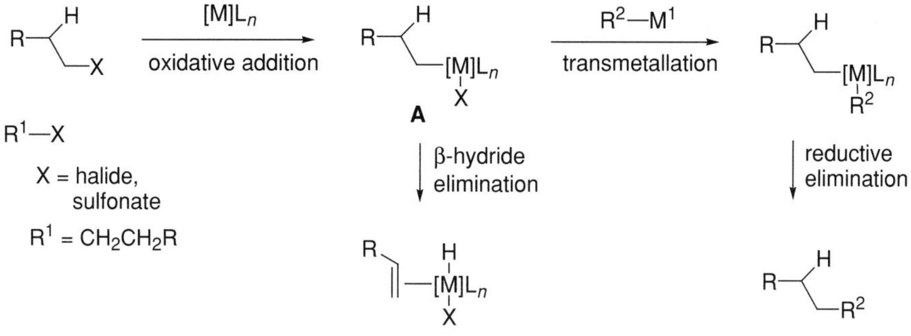

Scheme 5.13

5.2.1   Coupling of Grignard reagents

Copper-catalyzed cross-couplings between unactivated alkyl halides and organomagnesium compounds have been known since the late 1960s and probably are the most studied alkyl–alkyl coupling reactions[56]. The main drawback of this protocol was the need of stoichiometric amounts of copper salts for the formation of the organocuprate, and consequently this led to loss of 1 equiv. of the non-transferred alkyl group.

After Kumada and Tamao[57] and Corriu[58] independently reported the nickel(II) salts- and complexes-catalyzed cross-coupling reaction of Grignard reagents with aryl and alkenyl halides, the Pd-catalyzed reaction of Grignard reagents was first reported by Murahashi[59].

$$R^2MgX^1 \ + \ R^1{-}X^2 \ \xrightarrow{Pd(0) \ catalyst} \ R^2{-}R^1 \ + \ MgX^1X^2$$

Kumada coupling

The cross-coupling reaction of Grignard reagents with organic halides (alkenyl, aryl and heteroaryl halides or triflates) in the presence of Ni(II) complex is known as the **Kumada coupling**[57,58]. For example, cross-coupling of *m*-dichlorobenzene (**5.28**) and *n*-butylmagnesium bromide (**5.29**) in the presence of dichloro(1,2-bis(diphenylphosphino)ethane)nickel(II), [NiCl$_2$(dppe)], gave 94% *m*-di-*n*-butylbenzene[59] (**5.30**). Similarly, α-vinylnaphthalene (**5.33**) was obtained in 80% yield by the cross-coupling reaction[59] between vinyl chloride (**5.31**) and α-naphthylmagnesium bromide (**5.32**) in the presence of NiCl$_2$(dppe).

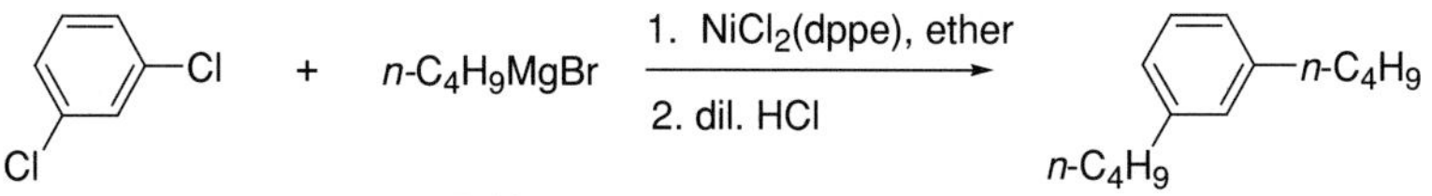

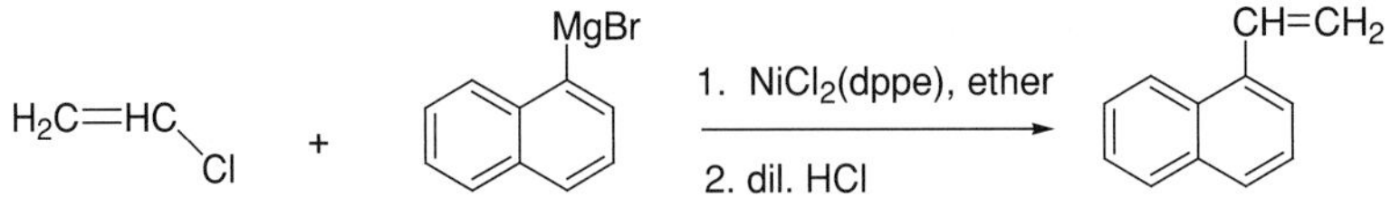

5.31   **5.32**   **5.33**

80%

dpe = 1, 2-Bis(diphenylphosphino)ethane or $Ph_2PCH_2CH_2PPh_2$

The advantage of this reaction is that numerous Grignard reagents are commercially available; those that are not commercially available can be readily prepared from the corresponding halides. Another advantage is that the reaction can often be run at room or lower temperature. A disadvantage of this method is the intolerance of many functional groups (such as OH, NH_2 and C=O) by the Grignard reagents.

It is reported recently that unactivated aryl chlorides can be used as a substrate in the Kumada coupling. For example, when 4-chlorotoluene (**5.34**) is reacted with phenyl-magnesium bromide (**5.35**) in the presence of $Pd_2(dba)_3$ and IPr·HCl [1,3-bis(2,6-diisopropylphenyl)imidazolium chloride] in dioxane–THF (tetrahydrofuran), the product 4-phenyltoluene (**5.36**) is isolated in 99% yield[60].

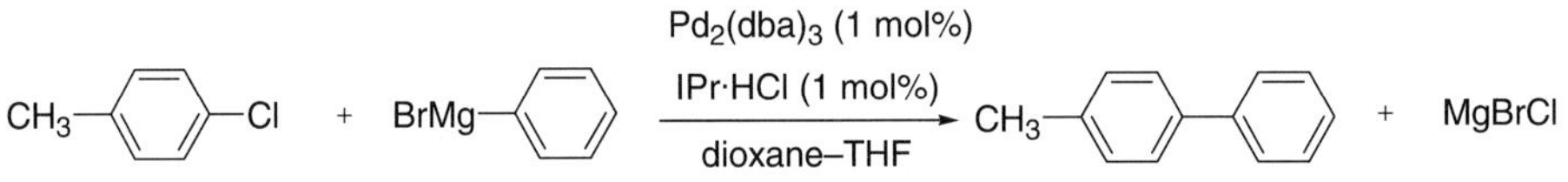

5.34   **5.35**   **5.36**

99%

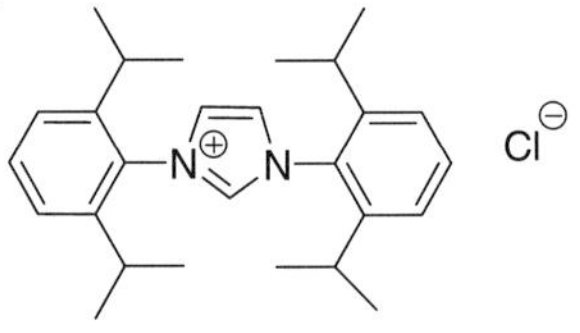

IPr·HCl

Mechanism: The intermediate diorganonickel complex **A** is formed by the reaction of dihalodiphosphine nickel (L_2NiX_2) with the Grignard reagent (R^2MgX^1). Complex **A** is converted into haloorganonickel complex **B** by an organic halide (R^1–X^2). Further, reaction of complex **B** with the Grignard reagent R^2MgX^1 forms a new diorgano complex **C**, which releases the cross-coupling product (R^1–R^2) on reaction with an organic halide (R^1–X^2) and the original complex **B** is regenerated to complete the catalytic cycle (Scheme 5.14).

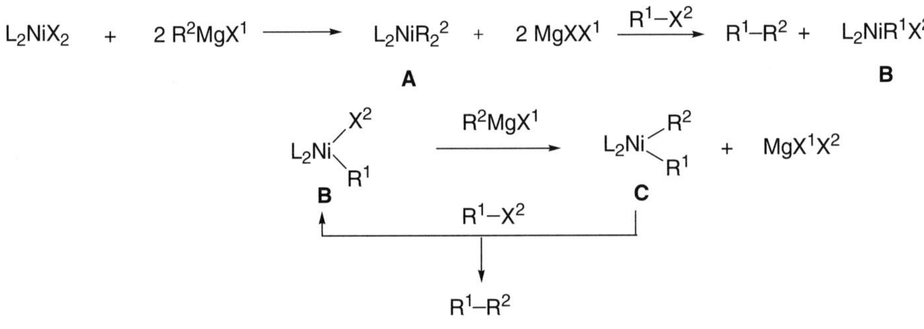

Scheme 5.14

Ni- and Pd-catalyzed cross-coupling reactions of halopurines with Grignard reagents can be used to introduce an aryl or alkyl group into 6- or 8-position.

R' = TMS; R = Me, allyl, aryl

R' = TBS; R = phenyl, alkyl

dppf = 1,1'-Bis(diphenylphosphino)ferrocene

The first cross-coupling reaction of simple aliphatic iodides with a variety of Grignard reagents in the presence of *in situ* generated Pd(0)(dppf) catalyst was reported in 1986. However, the mixture of alkane and alkene is obtained under these conditions. It was reported recently that the neopentyl iodide derivatives react with the Grignard reagent in the presence of a nickel catalyst such as $NiCl_2(dppf)$ to give the corresponding cross-coupling products[61].

R = Ph, H
R^2 = Me, Et

The best results were obtained when aromatic Grignard reagents[62] were employed in the presence of $NiCl_2(dppf)$. For example, neopentyl iodide (**5.37**) reacts with **5.35** to give 86% **5.38** in the presence of $NiCl_2(dppf)$.

| | 5.37 | | 5.35 | | | 5.38 |
| | | | | | | 86% |

Addition of 1,3-butadiene has recently been found[63] to remarkably increase the rate of Ni-catalyzed cross-coupling between Grignard reagents with alkyl chlorides, bromides and tosylates (R^1–X^2). For example, cross-coupling between *n*-decyl bromide (**5.39**) and *n*-butylmagnesium chloride (**5.29**) in the presence of isoprene (**5.40**) as additive and $NiCl_2$ gave tetradecane (**5.41**) in 92% yield. $Ni(acac)_2$ and $Ni(COD)_2$ also gave high yield of **5.41**.

$$R^1-X^2 \quad + \quad R^2MgX^1 \xrightarrow[\text{isoprene (1 equiv.)}]{\text{catalyst (3 mol\%)}} R^1-R^2$$

$$R^1 = \text{alkyl}; X^1 \neq X^2 = \text{Cl, Br, OTs}$$
$$R^2 = \text{alkyl, aryl}$$

$$n\text{-C}_{10}\text{H}_{21}\text{Br} \quad + \quad n\text{-BuMgCl} \xrightarrow{\text{NiCl}_2} n\text{-C}_{14}\text{H}_{30}$$

5.39   **5.29**   **5.41**

(5.40)   92%

Although alkyl (1° and 2°) and aryl Grignard reagents afforded the corresponding cross-coupling products, the alkenylmagnesium halide under similar conditions were found to be inactive.

Mechanism: Butadiene plays an important role in converting Ni(0) to Ni(II) species **A**, which does not oxidatively add but undergoes transmetallation with the Grignard reagent. NiCl$_2$ reacts with R^{2}MgX^1 and Ni(II) is reduced to Ni(0), which reacts with 2 equiv. of 1,3-butadiene to afford bis-π-allyl nickel complex **A**. Complex **A** is less reactive towards alkyl halides but reacts with the Grignard reagent. Thus, **A** undergoes transmetallation with R^{2}MgX^1 to form intermediate **B**. The oxidative addition of alkyl halides (R^1–X^2) to complex **B** yields dialkylnickel complex **C**, which on reductive elimination gives the coupling product R^1–R^2 and original complex **A** (Scheme 5.15).

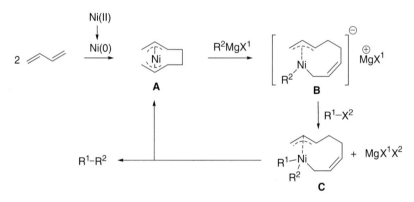

Scheme 5.15

5.2.2 Coupling of organostannanes

Stille coupling

The Stille coupling[64–66] between organostannanes (aryl and vinyl) and an electrophile (such as aryl halide) in the presence of palladium complex is a versatile method for the formation of carbon–carbon bonds.

$$\text{R} \overset{}{\bigcirc}\!\!-X \quad + \quad R^2{-}SnR_3 \xrightarrow[\text{base}]{\text{Pd catalyst}} \text{R} \overset{}{\bigcirc}\!\!-R^2$$

$$R^2 = \text{aryl and vinyl}$$

The moisture stability of the organostannanes and good functional group tolerance of the reaction make it most extensively used in coupling reactions. However, toxicity and low polarity of tin compounds are certain drawbacks of the use of the Stille reaction. The Suzuki coupling makes use of boronic acids and their derivatives, which is an improvement on the Stille coupling. In contrast to the Suzuki, Kumada, Heck and Sonogashira reactions which are carried out under basic conditions, the Stille reaction can be carried out under neutral conditions.

100%

Chloroaryls, which are electronically deactivated and thus resistant to enter to the oxidative addition, were coupled with phenyltributyltin ($PhSnBu_3$) by using tricyclohexylphosphine (Pcy_3) adducts of palladium in K_3PO_4 and 1,4-dioxane to yield the corresponding biaryls in good yield[67].

R = 4-OCH$_3$, 2-OCH$_3$, 4-CH$_3$, 2-CH$_3$

Chemoselective Stille cross-coupling of 2′-triflato-(Z)-2-stannyl-2-butenanilides **5.42** with aromatic or heteroaromatic iodides gives the 2′-triflato-(Z)-2-aryl(or heteroaryl)-2-butenanilides **5.43** in high yield[68]. The faster rate of the oxidative addition of iodides compared to triflates to Pd(0) is the origin of chemoselectivity observed in this reaction[69].

5.42   **5.43**

Mechanism: The Stille reaction is somewhat related to the Heck reaction of organic electrophiles; however, although the first step (i.e. the oxidative addition step) in both processes is identical, in the Heck reaction there is no transmetallation step. The Stille reaction involves a transmetallation step, a transfer of an organic group R^2 from Sn to Pd along with the coupling of two groups to give R^1–R^2. When there is more than one group attached to Sn, the order of transmetallation for different substituents is alkynyl > vinyl > aryl > allyl ~ benzyl ≫ alkyl.

The original mechanistic proposal for the Stille reaction is shown in Scheme 5.16. In the generalized mechanism, a Pd(0)L_n (L = PPh_3) complex was assumed to be the active catalytic species, which reacts with the organic electrophile R^1–X to form complex **A**. Transmetallation reaction with the organostannane is the slowest step and leads to the formation of complex **B**. A trans-to-cis isomerization gives complex **C**, which undergoes the reductive elimination to yield the organic product R^1–R^2 (Scheme 5.16). The most commonly used electrophiles in Stille couplings are organic iodides, bromides and triflates[70].

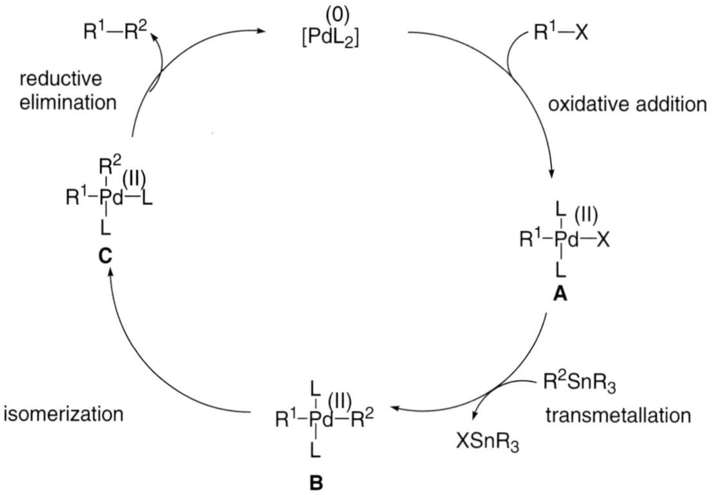

Scheme 5.16

Halopurines can undergo cross-coupling reactions with organotin reagents.

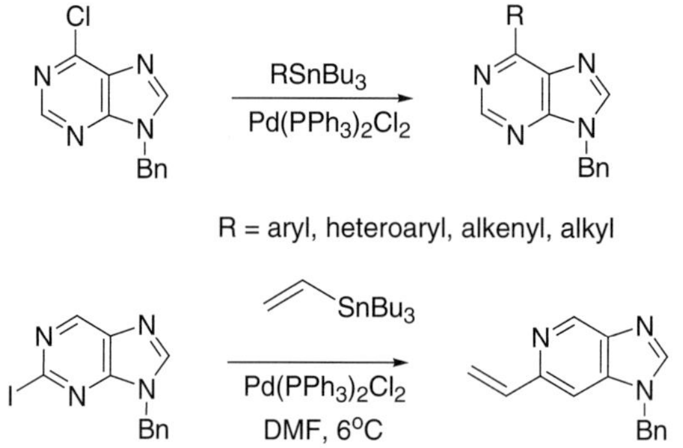

R = aryl, heteroaryl, alkenyl, alkyl

The Stille coupling can be used for the synthesis of oligoenes[71].

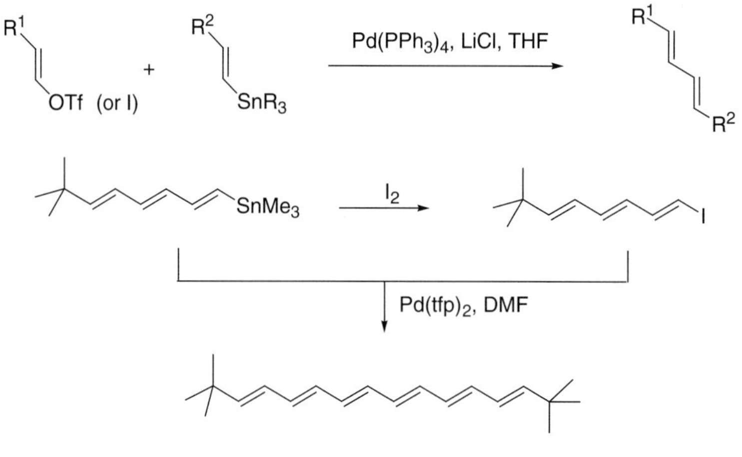

tfp = Tri-2-furylphosphine

5.2.3   Coupling of organoboranes

Suzuki–Miyaura coupling

The Suzuki–Miyaura reaction[72] is the Pd-catalyzed cross-coupling between organoboron compounds (aryl and vinyl boronic acid) and organohalides or triflates (aryl or vinyl) in the presence of a base. The reaction is very effective when aromatic iodides are utilized. The Suzuki–Miyaura cross-coupling reaction represents one of the most widely used processes for the synthesis of biaryls. The palladium catalyst $Pd(PPh_3)_4$ is most commonly used, but $Pd(PPh_3)_2Cl_2$ and $Pd(OAc)_2$ plus other phosphine ligands are also efficiently used. The higher cost and toxicity of organostannanes make the Suzuki–Miyaura coupling the preferred method.

Triflates (pseudohalides) instead of halides as well as boron esters instead of organoboronic acid may also be used.

Mechanism: The reaction proceeds first by the oxidative addition of organohalide to the Pd(0) complex to give a palladium(II) intermediate as in the case of Stille coupling. The Pd(II) complex then undergoes transmetallation with the base-activated boronic acid to give complex **B**. This is followed by reductive elimination to form the active Pd(0) species, HX and the cross-coupled product (Scheme 5.17).

Oxidative addition is often the rate-determining step and initially gives a cis complex that rapidly isomerizes to its *trans*-σ-palladium(II) complex. The reaction proceeds with the complete retention of configuration for alkenyl halides and with inversion for allylic and benzylic halides. The relative reactivity of leaving groups is $I^- > OTf^- > Br^- \gg Cl^-$.

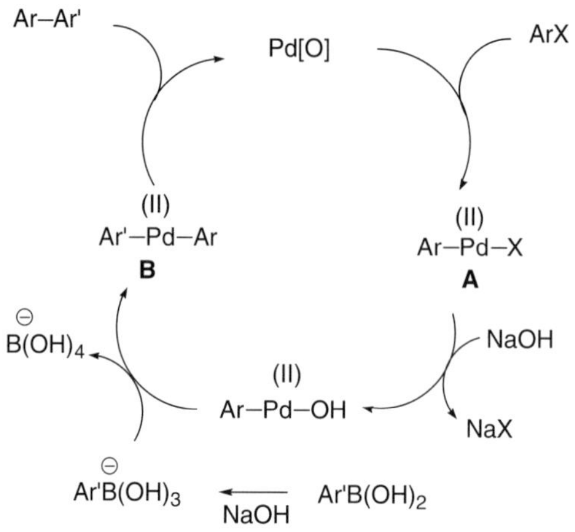

Scheme 5.17

A solventless Suzuki coupling reaction has been developed using both thermal and microwave-assisted methods. A potassium fluoride–alumina mixture is utilized along with palladium powder. The KF acts as a base[73].

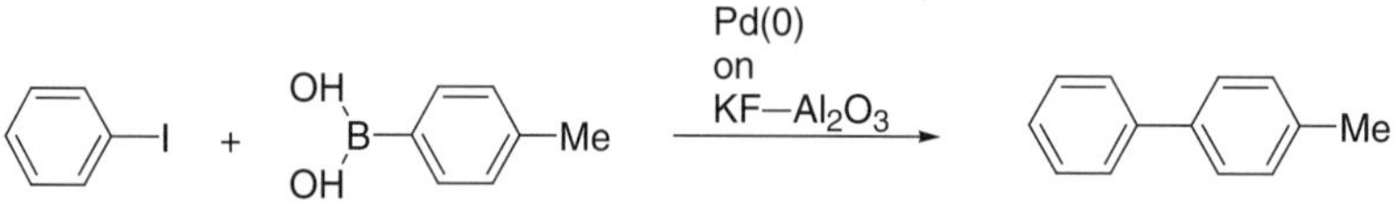

The transition metal activates the C–X bond in the oxidative addition step and normally the substrates have sp or sp^2 carbons at or immediately adjacent to an electrophilic centre. The reactivity of aliphatic C–X bond towards the oxidative addition with a transition metal is somewhat low. However, in 1992, Suzuki and co-workers[74] discovered that $Pd(PPh_3)_4$ can catalyze couplings of alkyl iodides with alkyl boranes at 60°C in moderate yields (50–71%). These conditions tolerated a wide variety of functional groups such as esters, ketals and cyanides.

$$C_{10}H_{21}I \ + \ (9\text{-BBN})\text{--Bu} \ \xrightarrow[\text{K}_3\text{PO}_4,\ \text{dioxane}]{\text{Pd(PPh}_3)_4} \ C_{14}H_{30}$$

$$CH_3I \ + \ (9\text{-BBN})\text{--}(CH_2)_{10}CO_2Me \ \xrightarrow[\text{K}_3\text{PO}_4,\ \text{dioxane}]{\text{Pd(PPh}_3)_4} \ CH_3(CH_2)_{10}CO_2Me$$

Charette synthesized polycyclopropane natural product by utilizing this approach, i.e. by coupling iodocyclopropanes to various boronic acids or esters[75].

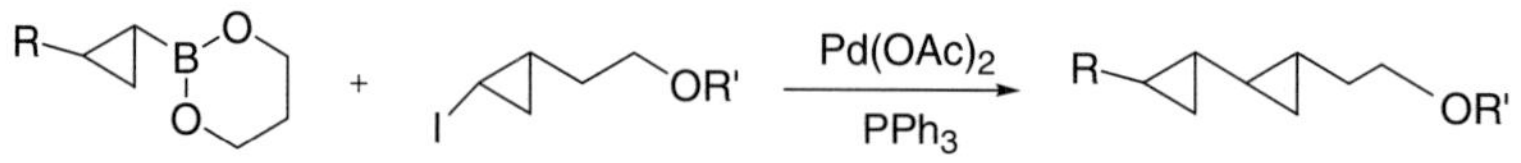

R = alkyl, alkoxy
R′ = H, benzyl

Stereospecific cross-couplings of *E*- and *Z*-tosylates **5.44** and **5.46** with aryl boronic acids in the presence of $PdCl_2(PPh_3)_2$ and aqueous Na_2CO_3 in THF gave the corresponding trisubstituted *E*- and *Z*–α,β-unsaturated esters[76] **5.45** and **5.47**, respectively.

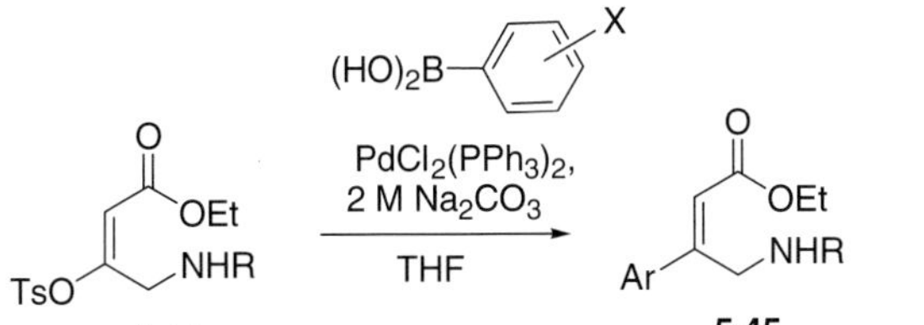

X = OMe, CN, CHO, F

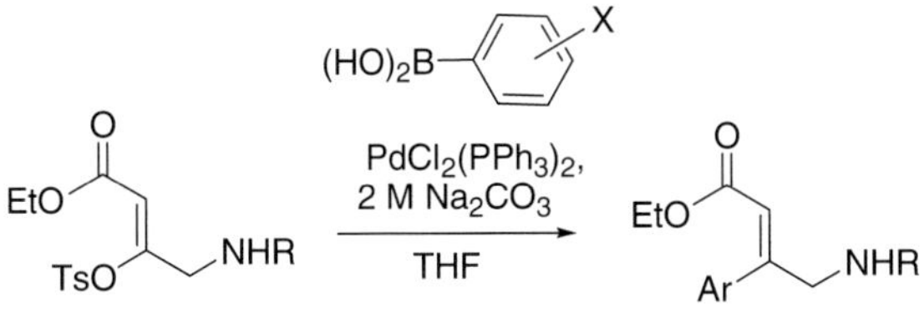

Ar = XC₆H₄

5.2.4 Coupling of organosilanes

The pioneering work of Hiyama has demonstrated that organosilanes[77] (suitably function-alized) in the presence of a nucleophilic activator can undergo Pd-catalyzed cross-coupling reactions. The chlorosilanes, fluorosilanes and alkoxysilanes are used to couple with a variety of electrophiles.

Hiyama coupling

It is a palladium-catalyzed cross-coupling reaction between organosilanes[78,79] (vinyl, ethynyl and allylsilanes) and organic halides (aryl, vinyl and allyl halides). Allylpal-ladium chloride dimmer [(η³-C₃H₅PdCl)₂] and either tris(diethylamino)sulfonium difluorotrimethylsilicate (TASF) or tetra-*n*-butylammonium fluoride (TBAF) are used as catalysts. Fluoride ion acts as an activator for the coupling, forming an intermediate hypervalent anionic silicon species, which can then transmetallate with palladium as a preliminary reaction to coupling.

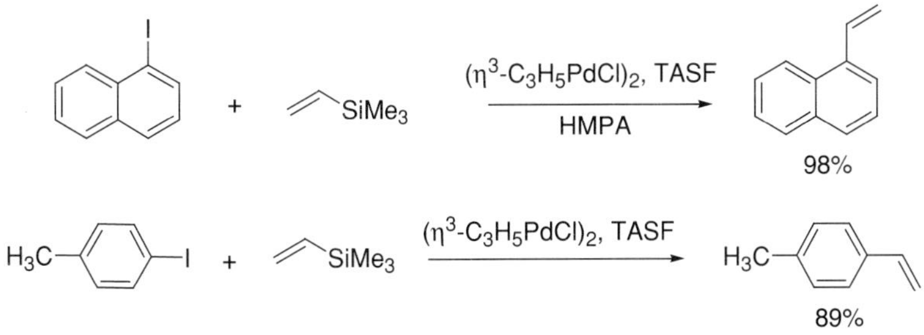

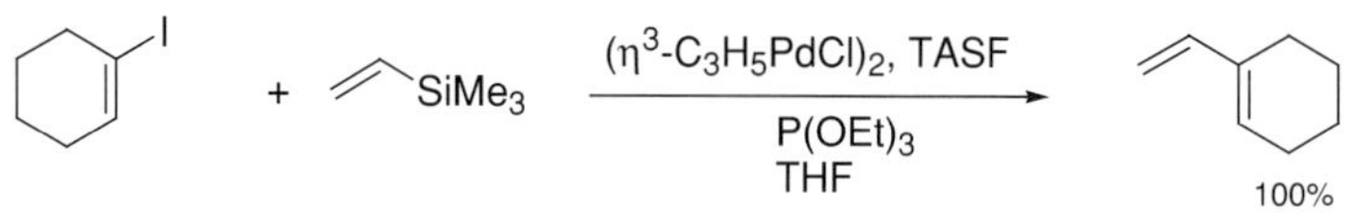

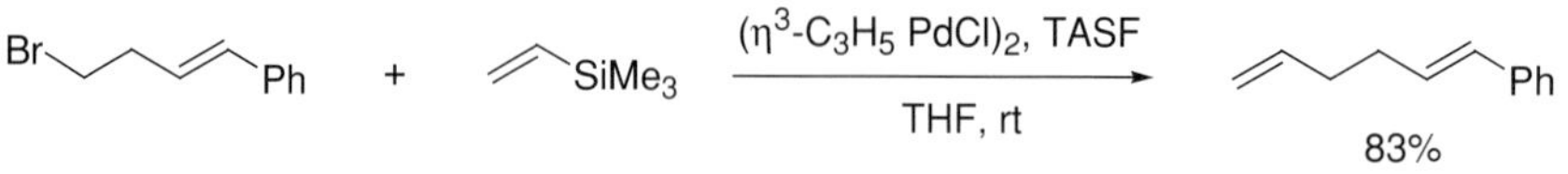

These coupling reactions possess several advantages as compared to the coupling reactions of organotin and organoboron, because organosilicon compounds are less toxic and less oxygen sensitive than the corresponding tin and boron compounds.

Stereospecificty and regioselectivity of the reaction are noteworthy. The reaction proceeds with the retention of the double bond geometry of the vinyl halide.

The Pd-catalyzed cross-coupling reaction of vinyl(2-pyridyl)silanes **5.48** with organic halides gave substituted vinyl(2-pyridyl)silanes **5.49** in high yields. The mechanism of this reaction involves the carbometallation pathways (Scheme 5.18).

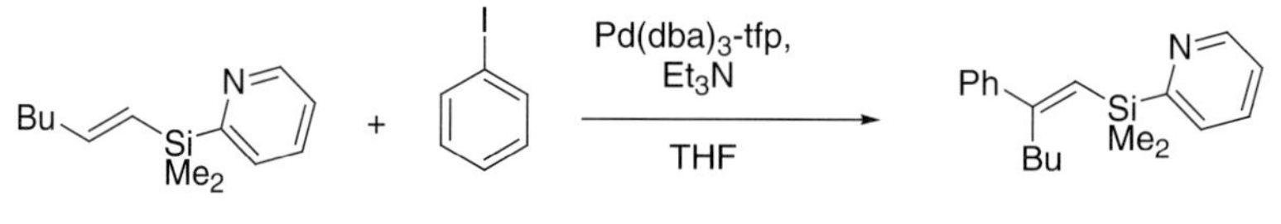

5.48 **5.49** 94%

tfp = Tri-2-furylphosphine

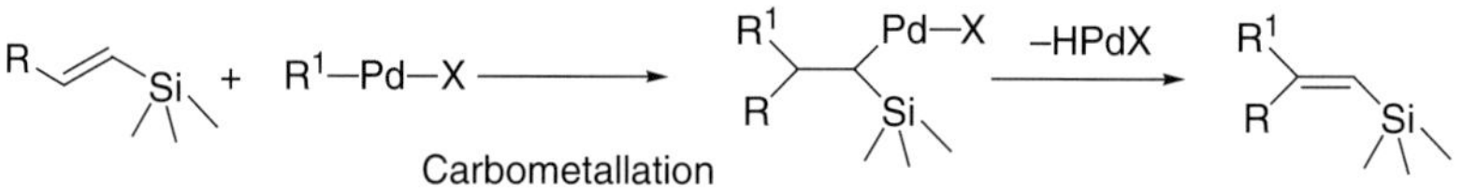

Scheme 5.18

However, when the Pd–TBAF system is used in the above reaction, the transmetallation from silicon to palladium can be accelerated by the fluoride ion and the reaction pathway is changed from carbometallation to transmetallation pathway (Scheme 5.19).

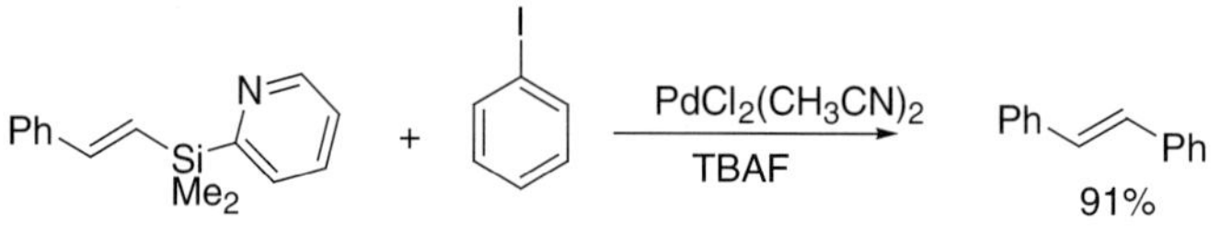

5.48

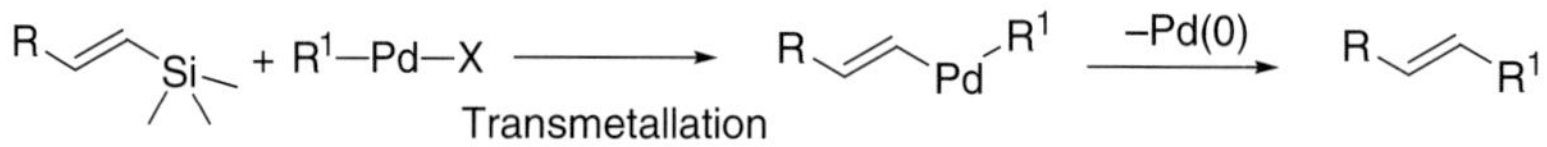

Scheme 5.19

The mechanism for this transformation is not clear, but one could speculate that the reaction proceeds via a transmetallation pathway similar to fluoride-induced silicon to palladium transmetallation mechanism.

5.2.5 Coupling of organocopper reagents

The most widely used Pd-catalyzed alkynylation for the synthesis of alkynes, the Sonogashira reaction, is the hybrid of alkyne version of the Heck reaction and Cu-promoted Castro–Stephens reaction[80] (Scheme 5.20).

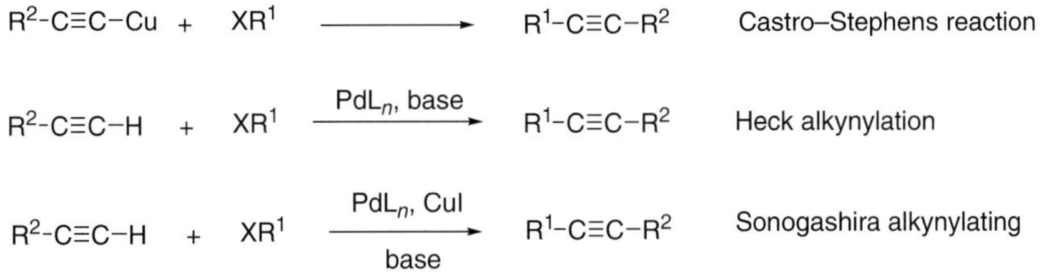

$$R^2\text{-C}{\equiv}\text{C-Cu} \;+\; \text{XR}^1 \quad\longrightarrow\quad R^1\text{-C}{\equiv}\text{C-}R^2 \qquad \text{Castro–Stephens reaction}$$

$$R^2\text{-C}{\equiv}\text{C-H} \;+\; \text{XR}^1 \;\xrightarrow{\text{PdL}_n,\ \text{base}}\; R^1\text{-C}{\equiv}\text{C-}R^2 \qquad \text{Heck alkynylation}$$

$$R^2\text{-C}{\equiv}\text{C-H} \;+\; \text{XR}^1 \;\xrightarrow[\text{base}]{\text{PdL}_n,\ \text{CuI}}\; R^1\text{-C}{\equiv}\text{C-}R^2 \qquad \text{Sonogashira alkynylating}$$

Scheme 5.20

Sonogashira coupling

The Pd-catalyzed cross-coupling of terminal alkynes with aryl and alkenyl halides in the presence of Cu(I) as co-catalyst to give arylalkynes and enynes is known as the Sonogashira coupling[81] (first time reported by Kenkichi Sonogashira *et al.*). Triethylamine or diethylamine is used as a solvent.

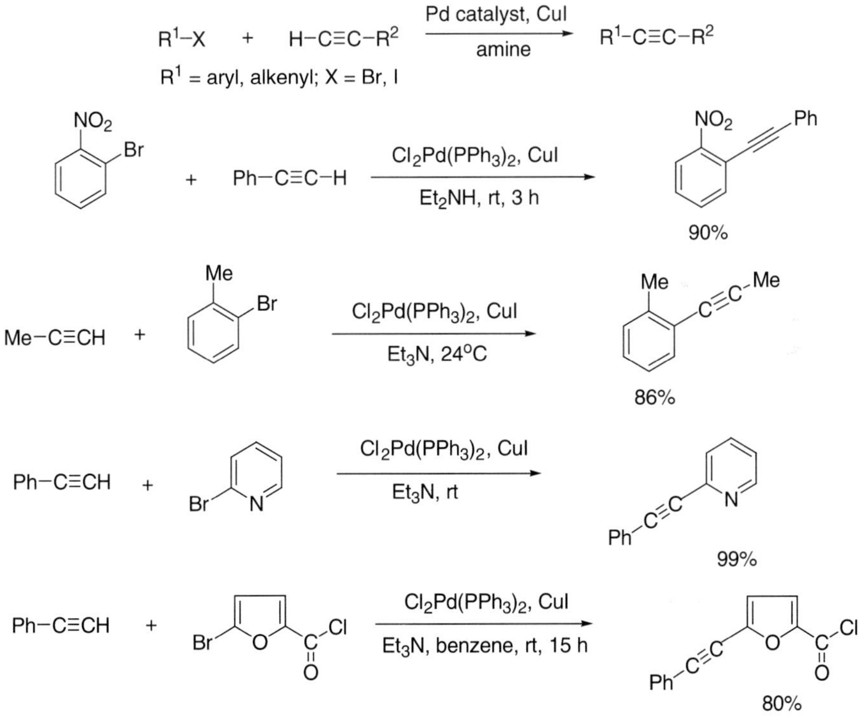

$$R^1\text{-X} \;+\; \text{H-C}{\equiv}\text{C-}R^2 \;\xrightarrow[\text{amine}]{\text{Pd catalyst, CuI}}\; R^1\text{-C}{\equiv}\text{C-}R^2$$

R^1 = aryl, alkenyl; X = Br, I

The most used catalyst in this reaction is $Cl_2Pd(PPh_3)_2$–CuI in Et_2NH or $Pd(PPh_3)_4$–CuI in R_2NH or R_3N. For example, arylacetylenes were prepared by the Pd-catalyzed reaction of arylbromides with terminal alkynes in the presence of CuI in triethylamine (Et_3N) base and THF as a solvent.

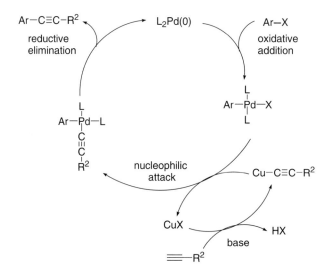

R = CHO, COMe, CO$_2$Me
R^2 = Me$_3$Si, Ph, *n*-Bu

Mechanism: The Pd complex such as Pd(PPh$_3$)$_4$ activates the organic halides by oxidative addition into the carbon–halogen bond. The copper(I) halides react with the terminal alkyne and produce copper acetylide, which acts as an activated species for the coupling reactions. The oxidative addition step is followed by the transmetallation step. The proposed catalytic cycle is shown in Scheme 5.21.

Scheme 5.21　Catalytic cycle of the Sonogashira coupling

5.2.6　Coupling of organozinc compounds

The use of transition metal catalysts, especially those containing Pd and Ni, has converted otherwise unreactive or sluggishly reactive organozincs into highly reactive reagents for their reactivity towards various common electrophiles. This includes carbonyl addition reactions of organozinc reagents, which are catalyzed by Pd complexes.

Fukuyama coupling
The coupling of organozinc reagents with thioesters in the presence of palladium catalyst is known as the Fukuyama coupling[82].

Mechanism: Oxidative addition of the thioester to Pd(0) complex and then transmetallation followed by reductive elimination gives the final product (Scheme 5.22).

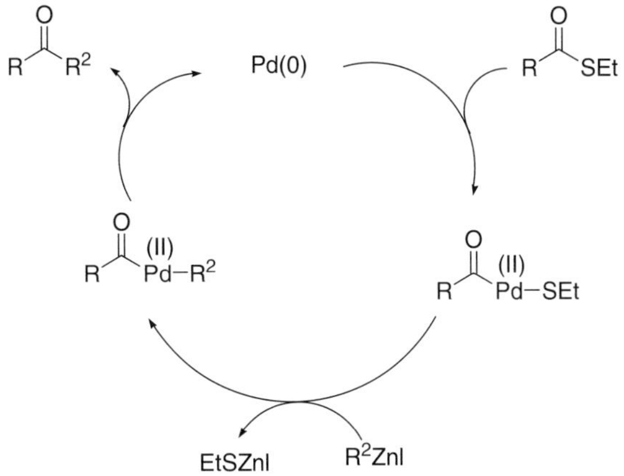

Scheme 5.22

Negishi coupling

Negishi *et al.*[83] reported Pd- or Ni-catalyzed cross-coupling of organozinc, organoaluminium and organozirconium compounds with organohalides (and triflates). The Ni-catalyst tetrakis(triphenylphosphine)nickel, Ni(PPh₃)₄, was generated *in situ* by the reaction of anhydrous Ni(acac)₂ with 1 equiv. of DIBAH (diisobutylaluminium hydride, DIBAL or DIBAL-H) in the presence of 4 equiv. of Ph₃P in THF. The Pd-catalyst was generated by the reaction of PdCl₂(PPh₃)₂ (1 mmol) with DIBAH (2 mmol) in THF.

$$\text{Ni catalyst} = \text{Ni(acac)}_2 + \text{DIBAH} + \text{PPh}_3 \ (1{:}1{:}4)$$
$$\text{Pd catalyst} = \text{PdCl}_2(\text{PPh}_3)_2 + \text{DIBAH} \ (1{:}2)$$

For example, (*E*)-1-octenylzinc chloride (**5.50**), generated *in situ* from (*E*)-1-octenyliodide, *tert*-butyllithium and dry ZnCl₂, on cross-coupling reaction with (*E*)-1-hexenyliodide (**5.51**) in the presence of Pd catalyst Pd(PPh₃)₄ gave (*5E,7E*)-5,7-tetradecadiene **5.52** in 95% yield (Scheme 5.23).

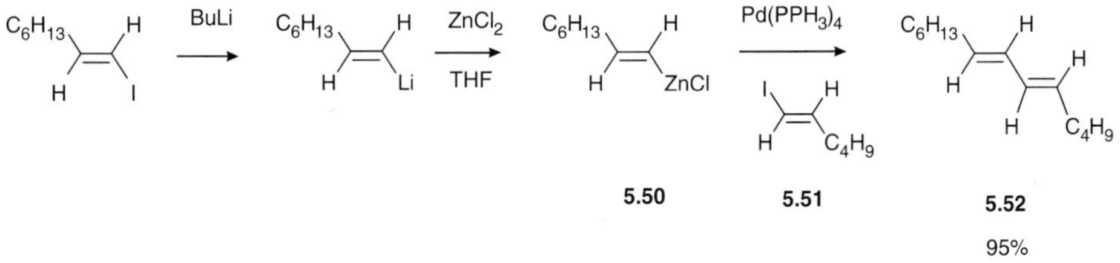

Scheme 5.23

The Ni-catalyzed reaction of (*E*)-1-alkenylalanes and (*E*)-1-alkenylzirconium derivatives with aryl halides (ArX) gave the corresponding cross-coupled products in good yields, as shown in Table 5.1.

Table 5.1 Ni-catalyzed reaction of (*E*)-1-alkenylalanes and (*E*)-1-alkenylzirconium derivatives with aryl halides (ArX)

(R²M)		(ArX)	Product
1-HexenylAl(*i*-Bu)₂	+	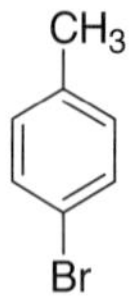 CH₃ ... Br	84%
1-HexenylAl(*i*-Bu)₂	+	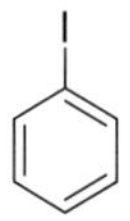 I	91%
1-HexenylZrCp₂–Cl	+	Br ... CN	92%
1-HexenylZrCp₂–Cl	+	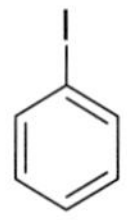 I	96%

In a similar manner, the Pd- or Ni-catalyzed reaction of (*E*)-1-alkenylalanes with alkenyl halides gave the corresponding dienes with high stereoisomeric purity (Scheme 5.24).

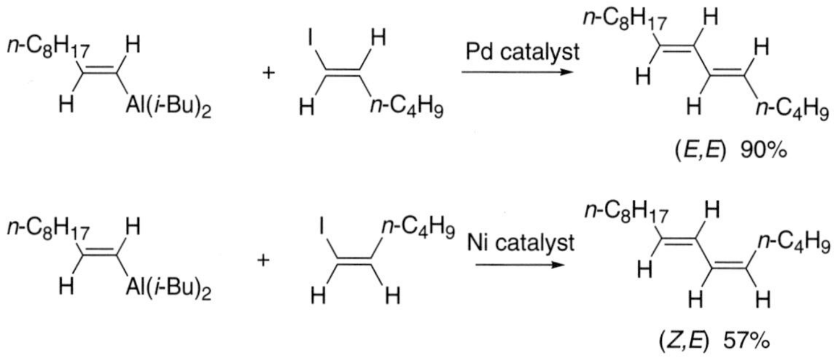

Scheme 5.24

Among all the alkenyl metals selected, Zn, Zr and Al give best results for the Pd-catalyzed cross-coupling reactions of alkenyl metals.

Mechanism: The Pd- or Ni-catalyzed alkenyl–alkenyl and alkenyl–aryl coupling reactions may proceed according to Scheme 5.25, involving the oxidative addition, transmetallation and reductive elimination.

The organic halide (R^1–X) is activated by a low-valent nickel Ni(0)L$_n$ to form a nickel(II)organic halide complex **A**. When organic bromides (R^1–Br) are used, the resulting L$_n$NiR^{1}Br complex **A** subsequently undergoes transmetallation with R^{2}ZnBr reagent to give a nickel(II)R^{2}R^1 complex **B**. The ZnBr$_2$ is obtained as a co-product. The nickel(II)RR^1 complex **B** further undergoes reductive elimination to give the final cross-coupled product R^2–R^1 and regenerates low-valent nickel. The generally accepted mechanism of the Negishi reaction is outlined in Scheme 5.25.

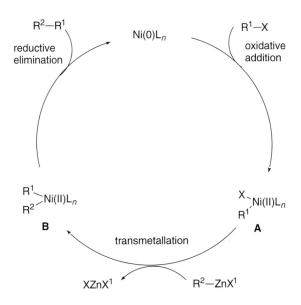

Scheme 5.25

The Negishi coupling is often advantageous over other cross-couplings, because organozinc reagents have a high tolerance of functional groups. This coupling allows the preparation of a wide range of coupling products and not restricted to the formation of biaryls only.

Organozinc reagents are usually generated and used *in situ* by transmetallation of Grignard or organolithium reagents with ZnCl$_2$. In addition, oxidative insertion of Zn(0) metal into some organohalides gives the corresponding organozinc reagents which can be coupled with aryl halides. For example, organozinc compounds obtained from unactivated alkyl bromides and chlorides undergo Pd- or Ni-catalyzed cross-coupling with aryl halides in high yields (Scheme 5.26)[84,85].

The Negishi reaction is utilized to construct substituted pyrimidines and purines.

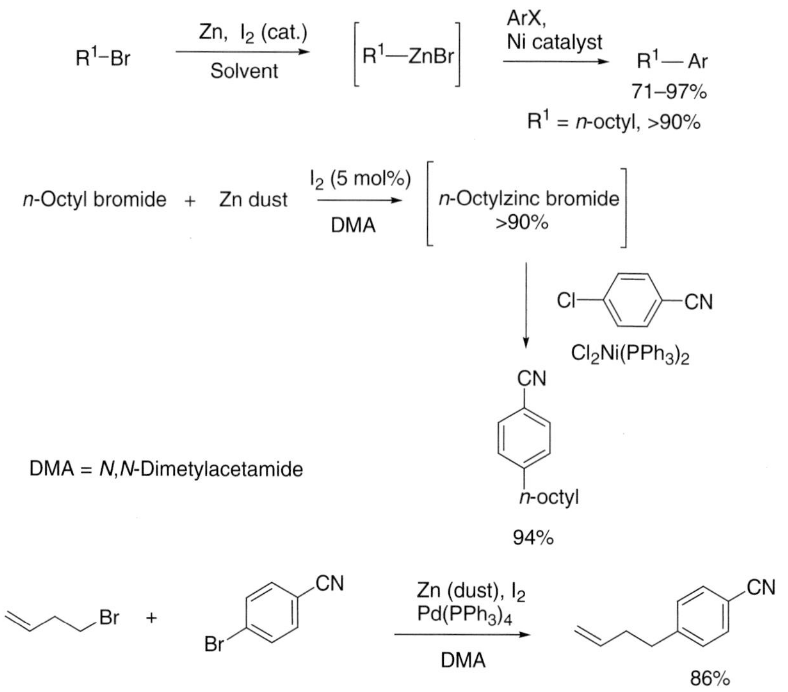

Scheme 5.26

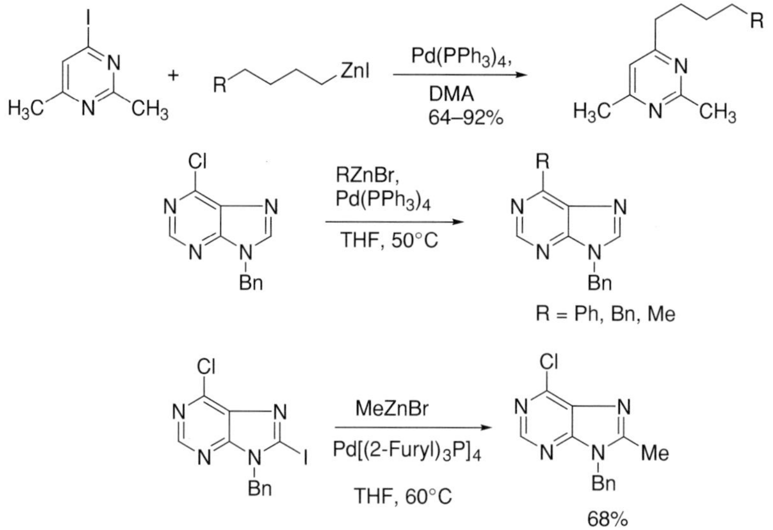

Negishi and co-workers also reported in 1978 Pd-catalyzed alkynylation reactions with alkynylmetals containing Mg, B, Al and Sn. But B and Sn versions have sometimes been known as the Suzuki and Stille alkynylation reactions, respectively. Negishi and co-workers[86] during 1977–1978 reported the Pd-catalyzed alkynylation reactions of alkynylzinc reagents with 1-halo-1-alkynes and aryl halides.

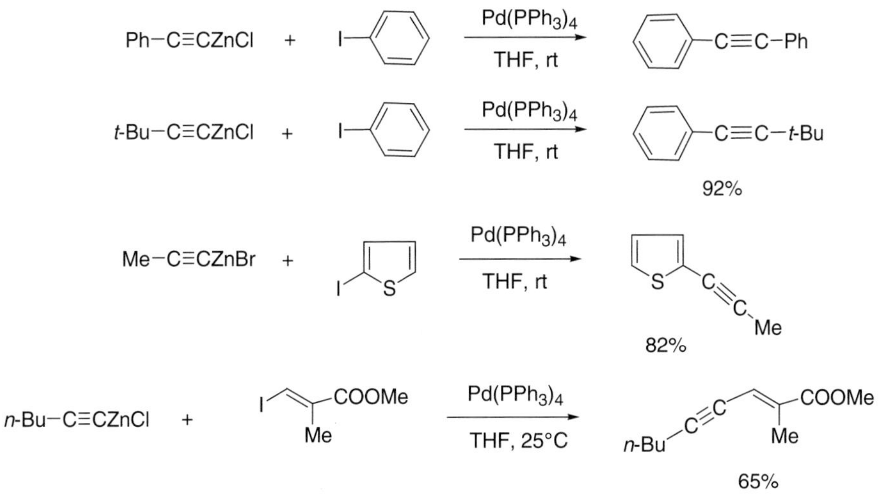

The insertion of (*E*)-1-chloro-1-litho-1,3-butadiene (**5.53**) into *n*-hexylzirconocene chloride (**5.54**) produces alkenylzirconium reagent **5.55**, which undergoes Pd(0)-catalyzed coupling with vinyl halides like **5.56**, **5.57** and allyl halides like **5.58** to give the corresponding coupling products (3*E*,5*E*)-4-hexyl-6-phenyl-1,3,5-hexatriene (**5.59**), (3*E*,5*E*)-4-hexyltetradecatriene (**5.60**) and (*E*)-4-hexyl-1,3,6-heptatriene (**5.61**), respectively. These organozirconium species are useful in organic synthesis for copper–nickel and palladium-catalyzed carbon–carbon bond forming reactions[87] (Scheme 5.27).

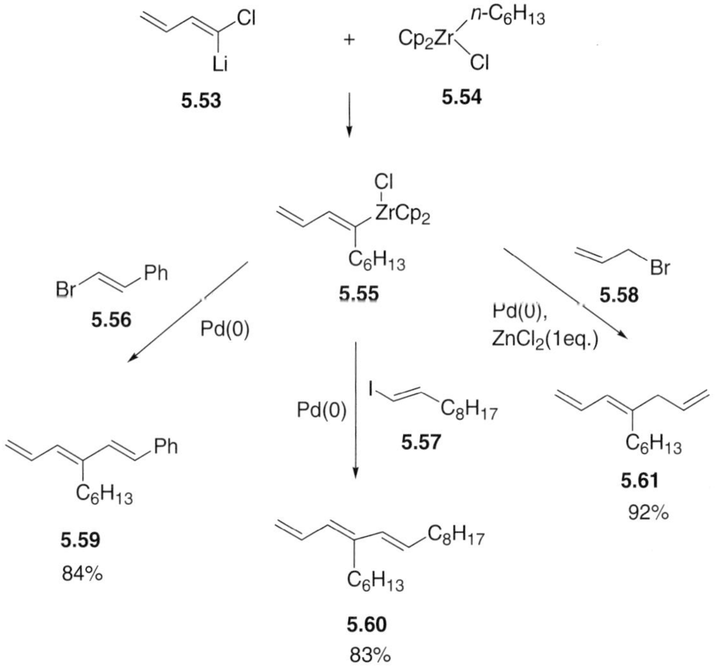

Scheme 5.27

Recent advances of the Negishi cross-coupling reaction have enabled chemists to catalytically couple two sp^3 carbon centres. These reactions provide the coupled alkane as the final product[88]. A remote double bond in primary halides also enabled the Ni-catalyzed cross-coupling reactions between two Csp^3 centres[89].

References

1. Urbanski, T., *Chem. Ber.*, **1976**, *12*, 191.
2. Heitbaum, M., Glorius, F. and Escher, I., *Angew. Chem., Int. Ed. Engl.*, **2006**, *45*, 4732.
3. Yin, L. and Liebscher, J., *Chem. Rev.*, **2007**, *107*, 133.
4. Cai, D., Payack, J. F., Bender, D. R., Hughes, D. L., Verhoeven, T. R. and Reider, P. J., *Org. Synth. Coll. Vol.*, **2004**, *10*, 112.
5. Cai, D., Payack, J. F., Bender, D. R., Hughes, D. L., Verhoeven, T. R. and Reider, P. J., *Org. Synth. Coll. Vol.*, **1999**, *76*, 6.
6. Noyori, R., *Science*, **1990**, *248*, 1194.
7. Higham, L. J., Clarke, E. F., Müller-Bunz, H. and Gilheany, D. G., *J. Org. Chem.*, **2005**, *690*(1), 211.
8. Kagan, H. B. and Dang-Tuan-Phat, *J. Am. Chem. Soc.*, **1972**, *94*, 6429.
9. Fryzuk, M. D. and Bosnich, B., *J. Am. Chem. Soc.*, **1977**, *99*, 6262.
10. Mizoroki, T., Mori, K. and Ozaki, A., *Bull. Chem. Soc. Jpn.*, **1971**, *44*, 581.
11. Heck, R. F., *J. Am. Chem. Soc.*, **1968**, *90*, 5518.
12. Heck, R. F. and Nolley, J. P., *J. Org. Chem.*, **1972**, *37*, 2320.
13. Bräse, S. and de Meijere, A., *Metal-Catalyzed Cross-Coupling Reactions* (eds F. Diederich and P. J. Stang) Wiley, New York, **1998**, Chapter 3.
14. Beletskaya, I. P. and Cheprakov, A. V., *Chem. Rev.*, **2000**, *100*, 3009.
15. Heck, R. F., *Comprehensive Organic Synthesis, Vol. 4* (ed. B. M. Trost) Pergamon, New York, **1991**, Chapter 4.3.
16. Heck, R. F., *Org. React.*, **1982**, *27*, 345.
17. Crisp, G. T., *Chem. Soc. Rev.*, **1998**, *27*, 427.
18. de Meijere, A. and Meyer, F. E., *Angew. Chem., Int. Ed. Engl.*, **1994**, *33*, 2379.
19. Jeffery, T., *Advances in Metal-Organic Chemistry, Vol. 5* (ed. L. S. Liebeskind) JAI, London, **1996**, pp. 153–260.
20. Cabri, W. and Candiani, I., *Acc. Chem. Res.*, **1995**, *28*, 2.
21. de Meijere, A. and Meyer, F. E., *Angew. Chem., Int Ed. Engl.*, **1994**, *33*, 2379.
22. Beletskaya, P. and Cheprakov, A. V., *Chem. Rev.*, **2000**, *100*, 3009.
23. Amatore, C. and Jutand, A., *J. Organomet. Chem.*, **1999**, 254.
24. Andersen, N. G., Parvez, M. and Keay, B. A., *Org. Lett.*, **2000**, *2*, 2817.
25. Kiely, D. and Guiry, P. J., *Tetrahedron Lett.*, **2002**, *43*, 9545.
26. Imbos, R., Minnaard, A. J. and Feringa, B. L., *J. Am. Chem. Soc.*, **2002**, *124*, 184.
27. Noyori, R., Okhuma, T., Kitamura, M., Takaya, H., Sayo, N., Kumobayashi, H. and Akuragawa, S., *J. Am. Chem. Soc.*, **1987**, *109*, 5856–5858. (Also see References 19–22 of Chapter 1.)
28. Sato, Y., Sodeoka, M. and Shibasaki M., *J. Org. Chem.*, **1989**, *54*, 4738.
29. Shibasaki, M., Boden, C. D. J. and Kojima, A., *Tetrahedron*, **1997**, *53*, 7371.
30. Dounay, A. B., Hatanaka, K., Kodanko, J. J., Oestreich, M., Overman, L. E., Pfeifer, L. A. and Weiss, M. N., *J. Am. Chem. Soc.*, **2003**, *125*, 6261.
31. Lebsack, A. L., Link, J. T., Overman, L. E. and Sterns, B. A., *J. Am. Chem. Soc.*, **2002**, *124*, 9008.
32. Heck, R. F., *Comprehensive Organic Synthesis, Vol. 4*, Pergamon, New York, **1991**, p. 585.
33. Williams, J. M. J., *Synlett*, **1996**, 705.
34. Trost. A. M., *Acc. Chem. Res.*, **1996**, *29*, 355.
35. Trost, B. M. and Vranken, D. L. V., *Chem. Rev.*, **1996**, *96*, 395.
36. Trost, B. M., Krueger, A. C., Bunt, R. C. and Zambrano, J., *J. Am. Chem. Soc.*, **1996**, *118*, 6521.
37. Longmire, J. M., Wang, B. and Zhang, X., *Tetrahedron Lett.*, **2000**, *41*, 5435.
38. Tsuji, J., Takahashi, H. and Morikawa, M., *Tetrahedron lett.*, **1965**, *6*, 4387.
39. Tsuji, J., Shimizu, I., Minami, I., Ohashi, Y., Sugiura, T. and Takahashi, K., *J. Org. Chem.*, **1985**, *50*, 1523.

40. Behenna, D. C. and Stoltz, B. M., *J. Am. Chem. Soc.*, **2004**, *126*, 15044.
41. Helmchen, G., Dahnz, A., Dubon, P., Schelwies, M. and Weihofen R., *Chem. Commun.*, **2007**, 675.
42. Kanayam, T., Yoshida, K., Miyabe, H. and Takemoto, Y., *Angew. Chem., Int. Ed.*, **2003**, *42*, 2054.
43. Kanayama, T., Yoshida, K., Miyabe, H., Kimachi, T. and Takemoto, Y., *J. Org. Chem.*, **2003**, *68*, 6197.
44. Tsou, T. T. and Kochi, J. K., *J. Am. Chem. Soc.*, **1979**, *101*, 7547.
45. Chodkiewicz, W. and Cadiot, P., *Compt. Rend.*, **1955**, *241*, 1055.
46. Castro, C. E. and Stephens, R. D., *J. Org. Chem.*, **1963**, *28*, 2163.
47. Stephens, R. D. and Castro, C. E., *J. Org. Chem.*, **1963**, *28*, 3313.
48. Diederich, F. and Stang, P. J. (eds) , *Metal-Catalyzed Cross-Coupling Reactions*, Wiley-VCH, New York, **1998**.
49. Murahashi, S., Yamamura, M., Yanagisawa, K., Mita, N. and Kondo, K., *J. Argo Chem.*, **1979**, *44*, 2408.
50. Lipshutz, B. H. and Sengupta, S., *Org. React.*, **1992**, *41*, 135.
51. Stille, B., *Angew. Chem., Int. Ed. Engl.*, **1986**, *25*, 508, and references therein.
52. Miyaura, N. and Suzuki, A., *Chem. Rev.*, **1995**, *95*, 2457.
53. Hatanaka, Y. and Hiyama, T., *J. Am. Chem. Soc.*, **1990**, *112*, 7793.
54. Hatanaka, Y. and Hiyama, T., *Synlett*, **1991**, 845.
55. Frisch, A. C. and Beller, M., *Angew. Chem., Int. Ed.*, **2005**, *44*, 674.
56. Lipshutz, B. H. and Sengupta, S., *Org. React.*, **1992**, *41*, 135.
57. Tamao, K., Sumitani, K. and Kumada, M., *J. Am. Chem. Soc.*, **1972**, *94*, 4374.
58. Corriu, R. J. P. and Masse, J. P., *Chem. Commun.*, **1972**, 144.
59. Yamamura, M., Mortitani, I. and Murahashi, S., *J. Organomet. Chem.*, **1975**, *103*, 91.
60. Kasatkin, A. and Whitby, R. J., *J. Am. Chem. Soc.*, **1999**, *121*, 9889.
61. Castle, P. L. and Widdowson, D. A., *Tetrahedron Lett.*, **1986**, *27*, 6013.
62. Yuan, K. and Scott, W. J., *Tetrahedron Lett.*, **1989**, *30*, 4779.
63. Terao, J., Watanabe, H., Ikumi, A., Kuniyasu, H. and Kambe, N., *J. Am. Chem. Soc.*, **2002**, *124*, 4222.
64. Stille, J. K., *Angew. Chem., Int. Ed. Engl.*, **1986**, *25*, 508.
65. Farina, V., *Pure Appl. Chem.*, **1996**, *68*, 73.
66. Farina, V., Krishnamurthy, V. and Scott, W. J., *The Stille Reaction*, Wiley, New York, **1998**.
67. Bedford, R. B., Cazin, C. S. J. and Hazelwood, S. L., *Chem. Commun.*, **2002**, 2608.
68. Dounay, A. B., Hatanaka, K., Kodanko, J. J., Oestreich, M., Overman, L. E., Pfeifer, L. A. and Weiss, M. M., *J. Am. Chem. Soc.*, **2003**, *125*, 6261.
69. Alcazar-Roman, L. M. and Hartwig, J. F., *Organometallics*, **2002**, *21*, 491.
70. Casado, A. L., Espinet, P. and Gallego, A. M., *J. Am. Chem. Soc.*, **2000**, *122*, 11771.
71. Kiehl, A., Eberhardt, A., Adam, M., Enkelmann, V. and Mullen, K., *Angew. Chem., Int. Ed. Engl.*, **1992**, *31*, 1588.
72. Miyaura, N. and Suzuki, A., *Chem. Rev.*, **1995**, *95*, 2457.
73. Kabalka, G. W., Pagni, R. M. and Hair, C. M., *Org. Lett.*, **1999**, *1*, 1423.
74. Ishiyama, T. A. S., Miyaura, N. and Suzuki, A., *Chem. Lett.*, **1992**, 691.
75. Charette, A. B. and De Freitas-Gil, R. P., *Tetrahedron Lett.*, **1997**, *38*, 2809.
76. Baxter, J., Steinhuebel, D., Palucki, M. and Davies, I. W., *Org. Lett.*, **2005**, *7*(2), 215.
77. Hatanaka, Y. and Hiyama, T., *Synlett*, **1991**, 845.
78. Hatanaka, Y. and Hiyama T., *J. Org. Chem.*, **1988**, *53*, 918.
79. Hiyama, T. and Hatanaka, Y., *Pure Appl. Chem.*, **1994**, *66*, 1471.
80. Negishi, E. and Anastasia, L., *Chem. Rev.*, **2003**, *103*, 1979.
81. Sonogashira, K., Tohada, Y. and Hagihara, N., *Tetrahedron Lett.*, **1975**, *16(50)*, 4467.
82. Tokuyama, H., Yokoshima, S., Yamashita, T., Lin, S., Li, L. and Fukuyama, T., *J. Braz. Chem. Soc.*, **1998**, *9*, 381.
83. Negishi, E., Takahashi, T., Baba, S., Horn, D. E. V. and Okukado, N., *J. Am. Chem. Soc.*, **1987**, *109*, 2393.
84. Negishi, E., *Acc. Chem. Res.*, **1982**, *15*, 340.
85. Huo, S., *Org. Lett.*, **2003**, *5*, 423.
86. Negishi, E., Okukado, N., King, A. O., Horn, D. E. V. and Spiegel, B. I., *J. Am. Chem. Soc.*, **1978**, *100*, 2254.
87. Matsushita, H. and Negishi, E., *J. Am. Chem. Soc.*, **1981**, *103*, 2882.
88. Anderson, T. A. and Vicic, D. A., *Organometallics*, **2004**, *23*(4), 623.
89. Giovannini, R., Studemann, T., Dussin, G. and Knochel, P., *Angew. Chem., Int. Ed. Engl.*, **1998**, *37*, 2387.

Chapter 6
Reduction

Reduction is one of the most synthetically useful reactions. The reduction process may be subdivided into three categories: the addition of H_2 (hydrogenation), removal of oxygen and gain of electrons. The reagents used for reduction may be divided into four classes: (i) hydrogen and catalyst (catalytic hydrogenation), (ii) metal hydrides, (iii) metal and hydrogen source and (iv) hydrogen transfer reagents.

6.1   Reduction of carbon–carbon double bond

For the reduction of non-polar double bonds such as a carbon–carbon double bond, a non-polar reagent is used. Among a wide number of reductive methods, catalytic hydrogenation is most common in chemistry.

6.1.1   Catalytic hydrogenation

Addition of hydrogen to a multiple bond is called **hydrogenation**. Although the overall hydrogenation reaction is exothermic, high activation energy prevents it from taking place under normal conditions. This restriction may be circumvented by the use of a catalyst.

   Platinum group of metals (Pt, Pd, Ru, Rh, Os and Ir) form a special category of catalysts. These catalysts can be classified into two categories: heterogeneous catalysts and homogeneous catalysts.

   Heterogeneous catalysis: The heterogeneous catalyst, either the finely dispersed metal itself or the metal adsorbed onto a support, remains in a separate phase during the course of reaction. Examples of hydrogenation of alkenes given below include reduction of oleic acid (**6.1**) with H_2 and Pd–C to octadecanoic acid (**6.2**) and cinnamyl alcohol (**6.3**) and Raney nickel in ethanol to 3-phenyl-1-propanol (**6.4**).

$$CH_3(CH_2)_7CH=CH(CH_2)_7COOH \xrightarrow[\text{5\% Pd–C}]{\text{H}_2} CH_3(CH_2)_{16}COOH$$

6.1   **6.2**

$$\text{(cinnamyl alcohol)} \xrightarrow[\substack{\text{Raney Ni,}\\ \text{EtOH}}]{\text{H}_2} \text{(3-phenyl-1-propanol)}$$

6.3   **6.4**

The brown platinum oxide $PtO_2 \cdot H_2O$, the **Adams catalyst**, when treated with H_2 gives a very finely divided black suspension of metal. The reagent is used in acetic acid or ethanol for the hydrogenation of alkenes.

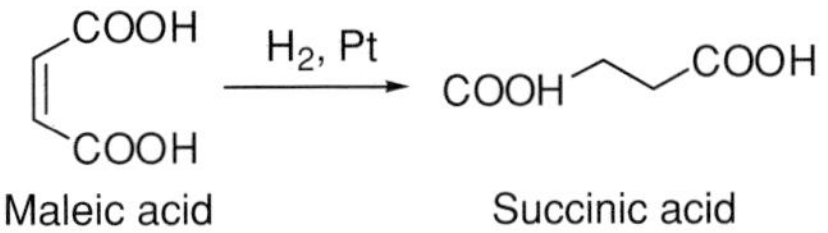

Maleic acid   Succinic acid

Mechanism: The atoms on the surface of metal are different from those buried in the body of the solid because they cannot satisfy their tendency to form strong metal–metal bonds. Some heterogeneous catalysts such as platinum or palladium or a finely divided form of nickel known as **Raney nickel** can satisfy a portion of their combining power by binding hydrogen atoms and/or substrate onto the surface. The details of the mechanism are still relatively unclear, but it is known that catalytic hydrogenation takes place as depicted in Scheme 6.1. First, the hydrogen and alkene molecules are adsorbed on the surface of the catalyst and then probably forming metal–hydrogen σ-bond, and then the interaction of π and π^* orbitals of alkene with appropriate orbitals of metal takes place. Next, two hydrogen atoms shift from the metal surface to the carbons of the double bond one after another. The resulting saturated hydrocarbon, which is more weakly adsorbed, leaves the catalyst surface. Although the hydrogen atoms are transferred one at a time, this reaction is believed to be fast enough that both of these atoms usually end up on the same side of the carbon–carbon double bond, i.e. *syn*-addition (in contrast to *anti*-addition when atoms add to the opposite face of the double bond).

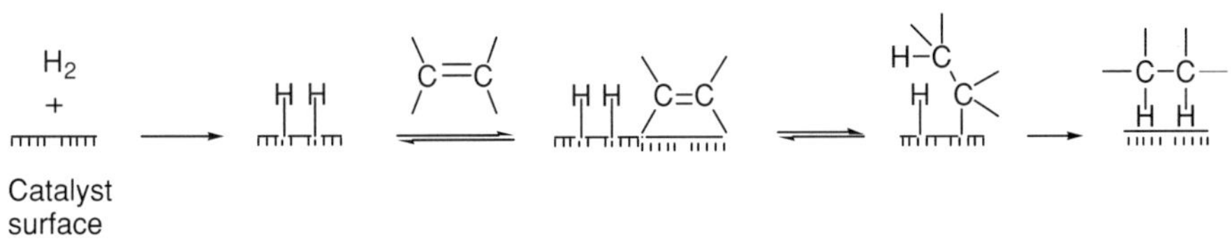

Scheme 6.1   Mechanism of hydrogenation on the surface of metal catalyst

Reduction of 1,2-dimethyl-1-cyclopentene (**6.5**) with hydrogen and Ni or Pt catalyst gives stereoselective product, *cis*-1,2-dimethylcyclopentane (**6.6**).

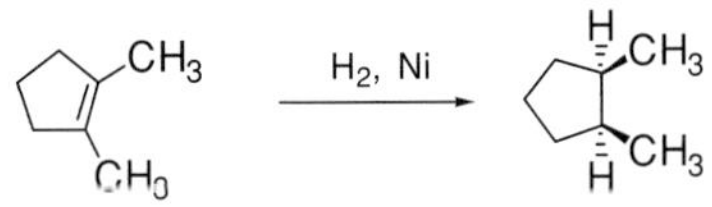

In some cases, hydrogenation catalysts may also cause double bond migration and *cis–trans*-isomerization of the double bond prior to hydrogen addition; in that case stereoselectivity may be uncertain.

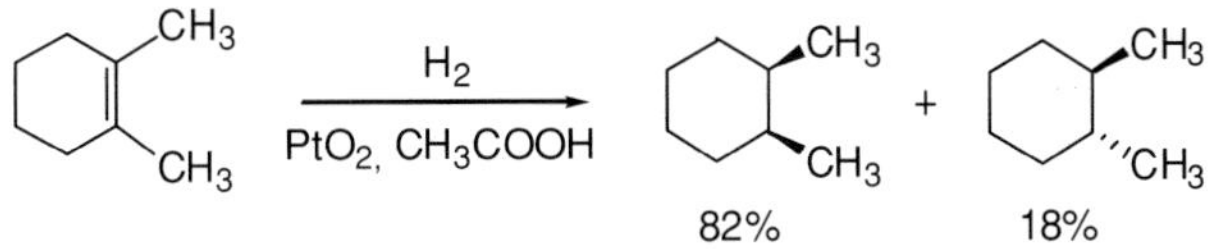

Catalytic hydrogenation can be used for the selective reduction of a carbon–carbon double bond in the presence of other functional groups such as a carbonyl group or an aromatic ring. Selective reduction of one double bond in (R)-limonene (**6.7**), which contains two double bonds, gives (R)-carvomenthene (**6.8**) by hydrogenation over Ni metal.

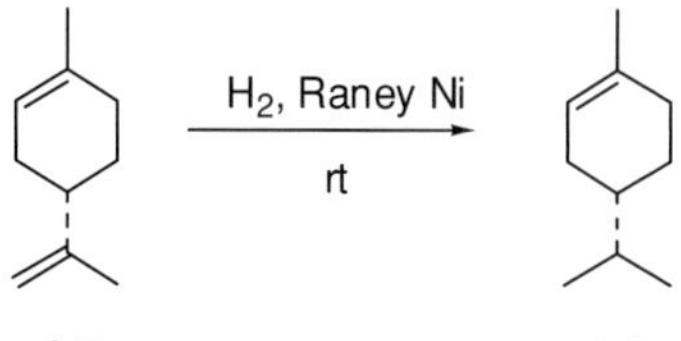

6.7   **6.8**

Homogeneous catalysis: Homogeneous catalysts or soluble metal complexes are present together with the reactants in a single phase. Hydrogenation with a homogeneous catalyst can be carried out under mild conditions with high selectivities. Platinum group metal complexes, the **Wilkinson catalyst**[1] $[Rh(PPh_3)_3]^+Cl^-$ (**6.9**) **and the Vaska catalyst**[2] $[(PPh_3)_2IrCO]^+Cl^-$, are excellent catalysts used for homogeneous hydrogenation. The **Wilkinson catalyst** is useful for the stereospecific reduction of unhindered carbon–carbon double bonds even in the presence of other groups such as keto, nitrite, nitro and sulfide.

$$n\text{-}C_8H_{17}CH{=}CH_2 \xrightarrow[D_2,\ C_6H_6]{\textbf{6.9}} n\text{-}C_8H_{17}CHDCH_2D$$

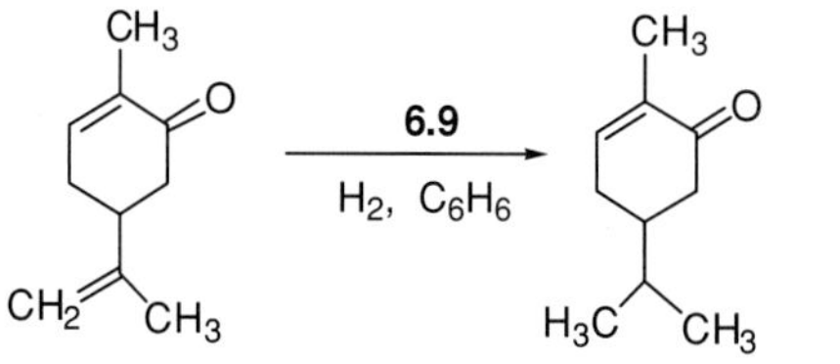

Carvone   Carvotanacetone

Mechanism: The Wilkinson catalyst (**6.9**), a 6-electron complex, loses one or two triphenylphosphine ligands and converts into a 14- or 12-electrons complex. The activation of hydrogen occurs by uptake on the metal complex catalyst via an **oxidative addition**. This is followed by π-complexation of alkene to metal. **Intramolecular hydride transfer** and subsequent **reductive elimination** release the alkane and complete the cycle (Scheme 6.2).

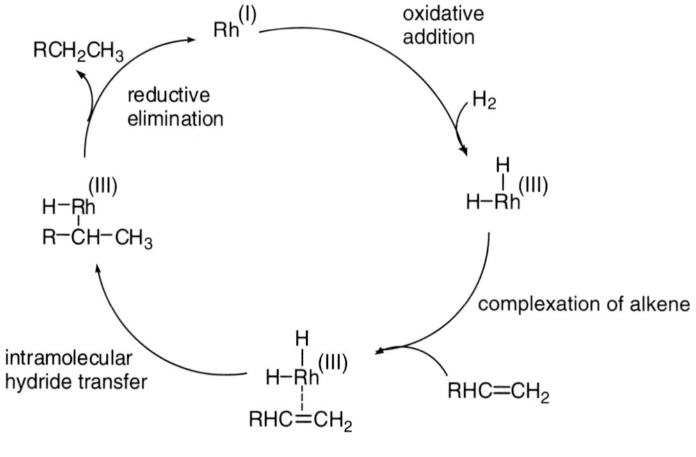

Scheme 6.2 Outline of catalytic hydrogenation cycle

Asymmetric catalytic hydrogenation: The metal is combined with an appropriate optically active ligand to form the optically active (or chiral) catalyst. The use of chiral catalysts to effect the asymmetric hydrogenation of prochiral alkene substrates with high optical yields represents one of the most impressive achievements in catalytic selectivity. The steric factors also determine the orientation of the substrate on the catalyst surface and hence control the stereochemistry of the hydrogenation. The most important chiral ligands used to prepare active chiral catalysts from platinum group metals are the chiral diphosphine ligands such as DIPAMP (**5.2**), DIOP (**5.3**) and CHIRAPHOS (**5.4**). Noyori and co-workers[3,4] successfully used diphosphine chiral ligands (*S*)-**BINAP** and (*R*)-**BINAP** [2,2′-bis(diphenylphosphino)-1,1′-binaphthyl] (**6.10** and **6.11**) in several homogeneous asymmetric catalysis reactions. BINAP ligands (**6.10** and **6.11**) can be prepared from the corresponding BINOL (1,1′-bi-2-naphthol). For example, (*R*)-BINAP is prepared from (*R*)-BINOL (**6.12**) via its bistriflate derivatives or via Grignard reagent (Scheme 6.3). BINAP is chiral due to restricted rotation about the single bond binding the two naphthyl groups. Therefore, the angle made by the two π planes is fixed to approximately 90° and two separate enantiomers exist. Both the (*R*)- and (*S*)-BINAP enantiomers (**6.10** and **6.11**) are commercially available.

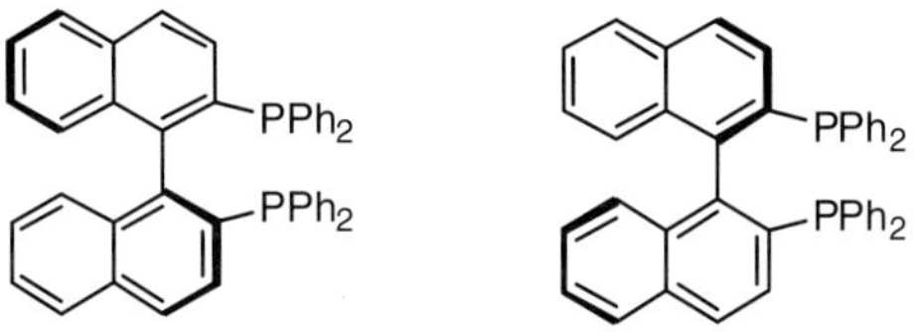

(*S*)-(–)-BINAP (**6.10**)

(*S*)-(–)-2,2′-Bis(diphenyl phosphino)-1,1′-binaphthyl

(*R*)-(+)-BINAP (**6.11**)

(*R*)-(+)-2,2′-Bis(diphenyl phosphino)-1,1′-binaphthyl

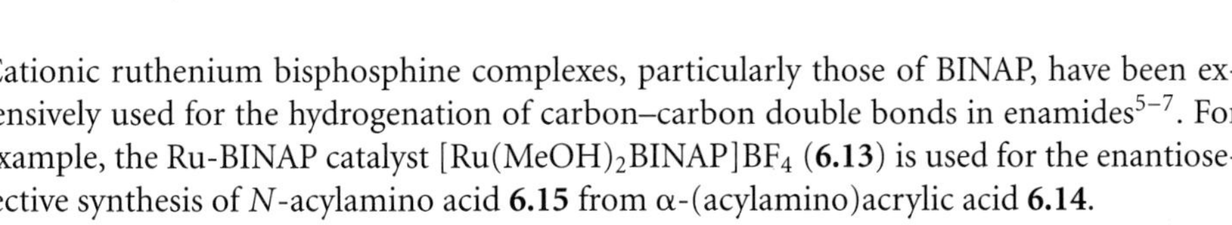

6.12

6.11

Scheme 6.3 Synthesis of (*R*) BINAP (**6.11**) from (*R*)-BINOL (**6.12**)

Cationic ruthenium bisphosphine complexes, particularly those of BINAP, have been extensively used for the hydrogenation of carbon–carbon double bonds in enamides[5-7]. For example, the Ru-BINAP catalyst [Ru(MeOH)$_2$BINAP]BF$_4$ (**6.13**) is used for the enantioselective synthesis of *N*-acylamino acid **6.15** from α-(acylamino)acrylic acid **6.14**.

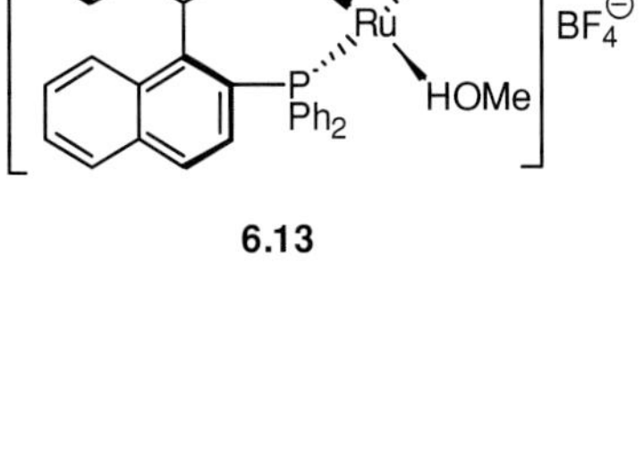

6.13

Reduction of the enamide double bond of α-(acylamino)acrylic acid **6.16** with Rh complex of diphosphine ligand DIPAMP (**5.2**) is an important step in the synthesis of L-DOPA (**6.17**), used for the treatment of Parkinson's disease.

N-Benzoyl-(*S*)-leucine (**6.19**), *N*-benzoyl-(*S*)-phenylalanine (**6.15**) and *N*-acetyl-(*S*)-phenylalanine (**6.21**) amino acids are synthesized in 94, 92 and 98% ee from **6.18**, **6.14** and **6.20**, respectively, by the asymmetric hydrogenation of the corresponding α-acylamino acrylic acids[8,9] in the presence of a chiral catalyst, **6.22**.

	R	R¹	
6.18:	Ph	CH(CH₃)₂	**6.19**
6.14:	Ph	Ph	**6.15**
6.20:	CH₃	Ph	**6.21**

6.22 = [Rh(*R*,*R*)-1,2-bis{*N*-methyl(diphenylphosphino)amino} cyclohexane-1,5-cyclohexadiene] hexafluorophosphate complex

Allylic alcohols such as geraniol (**6.23**) and nerol (**6.25**) are reduced to (*R*)-citronellol[10] (**6.24**) and (*S*)-citronellol (**6.26**), respectively, by a (OAc)₂RuBINAP complex without *E/Z* isomerization of the double bond with high chemical and optical yields (Scheme 6.4).

6.1.2   Hydrogen transfer reagents

A non-catalytic procedure for the *syn*-addition of hydrogen to alkenes makes use of the unstable compound diimide (N_2H_2). This strongly exothermic reaction is favoured by the elimination of nitrogen gas. The diimide reagent must be freshly generated in the reaction

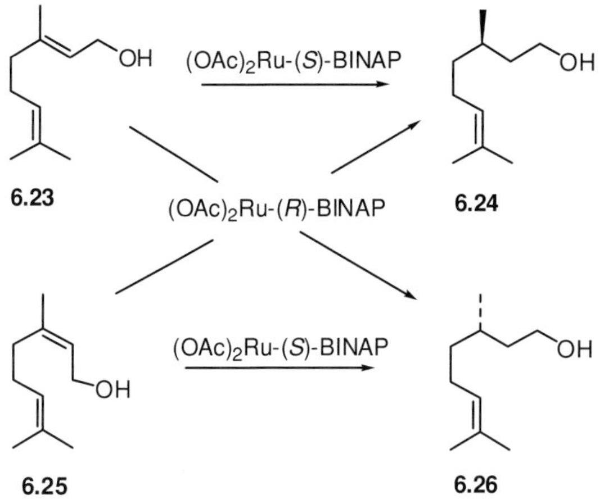

Scheme 6.4

system, usually by the oxidation of hydrazine, and may exist as *cis–trans*-isomers; only the *cis*-isomer serves as a reducing agent.

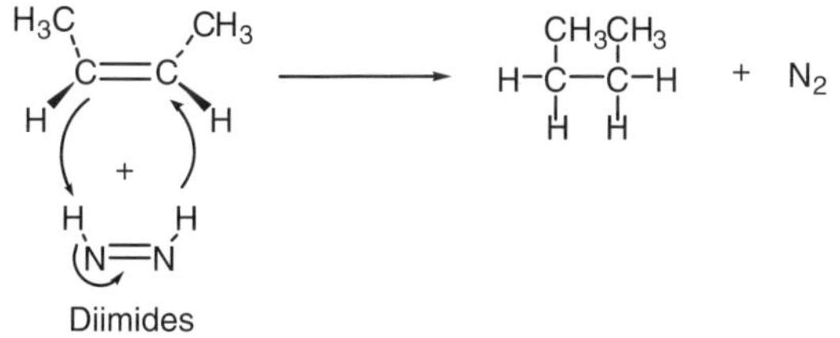

Diimides

$[Ir(COD)Cl]_2$ and 2-propanol, as an *in situ* source of hydrogen, are used for the carbon–carbon double bond hydrogenation of an α,β-unsaturated ketone[11].

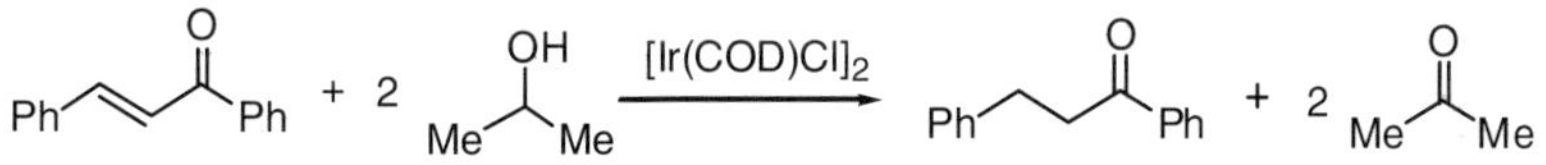

COD=*cis, cis*- 1, 5-Cyclooctadiene

6.2 Reduction of acetylenes

6.2.1 Catalytic hydrogenation

Reduction of alkynes with hydrogen on a metal catalyst (usually Ni, Pt and Ru) gives the corresponding alkanes.

$$\underset{\text{2-Butyne}}{CH_3{-}C{\equiv}C{-}CH_3} \xrightarrow[H_2]{\text{Pt, Ni or Pd}} \underset{\text{2-Butene}}{CH_3{-}CH{=}CH{-}CH_3} \xrightarrow[H_2]{\text{Pt, Ni or Pd}} \underset{\text{Butane}}{CH_3{-}CH_2{-}CH_2{-}CH_3}$$

However, by partially poisoning the palladium catalyst supported on calcium carbonate with lead acetate and quinoline (the **Lindlar catalyst**[12]) it is possible to reduce alkynes to alkenes. Once again, the reaction is stereoselective, adding both hydrogen atoms from the same side of the carbon–carbon bond to form the *cis*-alkene.

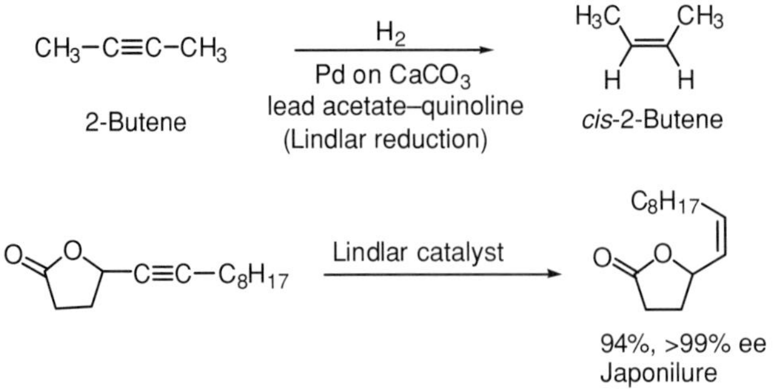

$$CH_3-C{\equiv}C-CH_3 \xrightarrow[\substack{\text{Pd on CaCO}_3 \\ \text{lead acetate–quinoline} \\ \text{(Lindlar reduction)}}]{H_2}$$

2-Butene

H$_3$C   CH$_3$

H   H

cis-2-Butene

$$\xrightarrow{\text{Lindlar catalyst}}$$

C$_8$H$_{17}$

94%, >99% ee
Japonilure

6.2.2 Dissolving metals

Alkynes are reduced to *trans*-alkenes with Na or Li and liquid NH$_3$ that contains stoichiometric amount of alcohol. This reduction, known as **Birch reduction**[13,14], is highly selective as no saturated product is formed and completely stereospecific as only *trans*-alkene is formed.

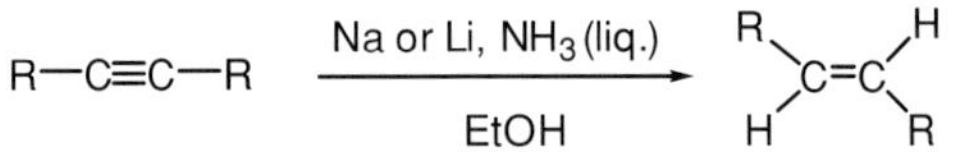

$$R-C{\equiv}C-R \xrightarrow[\text{EtOH}]{\text{Na or Li, NH}_3\,\text{(liq.)}}$$

R   H

H   R

Mechanism: Lithium metal donates an electron to the carbon–carbon triple bond. The radical anion **A** accepts a proton to give a vinylic radical **B**. Transfer of another electron gives a vinylic anion **C** (*trans*-vinylic anion is more stable than the *cis*-vinylic anion), which on protonation gives *trans*-alkene (Scheme 6.5).

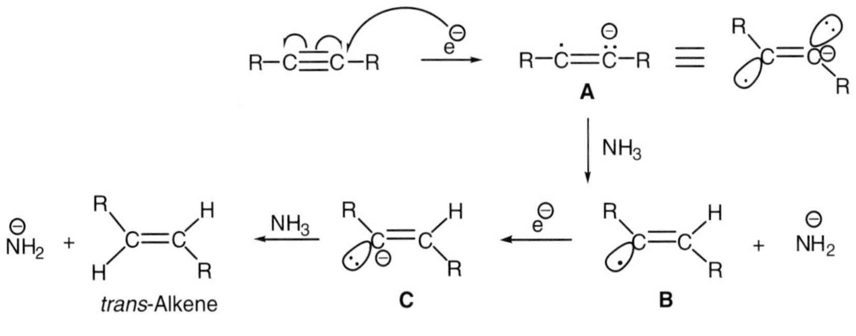

Scheme 6.5

6.2.3 Metal hydrides

Diisobutylaluminum hydride[15] (*i*-Bu$_2$AlH–H$_2$O) reduces alkynes to alkenes. LiAlH$_4$ can also reduce triple bonds in unsaturated alcohols.

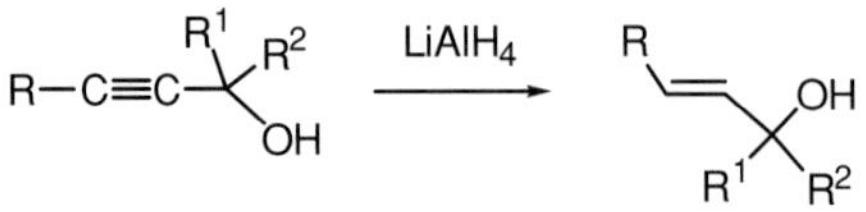

$$R-C{\equiv}C-\underset{\underset{OH}{\overset{R^2}{|}}}{\overset{R^1}{|}} \xrightarrow{\text{LiAlH}_4}$$

R

OH

R^1 R^2

Alkynes are reduced preferentially to alkenes with preferred *Z*-selectivity with molar mixtures of LiAlH$_4$ and several transition metal chlorides such as TiCl$_4$.

73% 16% 11%

Completely stereospecific *trans*-reduction of acetylenic alcohols to *E*-allyl alcohol is reported with sodium bis(2-methoxyethoxy)aluminium dihydride[16] (SMEAH or Red-Al), where LiAlH$_4$ in various solvents is less selective.

3-Trimethylsilyl-2-propyn-1-ol

(*E*)-3-Trimethylsilyl-2-propen-1-ol
70%

Chan stereospecific reduction[17] of acetylenic alcohol **6.27** gives *E*-allylic alcohol **6.28** by means of SMEAH in 83% yield.

6.27

6.28
83%

6.2.4 Hydroboration–protonation

Alkynes are also reduced to *cis*-alkenes by the hydroboration–protonation method[18].

1. disamylborane

2. HOAc

6.3 Reduction of benzene and derivatives

6.3.1 Catalytic hydrogenation

Catalytic hydrogenation of benzene under pressure by using Raney Ni as a catalyst results in the addition of three molar equivalents of hydrogen. First, benzene is converted into cyclohexadiene, which is reduced to cyclohexene. The hydrogenation of cyclohexadiene and cyclohexene is faster than the hydrogenation of benzene (aromatic compound). Similarly, catalytic hydrogenation of naphthalene in the presence of a Ni catalyst gives tetralin and then decalin.

Naphthalene   Tetralin   Decalin

6.3.2   Birch reduction

Birch reduction of aromatic compounds involves reaction with an electron-rich solution of alkali metal lithium or sodium in liquid ammonia (sometimes called metal ammonia reduction). Usually a proton donor such as *tert*-butanol or ethanol is used to avoid the formation of excess amount of LiNH$_2$ or NaNH$_2$. The major product is normally a 1,4-diene. This reaction is related to the reduction of alkynes to *trans*-alkenes[19,20] (section 6.2.2).

1,4-Cyclohexadiene

Naphthalene   1,4-Dihydronaphthalene

Anthracene   9,10-Dihydroanthracene

Phenanthrene   9,10-Dihydrophenanthrene

Mechanism: The mechanism of Birch reduction involves stepwise addition of two electrons to the benzene ring; each electron addition being followed by a protonation. The initial electron addition gives a radical anion **A** for which many resonance contributors may be written. Following the delivery of a proton by the weak acid ammonia or EtOH, the resulting delocalized radical **B** accepts a second electron to form a cyclohexadienyl anion **C**. The anion **C** generated by the second electron addition is delocalized over three carbon atoms, and is protonated on the central carbon (Scheme 6.6). The isolated (unconjugated) double bonds in the product are not reduced at the low temperatures of refluxing liquid ammonia ($-33°$C).

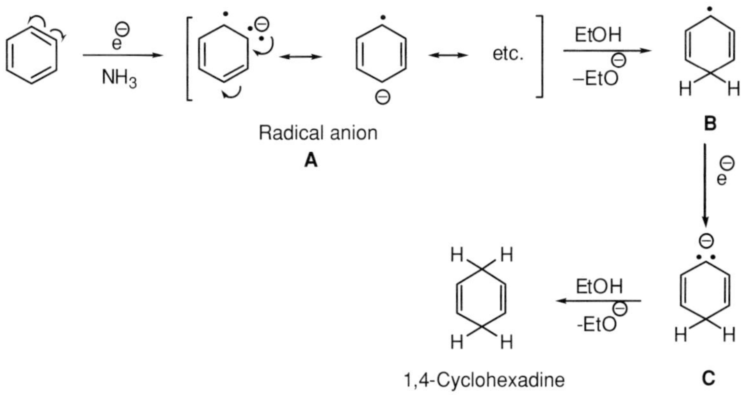

Scheme 6.6

When substituents are present, they may influence the regioselectivity of the Birch reduction. The product is determined by the site of the first protonation, since the second protonation is nearly always opposite (*para* to) the first. Electron-donating substituents such as ethers and alkyl groups favour protonation at an unoccupied site *meta* to the substituent whereas electron-attracting substituents such as carboxyl favour *para* protonation. This can be explained by the stability of the intermediates **6.29** and **6.30** formed in both cases.

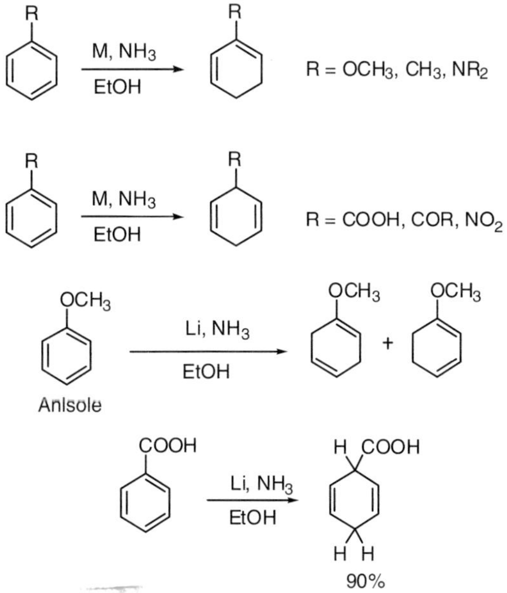

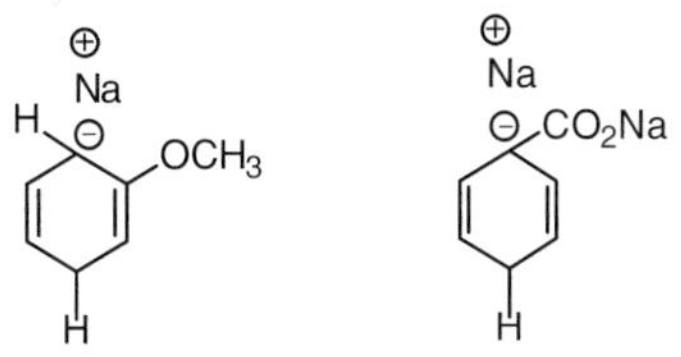

Reduction of aryl ethers is a particularly useful application of the Birch reduction. Thus, methoxybenzenes are reduced to 1,4-dienes, as expected, but one of the double bonds is enol ether and is readily hydrolyzed, followed by rearrangement to give the corresponding α,β-unsaturated-ketone[21].

6.4 Reduction of carbonyl compounds

Reduction of aldehydes and ketones usually occurs by the addition of hydrogen across the carbon–oxygen double bond to yield alcohols, but reductive conversion of a carbonyl group to a methylene group requires complete removal of the oxygen, and is called **deoxygenation**.

Although catalytic hydrogenation in the presence of H_2 and a catalyst such as Pt, Pd, Ni or Ru, reaction with diborane, and reduction by lithium, sodium or potassium in hydroxylic or amine solvents have all been reported to convert carbonyl compounds into alcohols, the most common reagents used for the reduction of carbonyl compounds are hydride donors.

6.4.1 Catalytic hydrogenation

Aldehydes are easily hydrogenated to alcohols but ketones are more difficult to reduce because of steric hindrance. Hydrogenolysis is a problem with the catalytic reduction of carbonyls, particularly when linked to aromatic systems. Pd and H_2 reduce alkenes faster than carbonyls. Metal catalyst Pt is commonly used for the reduction of carbonyls. For example, the **Adams catalyst** (PtO_2) reduces 2-naphthaldehyde (**6.31**) to **6.32** in 80% when used with $FeCl_3$ as a promoter. When excess of the promoter is used the product is 2-methylnaphthalene (**6.33**), which is also obtained by the reduction of **6.31** with Pd on $BaSO_4$ and H_2.

Rosenmund reduction[22,23], i.e. catalytic hydrogenation, of acid chlorides with poisoned Pd–$BaSO_4$ (in the presence of sulfur and quinoline poison) afforded aldehyde.

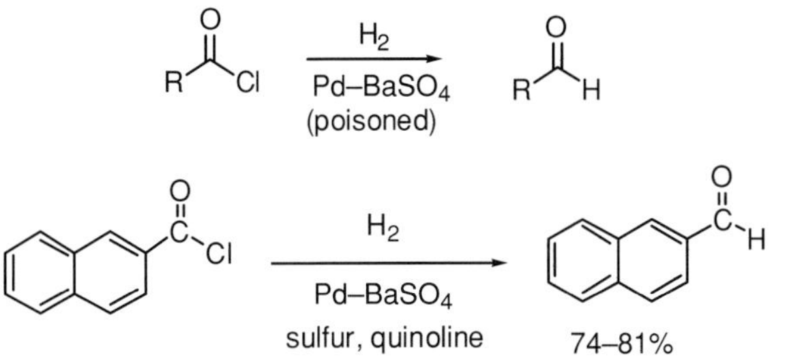

The mechanism of Rosenmund reduction is shown in Scheme 6.7.

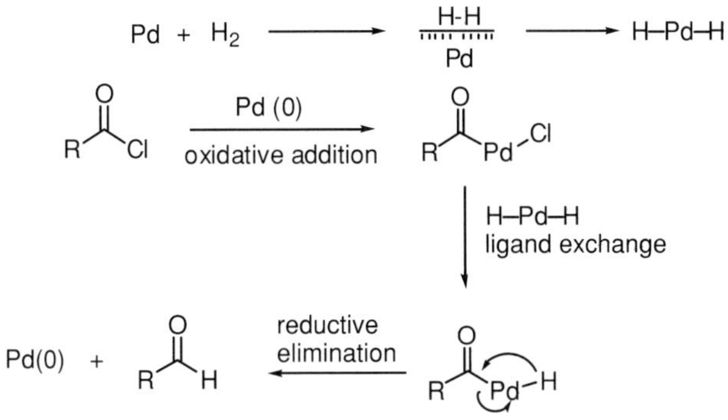

Scheme 6.7

Reduction of carbonyl group to methylene via thioacetals: In contrast to the Clemmensen reduction (section 6.4.3) and Wolff–Kishner reduction (section 6.4.4), this method does avoid treatment with strong acid or base but requires two separate steps. The first step is to convert the aldehyde or ketone into a thioacetal. The second step involves refluxing an acetone solution of the thioacetal over a Raney nickel. This reduction method is known as **Mozingo reduction**. Hydrazine can also be used in the second step.

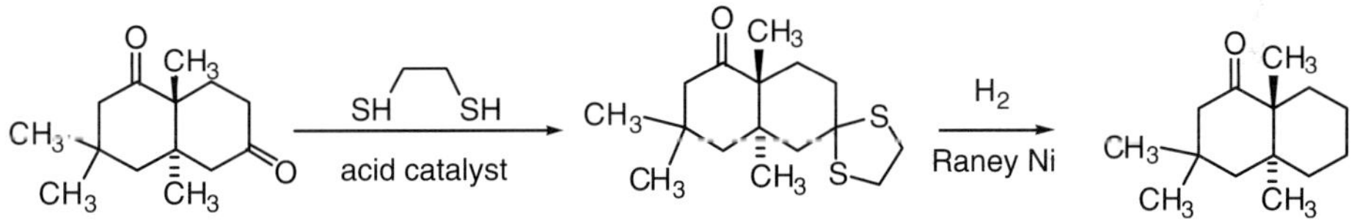

Reduction of carbonyl group to amines via imines (reductive alkylation): Hydrogenation of imines formed by the condensation of an amine with a ketone or aldehyde (without the isolation) is known as reductive alkylation.

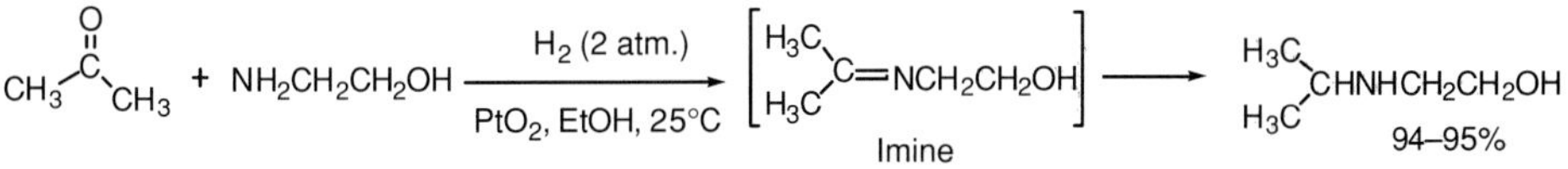

Sodium cyanoborohydride ($NaBH_3CN$) can also be used for the reductive alkylation.

Asymmetric reduction: The ruthenium(II)–BINAP catalysts developed by Noyori's group in 1980s were the most successful for the asymmetric hydrogenation of functionalized ketones such as α-ketoesters, α-hydroxyketones and α-aminoketones because the second functional group coordinates with the ruthenium metal centre. The chelates were believed to be necessary for high enantioselectivity.

The oxygen of the ketone acts as one donor ligand, and the other group (–NH₂, –OH or C=O) completes the chelation. The chelate ring can be either five or six membered. The reduction of β-diketone **6.34** and β-hydroxyketone **6.36** gives diols **6.35** and **6.37** respectively. β-Ketoester **6.38** and α-ketoester **6.40** give β-hydroxy ester **6.39** and α-hydroxy ester **6.41** respectively. α-Amino acid (**6.42**) is reduced to β-hydroxy amine **6.43**. These hydrogenation processes in the presence of Ru(X)₂ (*R*)- or (*S*)-BINAP are both highly enantioselective and diastereoselective.

6.4.2   Metal hydrides

The reduction of carbonyl compounds with metal hydride reagents can be viewed as nucleophilic addition of hydride to the carbonyl group. Addition of a hydride anion to an aldehyde or ketone produces an alkoxide anion, which on protonation gives the corresponding alcohol. Aldehydes give 1°-alcohols and ketone gives 2°-alcohols.

Two practical sources of hydride donors are lithium aluminium hydride (LiAlH₄) and sodium borohydride (NaBH₄).

$$4\ LiH\ +\ AlCl_3 \longrightarrow LiAlH_4\ +\ 3\ LiCl$$

$$4\ NaH\ +\ B(OMe)_3 \longrightarrow NaBH_4\ +\ 3\ MeONa$$

Several selective reagents are obtained by the modification[24] of $LiAlH_4$ and $NaBH_4$. The super hydrides such as lithium triethylborohydride are the most powerful reducing agents.

$$BEt_3\ +\ LiH \xrightarrow{\ THF,\ 65°C,\ 15\ min\ } LiBHEt_3$$

Both $LiAlH_4$ and $NaBH_4$ reduce polar carbonyl groups in aldehydes and ketones, but normally do not reduce simple alkenes or alkynes.

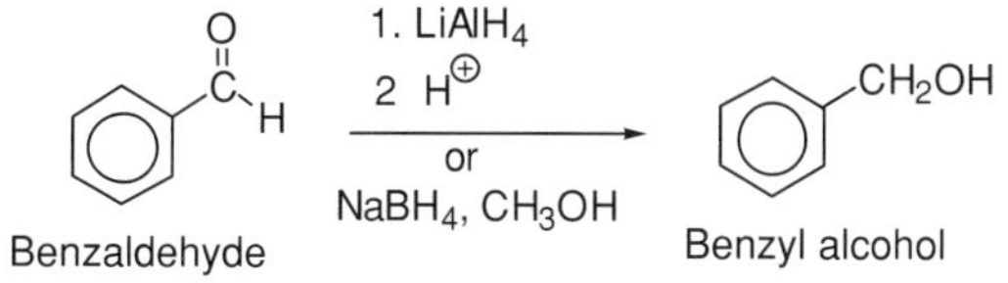

Benzaldehyde → Benzyl alcohol

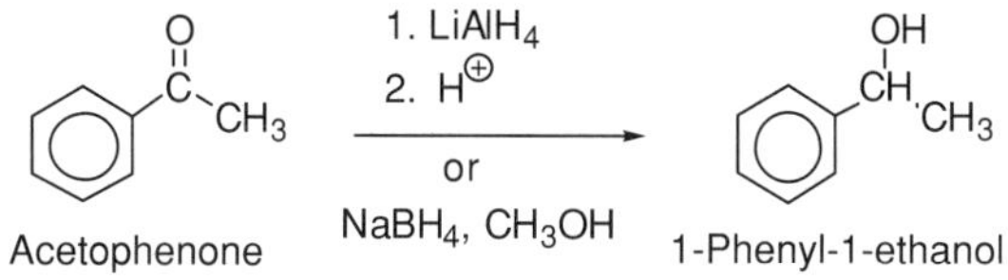

Acetophenone → 1-Phenyl-1-ethanol

Mechanism: Addition of a hydride anion to an aldehyde or ketone produces an alkoxide anion, which on protonation yields the corresponding alcohol.

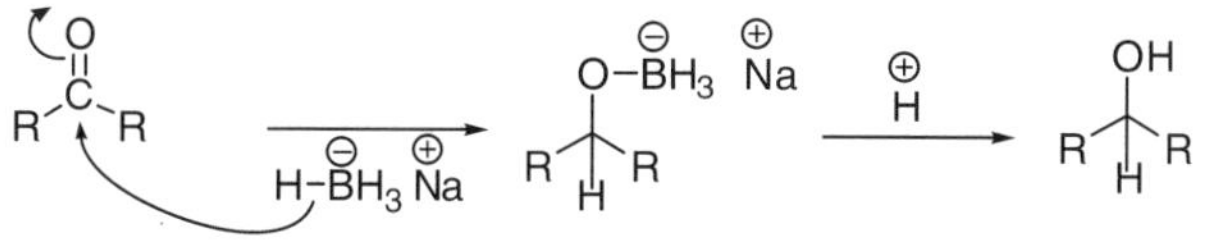

In the $LiAlH_4$ reduction, the resulting **alkoxyaluminate intermediates** are insoluble and need to be hydrolyzed (with care) before the alcohol product can be isolated (Scheme 6.8). In many cases, water can hydrolyze **alkoxyaluminate intermediates** but saturated ammonium chloride or dilute hydrochloric acid are sometimes necessary. In the borohydride reduction, the hydroxylic solvent system achieves this hydrolysis automatically.

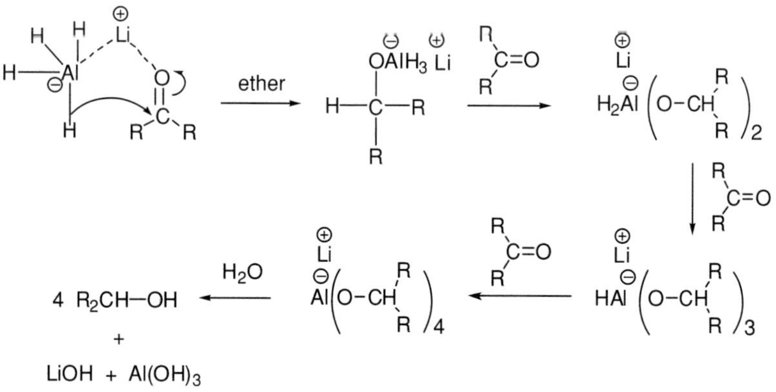

Scheme 6.8

Lithium aluminium hydride is highly reactive and can also reduce carboxylic acids, acid chlorides, anhydrides, esters, lactones, amides, lactams, imines, nitriles and nitro group. For example, $-COCl$, $-CO_2H$, $-CO_2Et$, $-CHO$ and $>CO$ are reduced to $-CH_2OH$ or $>CHOH$, provided the correct solvent is used.

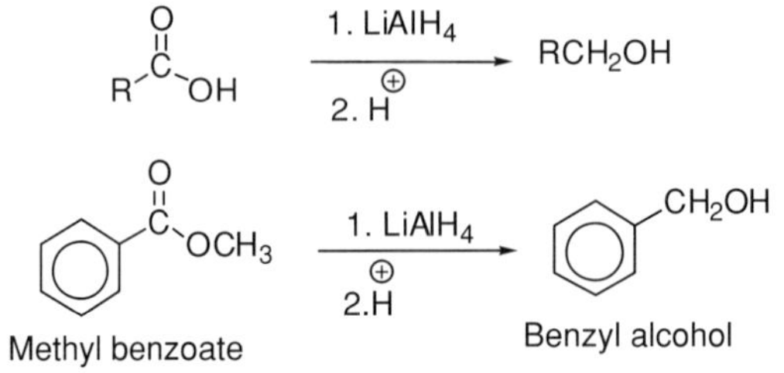

Methyl benzoate   Benzyl alcohol

Mechanism: The mechanism of acids and esters reduction with $LiAlH_4$ is shown in Schemes 6.9 and 6.10, respectively. The acidic hydrogen in acid reacts first. Then the reduction of carbonyl group proceeds via the usual alkoxyaluminate intermediate.

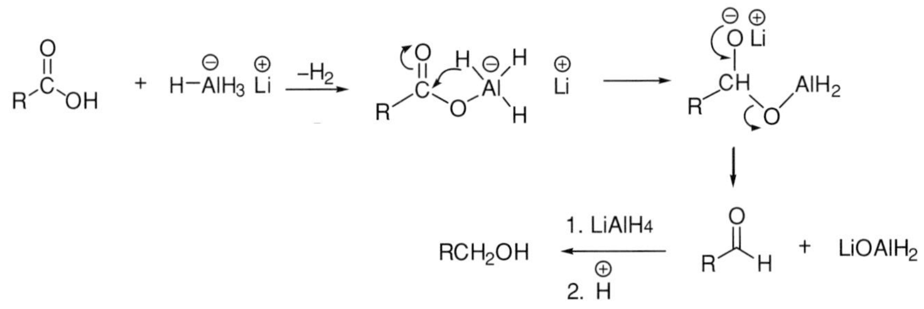

Scheme 6.9

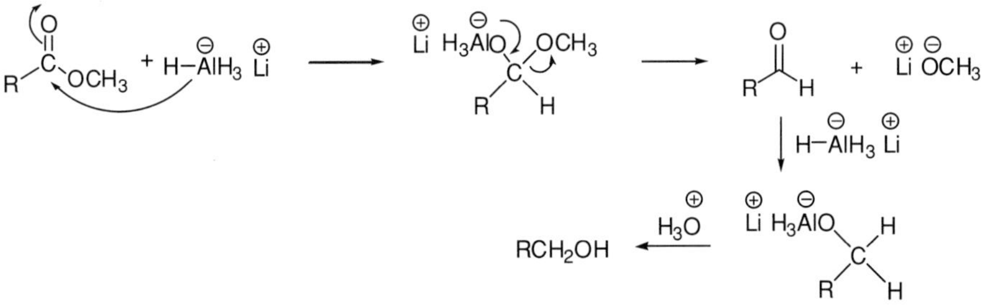

Scheme 6.10

AlH_3 also reduces acids to alcohols.

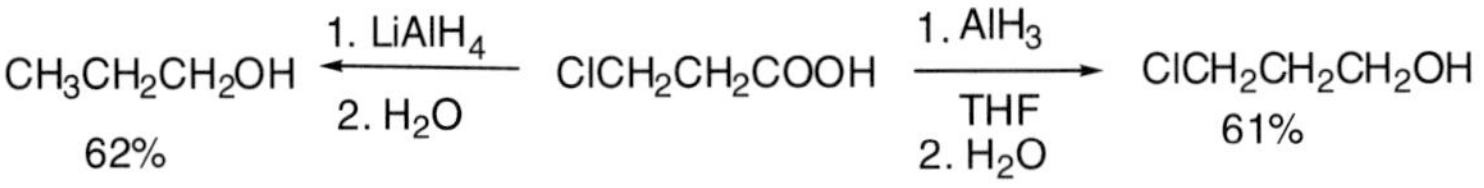

$CH_3CH_2CH_2OH$ ←(1. LiAlH_4 / 2. H_2O)— $ClCH_2CH_2COOH$ —(1. AlH_3 / THF / 2. H_2O)→ $ClCH_2CH_2CH_2OH$
62%   61%

Lithium aluminium hydride reduces amides to amines. Mechanism of LiAlH$_4$ reduction of amides is given in Scheme 6.11.

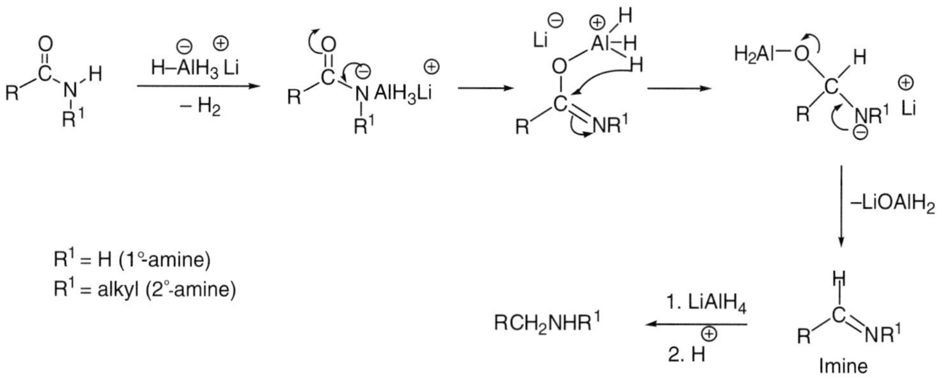

R^1 = H (1°-amine)
R^1 = alkyl (2°-amine)

Scheme 6.11

Reduction of acetanilide with tetra-*n*-butylammonium borohydride in the presence of dichloromethane followed by treatment with HCl gives 74% *N*-ethylaniline hydrochloride[25].

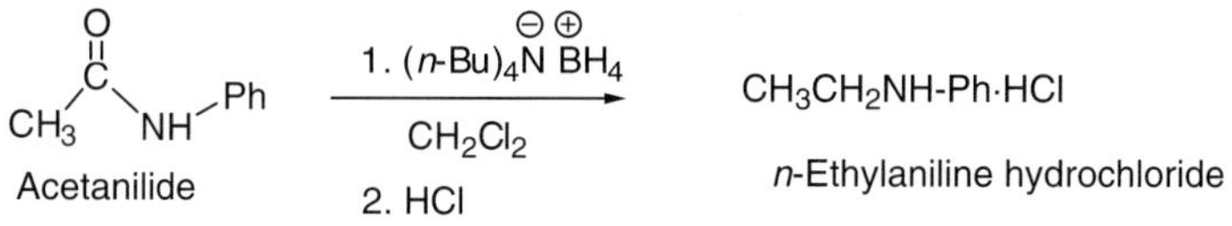

NaBH$_4$ reduces aldehydes or ketones in the presence of esters, amides, nitro and R–X groups in protic solvents.

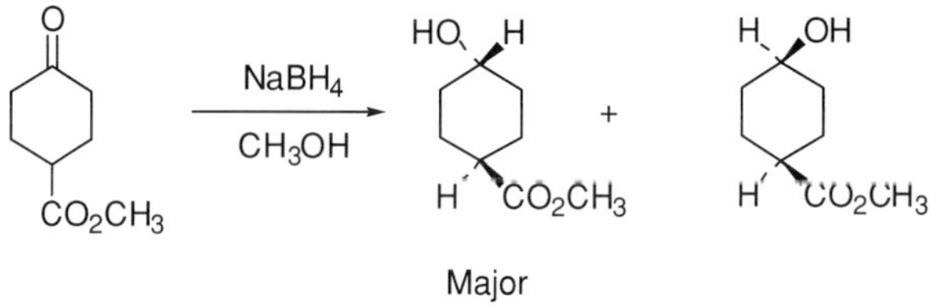

Sodium cyanoborohydride (NaBH$_3$CN), prepared from sodium borohydride and HCN, is less reactive and more selective. It reduces aldehydes and ketones only at low pH (3–4) and does not reduce acid chlorides and esters.

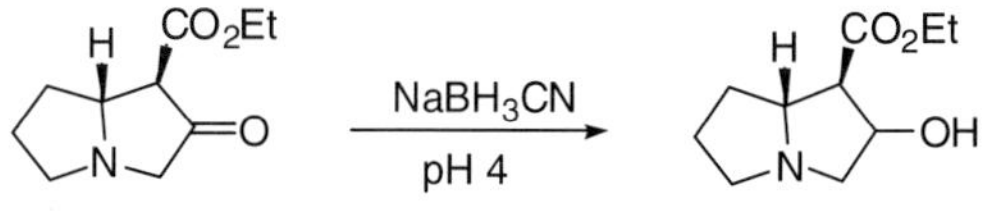

The lithium aluminium hydride reductions often proceed at room temperature or below and are usually rapid and free from side reactions. The compound to be reduced is added slowly to an excess of the reagent suspended or dissolved in ether (**normal addition**). Selective reduction of polar groups in the presence of other reducible functions can frequently be achieved by an **inverse addition method**: the reagent is added slowly to the substance to be reduced, so that the reagent is never present in excess. Thus, by **inverse addition** cinnamaldehyde (**6.44**) is reduced to cinnamyl alcohol (**6.3**). Normal addition gives dihydrocinnamyl alcohol (**6.4**).

Reduction of acid chlorides to aldehydes: One of the most useful synthetic transformations in organic synthesis is the conversion of an acid chloride to the corresponding aldehyde without over-reduction to the alcohol. Until recently, this type of selective reduction was difficult to accomplish and was most frequently effected by catalytic hydrogenation (the Rosenmund reduction; section 6.4.1). However, in the past few years, several novel reducing agents have been developed to accomplish the desired transformation. Among the reagents that are available for the partial reduction of acyl chlorides to aldehydes are bis(triphenylphosphine)cuprous borohydride[26–28], sodium or lithium tri-*tert*-butoxyaluminium hydride[29], complex copper cyanotrihydridoborate salts[30], anionic iron carbonyl complexes[31,32] and tri-*n*-butyltin hydride in the presence of tetrakis(triphenylphosphine)palladium(0)[33].

By using sodium borohydride in *N,N*-dimethylformamide solution containing a molar excess of pyridine as a borane scavenger, direct conversion of both aliphatic and aromatic acid chlorides to the corresponding aldehydes[34] can be achieved in >70% yield with 5–10% alcohol formation.

R = Ph, Ar, alkyl

Selective reduction of functional groups can be achieved by chemical modification of the LiALH$_4$ for example, lithium tri(t-butoxy)aluminium hydride [LiAIH(t-OBu)$_3$] is a more selective reagent, and reduces aldehydes and ketones, but slowly reduces esters and epoxides. Nitriles and nitro groups are not reduced by this reagent. Carboxylic acids can be converted into the aldehyde via acid chloride with lithium tri(*tert*-butoxy)aluminium hydride at a *low temperature* ($-78°C$). The nitro compounds are not reduced under this condition. Thus, selective reduction of 3,5-dinitrobenzoic acid (**6.45**) to 3,5-dinitrobenzaldehyde (**6.47**) can be achieved in two steps. First, **6.45** is converted into 3,5-dinitrobenzoyl chloride (**6.46**) and then LiAlH(t-OBu)$_3$ reduction of **6.46** gives **6.47**.

It is also possible to oxidize the alkoxyaluminium intermediate (without isolation), formed by the reaction of acid chlorides with half equivalent of $LiAlH_4$ at $0°C$, with PCC (pyridinium chlorochromate) or PDC (pyridinium dichromate) at room temperature[35].

R = Ph, Ar, alkyl

Lithium amino borohydride ($LiBH_3NR_2$) is powerful and selective reagent with reactivity comparable to $LiAlH_4$. It reduces aldehydes, ketones, esters and amides to alcohols. But sterically hindered lithium amino borohydrides convert amides to amines.

Diisobutylaluminium hydride reduces esters and ketones to alcohols at ordinary temperature. But at low temperature, it is used for the preparations of aldehydes from esters.

Red-Al can reduce esters to aldehyde in the presence of N-methyl-2-pyrrolidinone (NMP), whereas without the presence of an amine only dicarbinol results[36].

Lithium triethoxyaluminohydride [LTEAH, $Li(EtO)_3AlH$], formed by the reaction of 1 mol of $LiAlH_4$ solution in ethyl ether with 3 mol of ethyl alcohol or 1.5 mol of ethyl acetate, also reduces aromatic and aliphatic tertiary amides to corresponding aldehydes.

Useful modifications in the properties of lithium aluminium hydride are also effected with aluminium chloride and other Lewis acids[37–41]. $AlCl_3$ is added to $LiAlH_4$ in various proportions; the mixture becomes less reactive thus more specific than $LiAlH_4$. For example, a (1:1) mixture of $AlCl_3$ and $LiAlH_4$ reduces esters, aldehydes and ketones to alcohols, but C–X bonds and nitro groups are unaffected.

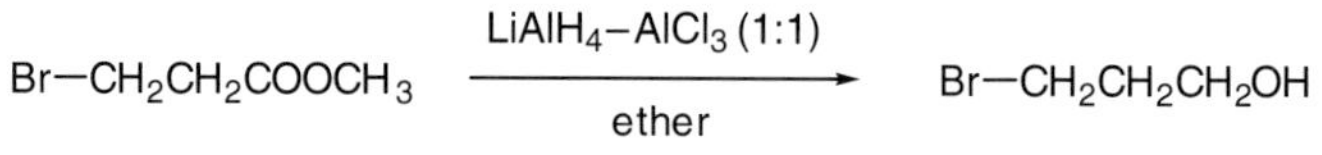

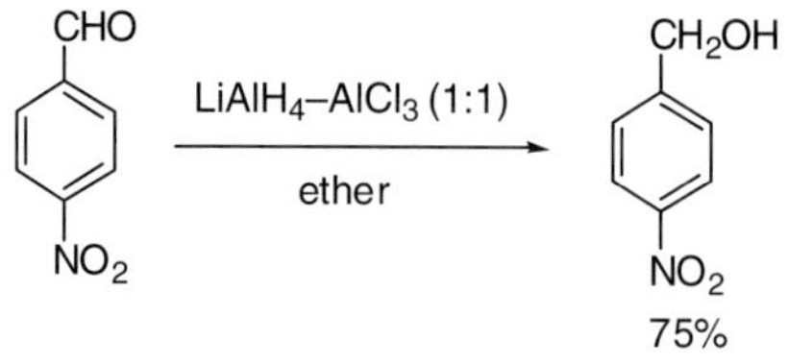

$$Br-CH_2CH_2COOCH_3 \quad \xrightarrow[\text{ether}]{\text{LiAlH}_4-\text{AlCl}_3 \text{ (1:1)}} \quad Br-CH_2CH_2CH_2OH$$

Diborane, B_2H_6, also reduces many carbonyl groups. In contrast to the metal hydride reagents, diborane is a relatively electrophilic reagent, as witnessed by its ability to add to carbon–carbon double bonds.

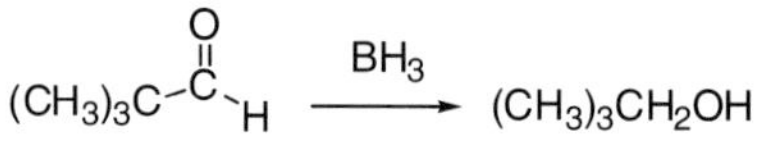

Borane is a Lewis acid and attacks electron-rich centres. Thus, the reduction of carbonyl group with borane takes place by addition of the borane (electrophile) to the oxygen atom (nucleophile) (Scheme 6.12).

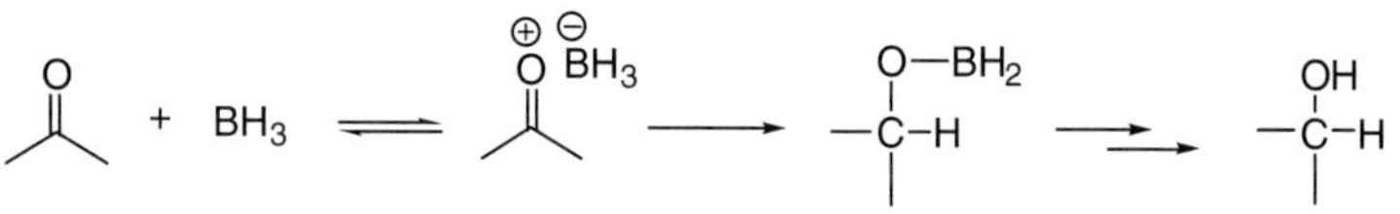

Scheme 6.12

Stereoselectivity of carbonyl reduction

The **asymmetric reduction** of carbonyl groups to form enantiopure secondary alcohols is a reaction of fundamental importance in modern synthetic chemistry. The carbonyl group is prochiral centre (if different groups are attached to carbonyl carbon) and can be attacked by hydride from *re* or *si* face to generate a racemic product if there are no additional chiral centres in the molecule.

The *re* and *si* faces can be assigned using a procedure similar to that used to assign absolute configurations R and S.

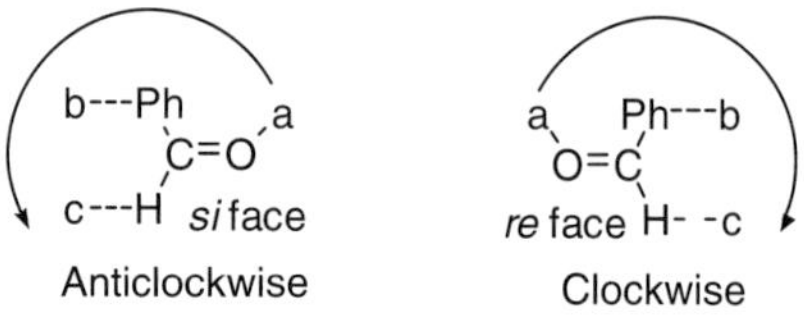

1. The carbonyl group is drawn in the plane of the paper.
2. Priority is assigned to all the groups (three) attached to the carbonyl carbon.
3. If the sequence a, b and c is anticlockwise it is the *si* face, and if it is clockwise it is the *re* face.

4. When addition to *re* and *si* faces produces enantiomers (Scheme 6.13) the faces *re* and *si* are said to be enantiotropic.

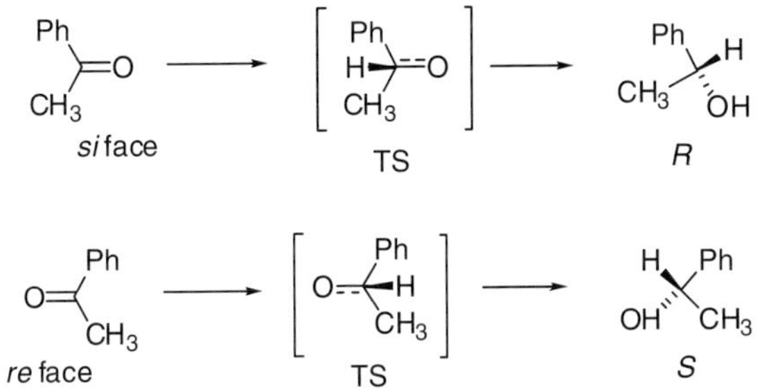

Scheme 6.13 Enantiomeric transition states lead to racemic products

5. When the nucleophile (H⁻ in this case) attacks from *si* face, enantiomer *R* is formed (Scheme 6.13).
6. When the nucleophile (H⁻ in this case) attacks from *re* face, enantiomer *S* is formed (Scheme 6.13).
7. The *re* and *si* faces are diastereotropic if the molecule is chiral and the products of addition to the two faces are diastereomers (Schemes 6.15 and 6.16).
8. In asymmetric synthesis, one of the enantiomers is formed in excess and the reaction is said to be enantioselective.

There are several ways for achieving asymmetric reduction of aldehydes and ketones. For example, a chiral catalyst or chiral reagent can be used for the enantioselective reduction.

Morrison and **Mosher** reviewed the use of chiral reagents to get the enantioselective reductions. These chiral reagents can be prepared from natural products that contain chiral centres of known absolute stereochemistry and reducing agent such as $LiAlH_4$. For example, reduction of acetophenone with $LiAlH_4$ in the presence of chiral reagents (**6.48** and **6.49**) generated from cinchona alkaloid, (−)-quinine and (+)-quinidine, gave (*R*)-alcohol in 48% ee and (*S*)-alcohol in 23% ee, respectively.

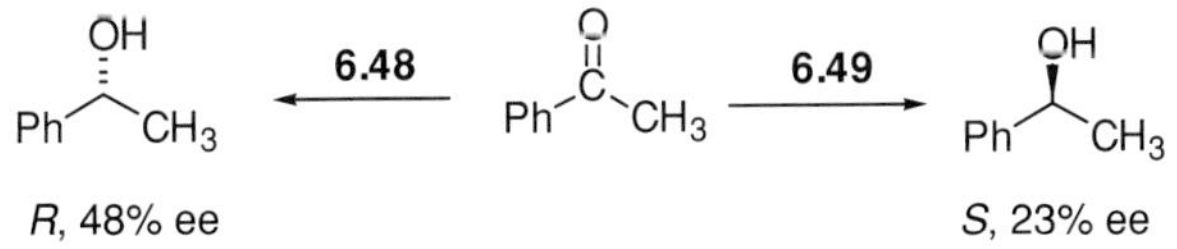

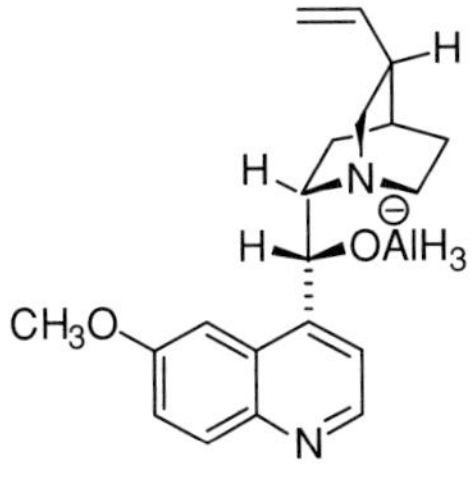

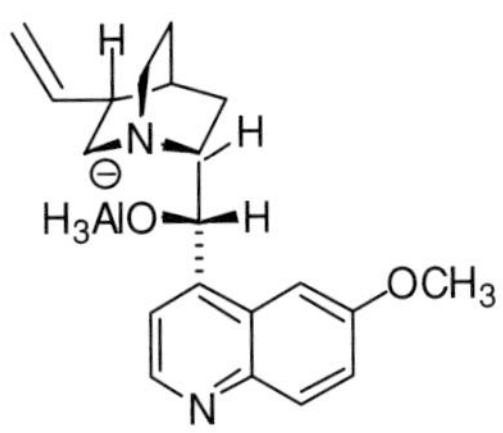

However, the reduction of ketones containing a chiral carbon generates diastereomers. If the stereogenic centre is adjacent to the carbonyl group, the chirality will exert an influence on the approach of the incoming reagent.

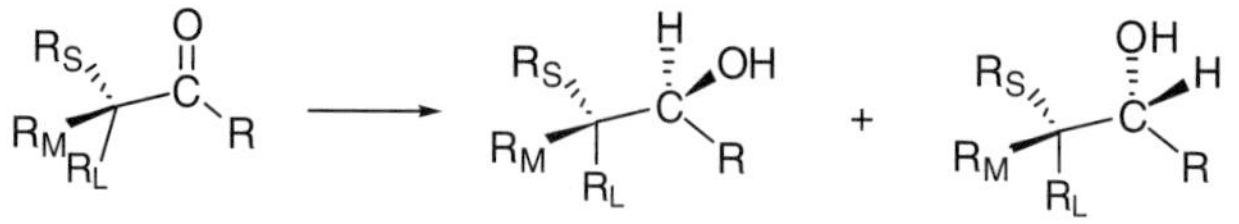

The extent of asymmetric induction in the above cases is suggested by **Cram's rule (Cram's model)**. The chiral centre adjacent to the carbonyl in ketones has three groups: R_S (small substituent), R_M (middle-sized substituent) and R_L (large substituent). The ketone assumes a predominant rotamer in which the carbonyl group is staggered between the medium and smallest substituents on the α-carbon or the large group is *syn* to the R group attached to the carbonyl group. The carbonyl group is attacked from the less hindered side (Scheme 6.14). The greater the difference in steric bulk between small and medium substituents, the greater the selectivity.

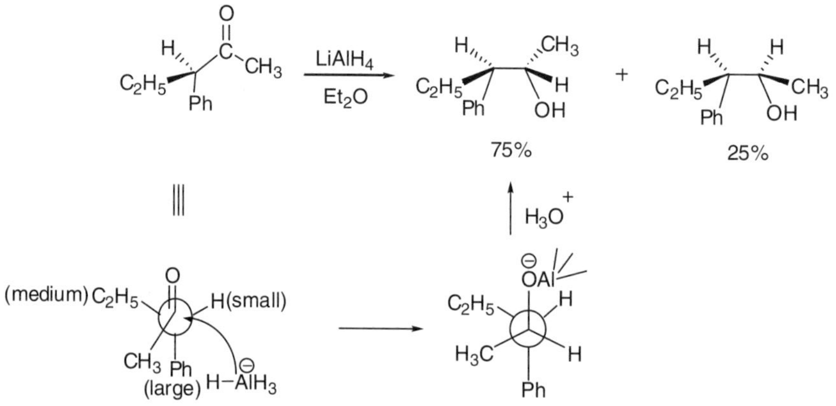

Scheme 6.14

However, the reducing agent may influence the conformation of ketone, and thereby the diastereoselectivity. If the small and medium substituents are close in size, then this model fails to predict the exact selectivity. This model also assumes less interaction between the large group and R group, which is not entirely correct.

   The **Felkin–Ahn model** assumed that the attack of nucleophile takes place antiperiplanar to a neighbouring σ bond of the large (R_L) group. Further, the attack shown in Scheme 6.15 is favoured over the attack shown in Scheme 6.16. In the latter, the nucleophile must pass close to the medium-sized group (R_M), whereas in the first case the nucleophile interacts with the smallest group.

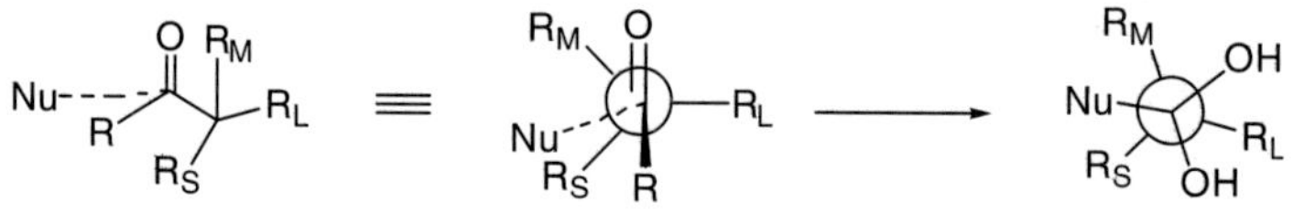

Scheme 6.15

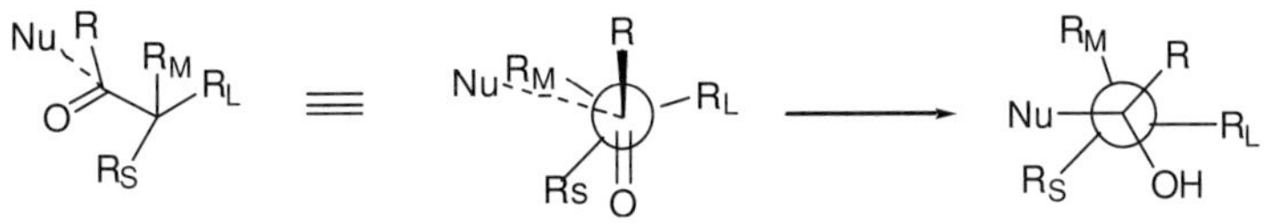

Scheme 6.16

When a group adjacent to the carbonyl group possesses a lone pair (usually either an oxygen or nitrogen substituent), then the reagent may undergo chelation (Scheme 6.17). Because of this chelation, the conformation is locked.

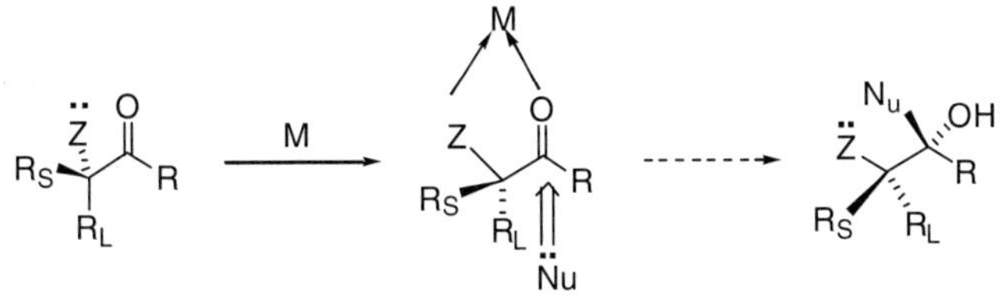

Scheme 6.17 Nucleophile attacks from less hindered face in chelation control carbonyl

The stereoselectivity in such cases is usually very high.

Prediction for the reduction of cyclic ketones on the basis of the **Cram, Karabatsos** and **Felkin–Ahn models** is usually unreliable and a simple model has yet to emerge.

The stereochemical product ratio for the reduction of cyclic ketones by hydrides is affected by the structure of the cyclic ketone and the nature of the hydride used. The reduction of substituted cyclohexanones avoids product interconversion by conformational ring flip because of conformationally locked cyclohexanones. In such cases, axial attack is preferred over equatorial attack. 4-*tert*-Butylcyclohexanone (**6.50**) is reduced by NaBH$_4$ and by LiAlH$_4$ to give 86% and 92% of *trans*-4-*tert*-butylcyclohexanol (**6.51**), respectively. Hindered hydrides such as *t*-Bu$_3$BHLi show more selectivity.

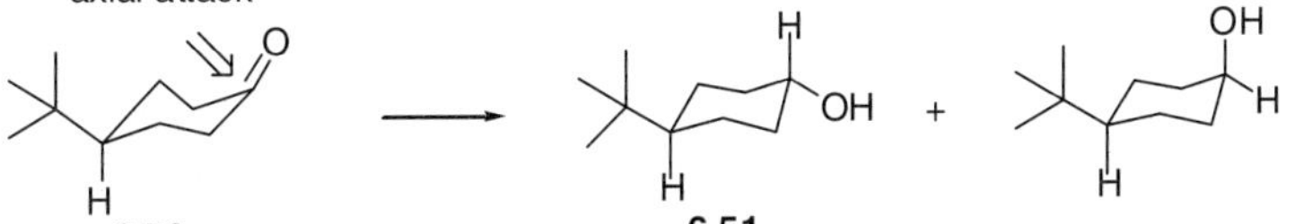

	6.51	
NaBH$_4$	86%	14%
LiAlH$_4$	92%	8%
t-Bu$_3$BHLi	> 99%	0%

But when steric environment on one side of the carbonyl group is different from the other side, then the hindered hydride reagents approach from the less hindered side as shown below.

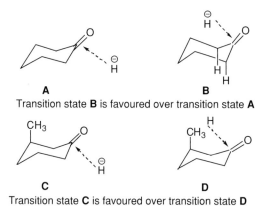

The mechanism of reduction of cyclic ketones by LiAlH$_4$ and NaBH$_4$ is quite different. The LiAlH$_4$ reduction involves reactant-like transition states (Scheme 6.18) and NaBH$_4$ reductions involve product-like transition states (Scheme 6.19). The LiAlH$_4$ reductions favour equatorial attack if bulky axial groups are present at C-3 and C-5 because of steric factors. Some non-steric factors which also favour equatorial attack can be explained by **Felkin–Ahn** rationalization on the basis of either torsional strain or the need for antiperiplanarity.

A

B

Transition state **B** is favoured over transition state **A**

C

D

Transition state **C** is favoured over transition state **D**

Scheme 6.18   Rationalization of stereochemistry in LiAlH$_4$ reduction

Thus, the axial substituents in the 3-position with respect to the carbonyl group disfavours axial attack.

Up to 60% de

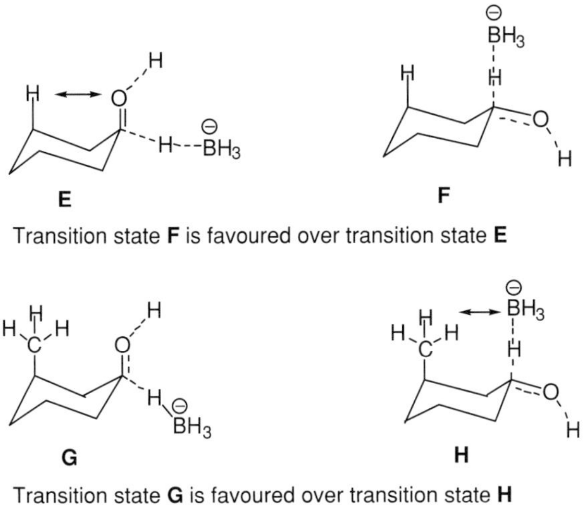
E
F
Transition state **F** is favoured over transition state **E**
G
H
Transition state **G** is favoured over transition state **H**

Scheme 6.19 Rationalization of stereochemistry in $NaBH_4$ reduction

Chiral boranes reduce carbonyl to alcohols with high asymmetric induction.

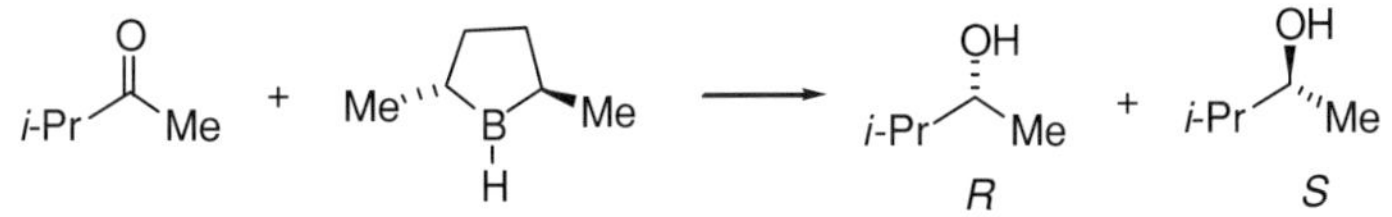
i-Pr
Me
Me
R
S

Enantioselective reduction of ketones by boranes and an enantiomeric catalyst oxazaborolidine (the **CBS catalyst**) is known as the Corey, Bakshi and Shibata method[42,43]. Both enantiomers of 2-methyl-CBS-oxazaborolidine (**6.52** and **6.53**) are used for the reduction of prochiral ketones, imines and oximes to produce chiral alcohols, amines and amino alcohols in excellent yields and enantiomeric excesses.

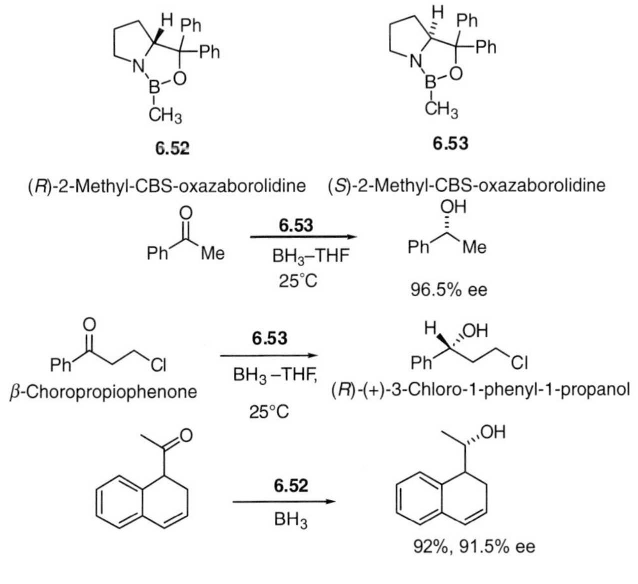
6.52
6.53
(*R*)-2-Methyl-CBS-oxazaborolidine
(*S*)-2-Methyl-CBS-oxazaborolidine
6.53
BH3–THF
25°C
96.5% ee
6.53
BH3 –THF,
25°C
β-Choropropiophenone
(*R*)-(+)-3-Chloro-1-phenyl-1-propanol
6.52
BH3
92%, 91.5% ee

Mechanism: The mechanism of oxidation[44] of carbonyl compounds with **6.53** is given in Scheme 6.20.

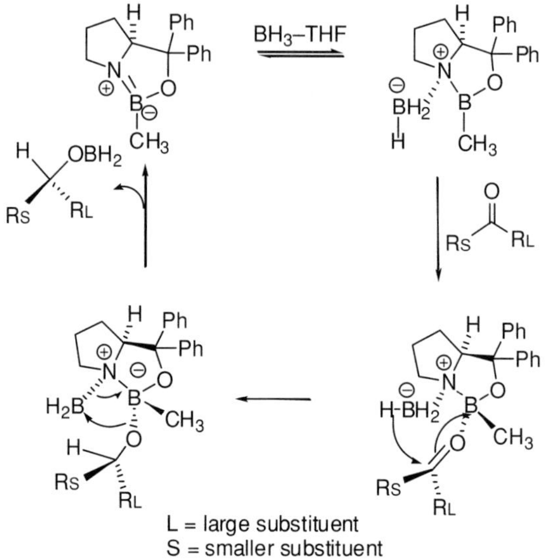

L = large substituent
S = smaller substituent

Scheme 6.20

Midland and co-workers[45] demonstrated that certain β-alkyl-9-borabicyclo [3.3.1] nonanes can reduce benzaldehyde to benzyl alcohols under mild conditions. Several optically active terpenes such as (+)-α-pinene (**6.54**), (−)-β-pinene (**6.55**), (−) camphene (**6.56**) and (+)-3-carene (**6.57**) have been used to prepare asymmetric β-alkyl-9-BBN reagents.

6.54	6.55	6.56	6.57
(+)-α–Pinene	(−)-β-Pinene	(−)-Camphene	(+)-3-Carene

For example, **6.58** is prepared by the reaction of **6.54** with 9-BBN.

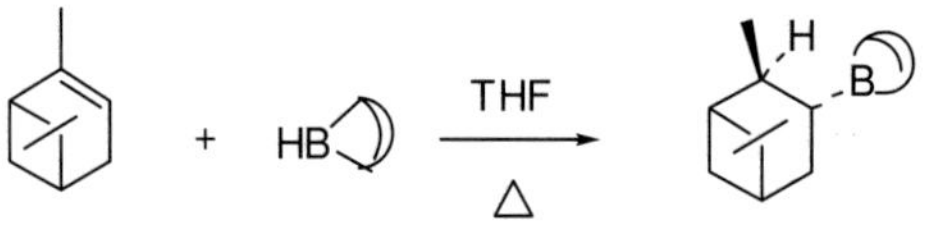

α-pinene   9-BBN   β–IPC-9-BBN
6.54   (β-3-Pinanyl-9-BBN)
6.58

These chiral reagents can transfer hydride from a chiral centre of the alkyl group to a new chiral centre of the carbonyl group, thus inducing optical activity into the reduced product, alcohol. For example, **6.58** transfers its β-hydrogen for the asymmetric reduction of the carbonyl compound and liberates 1 equiv. of α-pinene (**6.54**), which can be recycled.

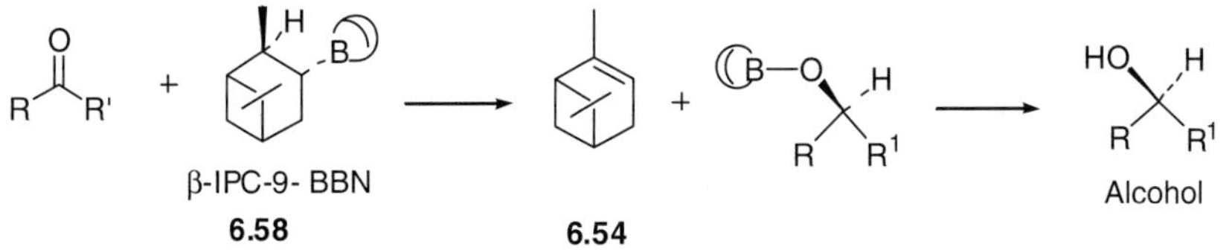

β-IPC-9- BBN

6.58   **6.54**   Alcohol

Reduction of benzaldehyde-1-D with chiral β-alkyl-9-BBN reagents **6.58**, **6.59**, **6.60** and **6.61** obtained from optically active terpenes such as (+)-α-pinene (**6.54**), (−)-β-pinene (**6.55**), (−)-camphene (**6.56**) and (+)-3-carene (**6.57**), respectively, is shown below in Scheme 6.21.

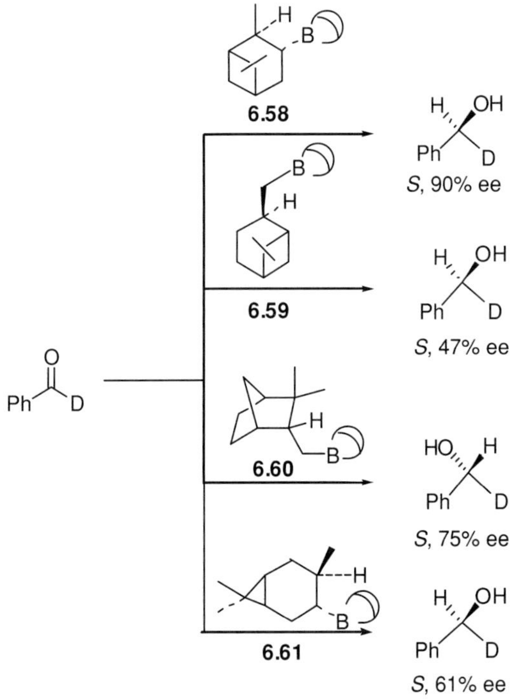

6.58

6.59

6.60

6.61

S, 90% ee

S, 47% ee

S, 75% ee

S, 61% ee

Scheme 6.21

Hydroboration of (+)-α-pinene **6.54** affords diisopinocampheylborane [(IPC)$_2$BH] (**6.62**).

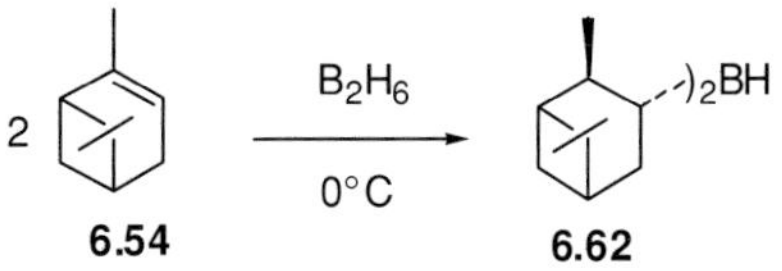

2   **6.54** $\xrightarrow[\text{0°C}]{B_2H_6}$ **6.62**

In the case of **6.62**, the hydride transfer takes place one atom away from the chiral centre. This factor is responsible for low optical purities observed in the reduction of ketones using (IPC)$_2$BH, **6.62** or (IPC)$_2$BD (Scheme 6.22).

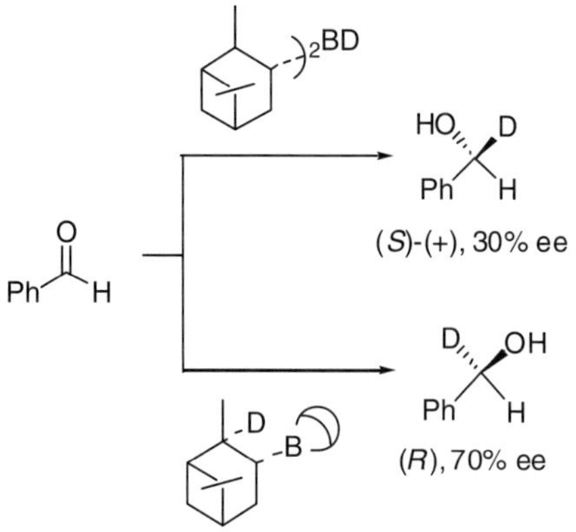

Scheme 6.22

Lithium β-3-pinanyl-9-boratabicyclo [3.3.1] nonyl hydride (**6.63**), which contains an asymmetric alkyl group, reduces ketones to the corresponding optically active alcohols at −78°C. The hydride transfer from this reagent takes place one atom away from the chiral centre, as compared to **6.58** in which the hydride is directly transferred from a chiral centre by a cyclic mechanism. Thus, the selectivity achieved in this reduction is also lower (Scheme 6.23).

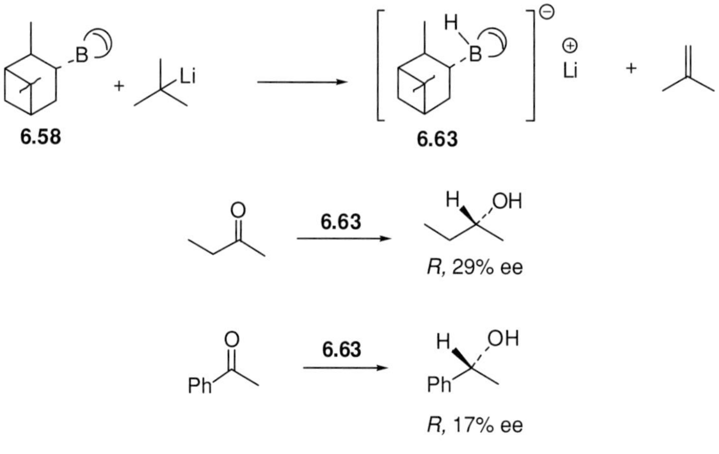

Scheme 6.23

(IPC)$_2$BCl (**6.64**) also transfers β-hydrogen and librates 1 equiv. of α-pinene. The reagent is best used for the asymmetric reduction of arylalkyl ketones, α-*t*-alkyl ketones and α-halo ketones.

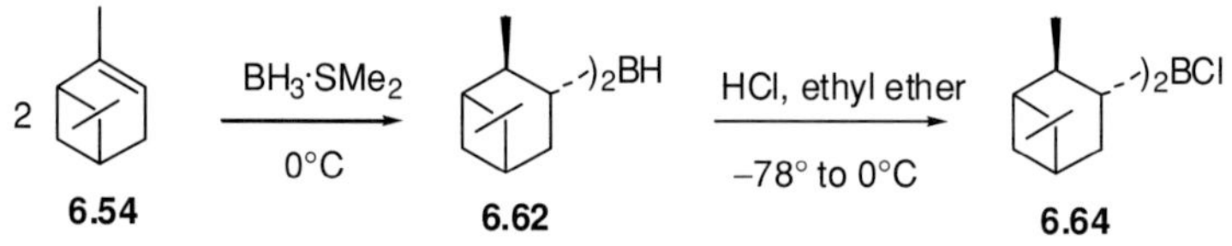

Silanes (H_2SiPh_2) also reduce[46] ketones asymmetrically in the presence of the chiral catalyst **6.65**.

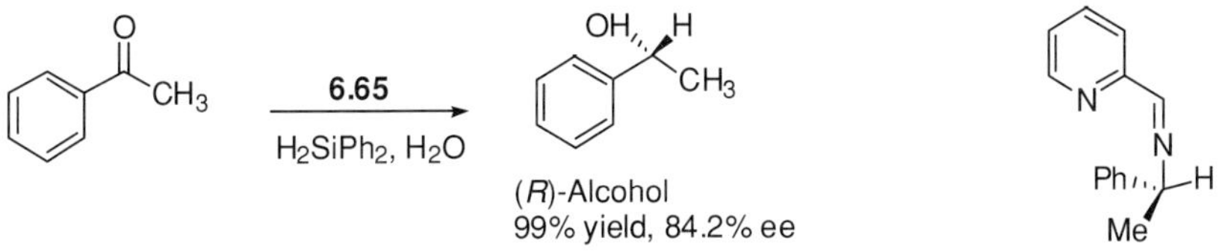

(*R*)-Alcohol
99% yield, 84.2% ee

6.65

The stereochemical result of the Lewis acid-mediated reduction of the carbonyl group in α-alkyl-β-ketoesters depends on Lewis acid employed, reducing agent and solvent. For example, reduction of **6.66** with BH_3–Py in the presence of $TiCl_4$ in CH_2Cl_2 solvent yields the *syn*-isomer **6.67** (with diastereomeric *syn/anti* excesses higher than 95/5). But reduction of **6.66** with lithium triethylborohydride ($LiEt_3BH$) in the presence of $CeCl_3$ in THF (tetrahydrofuran) solvent gives *anti*-isomer **6.68** (with diastereomeric *syn/anti* excesses higher than 90/10). The *syn* means that R^2 and OH are on the same side of the molecule on staggered conformation, whereas in *anti*-alcohol the OH and R^2 groups are on the opposite side of the molecule on staggered conformation (Scheme 6.24).

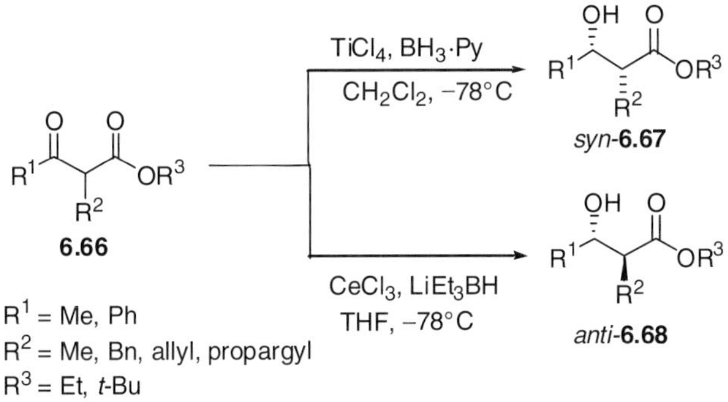

Scheme 6.24

The change in stereoselectivity observed in the reduction of β-ketoesters could be rationalized on the basis of different chelating ability of the Lewis acid used in the reaction. In this case, $TiCl_4$ is a strong chelating agent than $CeCl_3$. According to the **Felkin–Ahn model**, the relative conformation for the addition of nucleophile (in this case hydride) to the carbonyl group has the bonds to the L, M and S substituents staggered relative to the carbonyl function, and the hydride will attack from the side of the plane containing the small group. But when Lewis acid is present, it coordinates to the oxygen atom of the carbonyl group. The chelation is strong when a substituent with a heteroatom able to coordinate with the metal is placed next to the carbonyl group. This is explained by the **Cram cyclic model** in which the hydride again attacks from the less hindered side.

When TiCl₄ is used as a Lewis acid, the TiCl₄ complex can be represented as an equilibrium between the conformations **A** and **B**, where **A** is more stable than **B**. Thus, the hydride ion attacks the **A** conformation from less hindered side leading to *syn*-alcohol **6.67** (with high diastereoselectivity) (Scheme 6.25).

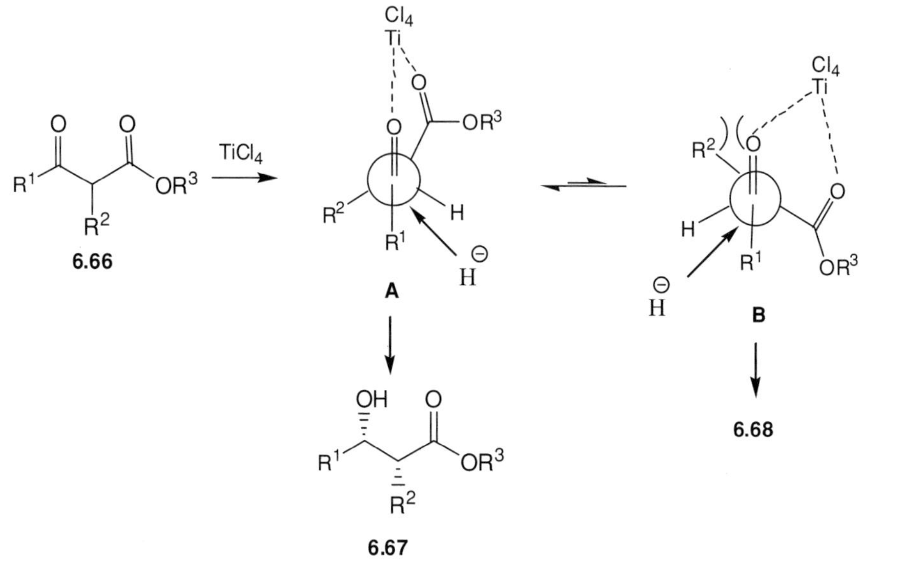

Scheme 6.25

However, CeCl₃ is a poor chelating Lewis acid. The stereochemistry of the reaction is reversed in this case, which can be explained on the basis of an open-chain **Felkin–Ahn model**. The *anti*-alcohol 6.68 is produced by the attack of hydride from the less hindered side to the most stable conformation, **C** of CeCl₃ complex. The **D** conformation is less stable than **C** (Scheme 6.26).

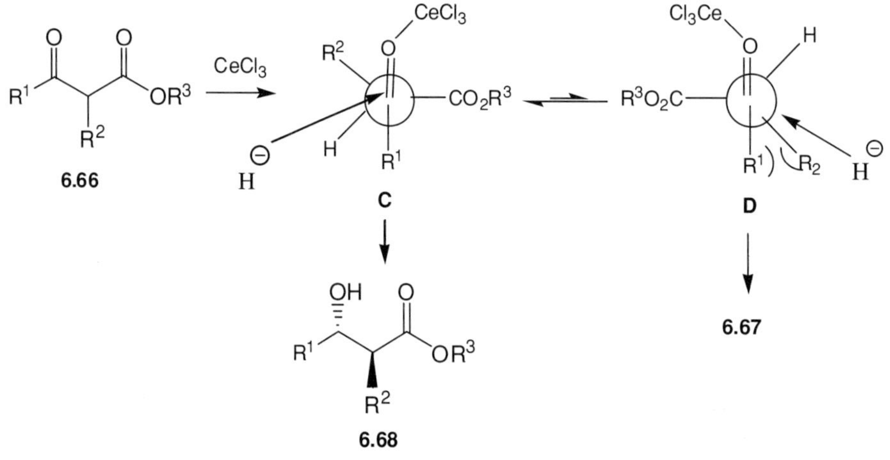

Scheme 6.26

6.4.3 Metal and proton source

The ability of certain metals to donate electrons to electrophilic or unsaturated functional groups has proven useful in several reductive procedures. The facility with which these metals donate electrons is given by their standard reduction potentials.

$$Li(Li^+)\ K(K^+)\ Na(Na^+)\ Mg(Mg^{2+})\ Al(Al^{3+})\ Ti(Ti^{3+})\ Zn(Zn^{2+})\ Fe(Fe^{3+})\ Sn(Sn^{2+})$$
$$-3.05\ -2.93\ -2.71\quad -2.37\qquad -1.66\quad -1.21\quad -0.76\quad -0.44\quad -0.14$$

From the values of potentials, the qualitative order of their reducing power is as follows: $Li > K > Na > Mg > Al = Ti > Zn > Fe > Sn$.

Carbonyl groups and conjugated π-electron systems are reduced by metals such as Li, Na and K, usually in liquid ammonia solution. Other reactive metals such as zinc and magnesium reduce aldehydes and ketones in the presence of a proton source.

Clemmensen reduction: Aldehydes and ketones are reduced by heating with a solution of zinc metal in mercury (zinc amalgam) and hydrochloric acid.

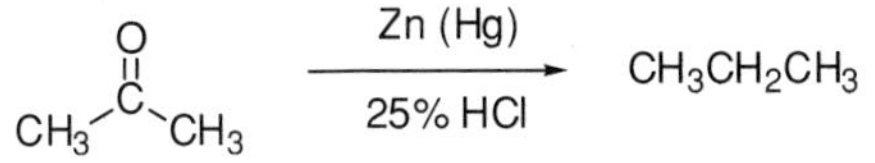

First, functional substituents α to the carbonyl group such as hydroxyl, alkoxyl and halogens are reduced, and then the resulting unsubstituted aldehyde or ketone is reduced to the parent hydrocarbon.

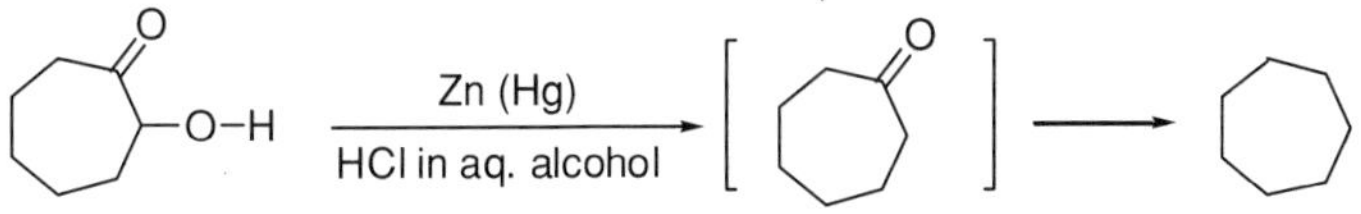

As the metal dissolves, it gives up two electrons which reduces the C=O bond. The mercury alloyed with the zinc does not participate in the reaction; it serves only to provide a clean active metal surface. The mechanism of reduction[47] is not very clear, but it is believed to proceed as shown in Scheme 6.27.

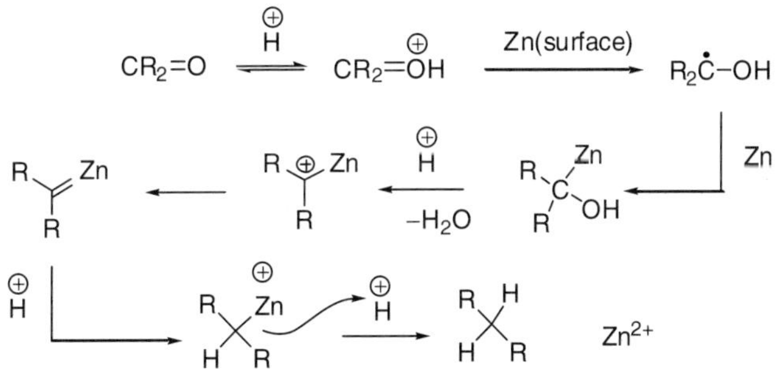

Scheme 6.27

Reduction with alkali metals: The solvents used for alkali metal reductions include hydrocarbons, ethers and, most commonly, liquid ammonia. Alcohols may also be used, but usually as co-solvents, since they react vigorously with these metals. Aldehydes are not usually reduced in this manner, because they react with ammonia to form unreactive imine condensation products.

Mechanism: Lithium, sodium or potassium reduce ketones by a *one-electron transfer* that generates a *radical anion* **A** known as a *ketyl*. Once such a reactive species is formed, it may react further by several modes, as described in Scheme 6.28. If a proton source is present, the *ketyl* undergoes carbon protonation. The resulting oxy radical **B** adds another electron to generate an alkoxide salt **C**, which takes proton to yield alcohol **D**.

Alternatively, ketyls **A** may dimerize to pinacol salts **E**. Isolation of pinacol products **F** requires further protonation by acids at least as strong as water or ethanol. The H^+ notation refers to any of the several possible proton sources, including ammonia, alcohols and the ammonium cation (a strong acid in the liquid ammonia system) (Scheme 6.28).

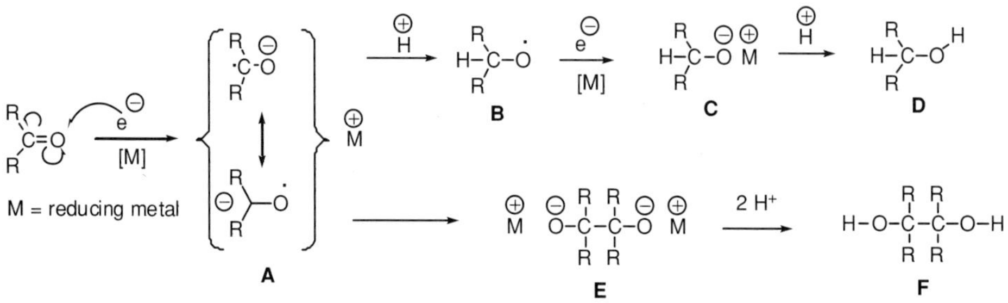

Scheme 6.28

Reduction of benzophenone in liquid ammonia gives both alcohol and pinacol products. The *ketyl* intermediate in this reaction is stabilized by phenyl substituents, and competitive carbon atom protonation and dimerization generate alkoxide salts that remain in solution until hydrolyzed prior to product isolation. Benzophenone (diphenyl ketone) forms a deep blue ketyl which is stable in solvents that lack acidic hydrogens such as hydrocarbons and ethers. It is widely used as an indicator of oxidizing or acidic impurities during the purification of such solvents.

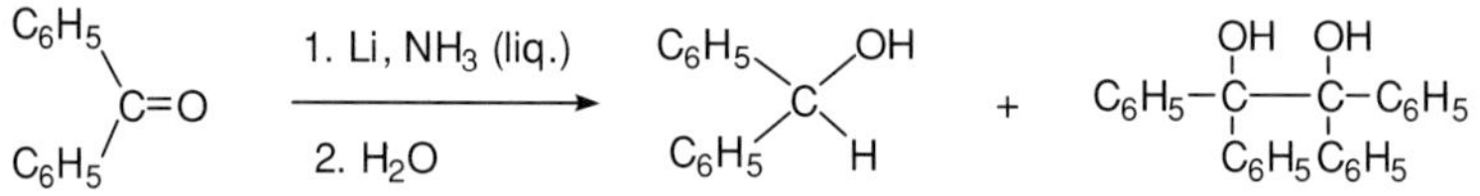

When pinacol products are desired, a less reactive metal such as magnesium is often used, and best results have been achieved when the metal is activated by amalgamation (alloyed with mercury) and Lewis acids are present. Hydrolysis of metal alkoxides releases the product.

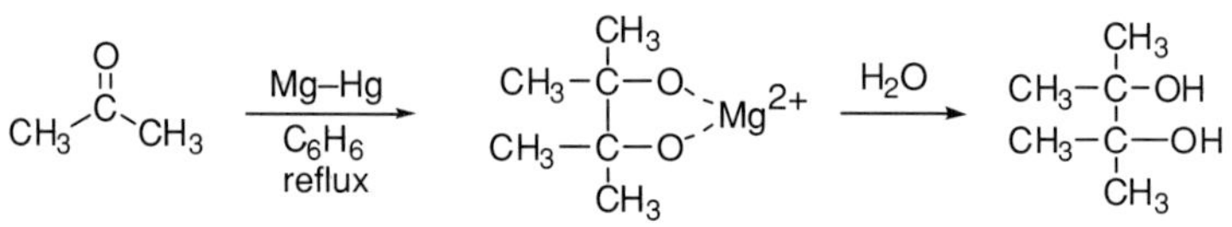

Titanium forms sufficiently strong bonds to oxygen to remove the oxygen from the dimeric species as titanium dioxide; this results into an alkene rather than a diol (see 'McMurry alkenation' under section 4.3.4).

Ester functions also undergo similar reduction on treatment with sodium. The most useful reaction of this kind is the **acyloin condensation**. To avoid protonation at carbon, this reaction is normally carried out in hydrocarbon solvents. The acyloin condensation creates α-hydroxy ketones.

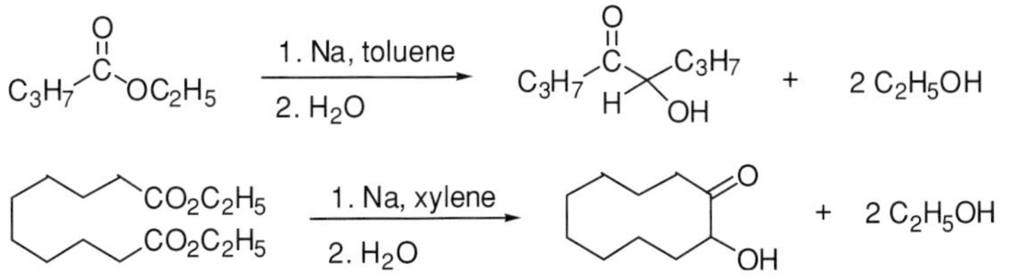

When a proton source such as ethanol is used, sodium reduces esters, aldehydes and ketones to alcohols and the reduction is known as **Bouveault–Blanc reduction**[48]. But metal hydride reduction of carbonyl compounds gives higher yields of alcohols.

Mechanism: Mechanism of esters reduction involves the electron transfer from metal, followed by protonation by EtOH as shown in Scheme 6.29.

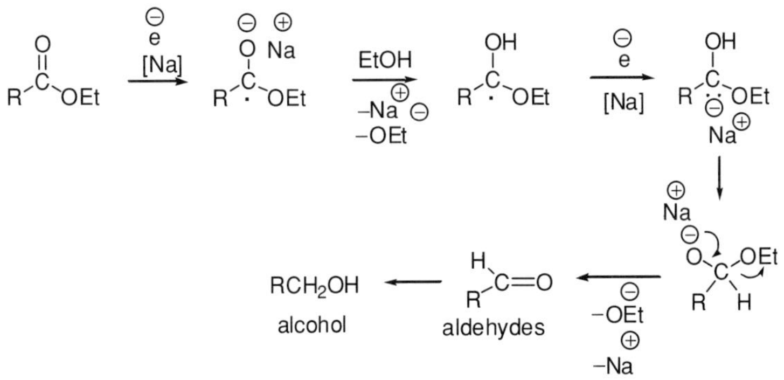

Scheme 6.29

6.4.4 Hydrogen transfer reagents

Wolff–Kishner reduction: The reduction of ketones to a methylene group via the decomposition of hydrazones of ketones is known as **Wolff–Kishner reduction**. The ketones are heated with NH_2NH_2 in aqueous KOH. The variation of this reaction is **Huang-Minlon**[49] where hydrazone is heated with alkali in ethylene glycol at 200°C. For example, cyclopentanone is reduced to cyclopentane on heating with NH_2NH_2 and KOH in diethylene glycol (DEG) solution. An aldehyde is similarly reduced to a methyl group.

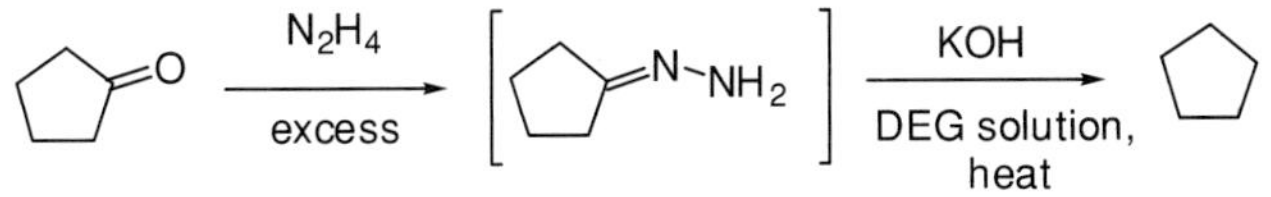

Mechanism: The reaction involves converting a ketone to the corresponding hydrazone **A**, which undergoes a base-catalyzed double bond migration (tautomerization) of the initially formed hydrazone to an azo-isomer **B**, and the loss of N_2 then follows to give carbanion **C**. Finally, an alkane derivative is formed by protonation of carbanion **C** (Scheme 6.30).

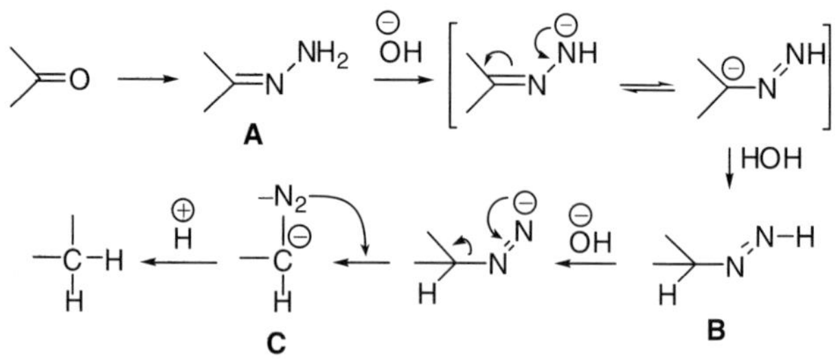

Scheme 6.30

The strongly basic conditions used in this reaction preclude its application to base-sensitive compounds.

Asymmetric reduction: Noyori's group in the 1990s developed a second generation of catalysts based on a ruthenium metal centre bearing a chiral diphosphine and a chiral diamine ligand. With these catalysts, asymmetric hydrogenation of ketones is possible without the presence of a secondary functional group. For example, several alcohols were obtained from unfunctionalized ketones in the presence of a base such as t-BuOK by using the **Noyori second-generation ruthenium catalyst 6.69**. The mechanistic studies confirmed that the catalytic reaction took place without the direct coordination of the substrate to the metal centre. The catalyst **6.69** is acting as a bifunctional scaffold for anchoring the substrate and transferring the hydride[50,51].

In the transfer hydrogenation, the hydride donor such as **2-propanol or formic acid** generates a metal hydride (ruthenium hydride in this case). The metal hydride selectively transfers the hydride to ketone via a bifunctional mechanism related to the one that operates for hydrogenation. The metal hydride is regenerated *in situ* from the hydrogen donor (Scheme 6.31).

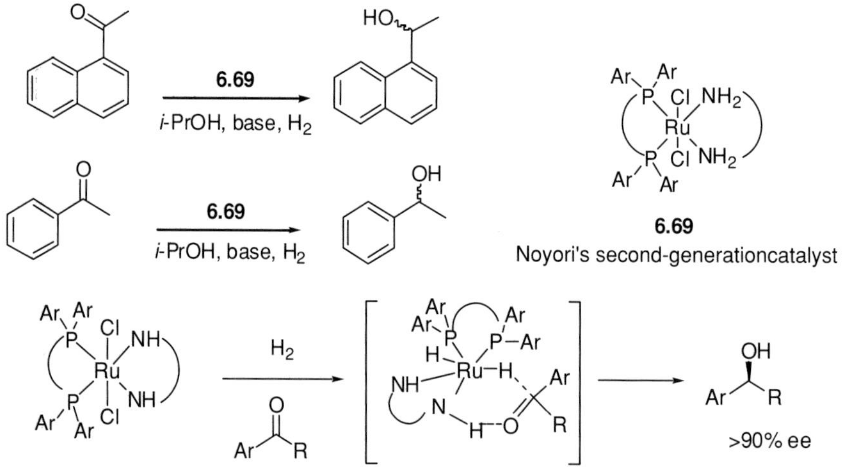

Scheme 6.31

The catalyst bearing BINAP (**6.10** and **6.11**) (developed by Noyori's group) or P-Phos ligands **6.70** (developed by Professor Albert Chan; which is more active and selective than the analogous BINAP ligand) and ancillary ligands such as 1,2-diamines, DAIPEN (**6.71**) and DPEN (**6.72**) or 1,4-diamine, **6.73** could be useful for the hydrogenation of some difficult substrates. For example, tetralone is efficiently reduced using the ruthenium catalyst **6.74**. Similarly, hydrogenation of sterically hindered aromatic ketones such as isobutyrophenone is also possible using the Ru catalyst **6.75**.

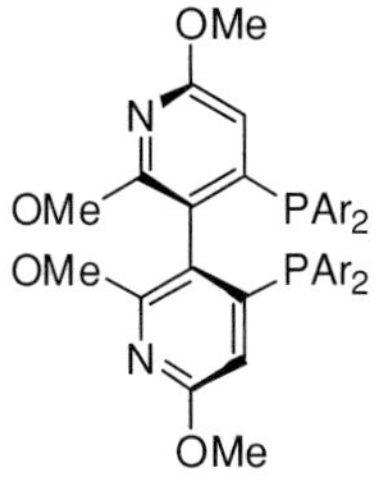

6.70

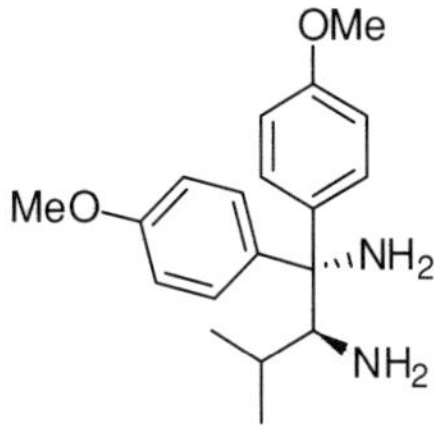

(S)-DAIPEN	(S,S)-DPEN	
6.71	**6.72**	**6.73**
1,2-Diamine ancillary ligand		1,4-Diamine ancillary ligand

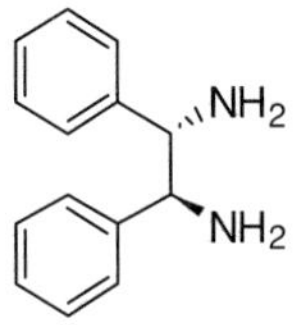

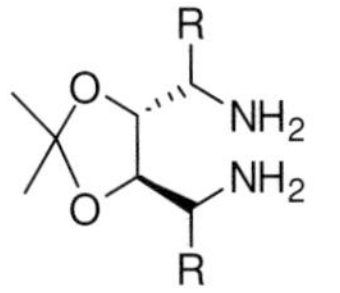

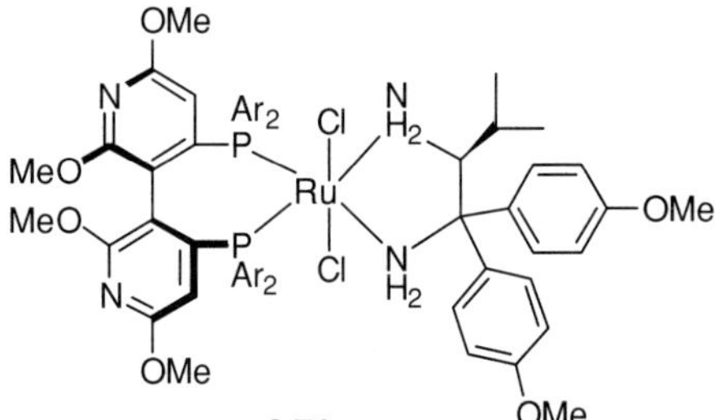

6.74 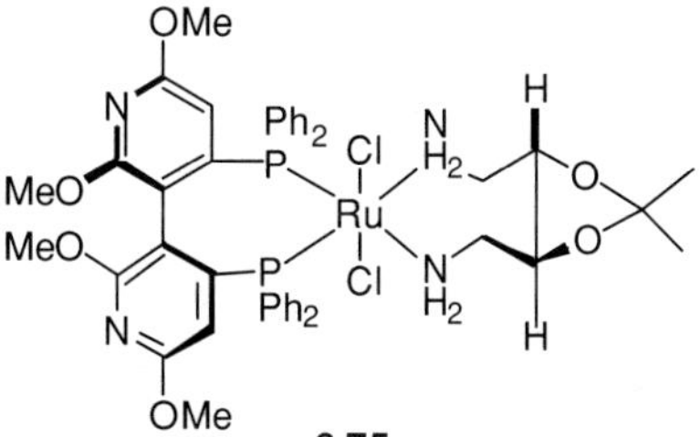　　　　　　**6.75**

Ru catalysts with a P-Phos and a 1,2 - or 1,4-diamine ancillary ligand

The catalyst (sulfonyl-diamine)RuCl(arene) **6.76**, without the expensive phosphine ligand, and the hydrogen donor such as 2-propanol or formic acid in the presence of a base are used for the asymmetric transfer hydrogenation of ketones[52–55].

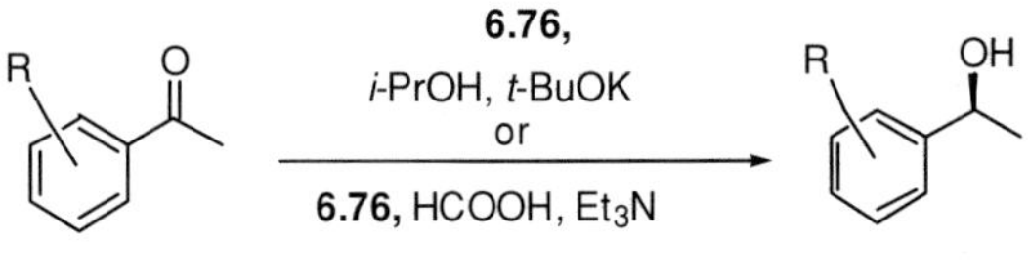

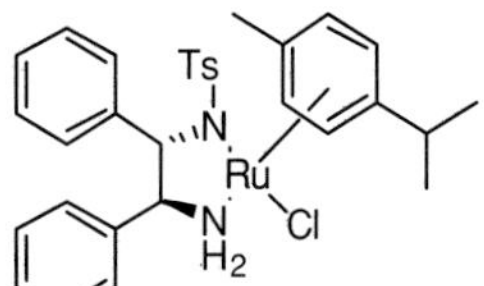

6.76

[Ts -DPEN RuCl(*p*-cymene)]

catalyst

With these homogeneous catalysts, it is possible to selectively reduce the C=O group without any competitive reduction of the carbon–carbon double bond, nitro group, arenes and aryl–halide bond.

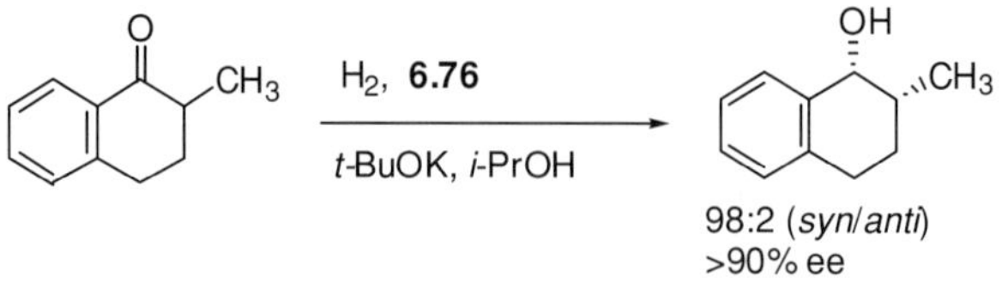

98:2 (*syn/anti*)
>90% ee

The enantioselective reduction of acetophenone using pyrrolidine–oxazoline-derived ligand[56] **6.77** and [IrCl(COD)]$_2$ gave optically pure alcohol (41–49% ee).

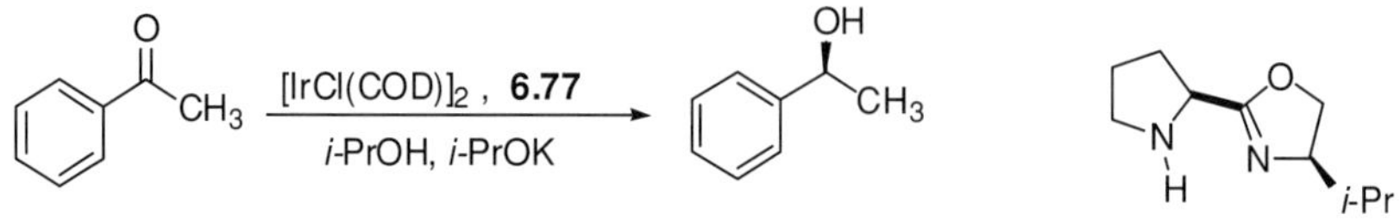

6.77

6.5 Reduction of α, β-unsaturated aldehydes and ketone

The α,β-unsaturated carbonyl compounds can be selectively reduced to either corresponding unsaturated alcohols (1,2-reduction) or saturated carbonyl compounds (1,4-reduction) by various methods[57].

6.5.1 Catalytic hydrogenation

Ru–BINAP complexes **6.78** and **6.79** can be used for the 1,2-reduction of α,β-unsaturated carbonyl compounds.

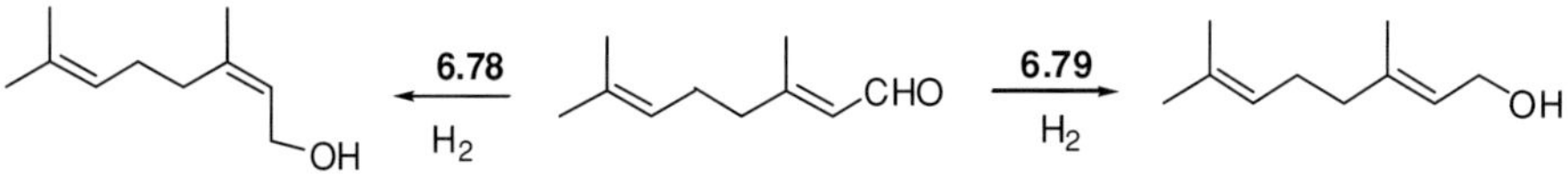

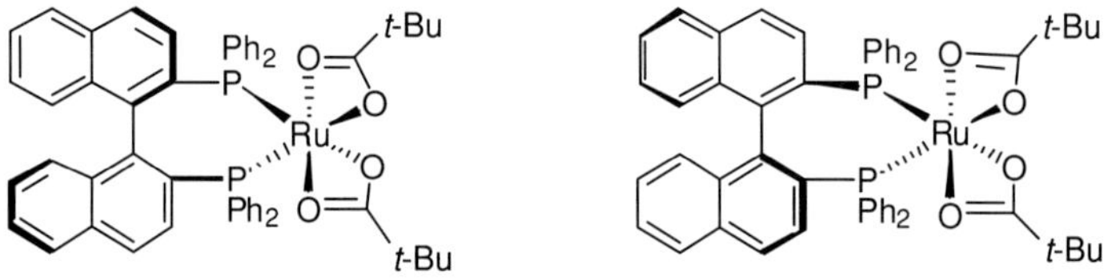

Ru(*R*-BINAP)(OPiv)$_2$
6.78

Ru(*S*-BINAP)(OPiv)$_2$
6.79

6.5.2 Hydride reagents

Carbon–carbon double and triple bonds are generally unaffected with hydride reagents except when they are α,β- to a polar group such as the carbonyl group. Reduction of α,β-unsaturated ketones with metal hydride reagents (LiAlH₄ and NaBH₄) sometimes leads to a saturated alcohol (1,4 reduction), especially with sodium borohydride. However, DIBAH and 9-BBN give exclusive 1,2 reduction (C=O reduction).

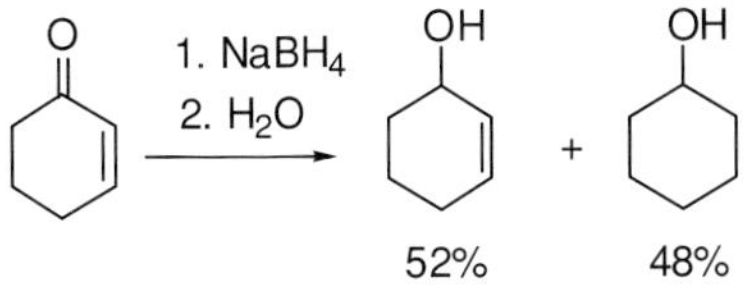

52%   48%

Mechanism: 1,4-Reduction product is formed by an initial conjugate addition of hydride to the β-carbon atom. The enolate thus formed undergoes ketonization, and reduction of the resulting saturated ketone gives saturated alcohol (Scheme 6.32).

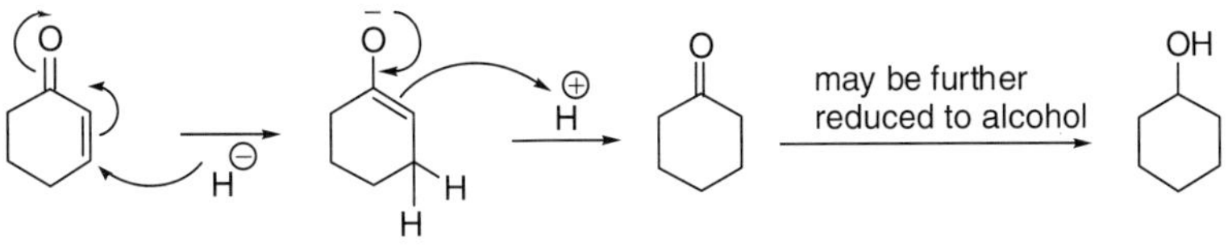

Scheme 6.32

If the R group in α,β-unsaturated ketones is small or R=H, i.e. in α,β-unsaturated aldehydes, then 1,2-addition predominates.

AlH₃ reduces 2-cyclopentenone to give unsaturated alcohol as the major product (1,2-reduction). The Lewis acid, AlH₃, first complexes with the carbonyl oxygen and then hydride is transferred to the carbonyl group by a four-centred transition state. However, LiAlH₄ forms very weak coordination with the carbonyl group; thus, both 1,2-reduction and 1,4-reduction occur. However, at −70°C LiAlH₄ undergoes 1,2-reduction of 2-cyclopentenone to give unsaturated alcohol as the major product.

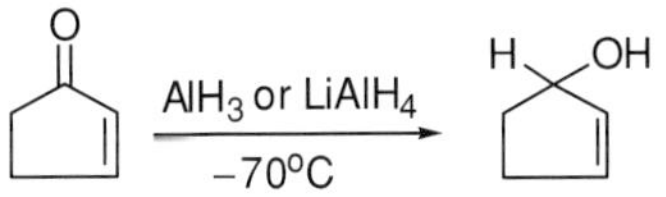

Cerium(III) chloride is an efficient catalyst for the regioselective 1,2-reduction of α,β-unsaturated ketones by sodium borohydride in methanol solution. The cerium-NaBH₄ reagent is known as the **Luche reagent**. Metal salts activate the ketone group by coordinating with the oxygen of the carbonyl group, which easily undergoes addition reactions[58–60].

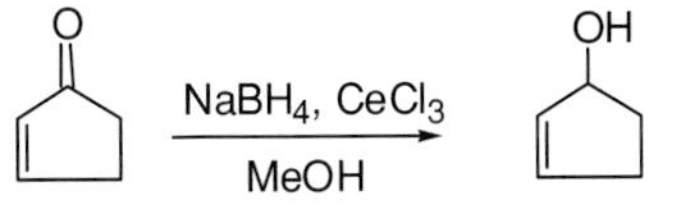

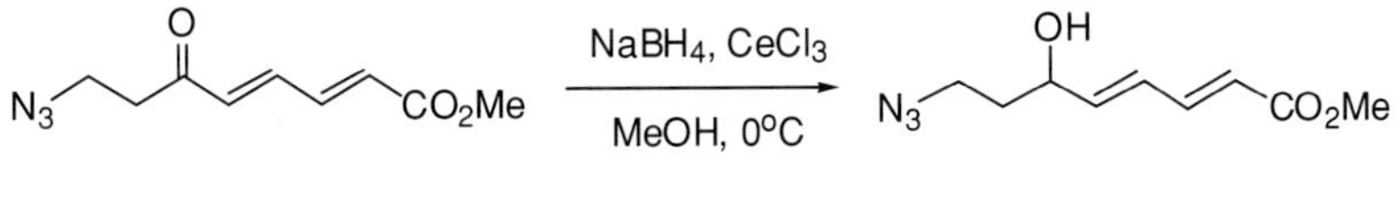

6.5.3 Dissolving metals

Under typical Birch reduction conditions α,β-unsaturated aldehydes, ketones and carboxylic esters are reduced to enolates, which can be trapped regiospecifically by reactive alkylating agents or other electrophiles[61,62].

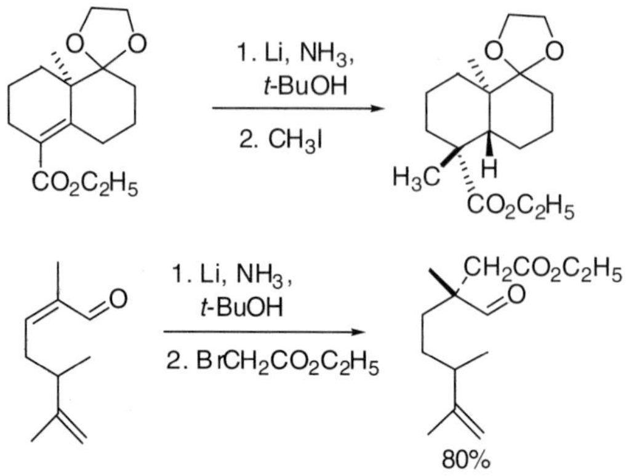

Mechanism: Because the π-electron systems of the two functional groups in α,β-unsaturated ketone are conjugated, the radical anion **A** formed by electron addition from a reducing metal is resonance stabilized. The usual fate of the **A** is protonation (or other electrophilic bonding) at the β-carbon atom. This creates an enoxy radical **B** which immediately accepts an electron to form an enolate anion **C**. Protonation or alkylation of this enolate species then gives a saturated ketone **D or E**, which may be isolated or further reduced depending on the reaction conditions (Scheme 6.33).

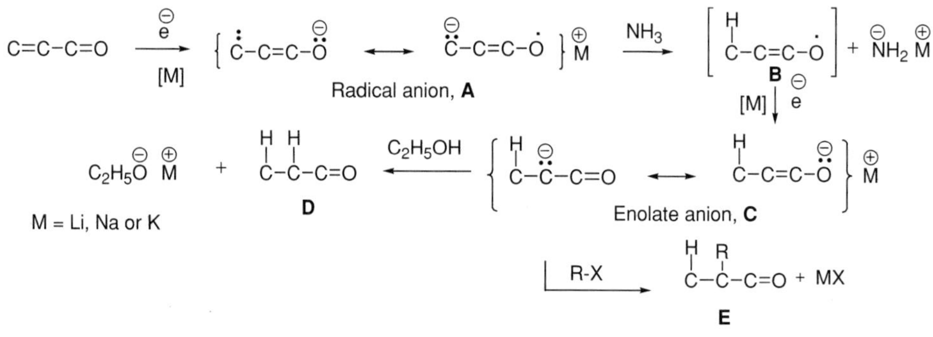

Scheme 6.33

If the lithium reduction is carried out in liquid ammonia without any acidic co-solvents, the enolate anion **C** is stable and remains unchanged until an electrophilic reagent such as R–X is added. If an acidic co-solvent such as ethanol is present, the enolate anion **C** is protonated, and the resulting ketone **D** is then reduced to an alcohol. Although the radical anion intermediate **C** usually undergoes protonation at the β-carbon, this is not a fast reaction in liquid ammonia.

6.6 Reduction of nitro, *N*-oxides, oximes, azides, nitriles and nitroso compounds

6.6.1 Catalytic hydrogenation

Hydrogenation of **nitriles** in the presence of a metal catalyst (such as rhodium, platinum or palladium) can produce primary, secondary or tertiary amines via an intermediate imine. Indeed, aldehydes and alcohols can be the reduced product if hydrogen supply is restricted or a sufficient amount of water is present in the reaction mixture. The **aromatic nitro and nitroso** groups are hydrogenated to the corresponding amine over the metal catalyst, Pd or Pt. The **aliphatic nitro** compounds are less easily reduced as the amine formed poisoned the catalyst. Cloronitrobenzene can be hydrogenated to chloroaniline in the presence of a highly selective 1% Pt–C catalyst.

6.6.2 Metal hydrides

The reduction of nitriles and amides to the amines is a very important, but not so easy, organic transformation. Several methods for the reduction of nitriles[63] and amides[64–67] with complex metal hydrides have been reported so far.

LiAlH$_4$ reduction of nitriles gives amines.

Mechanism: Reduction of nitriles with LiAlH$_4$ produces iminoalanate **A**, which undergoes shift of H and migration of double bond to form iminoalanate **B**. Hydrolysis of iminoalanate **B** gives an amine (Scheme 6.34).

LiBH$_3$N(CH$_3$)$_2$ can also reduce nitriles to amine.

Another convenient and efficient method for the reduction of nitriles to the amines is with tetra-*n*-butylammonium borohydride in refluxing dichloromethane. The chemospecificity of tetra-*n*-butylammonium borohydride towards organic cyano and amide compounds

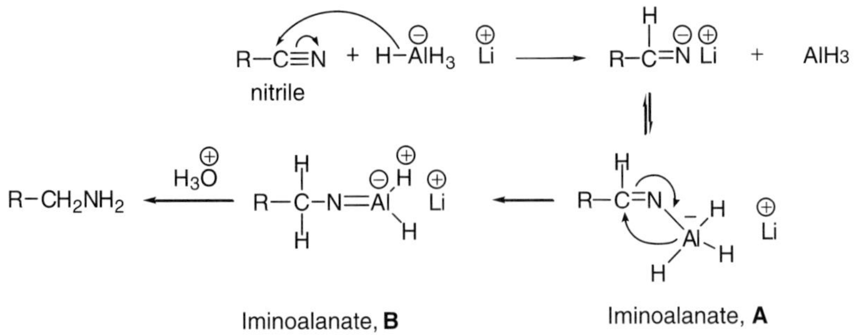

Scheme 6.34

was observed. The other functional groups, e.g., ester, nitro and halogen attached to the aromatic ring are not affected. For example, reduction of *p*-toluonitrile (**6.80**) with tetra-*n*-butylammonium borohydride in the presence of dichloromethane, followed by treatment with HCl, gives 87% *p*-methylbenzylamine hydrochloride (**6.81**).

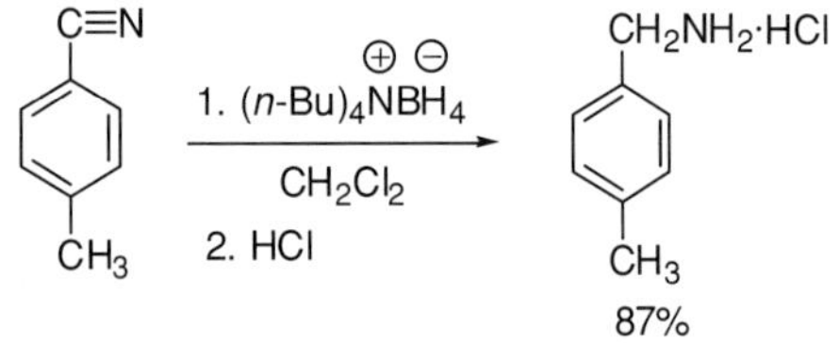

However, with less reactive hydrides such as $LiAlH(OEt)_3$ and DIBAL, the iminoalanate **A** and **B** are not formed (Scheme 6.34). The intermediate iminium salt can be hydrolyzed to RCHO. Red-Al [sodium bis(2-methoxyethoxy)aluminium hydride] can also be used to get even better yield of aldehyde.

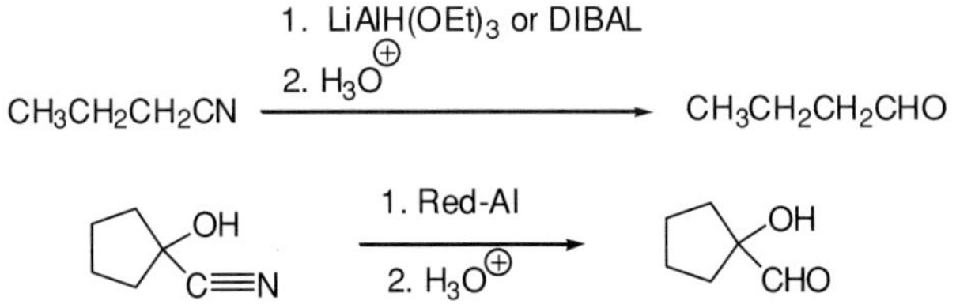

Red-Al = Sodium bis(2-methoxyethoxy)aluminium hydride

The iminoalane **A** obtained by the reaction of nitrile with DIBAL (diisobutylaluminium hydride) on hydrolysis gives the aldehyde via imine **B** (Scheme 6.35).

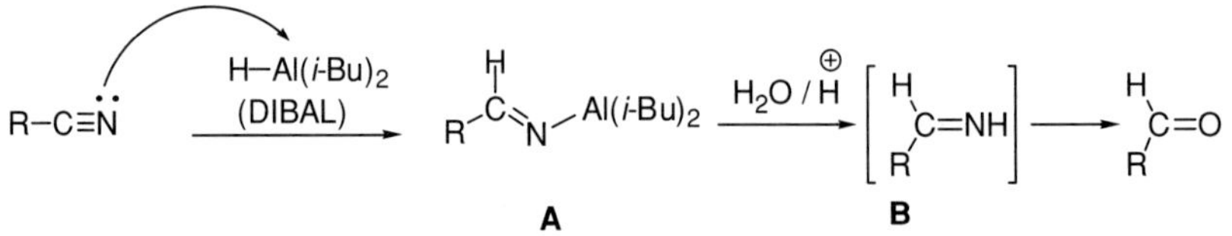

Scheme 6.35

The –CH=NOH and –CH$_2$N$_3$ groups are also reduced to –CH$_2$NH$_2$ by LiAlH$_4$. The primary aliphatic nitro compounds are reduced to amines with the loss of oxygen. However, nitrobenzenes are not reduced to anilines, but rather a diazo compound is formed.

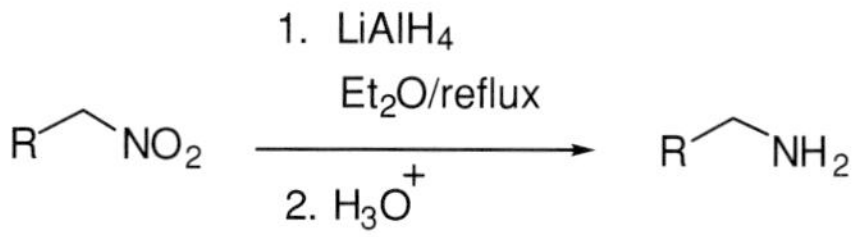

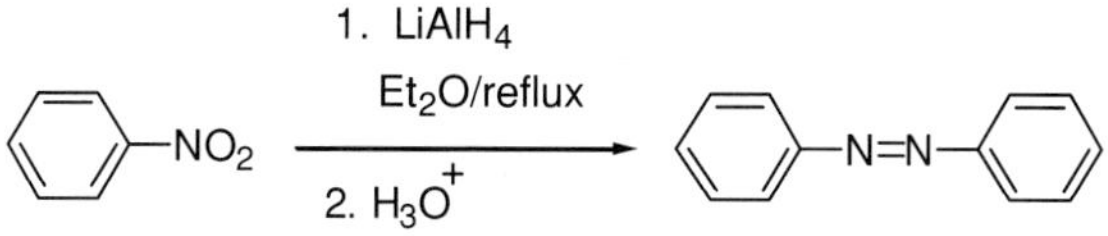

Sodium cyanoborohydride reduces enamines to amines. Diborane reduces nitriles to amines but the nitro group remains unaffected.

6.6.3   Metal and proton source

Aromatic nitro compounds are reduced by the iron or tin in hydrochloric acid to aromatic amines.

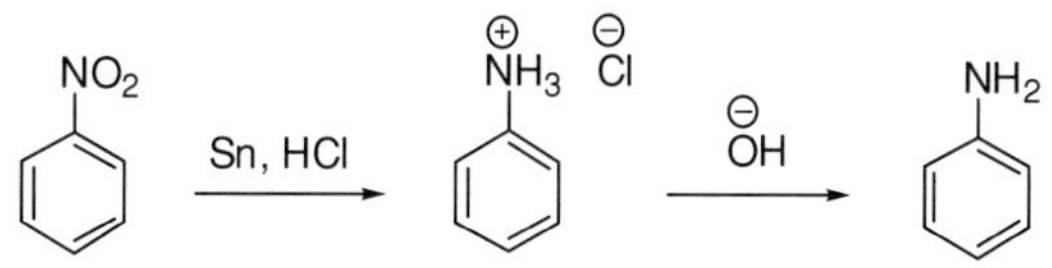

Mechanism: Mechanism of nitrobenzene reduction to aniline is shown in Scheme 6.36.

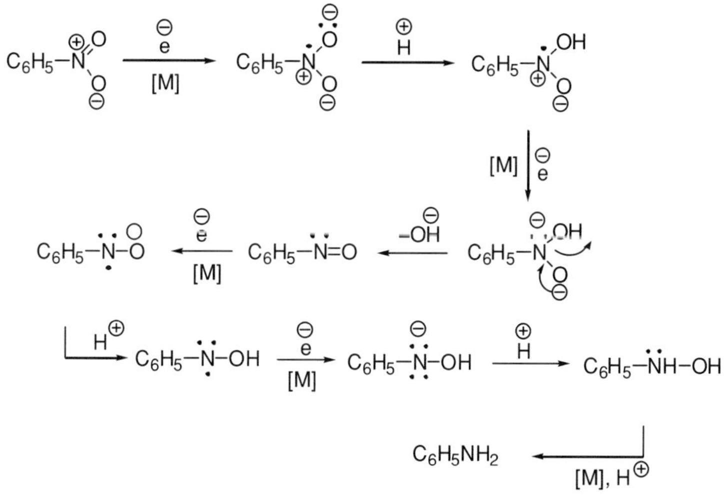

Scheme 6.36

Under alkaline conditions, the amine may react with nitro compound to give azoxybenzene, azobenzene and hydrazobenzene (Scheme 6.37).

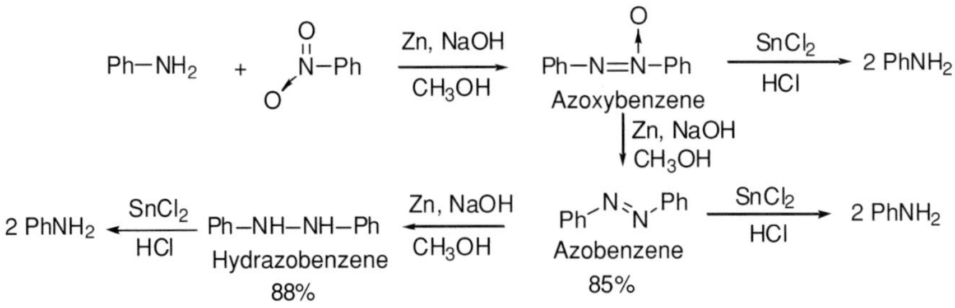

Scheme 6.37

m-Dinitrobenzene can be reduced stepwise by sodium sulfide to *m*-nitroaniline and then to *m*-phenylenediamine.

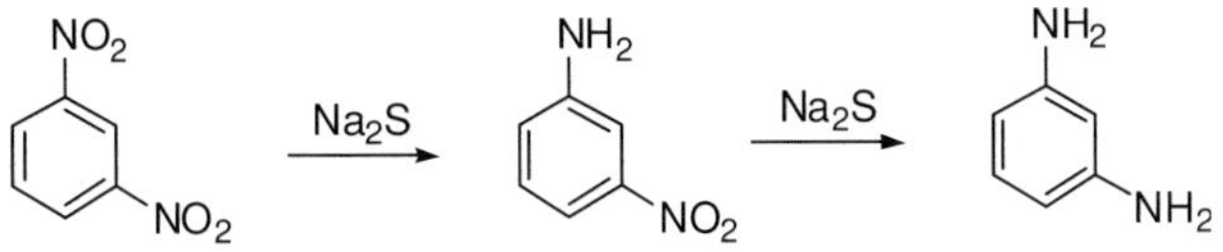

Nitrosobenzene and oximes are also reduced to the corresponding amine with metal and proton source.

$$C_6H_5NO \xrightarrow{Fe, HCl} C_6H_5NH_2$$

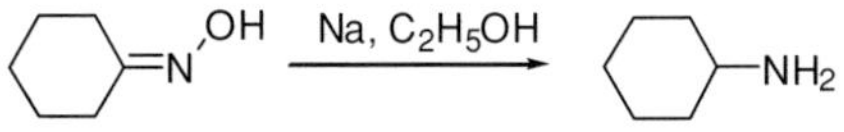

6.6.4 Triphenylphosphine

The affinity of trivalent phosphorus for oxygen (and sulfur) has been put to use in many reaction systems. For example, triphenylphosphine takes oxygen from *N*-oxides to form amines and triphenylphosphine oxide, which is a very stable polar compound, and in most cases it is easily removed from the other products.

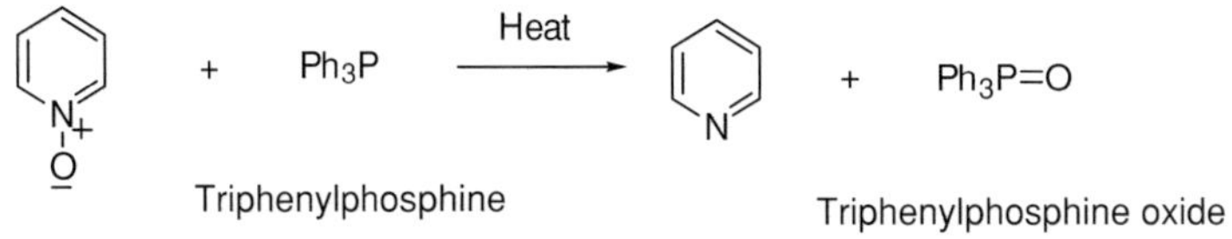

Triphenylphosphine   Triphenylphosphine oxide

Azides are converted into amines by the **Staudinger reaction**. It is a very mild azide reduction with triphenylphosphine[68].

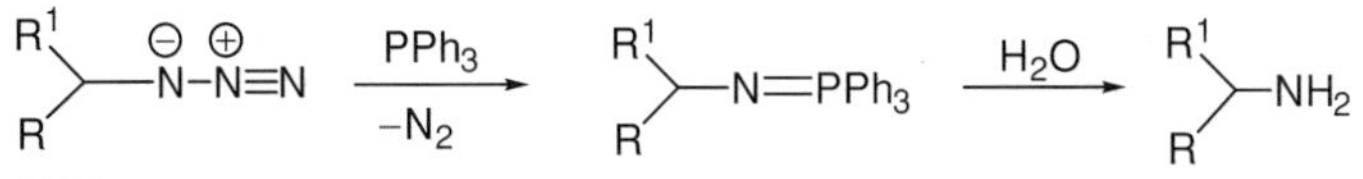

Mechanism:[69,70] Triphenylphosphine reacts with the azide to generate a phosphazide **A**, which loses N_2 via a cyclic transition state to form an iminophosphorane **B**. Aqueous work

up of the reaction mixture leads to the amine and very stable phosphine oxide (Scheme 6.38).

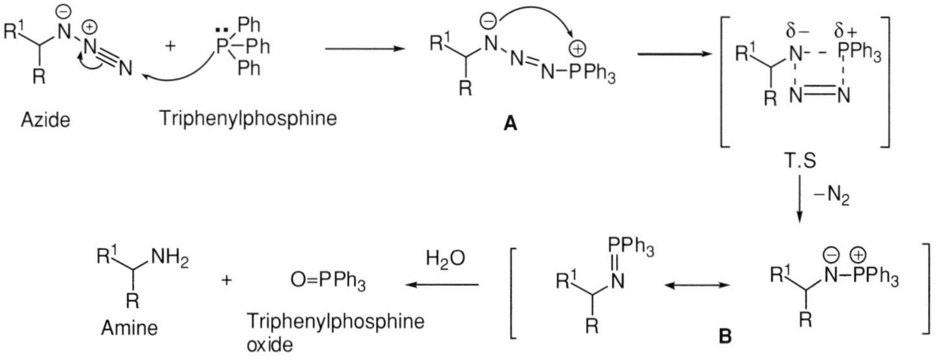

Scheme 6.38

6.7 Hydrogenolysis

The cleavage of a single bond (allylic or benzylic C–O and C–N, C–X, C–S, N–N, N–O, O–O bonds, opening of cyclopropane, epoxides and aziridine by catalytic hydrogenation) is known as **hydrogenolysis**. **Dehalogenation** is favoured by basic conditions in the presence of a metal catalyst, Pd. The ease of dehalogenation decreases in the order I > Br > Cl > F.

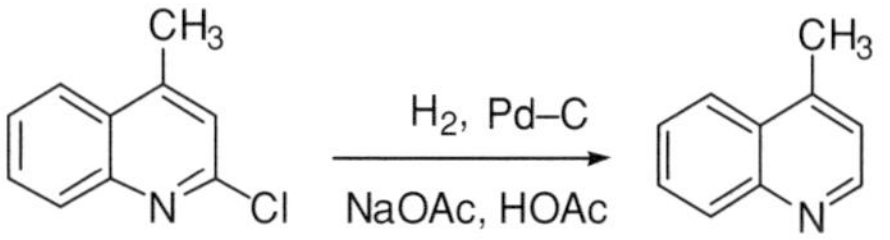

Hydrogenation of thioacetals with Raney Ni and H_2 breaks the C–S bonds and leads to desulfurization.

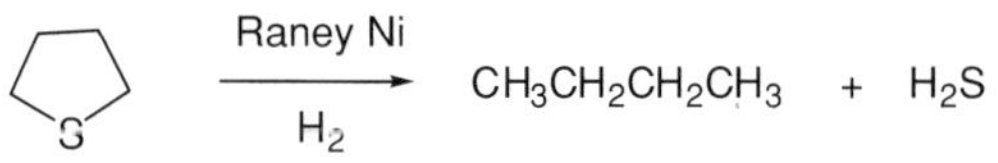

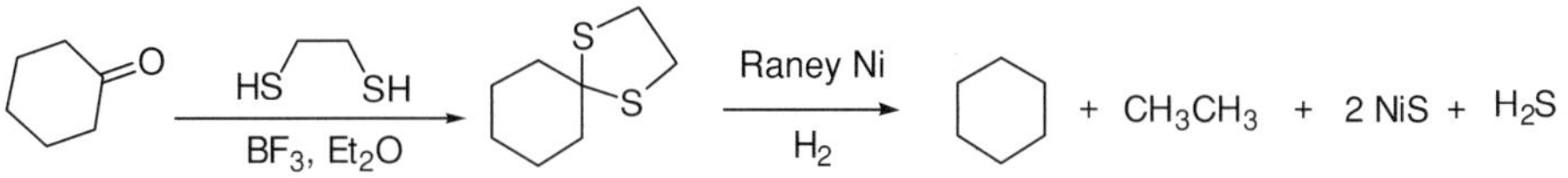

The metal and proton source system such as Zn and AcOH reduces 2-haloketones to ketones.

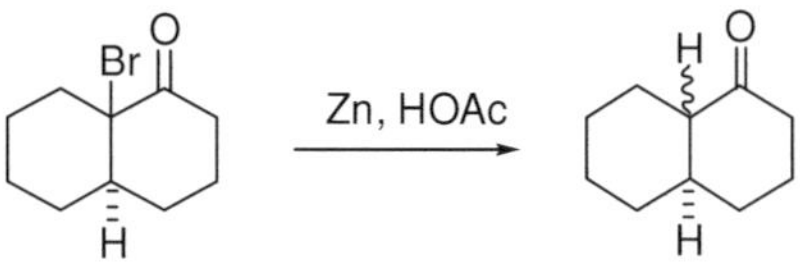

Primary and secondary halides can be reduced to hydrocarbons by metal hydrides. Tin hydride, $(n\text{-}Bu)_3SnH$ cleaves some C–X bonds.

97% both isomers

Red-Al [sodium bis(2-methoxyethoxy)aluminium hydride] reduces aliphatic halides and aromatic halides to hydrocarbons. Reductive dehalogenation of alkyl halides is most commonly carried out with super hydride. Epoxide ring can also be opened by super hydride.

Sulfonate esters are also reduced to hydrocarbon with $LiAlH_4$ via cleavage of the C–O bond. For example, menthol tosylate (**6.82**) gave the menthane (**6.83**) in 60% yield.

6.82   **6.83**

References

1. Osborn, J. A., Jardine, F. H., Young, J. F. and Wilkinson, G., *J. Chem. Soc.*, **1996**, 1711.
2. Vaska, L. and Diluzio, J. W., *J. Am. Chem. Soc.*, **1962**, *84*, 679.
3. Noyori, R., *Angew. Chem., Int. Ed. Engl.*, **2002**, *41*, 2008.
4. Noyori, R., Kitamura, M. and Ohkuma, T., *Proc. Natl. Acad. Sci. U.S.A.*, **2004**, *101*, 5356. (Also see references 21 and 22 of Chapter 1, and references 27 and 28 of Chapter 5.)
5. Ohta, T., Takaya, H., Kitamura, M., Nagai, K. and Noyori, R., *J. Org. Chem.*, **1987**, *52*, 3174.
6. Noyori, R., Ohta, M., Hsiao, Y., Kitamura, M., Ohta, T. and Takaya, H., *J. Am. Chem. Soc.*, **1986**, *108*, 7117.
7. Takaya, H., Ohta, T., Sayo, N., Kumobayashi, H., Akutagawa, S., Inoue, S., Kasahara, I. and Noyori, R., *J. Am. Chem. Soc.*, **1987**, *109*, 1596.
8. Kashibaia, K., Hanaki, K. and Fujia, J., *Bull. Chem. Soc. Jpn.*, **1980**, *53*, 2275.
9. Onuma, K., Ito, T. and Nakamura, A., *Bull. Chem. Soc. Jpn.*, **1981**, *53*, 2016.
10. Gramatica, P., Manitto, P., Ranzi, B. M., Delbianco, A. and Francavilla, M. *Experientia*, **1982**, *38*, 775.
11. Camu, A., Mestroni, G. and Zassinovich, G., *J. Organomet. Chem.*, **1980**, *184*, 10.
12. Papillon, J. P. N. and Taylor, R. J. K., *Org. Lett.*, **2002**, *4*, 119.
13. Blomquist, A. T., Liu, L. H. and Bohrer, J. C., *J. Am. Chem. Soc.*, **1952**, *74*, 3643.
14. Birch, A. J. and Smith, H., *Q. Rev. (London)*, **1958**, *12*, 17.
15. Granitzer, W. and Stütz, A., *Tetrahedron Lett.*, **1979**, *20*(34), 3145.
16. Jone, T. K. and Denmark, S. E., *Org. Synth.*, **1990**, *7*, 524.
17. Chan, K. K., Specian, A. C. and Saucy, J. R., *J. Org. Chem.*, **1978**, *43*, 3435.
18. Gollnick, K. and Hartmann, H., *Tetrahedron Lett.*, **1982**, *23*, 2651.
19. Birch, A. I., *J. Chem. Soc.*, **1944**, 430.; 1945, 809.; 1946, 593.; 1949, 2531.
20. Webster, F. X., *Synthesis*, **1987**, 923.
21. Subba Rao, G. S. R., *Pure Appl. Chem.*, **2003**, *75*(10), 1443.
22. Rosenmund, K. W., *Ber. Chem.*, **1918**, *51*, 585.
23. Mossettig, E. and Mozingo, R., *Org. React.*, **1948**, *4*, 362.

24. Hajos, A., *Complex Hydrides and Related Reducing Agents in Organic Chemistry*, Elsevier, Amsterdam, **1979**.
25. Wakamatsu, T., Inaki, H., Ogawa, A., Watanabe, M. and Ban, Y., *Heterocycles*, **1980**, *14*(10), 1437.
26. Fleet, G. W. J., Fuller, C. J. and Harding, P. J. C., *Tetrahedron Lett.*, **1978**, *19*(16), 1437.
27. Sorrell, T. N. and Spillane, R. J., *Tetrahedron Lett.*, **1978**, *19*(28), 2473.
28. Sorrell, T. N. and Pearlman, P. S., *J. Org. Chem.*, **1980**, *45*, 3449.
29. Brown, H. C. and Rao, B. C. S., *J. Am. Chem. Soc.*, **1958**, *80*, 5377.
30. Hutchins, R. O. and Markowitz, M., *Tetrahedron Lett.*, **1980**, *21*, 813.
31. Watanabe, Y., Mitsudo, T., Tanaka, M., Yamamoto, K., Okajima, T. and Takegami, Y., *Bull. Chem. Soc. Jpn.*, **1971**, *44*, 2569.
32. Cole, T. E. and Pettit, R., *Tetrahedron Lett.*, **1977**, *18*(9), 781.
33. Four, P. and Guibe, F., *J. Org. Chem.*, **1981**, *46*, 4439.
34. Babler, J. H., *Synth. Commun.*, **1982**, *12*(11), 839.
35. Cha, J. S. and Chun, J. H., *Bull. Korean Chem. Soc.*, **2000**, *21*(4), 375.
36. Hagiya, K., Mitsui, S. and Taguchi, H., *Synthesis*, **2003**, 823.
37. Brown, H. C. and Krishnamurthy, S., *Tetrahedron*, **1979**, *35*, 567.
38. Walker, E. R. H., *Chem. Soc. Rev.*, **1976**, *5*, 23.
39. Gaylor, N. G., *Complex Metal Hydrides*, Wiley-Interscience, New York, **1956**, p. 107.
40. Nystrom, R. F. and Brown, W. G., *J. Am. Chem. Soc.*, **1947**, *69*, 1197.
41. Hill, A. J. and Nason, E. H., *J. Am. Chem. Soc.*, **1924**, *46*, 2236.
42. Corey, E. J., Shibata, S. and Bakshi, R. K., *J. Org. Chem.*, **1988**, *53*, 2861.
43. Corey, E. J. and Link, J. O., *Tetrahedron Lett.*, **1989**, *30*, 6275.
44. Koning, C. B., Giles, R. G. F., Green, I. R. and Jahed, N. M., *Tetrahedron Lett.*, **2002**, *43*(23), 4199.
45. Midland, M. M., Tramontano, A. and Zderic, A., *J. Organomet. Chem.*, **1977**, *134*, C17.
46. Brunner, H., Riepl, G. and Neitzer, H., *Angew. Chem., Int. Ed. Engl.*, **1983**, *22*, 331.
47. DiVona, M. L. and Rosnati, V., *J. Org. Chem.*, **1991**, *56*, 4269.
48. Chaussar, J., *Tetrahedron Lett.*, **1987**, *28*, 1173.
49. Huang-Minlon, *J. Am. Chem. Soc.*, **1946**, *68*(12), 2487.; Huang-Minlon, *J. Am. Chem. Soc.*, **1949**, *71*(10), 3301.
50. Doucet, H., Ohkuma, T., Murata, K., Yokozawa, T., Kozawa, M., Katayama, E., England, A. F., Ikariya, T. and Noyori, K., *Angew. Chem., Int. Ed. Engl.*, **1998**, *37*, 1703.
51. Noyori, R. and Ohkuma, T., *Angew. Chem., Int. Ed.*, **2001**, *40*, 40.
52. Fujii, A., Hashiguchi, S., Uematsu, N., Ikariya, T. and Noyori, R., *J. Am. Chem. Soc.*, **1996**, *118*, 2521.
53. Uematsu, N., Fujii, A., Hashiguchi, S., Ikariya, T. and Noyori, R., *J. Am. Chem. Soc.*, **1996**, *118*, 4916.
54. Hashiguchi, S., Fujii, A., Haack, K.-J., Matsumura, K., Ikariya, T. and Noyori, R., *Angew. Chem., Int. Ed.*, **1997**, *36*, 288.
55. Haack, K.-J., Hashiguchi, S., Fujii, A., Ikariya, T. and Noyori, R., *Angew. Chem., Int. Ed.*, **1997**, *36*, 285.
56. McManus, H. A., Barry, S. M., Anderson, P. G. and Guiry, P. J., *Tetrahedron*, **2004**, *60*(15), 3405.
57. Larock, R. C., *Comprehensive Organic Transformation*, Wiley-VCH, New York, **1989**, pp. 8–17.
58. Luche, J. L., *J. Am. Chem. Soc.*, **1978**, *100*(7), 2226.
59. Luche, J. L. and Gemal, A. L., *J. Am. Chem. Soc.*, **1979**, *101*, 5848.
60. Morisso, F. D. P., Wagner, K., Hörner, M., Burrow, R. A., Bortoluzzi, A. J. and Costa, V. E. U., *Synthesis*, **2000**, 1247.
61. Taschnu, M. J. and Shahripour, A. J., *J. Am. Chem. Soc.*, **1985**, *107*, 5570.
62. Ling, T., Chowdhury, C., Kramu, B. A., Vong, B. G., Palladino, M. A. and Theodorakis, E. A., *J. Org. Chem.*, **2001**, *66*, 8843.
63. Umino, N., Iwakuma, T. and Itoh, N., *Tetrahedron Lett.*, **1976**, *17*(33), 2875, and references cited therein.
64. Umino, N., Iwakuma, T. and Itoh, N., *Tetrahedron Lett.*, **1976**, *17*(10), 763, and references cited therein.
65. Tsuda, Y., Sano, T. and Watanabe, H., *Synthesis*, **1977**, 652.
66. Kuehne, M. E. and Shannon, P. J., *J. Org. Chem.*, **1977**, *42*, 2082.
67. Basha, A. and Rahman, A., *Experientia*, **1977**, *33*, 101.
68. Thomas, S., Collins, C. J., Cuzens, J. R., Spiciarich, D., Goralski, C. T. and Singaram, B., *J. Org. Chem.*, **2001**, *66*, 1999.
69. Tian, W. Q. and Wang, Y. A., *J. Org. Chem.*, **2004**, *69*(13), 4299.
70. Lin, F. L., Hoyt, H. M., Halbeek, H. V., Bergman, R. G. and Bertozzi, C. R., *J. Am. Chem. Soc.*, **2005**, *127*(8), 2686.

Chapter 7
Oxidation

7.1 Oxidation of alcohols

Oxidation of alcohols to aldehydes or ketones is one of the most useful transformations in organic chemistry. Simple 1°- and 2°-alcohols in the gaseous state lose hydrogen (**dehydrogenation reaction**) when exposed to a hot copper surface. However, in solution phase alcohol oxidations are carried out using reactions in which the hydroxyl hydrogen is replaced by an atom or group that is readily eliminated together with the α-hydrogen. The decomposition of 1°- and 2°-alkyl hypochlorites is an example of such a reaction.

$$RCH_2OH \ + \ Hot \ Cu \longrightarrow RCHO \ + \ H_2$$

$$RCH_2OCl \ + \ Base \longrightarrow RCHO \ + \ H\text{-}Cl$$

In a similar manner, toluene-*p*-sulfonate derivative of alcohols on nucleophilic displacement with trimethylamine *N*-oxide followed by treatment with a base gives the carbonyl compound and trimethylamine.

7.1.1 Chromium(VI)

The most common reagent for the oxidation of 1°- and 2°-alcohols is chromic acid (H_2CrO_4). It is difficult to stop the reaction at the aldehyde stage; thus, primary alcohols are oxidized to carboxylic acids. In some cases, good yield of an aldehyde can be obtained by removing the aldehyde from the reaction mixture.

Aldehyde is isolated if removed by continuous distillation

However, the secondary alcohols give ketones.

In aqueous medium, chromium trioxide (CrO_3) exists in equilibrium with several Cr(VI) species such as H_2CrO_4, $Cr_2O_7^{2-}$, $H_2Cr_2O_7$, $H_2Cr_2O_7^-$, CrO_4^{2-} and $HCrO_4^-$. The

268

dominant form of Cr(VI) species formed in the aqueous solution depends on the pH value, solvent and concentration. At high dilution, **dichromate anion** $(Cr_2O_7^{2-})$ is the major species.

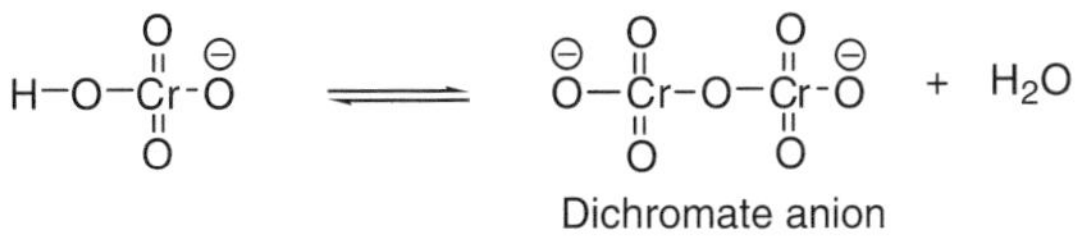

Dichromate anion

At high concentration, $(CrO_3)_n$ and chromic acid (H_2CrO_4) are the main species. As alcohol is oxidized to carbonyl compound, Cr(VI) is reduced to Cr(III). However, the detail mechanism shows that Cr(V) and Cr(IV) species are also involved in the oxidation of alcohols. The Cr(VI) species are usually $HCrO_4^-$ and CrO_3 and the Cr(IV) is usually $HCrO_3^-$. The balanced equation for the oxidation of alcohols with chromic acid involves both acid catalyst and water as shown below:

Mechanism: The mechanism of the alcohol oxidation with Cr(VI) is outlined in Scheme 7.1, and involves the formation of chromate ester. The base removes the proton and Cr species leaves in an intermolecular process (**A**); however, an intramolecular process (**B**) may also operate. The Cr(IV) ions in H_2CrO_3 or $HCrO_3^-$ are converted back to Cr(III) ions. It is believed that part of alcohol molecules are oxidized by the free radical mechanism.

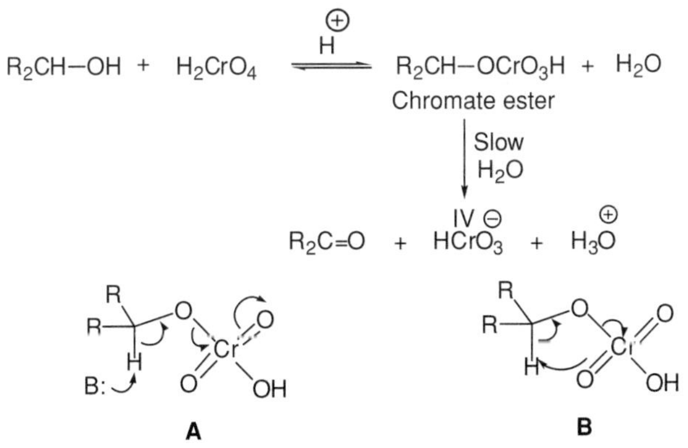

Scheme 7.1

In the alcohol oxidation to carbonyl compounds several Cr(VI) reagents can be used such as CrO_3 in aqueous acetic acid or in other solvents along with catalytic amount of mineral acid, sodium dichromate in aqueous acetone and mineral acid or base as catalyst, sodium dichromate in acetic acid, CrO_3–pyridine complex and *tert*-butyl chromate.

Jones reagent: Jones reagent[1,2] $(CrO_3, H_2SO_4, H_2O,$ acetone) is used for the oxidation of secondary alcohols to ketones. For example, isoborenol (**7.1**) is oxidized by CrO_3 and H_2SO_4 to camphor (**7.2**). A few primary alcohols have been oxidized to aldehydes by this reagent, but strong aqueous acidic conditions allow the oxidation of primary alcohols to carboxylic acids.

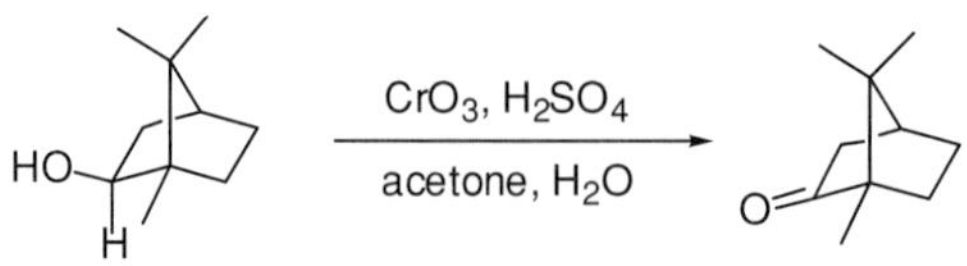

7.1   **7.2**

The conditions of Jones oxidation are compatible with complex organic compounds containing functional groups such as esters, ketones, amides and alkenes. For example, ethyl 3-hydroxy-4-pentenoate (**7.3**) on oxidation with Jones reagent gave ethyl 3-oxo-4-pentenoate (Nazarov's reagent)[3,4] (**7.4**).

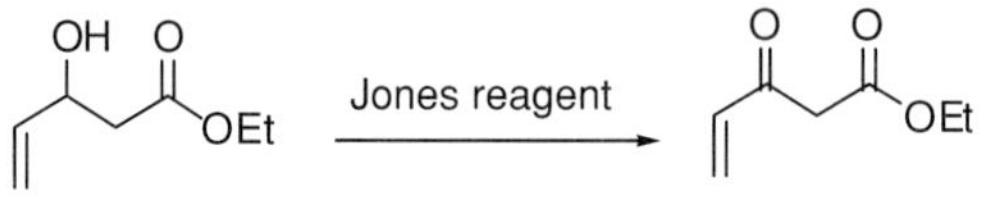

7.3   **7.4**

Mechanism: The mechanism of the Jones oxidation proceeds through a chromate ester intermediate which undergoes an E2-like elimination to the carbonyl product (Scheme 7.2). Since chromate reagents are of dark orange-red colour (Cr in VI oxidation state) and Cr(III) compounds are normally green, the progress of these oxidations is easily observed. Indeed, this is the chemical transformation on which the **Breathalizer test** is based.

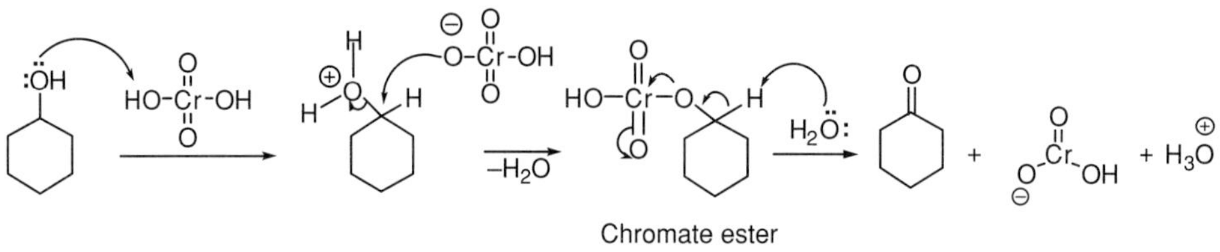

Chromate ester

Scheme 7.2

The oxidation of benzyl alcohols to benzaldehydes, which is difficult with Jones reagent, can be achieved by using silica gel-supported[1,2] Jones reagent in CH_2Cl_2.

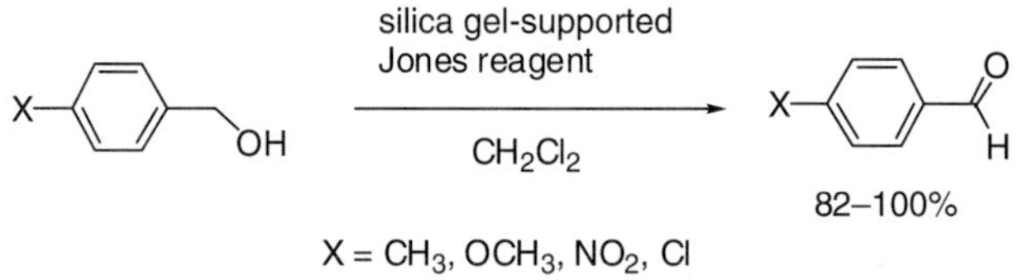

82–100%

$X = CH_3, OCH_3, NO_2, Cl$

CrO_3–pyridine complex: The oxidation of alcohols to aldehydes and ketones with CrO_3 in the presence of pyridine solvent is known as the **Sarett oxidation**[5]. Sarett used the CrO_3–pyridine complex in the synthesis of steroids. Although primary alcohols give poor yield, benzylic, allylic and secondary alcohols give good yields with Sarett reagent.

$$C_6H_5CH{=}CH{-}CH_2OH \xrightarrow{\quad CrO_3\text{–pyridine} \quad} C_6H_5CH{=}CH{-}CHO$$

Cinnamyl alcohol   Cinnamaldehyde
                                          80%

Collins reagent: It is difficult to isolate the products from the pyridine solvent; thus, the reagent formed by mixing CrO_3 and pyridine is first removed and then added to CH_2Cl_2 (**Collins reagent**). This reagent gives good yields of oxidation of primary alcohols to aldehydes[6].

$$CrO_3 \text{ (anhydrous)} + \text{Pyridine (anhydrous)} \longrightarrow$$

2 Py·CrO$_3$ in CH$_2$Cl$_2$
(Collins reagent)

Pyridinium chlorochromate (PCC): Corey and Suggs[7] prepared PCC by mixing CrO_3 with pyridine in HCl. PCC is used for the oxidation of primary and secondary alcohols in CH_2Cl_2. This reagent is less efficient than **Collins reagent** for the oxidation of allyl alcohols.

$$CrO_3 + \text{Pyridine} \xrightarrow{HCl} \text{PCC}$$

$$\underset{R^1}{\overset{R}{\diagup}}CHOH \xrightarrow{PCC} \underset{R^1}{\overset{R}{\diagup}}{=}O$$

$$CH_3(CH_2)_5CH_2OH \xrightarrow[CH_2Cl_2]{PCC} CH_3(CH_2)_5CHO$$

1-Heptanol   Heptanal
                                  93%

Pyridinium dichromate (PDC): The problem caused by the acidic nature of PCC is largely removed by using the neutral reagent **PDC**[8,9]. For example, geraniol (**7.5**) was oxidized to geranial (**7.6**) in 92% yield by PDC in DMF (dimethylformamide) solvent.

$$\xrightarrow[\text{DMF}]{\text{(PDC)}}$$

7.5   **7.6**

But non-conjugate alcohols such as citronellol (**7.7**) on treatment with PDC in DMF give the corresponding acid **7.8**; however, the aldehyde **7.9** is obtained with PDC in CH_2Cl_2.

$$\xleftarrow{\text{PDC, CH}_2\text{Cl}_2} \qquad \xrightarrow[25°C]{\text{PDC, DMF}}$$

7.9   **7.7**   **7.8**

CrO₃–dimethylpyrazole: CrO_3–3,5-dimethylpyrazole in methylene chloride is also used for the oxidation of 1°-alcohols to aldehydes and 2°-alcohols to ketones[10].

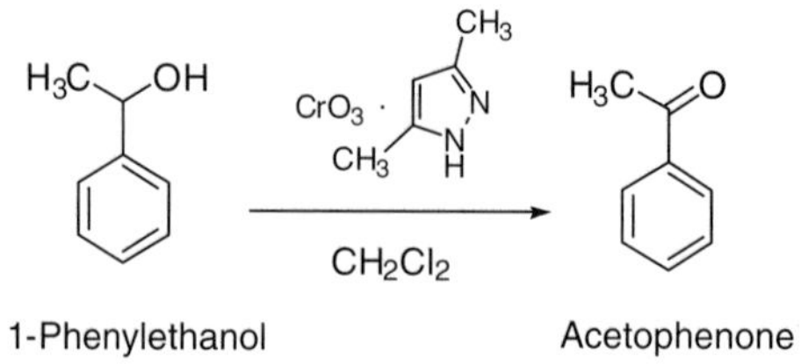

1-Phenylethanol   Acetophenone

The mechanism of the Sarett oxidation, Collins oxidation, with Corey's PCC and with PDC, follows a similar mechanism as shown in Scheme 7.1. The alcohol reacts with CrO_3 to give a chromate ester. Either a base (Py) removes a proton from the chromate ester to give an oxidized product (aldehyde or ketone) and $HCrO_3^-$ or a proton is transferred by the intramolecular mechanism to give an aldehyde or ketone and H_2CrO_3.

7.1.2 Potassium permanganate

The Cr(VI) oxidation of alcohols is acid catalyzed, but both under acidic or basic conditions permanganate effects the same transformation.

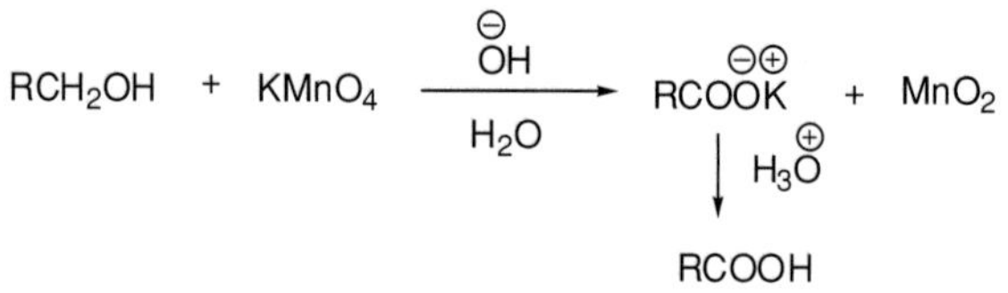

The base converts alcohol into an alkoxide ion, which loses H to give an aldehyde first, as shown in Scheme 7.3. The $HMnO_4^{2-}$ [Mn(V)] quickly disproportionates to Mn(IV) and Mn(VI). MnO_2 is the usual end product of permanganate oxidation in basic solution.

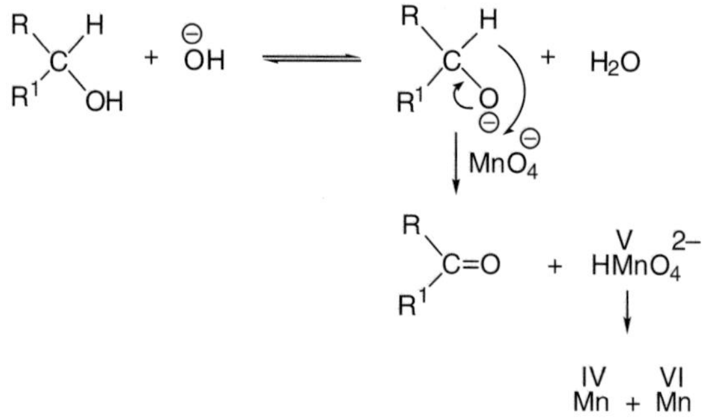

Scheme 7.3

$KMnO_4$ and 18-crown-6 (purple benzene) (**7.10**) are used to oxidize 1°-alcohols and aldehydes to carboxylic acids.

7.10

Barium permanganate ($BaMnO_4$) is also used for the oxidation of 1°- and 2°-alcohols to aldehydes and ketones, and no overoxidation is observed.

7.1.3 Manganese dioxide (MnO_2)

Manganese dioxide is extensively used as an oxidizing agent for the oxidation of allylic alcohols to the corresponding aldehydes. Benzylic and unactivated alcohols are also oxidized by MnO_2.

Benzyl alcohol   Benzaldehyde

Crotonyl alcohol   Crotonaldehyde
70%

Retinol (vitamin A)   Retinal

As seen in the last example, the configuration of the double bond is conserved in the reaction. When MnO_2 is used in combination with the Bestmann–Ohira reagent,[11a] activated alcohols are converted into terminal alkynes[11b].

Mechanism: The MnO_2 oxidation reaction is believed[12] to proceed by adsorption of alcohol on MnO_2 followed by coordination of both alcohol and MnO_2. Electron transfer then produces the radical, and Mn(IV) is reduced to Mn(III). Another electron transfer is followed by the release of adsorbed carbonyl product with the loss of water from $Mn(OH)_2$ (Scheme 7.4). An ionic mechanism is also proposed for this oxidation which involves the formation of manganate ester[13].

There are problems associated with the expensive disposal of toxic waste from metal-based oxidations of alcohols. Thus, the focus has been largely on catalytic reactions as typified by Ley and Griffith's tetrapropylammonium perruthenate oxidant (section 7.1.6). Completely metal-free oxidations have much potential for environment-friendly oxidations, particularly if the reagent can be recovered and recycled. The most common metal-free oxidation of alcohols are TEMPO/oxone or TEMPO/N-chlorosuccinimide oxidation[14], Dess–Martin periodane oxidation (section 7.1.5) and Swern oxidation (section 7.1.4) and its several variants.

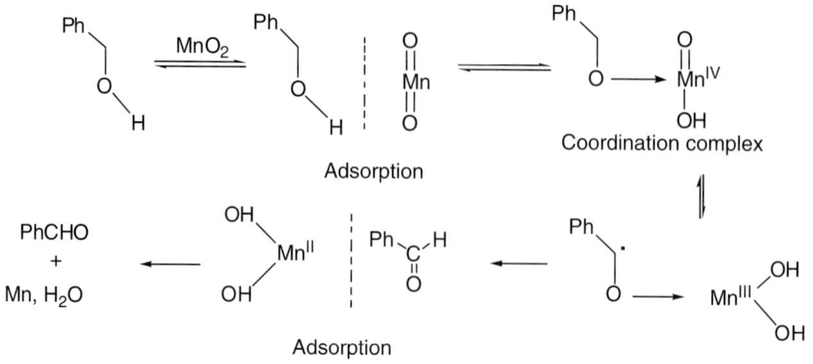

Scheme 7.4

7.1.4 Dimethylsulfoxide-mediated oxidations

A very mild oxidation procedure that tolerates a variety of other functional groups is the oxidation of alcohols to aldehydes and ketones by the dimethyl sulfoxide (DMSO). A DMSO solution of the alcohol is treated with one of the several *electrophilic dehydrating reagents*. The alcohol is oxidized to the aldehyde or ketone and DMSO is reduced to dimethyl sulfide (DMS), which is a volatile liquid (b.p. 37°C). Since pure DMSO freezes at 18°C and an oxidation reaction is carried out at −50°C or lower, a co-solvent such as methylene chloride or THF (tetrahydrofuran) is needed.

   Mechanism: The initial attack on the sulfoxide oxygen by *electrophile* (E$^+$) enhances the electrophilic character of the sulfur atom. Bonding of sulfur in **A** to the alcohol oxygen atom gives **B**. This is followed by removal of a proton by a base to give ylide **C** and then the elimination step, similar in nature to those proposed for other alcohol oxidations. In some cases triethyl amine is added to provide an additional base. A plausible general mechanism for this useful reaction is shown in Scheme 7.5.

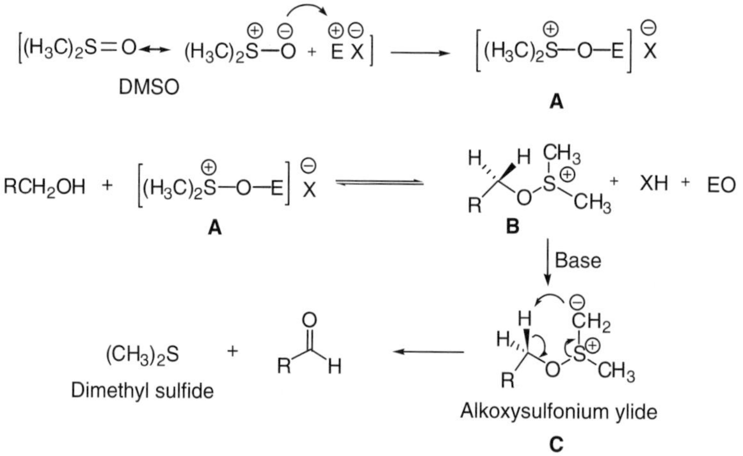

Scheme 7.5

Different *electrophiles* have been used to effect this oxidation. For example, in **Pfitzner–Moffatt oxidation**[15–17], alcohols are oxidized to aldehydes and ketones by the DMSO and DCC (dicyclohexylcarbodiimide). The resulting alkoxysulfonium ylide **C** rearranges to generate aldehydes and ketones (Scheme 7.6).

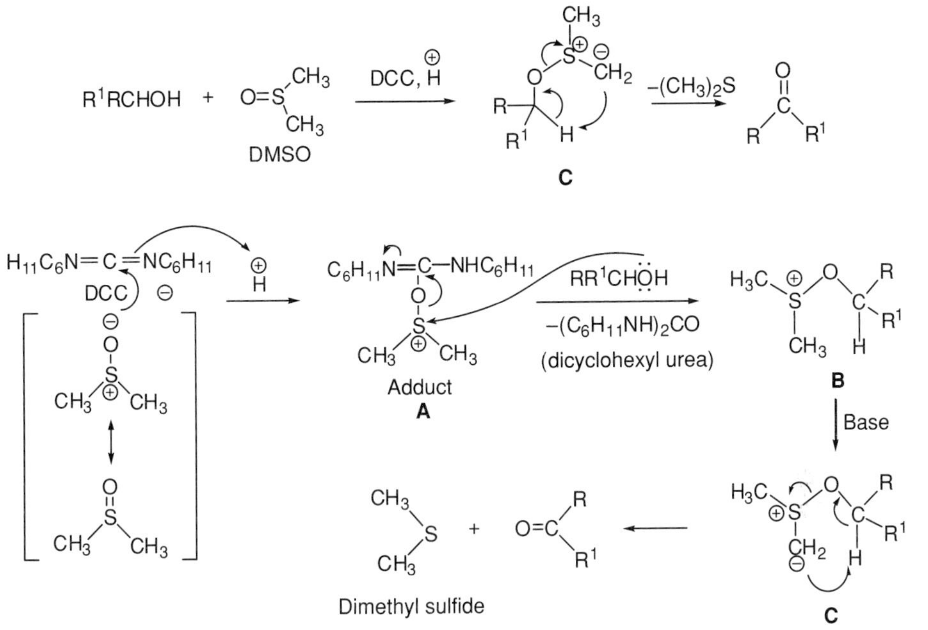

Scheme 7.6

But the **Pfitzner–Moffatt oxidation** is not in much use because the **Swern oxidation**[18] gives better yield and fewer side products. For example, methyl 12-hydroxydodecanoate (**7.11**) on treatment with DMSO and oxalyl chloride [$(COCl)_2$] in CH_2Cl_2 followed by treatment with triethylamine yields 87% methyl 12-oxododecanoate[19] (**7.12**).

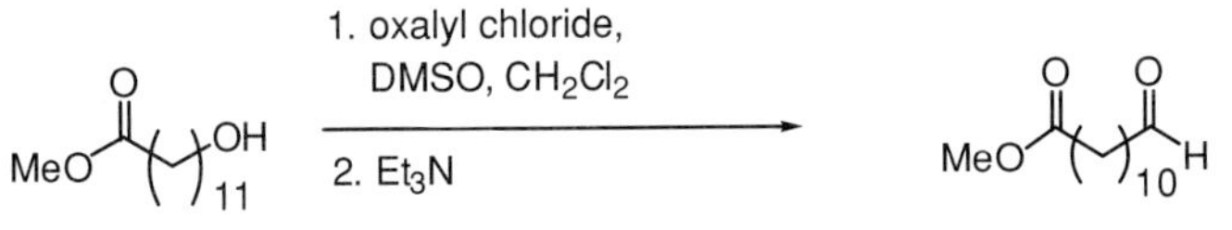

In the **Swern oxidation**, the reaction of oxalyl chloride with DMSO may generate chlorodimethylsulfonium chloride (**A**), which reacts with the alcohol to give alkoxysulfonium ion intermediate **B**. The base, typically triethylamine, deprotonates the alkoxysulfonium ion **B** to give the sulfur ylide **C**, which decomposes to give DMS and the desired aldehyde or ketone (Scheme 7.7). The temperature of this reaction is kept near −60 to −78°C.

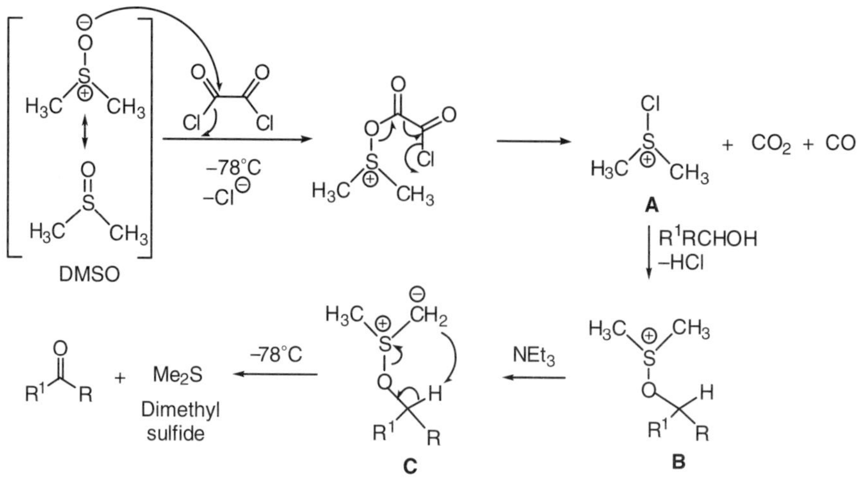

Scheme 7.7

The disadvantage of the Swern oxidation is the formation of side products from the **Pummerer rearrangement** (see section 1.6.1, Scheme 1.26). To avoid the side reactions in the Swern oxidation, the temperature is kept at −78°C, but when trifluoroacetic anhydride instead of oxalyl chloride is used the reaction can be warmed to −30°C. The use of diisopropylamine as a base stops side reactions.

The variation of the Swern oxidation is the **Corey–Kim oxidation**[20,21]. Treatment of DMS (Me₂S) with Cl₂ or N-chlorosuccinimide (NCS) forms dimethylchlorosulfonium ion (**A**), which reacts with alcohol to form sulfoxonium complex **B**. Treatment of sulfoxonium complex **B** with a base Et₃N gives the corresponding aldehyde or ketone via ylide **C** (Scheme 7.8).

Thus, 4-*tert*-butylcyclohexanol (**7.13**) on treatment with $(CH_3)_2S$ and NCS in toluene followed by the reaction with Et₃N yields 4-*tert*-butylcyclohexanone (**7.14**) in 100% yield.

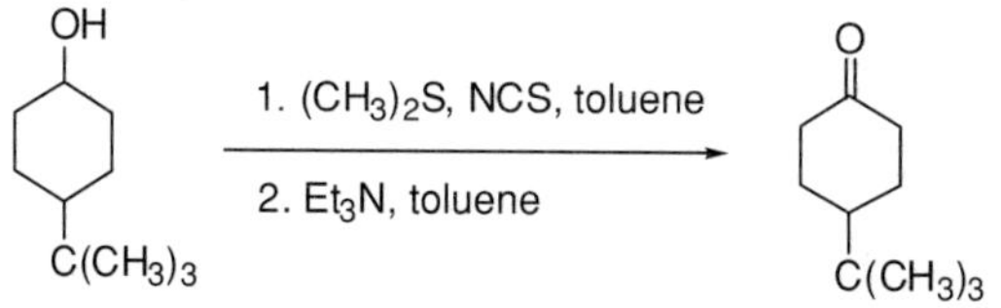

7.13   **7.14**

Because of the foul smell of DMS, several **odourless versions of Corey–Kim and Swern oxidations**[22,23] have been reported. For example, odourless dodecyl methyl sulfide (Dod-S-Me) is an alternative to DMS in Corey–Kim oxidations.

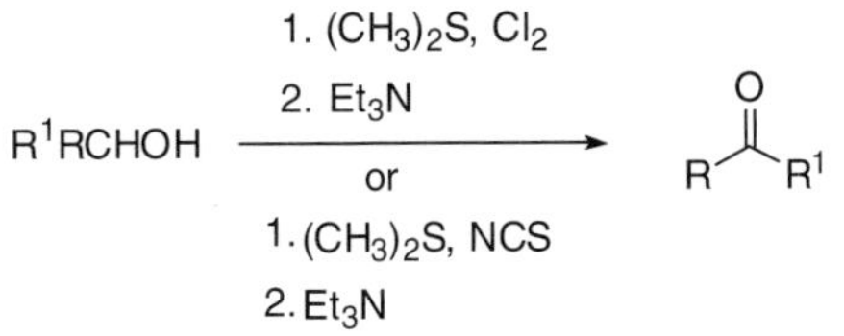

$$R^1RCHOH \quad \xrightarrow[\substack{or \\ 1.(CH_3)_2S,\ NCS \\ 2.\ Et_3N}]{\substack{1.\ (CH_3)_2S,\ Cl_2 \\ 2.\ Et_3N}} \quad R\overset{O}{\underset{}{\|}}R^1$$

Dimethyl sulfide

$$\xrightarrow[\substack{or \\ Cl_2}]{(NCS)} \quad \mathbf{A}$$

A → R^{1}RCHOH, −HCl

B → NEt$_3$ → **C** → −78°C →

$$\overset{O}{\underset{R^1 \quad R}{\|}} + Me_2S$$

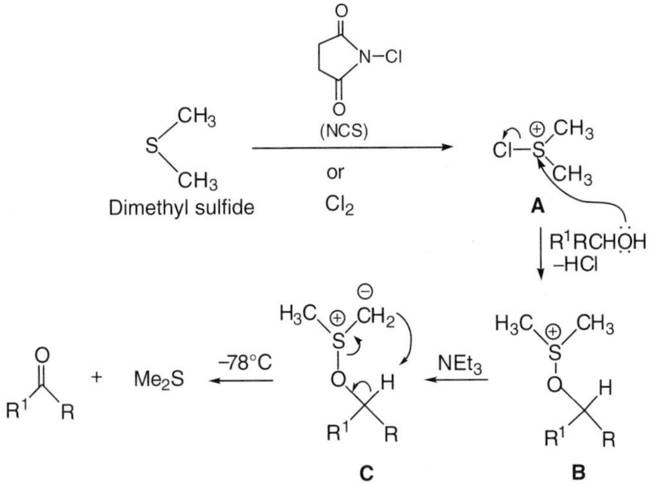

Scheme 7.8

Alcohols $\xrightarrow[\substack{2.\ Et_3N\ (5\ equiv.),\ -40°C}]{\substack{1.\ Dod\text{-}S\text{-}Me\ (3\ equiv.) \\ NCS\ (3\ equiv.)}}$ Aldehydes or ketones

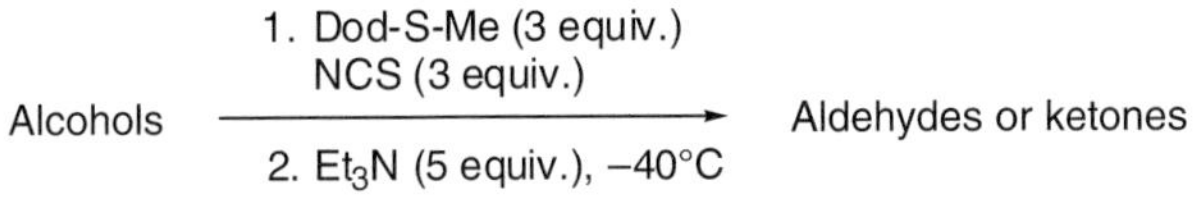

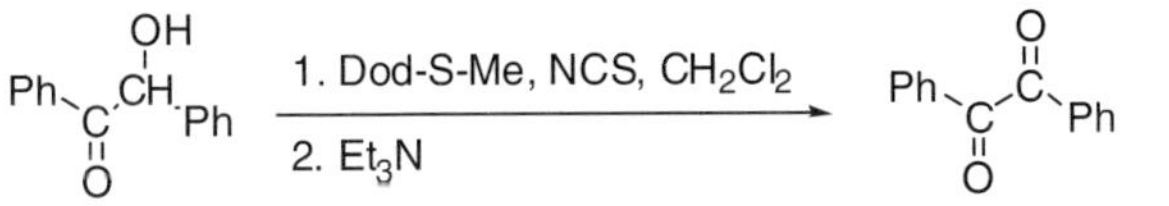

Benzoin
(2-Hydroxy-1,2-diphenylethan-1-one) $\xrightarrow[\substack{2.\ Et_3N}]{\substack{1.\ Dod\text{-}S\text{-}Me,\ NCS,\ CH_2Cl_2}}$ Benzil 99%

4-Phenyl-3-buten-1-ol $\xrightarrow[\substack{2.\ Et_3N}]{\substack{1.\ Dod\text{-}S\text{-}Me,\ NCS,\ CH_2Cl_2}}$ 4-Phenyl-3-butenal 91%

Similarly, when dodecyl methyl sulfoxide ($C_{12}H_{25}SOCH_3$) is used in the Swern oxidation, the odourless Dod-S-Me by-product instead of foul-smelling DMS is formed.

This Dod-S-Me recovered from the reaction can be reoxidized to dodecyl methyl sulfoxide by sodium periodate.

The DMSO oxidation of alcohols can also be carried out without the use of a base. Thus, benzyl alcohols on oxidation with DMSO in the presence of acids typically HBr gave high yields of the corresponding benzaldehydes[24]. The presence of an electron-donating group on aromatic ring gives higher yield than when electron-withdrawing group is present on aromatic ring.

$R = H, CH_3, NO_2, Cl, OH, OCH_3$

71–96%

7.1.5   Dess–Martin periodinane (DMP)

The Dess–Martin reagent (**7.16**) belongs to a group of compounds containing a hypervalent iodine atom. It is prepared from *o*-iodobenzoic acid via **7.15**.

The oxidation of alcohols to aldehydes or ketones by periodane has several advantages over chromium and DMSO-based oxidants because of its shorter reaction times, higher yields and simplified work up[25]. There is very little overoxidation to the carboxylic acid. It is a practical reagent for the facile and efficient oxidation of benzylic and allylic alcohols. Saturated alcohols are slow in their reactions with it. It oxidizes alcohols in the presence of non-hydroxylic functional groups such as sulfides, enols, ethers, furans and 2°-amides. An example of the DMP oxidation is the oxidation of 3,4,5-trimethoxybenzyl alcohol (**7.17**) with **7.16** in CH_2Cl_2 to give 94% yield of 3,4,5-trimethoxybenzaldehyde (**7.18**).

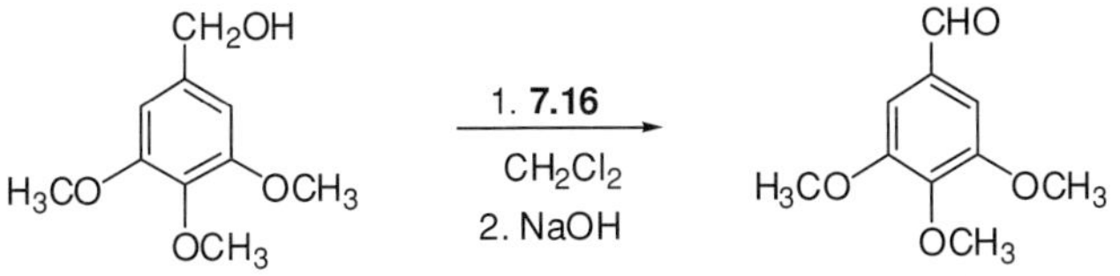

7.17 → **7.18**

Mechanism: A probable mechanism of the DMP oxidation involves the reaction of tri-acetoxyperiodinane called Dess–Martin periodinane (**7.16**) with alcohol (primary or sec-ondary) to give an alkoxydiacetoxyperiodinane **A**. Periodinane **A** loses one acetate and reacts with another alcohol molecule to form periodinane **B**, which releases an aldehyde or ketone and acetic acid, and **C** is formed. The periodane **A** may slowly lose AcOH, aldehyde or ketone and form **D** (Scheme 7.9).

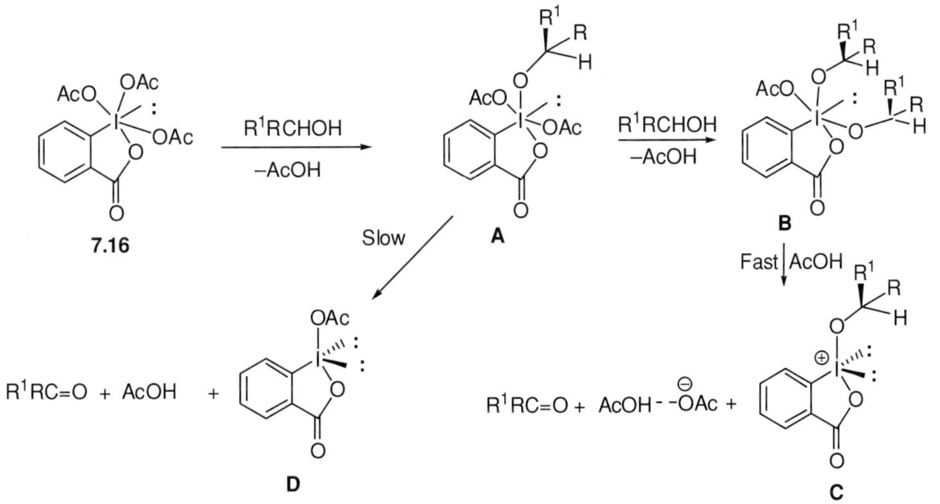

Scheme 7.9

o-Ioduxybenzoic acid (**7.15**) is also used for the oxidation of alcohols[26].

7.1.6 Tetra-*n*-propylammonium perruthenate (TPAP)

Oxidation of the alcohol 10-[3′-methoxy-2′-(methoxymethoxy)phenyl]decan-1-ol (**7.19**) with TPAP [(C$_3$H$_7$)$_4$NRuO$_4$] afforded the aldehyde 10-[3′-methoxy-2′-(methoxymethoxy)phenyl]decan-1-al (**7.20**) in quantitative yield[27]. The reagent TPAP is known as **Ley–Griffith's reagent**.

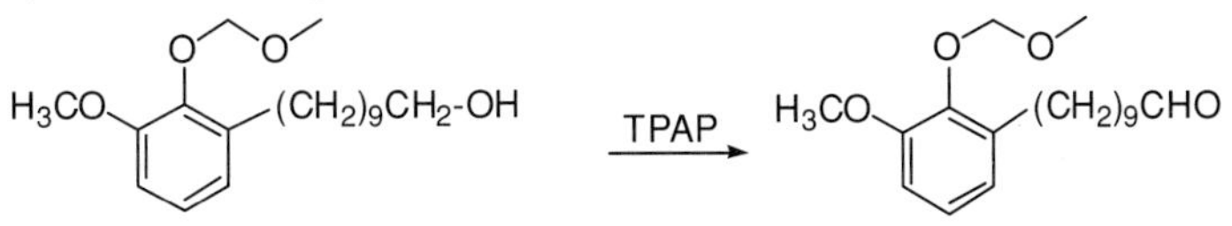

7.19 → **7.20**

In a similar manner, 2-propanol is oxidized to 2-propanone and cyclobutanol to cyclobutanone. N-Methylmorpholine-N-oxide (NMO) or oxygen may be used to reoxidize[28] the expensive TPAP catalyst. The water formed in the reaction is removed by molecular sieves, which prevents further oxidation of the aldehyde to the carboxylic acid. The presence of water increases an equilibrium concentration of the aldehyde hydrate, which can undergo further oxidation to carboxylic acid. The oxidation of **7.21** with TPAP in the presence of NMO gave **7.22** in good yield[29].

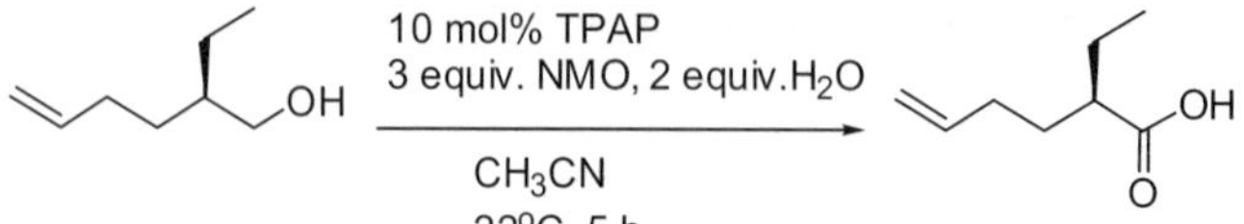

7.21   **7.22**

Mechanism: The mechanism of the oxidation of alcohols with TPAP corresponds to the non-radical $Cr(VI)$ oxidation. The perruthenate ion reacts with alcohols to form the ester of ruthenium(VII) acid **A**, which on elimination gives an aldehyde and reduced ruthenium(V) acid **B** (Scheme 7.10). It is believed that NMO reoxidizes the ruthenium(V) acid to perruthenate faster than the ruthenium(V) acid could attack on alcohol molecule.

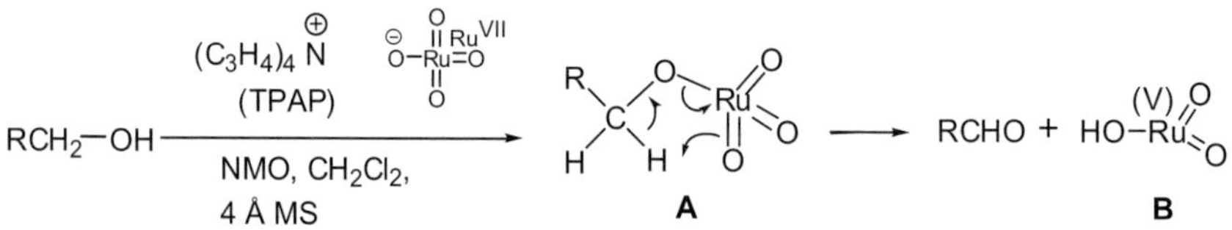

Scheme 7.10

7.1.7   Silver oxide and silver carbonate

Silver(I) oxide (Ag_2O), prepared *in situ* by mixing a solution of $AgNO_3$ and excess of NaOH, oxidizes primary alcohol into carboxylic acid. For example, 1-dodecanol is oxidized to dodecanoic acid[30] with Ag_2O.

$$CH_3(CH_2)_{10}CH_2OH \xrightarrow{Ag_2O} CH_3(CH_2)_{10}COOH$$

1-Dodecanol   Dodecanoicacid

Although **silver carbonate** although is not a very powerful oxidizing agent, it is extremely useful for the oxidation of alcohols to carbonyl compounds. The silver carbonate condensed on **Celite** is known as **Fetizon's reagent**[31], which oxidizes primary alcohols to aldehydes (e.g. **7.5** is converted into **7.6**) and secondary alcohols to ketones.

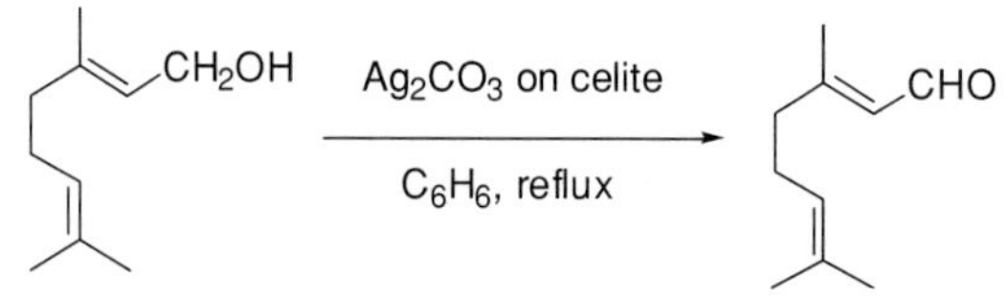

7.5   **7.6**

Ag_2CO_3–celite oxidizes selectively 2°-alcohols over 1°-alcohols.

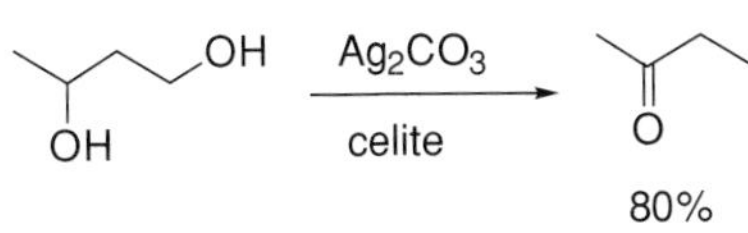

80%

Mechanism: The mechanism[32] involves the adsorption of alcohol on silver carbonate followed by formation of the protonated carbonyl. The carbonate ion acts as a hydrogen ion acceptor, generating carbonic acid, which decomposes to CO_2 and H_2O (Scheme 7.11).

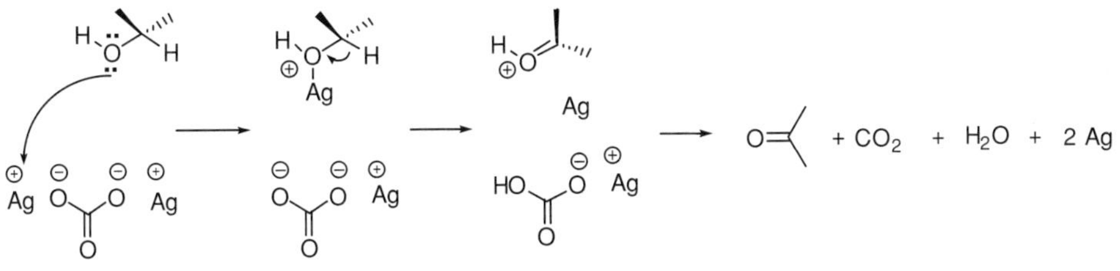

Adsorption

Scheme 7.11

Silver(II) oxide (AgO), prepared by the oxidation of silver nitrate ($AgNO_3$) with potassium pessulfate ($K_2S_2O_8$), oxidizes alcohols to carboxylic acids or aldehydes or ketones.

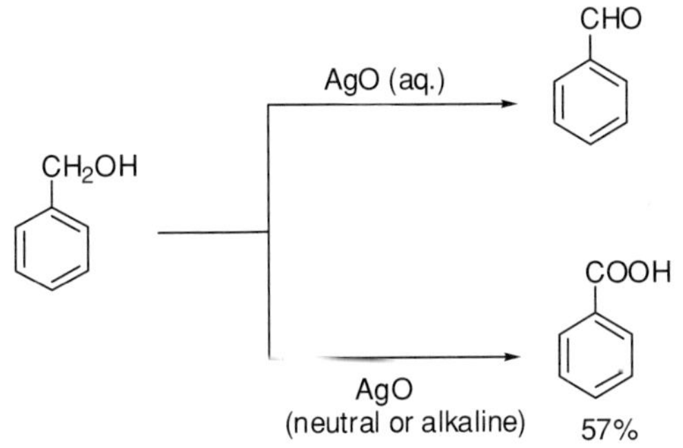

7.1.8 Oppenauer oxidation

The secondary alcohols are oxidized to ketones by refluxing with aluminium isopropoxide, $Al[OCH(CH_3)_2]_3$ [or $Al(O\text{-}iPr)_3$], or potassium *t*-butoxide, $KOC(CH_3)_3$ [or KO-*t*-Bu]. A ketone such as acetone used in the reaction as refluxing agent is reduced to alcohol, 2-propanol. The reaction is known as the **Oppenauer oxidation**. The reverse reaction known as the **Meerwein–Ponndorf–Verly reduction** is the reduction of ketones to alcohols in the presence of alcohol such as 2-propanol. Potassium *tert*-butoxide can be used for the oxidation of primary alcohols. Aluminium isopropoxide in acetone is particularly used for the oxidation of unsaturated alcohols. It can be used for the oxidation of primary alcohols but only in the presence of good hydrogen acceptor such as *p*-benzoquinone.

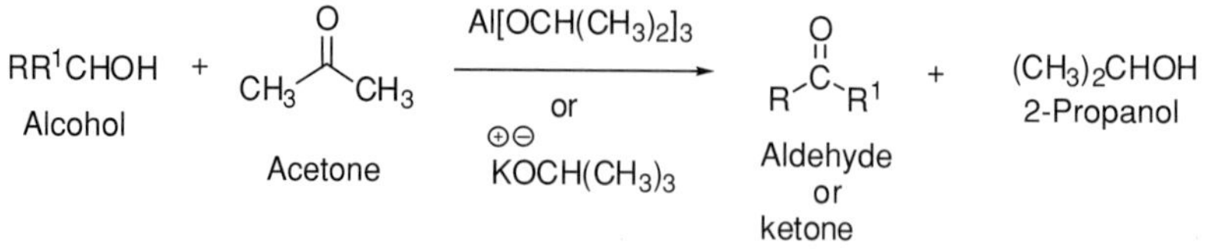

Mechanism: The aluminium alkoxide complex **A** (of starting alcohol and aluminium alkoxide) reacts with ketone (hydrogen acceptor) to form aluminium coordination complex **B**. Transfer of hydrogen takes place via a cyclic transition state to give the oxidized product ketone of the starting alcohol and aluminium alkoxide **C**, which releases 2-propanol (Scheme 7.12).

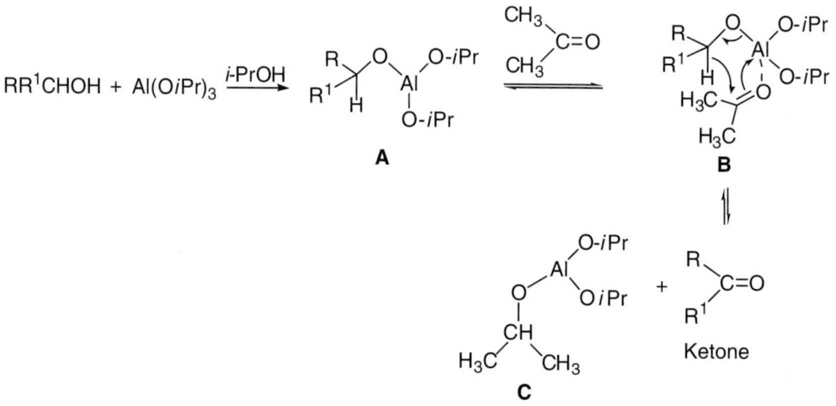

Scheme 7.12

The mechanism of the Oppenauer oxidation with potassium *tert*-butoxide is given in Scheme 7.13.

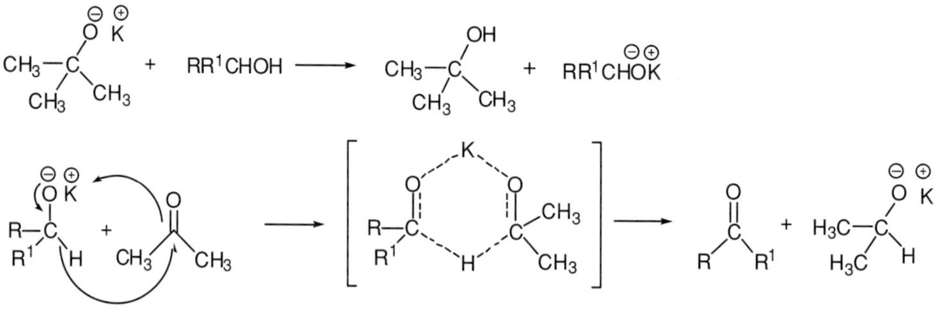

Scheme 7.13

7.2   Oxidation of aldehydes and ketones

Aldehydes are much more easily oxidized than ketones. **Tollen's, Benedict's** and **Fehling's** tests take this advantage of ease of oxidation of aldehydes by using Ag^+ (Tollen's reagent, i.e. $AgNO_3$, NH_4OH, NaOH and H_2O) or Cu^{2+} (Benedict's reagent, alkaline copper sulfate as its citrate complex, or Fehling's reagent, alkaline copper sulfate as its tartrate complex) oxidizing agents.

$$2\ Ag(NH_3)_2^{\oplus} + RCHO + 3\ \overset{\ominus}{O}H \longrightarrow 2\ Ag\downarrow + RCO\overset{\ominus}{O} + 4\ NH_3 + 2\ H_2O$$

$$2\ Cu^{2+}\ (\text{in complex}) + RCHO + 5\ \overset{\ominus}{O}H \longrightarrow \underset{Red}{Cu_2O\downarrow} + RCO\overset{\ominus}{O} + 3\ H_2O$$

The Cu^{2+} is used as a citrate or tartrate complex to avoid the formation of $Cu(OH)_2$. In Benedict's reagent, Na_2CO_3 is used, and in Fehling's reagent, NaOH is used. The strong oxidizing agents such as $K_2Cr_2O_7$ and H_2SO_4, PDC in DMF, potassium permanganate and mild oxidizing agents such as silver oxide (Ag_2O and AgO) oxidize aldehydes to carboxylic acids. Even the oxygen in air will slowly oxidize aldehydes to acids or peracids, most likely by a radical mechanism.

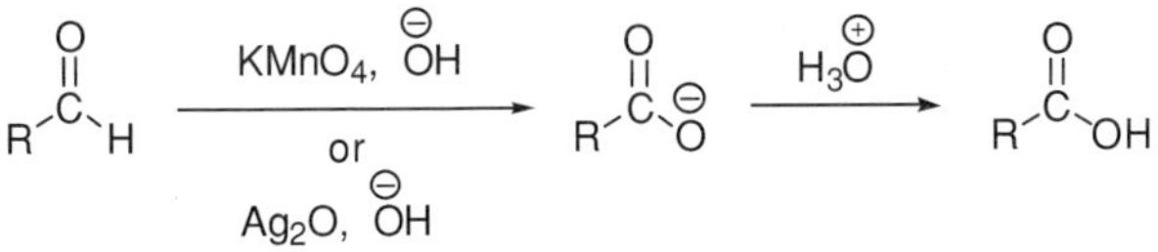

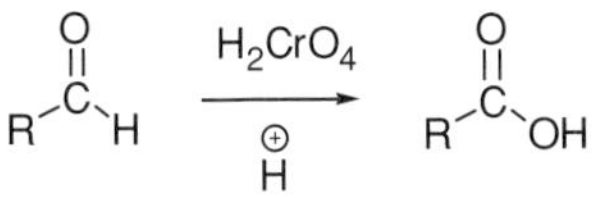

Mechanism: The mechanism of the Cr(VI) oxidation of aldehydes has been studied in detail in Scheme 7.14. A hydrate of aldehyde **A** is formed first, which reacts with chromium species to form a chromate ester **B**. A base abstracts a proton from the chromate ester **B** and Cr species leaves (E2 elimination) to give carboxylic acid.

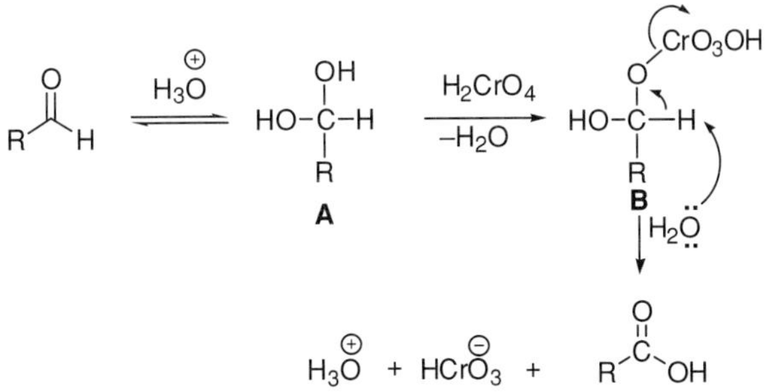

Scheme 7.14

Ketones are oxidized by strong oxidizing agents such as alkaline $KMnO_4$. HNO_3 also oxidizes ketone to carboxylic acid with fewer carbon atoms than the original ketone.

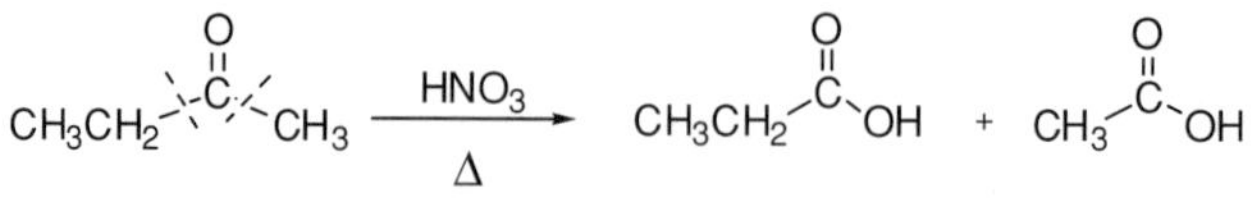

The cyclic ketones on treatment with an alkaline or acidic $KMnO_4$ give dicarboxylic acid. Adipic acid is commercially prepared by the oxidation of cyclohexanone. The oxidizing agent attacks on an enol form.

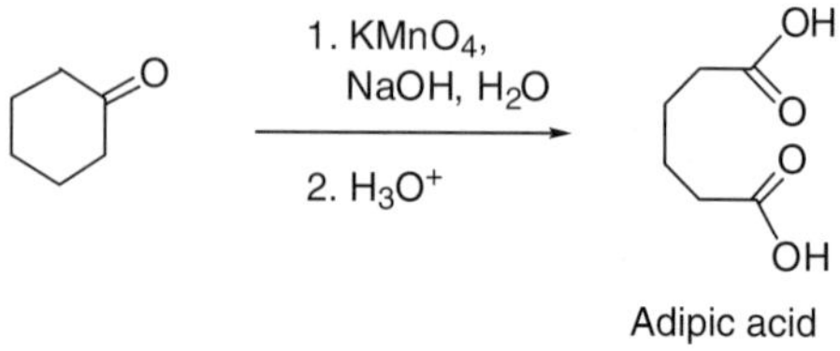

Adipic acid

Silver oxide: The mild oxidizing agent silver(I) oxide[33] (Ag_2O) also oxidizes aldehyde to carboxylic acid.

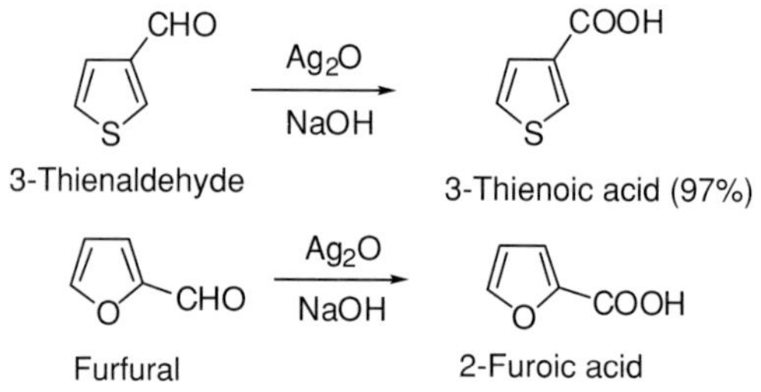

3-Thienaldehyde   3-Thienoic acid (97%)

Furfural   2-Furoic acid

Mechanism: The mechanism of the oxidation of aldehydes to carboxylic acid with silver oxide is given in Scheme 7.15.

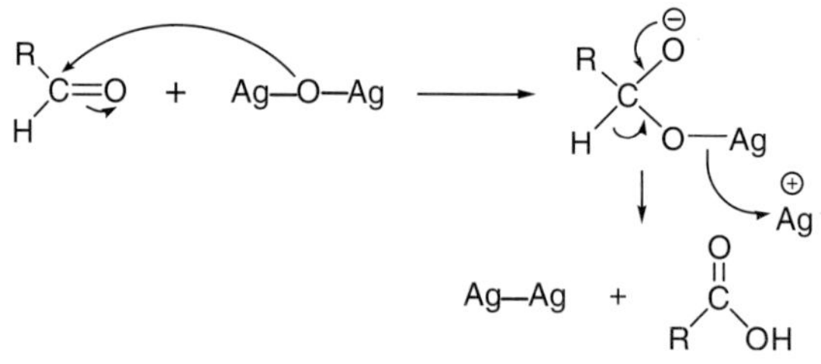

Scheme 7.15

The use of silver(II) oxide is less common because of its cost and limited availability. Corey[34] converted an benzaldehyde into the corresponding benzoic acid in 97% yield with

AgO and NaCN.

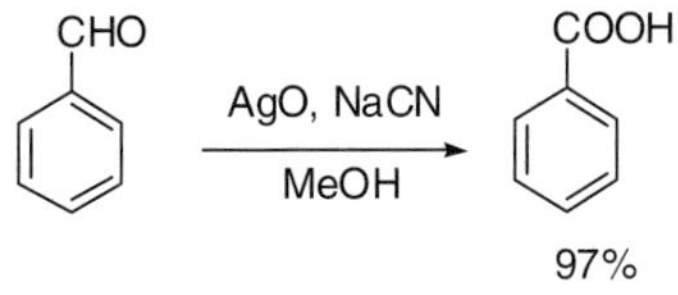
CHO
AgO, NaCN
MeOH
COOH
97%

Selenium dioxide (SeO_2) oxidation[35,36]: Selenium dioxide is an excellent oxidizing agent for the oxidation of allylic and benzylic C–H fragments to allylic or benzylic alcohol. It also oxidizes the aldehydes and ketones to 1,2-dicarbonyl compounds (i.e. oxidation of active methylene groups to carbonyl groups).

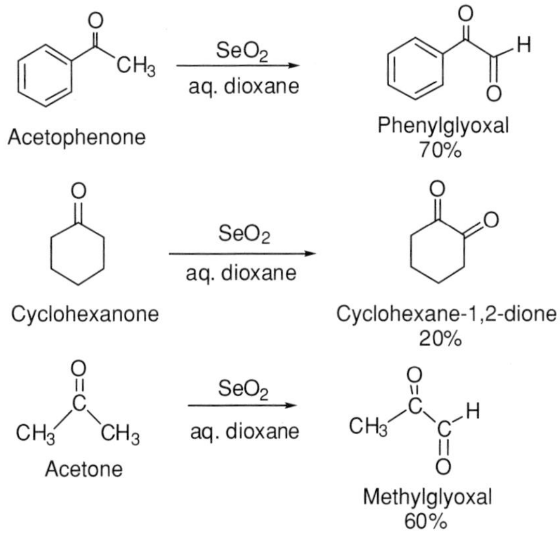
SeO2
aq. dioxane
Acetophenone
Phenylglyoxal
70%
Cyclohexanone
Cyclohexane-1,2-dione
20%
Acetone
Methylglyoxal
60%

Mechanism: The reaction of the enol form of the carbonyl compound **A** with selenium dioxide gives selenous enol ester **B**. The oxidative rearrangement of selenous enol ester **B** gives **C**. Loss of selenium and water from **C** gives the dicarbonyl compound (Scheme 7.16).

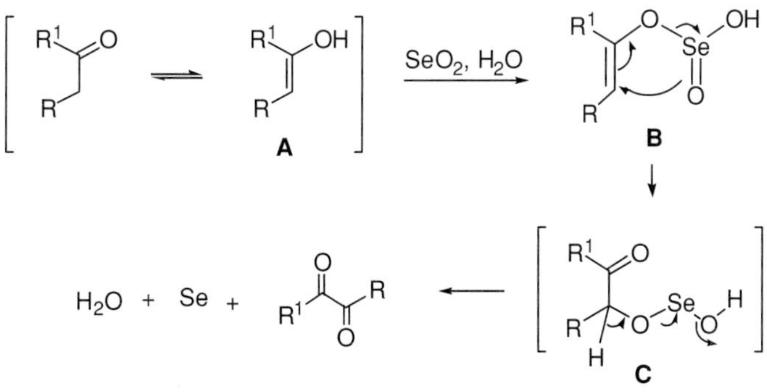
SeO2, H2O
A
B
C
H2O + Se +

Scheme 7.16

Baeyer–Villiger oxidation: Ketones (RCOR^1) are oxidized by peracids (or peroxy acids, RCO$_2$OH) to give esters (RCO$_2$R^1) and cyclic ketones give lactones.

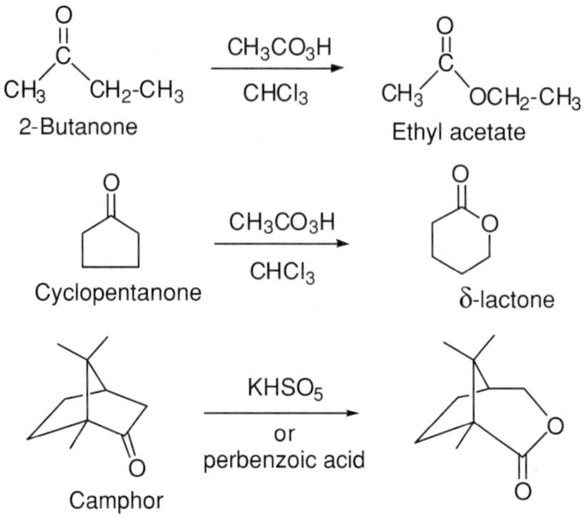

2-Butanone → Ethyl acetate

Cyclopentanone → δ-lactone

Camphor

7.2

Cycloheptanone and cyclooctanone on treatment with *m*-chloroperoxybenzoic acid (*m*-CPBA) in DCM (dichloromethylene) at room temperature give the corresponding lactones oxacyclooctan-2-one and oxacyclononan-2-one, in 47 and 49% yields, respectively[37,38].

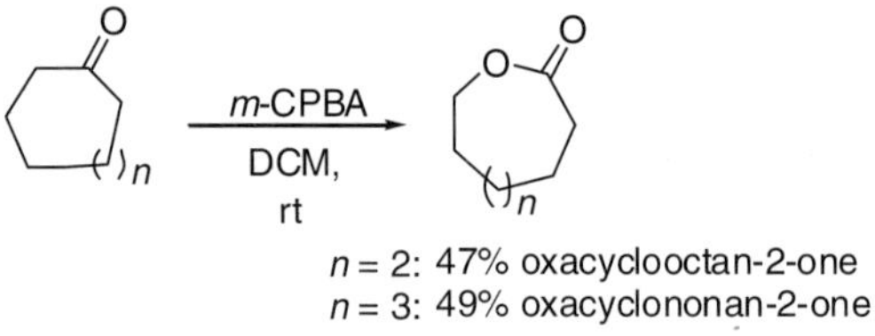

n = 2: 47% oxacyclooctan-2-one
n = 3: 49% oxacyclononan-2-one

Mechanism: The protonation of C=O bond with acid of carbonyl compound leads to **A**, which undergoes a nucleophilic attack by peroxy acid. The migration of one alkyl group to electropositive oxygen and elimination of carboxylate from perester **B** gives the protonated ester **C**. Finally, the base takes a proton from **C** and ester is formed (Scheme 7.17).

Asymmetric Baeyer–Villiger oxidation reaction: In 1994, Bolm *et al.*[39] used chiral Cu and Ni complexes (**7.23**) in catalytic amount with different oxidation systems.

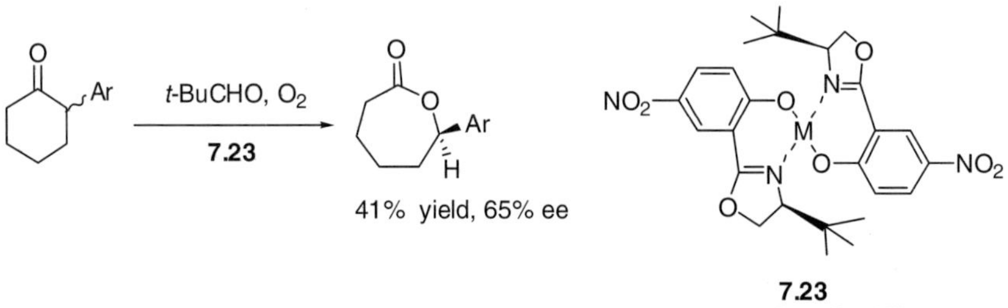

41% yield, 65% ee

7.23
(M = Cu, Ni)

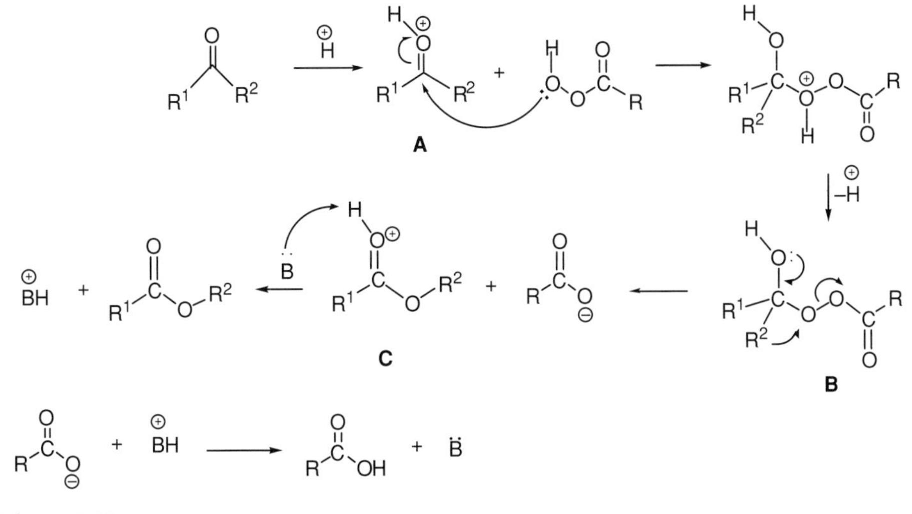

Scheme 7.17

Strukul and co-workers used chiral Pt catalyst (**7.24**) for the oxidation of 2-alkylcycloalkanones using hydrogen peroxide oxidant[40].

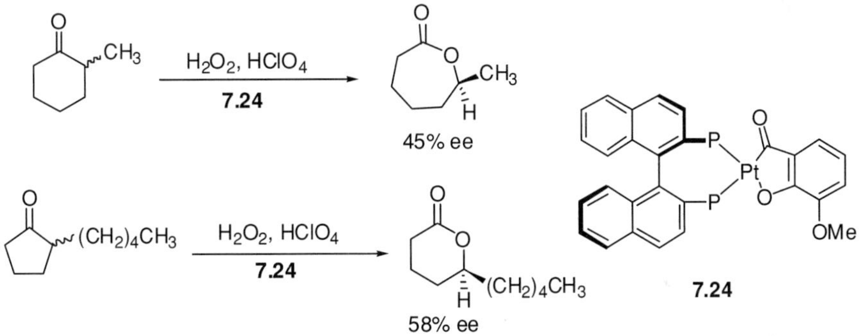

7.24

The most famous asymmetric oxidation catalyst, **Sharpless–Katsuki complex** [Ti(O-*i*Pr)$_4$, *t*-BuOOH and ester of tartaric acid], used for the asymmetric epoxidation of allylic alcohols[41,42] can also oxidize prochiral and racemic cyclobutanones **7.25** and **7.27** to enantiomerically enriched lactones[43] **7.26** and **7.28**, respectively.

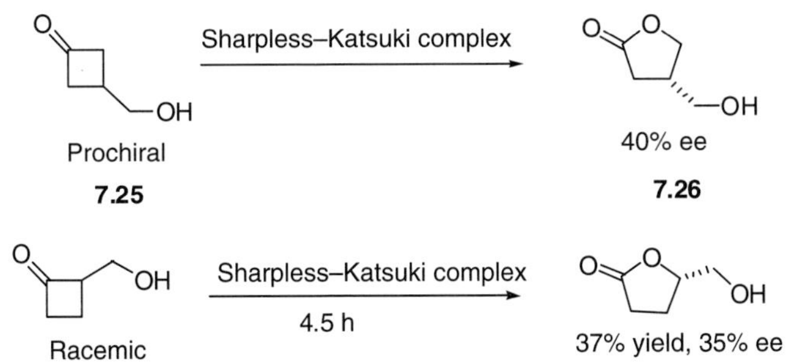

7.25 **7.26** **7.27** **7.28**

Lopp and co-workers modified the Sharpless–Katsuki Ti-complex [Ti(O-*i*Pr)$_4$, *t*-BuOOH and ester of tartaric acid ligand] by changing the tartaric acid ester by another chiral ligand such as TADDOL[44] (α,α,α',α'-tetraaryl-1,3-dioxolane-4,5-dimethanol, **7.29**).

Racemic   33% yield,41% ee

                                                                    7.29

7.3 Oxidation of phenols

Phenols are rather easily oxidized despite the absence of a hydrogen atom on the hydroxyl-bearing carbon. Among the coloured products from the oxidation of phenol by chromic acid is the dicarbonyl compound *p*-benzoquinone (also known as 1,4-benzoquinone or simply quinone). Dihydroxybenzenes, hydroquinone (**7.30**) and catechol (**7.32**) are oxidized to *p*-benzoquinone (**7.31**) and *o*-benzoquinone (**7.33**), respectively, by milder oxidants such as Jones reagent. **Fremy's radical** (**7.34**) is an excellent and very specific oxidizing agent for the oxidation of phenols to *o*- or *p*-benzoquinones. (*m*-Quinones do not exist.)

7.30   7.31

7.32   7.33

The oxidation of phenols with **Fremy's radical**, dipotassium nitrosodisulfonic acid [NO(SO$_3$K)$_2$, **7.34**], to hydroquinones is known as the **Teuber reaction**[45,46]. The disodium salt of nitrosodisulfonic acid [NO(SO$_3$Na)$_2$] is also used for the oxidation of phenols.

7.34

(Fremy's salt)

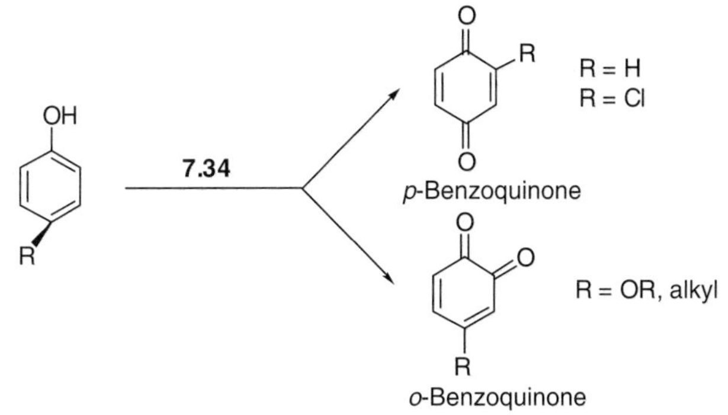

p-Benzoquinone

R = H
R = Cl

R = OR, alkyl

o-Benzoquinone

Mechanism: Fremy's radical (**7.34**) abstracts a hydrogen from phenol to give phenoxy radical, which is resonance stabilized. The resonating structures **A** or **B** then react with the second equivalent of Fremy's radical (**7.34**) to give either of the cyclohexadienone intermediates **C** or **D**, depending on the nature of R. Loss of $HN(SO_3K)_2$ gives the corresponding benzoquinones **E** or **F** (Scheme 7.18).

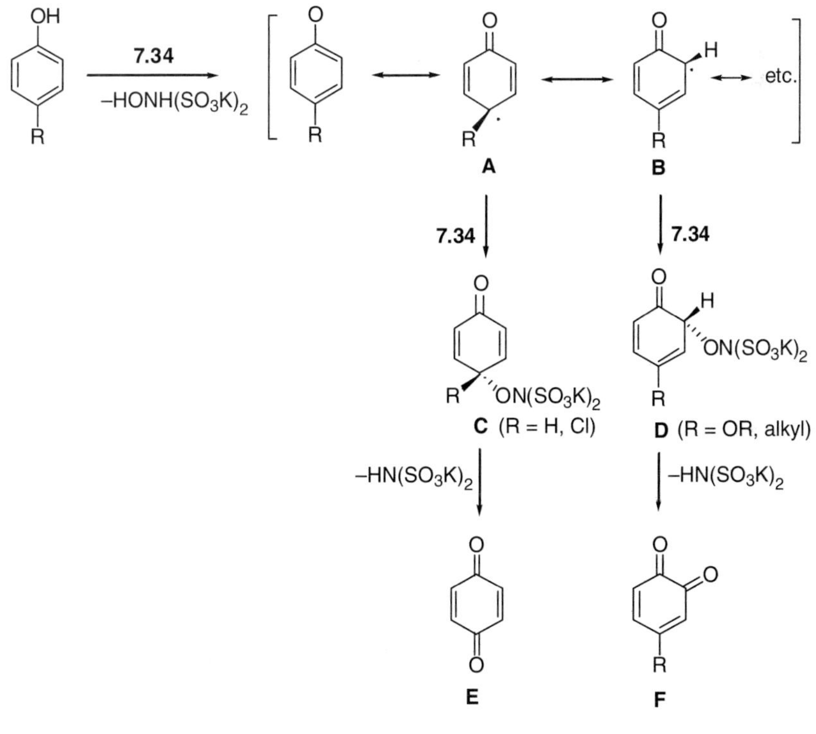

Scheme 7.18

For example, 3,4-dimethylphenol (**7.35**) is oxidized to 4,5-dimethyl-*o*-benzoquinone[47,48] (**7.36**) with Fremy's radical (**7.34**) and sodium dihydrogen phosphate. The oxidation of the phenol 2-[(*Z*)heptadec-10′-enyl]-6-methoxyphenol (**7.37**) with Fremy's radical produced irisquinone[27] (**7.38**) in 87% yield.

7.35   **7.36**

7.37   **7.38**
                                                                       87%

α-Naphthols are also oxidized by Fremy's radical to yield either *o*- or *p*-naphthoquinones, depending on the nature of R at position 4, *para* to the hydroxyl group. In some cases both are formed.

α-Naphthol

β-Naphthol is generally oxidized to give *o*-naphthoquinones through radical intermediates **A** and **B**, respectively.

β-Naphthol   *o*-Naphthoquinones

More stable   Less stable
      A   **B**

A cobalt complex salcomine (**7.39**) oxidizes substituted phenols but unsubstituted at *para* position, such as 2,6-di-*tert*-butylphenol (**7.40**), to give the corresponding *p*-quinone, 2,6-di-*tert*-butyl-*p*-benzoquinone (**7.41**).

7.40   **7.41**   **7.39**

The oxidation of phenol with potassium persulfate $(K_2S_2O_8)$ in an alkaline medium to *p*-hydroquinone is known as the **Elbs persulfate oxidation**[49]. Elbs used ammonium persulfate for the oxidation of 2-nitrophenol (**7.42**) to 2-nitro-*p*-hydroquinone (**7.43**), but the potassium salt is commonly used for the hydroxylation of phenols.

7.42   **7.43**

Mechanism[50]: The nucleophilic attack of phenoxide through *para* position of the aromatic ring carbon on the peroxide oxygen of the peroxydisulfate ion forms the intermediate persulfate **7.44**. Hydrolysis of **7.44** gives the *p*-hydroquinone **7.30**.

Phenol   **7.44**   **7.30**

7.4 Epoxidation

Oxidation reactions of alkenes with peracids (or peroxyacids, RCO_3OH) give cyclic ethers called **epoxides** or **oxiranes** (**Prilezhaev reaction**). The oxygen–oxygen bond of such peroxide derivatives is not only weak, but in this case is also polarized so that the acyloxy group is negative and the hydroxyl group is positive. The epoxidation reaction occurs in a single step with a transition state like a **butterfly** incorporating all the bonding events. Consequently, epoxidations by peracids always have *syn*-stereoselectivity and seldom give a structural rearrangement. Peroxybenzoic acid, *m*-CPBA (*m*-chloroperoxybenzoic acid) and magnesium monoperoxyphthalate are the most common peracids used for the epoxidation of alkenes.

Peroxybenzoic acid   *m*-CPBA   MMPP

Alkene   Peracid   TS   Epoxide (oxirane)

Peracid, CH_3CO_2OH in the presence of a catalyst, trimanganese complexes bearing bidentate nitrogen ligands [$Mn_3L_2(OAc)_6$] can also be used for the selective and efficient epoxidation of alkenes[51] under mild conditions.

$$cat. = [Mn_3L_2(OAc)_6]; \quad L = dipy, ppei$$

ppei = 2-Pyridinal-1-phenylethylimine

dipy = Dipyridyl

Asymmetric epoxidation[52]: The catalytic asymmetric epoxidation of alkenes has been the focus of many research efforts over the past two decades. The non-racemic epoxides are prepared either by enantioselective oxidation of a prochiral carbon–carbon double bond or by enantioselective alkylidenation of a prochiral C=O bond (e.g. via a ylide, carbene or the **Darzen reaction**). The **Sharpless asymmetric epoxidation** (SAE) requires allylic alcohols. The **Jacobsen epoxidation**[53,54] (using manganese–salen complex and NaOCl) works well with *cis*-alkenes and dioxirane method is good for some *trans*-alkenes (see Chapter 1, section 1.5.3).

The **SAE** is arguably one of the most important reactions discovered in the last 30 years. The SAE converts the double bond of allyl alcohols into epoxides with high enantioselective purity using a titanium tetraisopropoxide catalyst, Ti(O-*i*Pr)$_4$, chiral additive, either L-(+)-diethyl tartrate [(+)-DET, **7.45**] or D-(−)-diethyl tartrate [(−)-DET, **7.46**], and *tert*-butyl peroxide (*t*-BuOOH, TBHP (*t*-butylhydroperoxide)) as the source of the oxidant in stoichiometric amounts (see section 1.5, references 28–30 of Chapter 1).

Mechanism: The mechanism of the Sharpless epoxidation is quite complicated. Either enantiomer of diethyl tartrate (**7.45** or **7.46**) may be used and these affect the chirality of the epoxide formed. The enantiomer **7.45** induces epoxide formation on the bottom face and **7.46** induces epoxide formation on the top face (Scheme 7.19) if an allyl alcohol is drawn so that hydroxyl group is at the lower right.

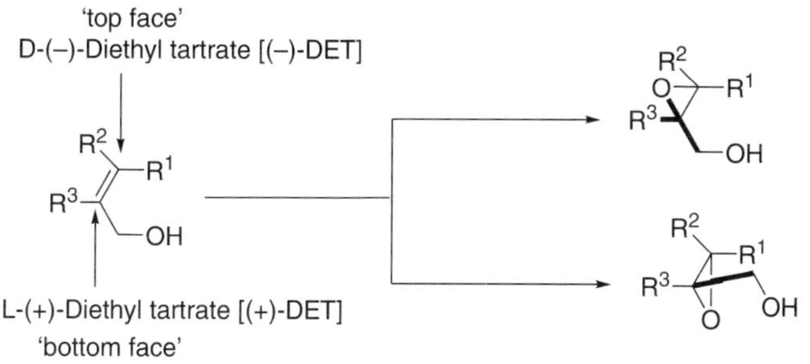

'top face'
D-(–)-Diethyl tartrate [(–)-DET]

L-(+)-Diethyl tartrate [(+)-DET]
'bottom face'

Scheme 7.19

Coordination of the chiral ligand DET and the oxidant source *t*-BuOOH to the metal centre forms the catalytically active species **7.47**. Several models for the stereocontrol of the Sharpless epoxidation are given, e.g. Sharpless model, Corey model and Hoffmann model. It is generally believed that the reaction proceeds through catalysis by a dimeric titanium species **7.47**, i.e. two metal centres are bridged via two oxygen ligands, giving the overall shape of two edge-fused octahedra. Coordination of the substrate can occur only in one orientation without causing severe steric interactions (Scheme 7.20).

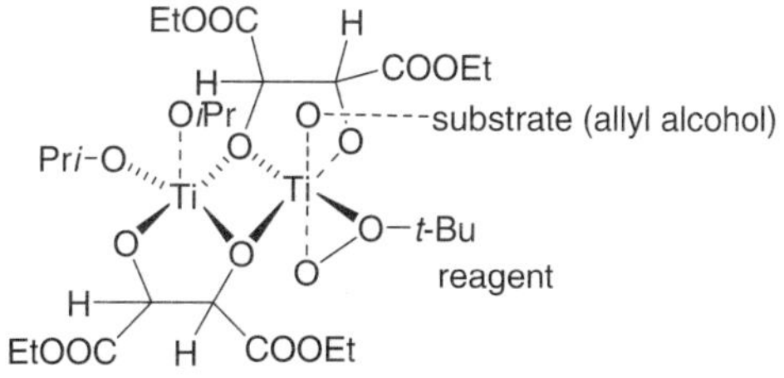

7.47

Ti(O-*i*Pr)$_4$ + DET $\longrightarrow$ Ti(DET)(O-*i*Pr)$_2$ + 2 *i*PrOH

2 Ti(DET)(O-*i*Pr)$_2$ $\longrightarrow$ [Ti(DET)(O-*i*Pr)$_2$]$_2$

[Ti(DET)(O-*i*Pr)$_2$]$_2$ $\downarrow$ *t*-BuOOH (reagent)

[Ti$_2$(DET)$_2$(O-*i*Pr)$_2$(*t*-BuOO)(allyl alcohol)] $\xleftarrow{\text{allyl alcohol}}$ [Ti$_2$(DET)$_2$(O-*i*Pr)$_3$ (*t*-BuOO)]

7.47

+ *i*-PrOH   + *i*-PrOH

Scheme 7.20

The direction of the attack is decided by the stereochemistry of diethyl tartrate in the reaction. The ester groups of tartrate molecules extend outward from the ring, thus limiting the manner in which the allylic alcohol and hydroperoxide can bind to the metal.

Sharpless and co-workers also used two new asymmetric epoxidating catalysts. The first of the new catalyst systems utilizes Ti(O-iPr)$_4$–tartramide ligand (**7.48**) in the 2:1 ratio.

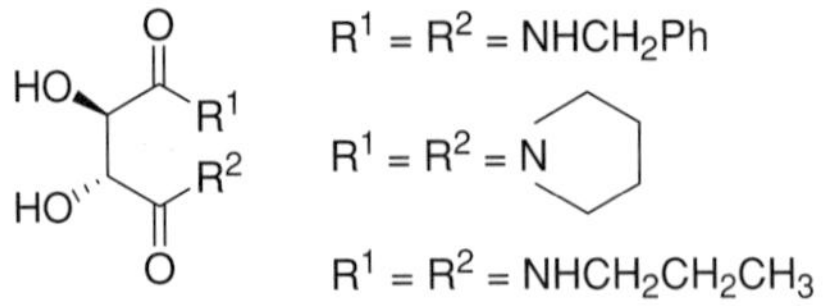

$R^1 = R^2 = NHCH_2Ph$

$R^1 = R^2 = N$

$R^1 = R^2 = NHCH_2CH_2CH_3$

7.48

By carefully controlling the amounts of ligand to titanium, the stereoselection of the epoxidating may be controlled as shown below in the epoxidation of (E)-α-phenylcinnamyl alcohol with various Ti(O-iPr)$_4$–tartramide complexes.

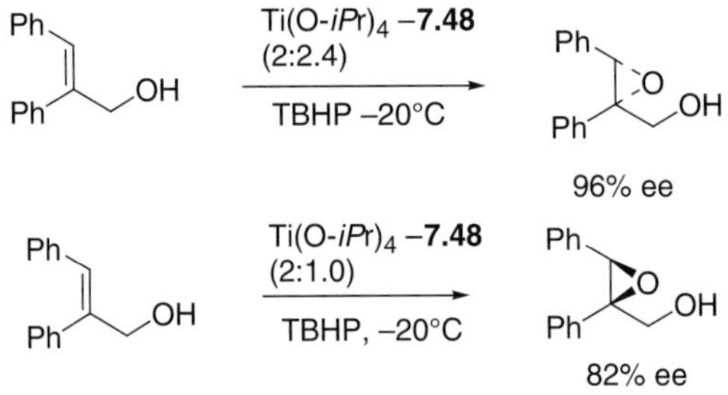

Ti(O-iPr)$_4$ –**7.48**
(2:2.4)
TBHP –20°C

96% ee

Ti(O-iPr)$_4$ –**7.48**
(2:1.0)
TBHP, –20°C

82% ee

The selectivity obtained with new systems is always opposite to the standard system. The second of the new catalyst utilizes TiCl$_2$(O-iPr)$_2$.

The epoxidation of the allyl alcohol **7.49** with Ti(O-tBu)$_4$, (+)-DET and TBHP (1:1:1) at −20°C gives the epoxide **7.50** in 51% as compared to 15% under standard conditions. But when the epoxidation is carried out with the new catalyst TiCl$_2$(O-iPr)$_2$, 68% of the epoxide **7.52** is formed via **7.51**.

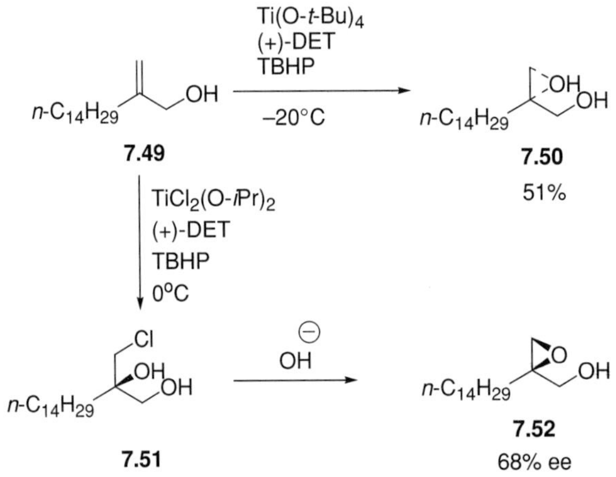

Ti(O-t-Bu)$_4$
(+)-DET
TBHP
−20°C

n-C$_{14}$H$_{29}$

7.49

7.50

51%

TiCl$_2$(O-iPr)$_2$
(+)-DET
TBHP
0°C

n-C$_{14}$H$_{29}$

7.51

$\ominus$
OH

n-C$_{14}$H$_{29}$

7.52

68% ee

Ruthenium complexes [RuCl$_2$(PNNP)] containing tetradentate hybrid ligands **7.53** with P- and N-donors (PNNP) catalyze the asymmetric epoxidation of unfunctionalized alkenes with aqueous H_2O_2 as primary oxidant. The epoxidation of styrene (**7.54**) affords the corresponding *cis*-epoxide, **7.55** exclusively, showing that the reaction is stereospecific[55].

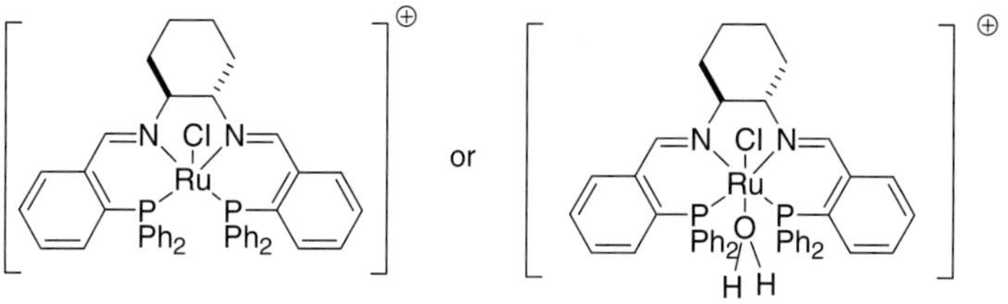

Catalyst (ruthenium complexes)
7.53

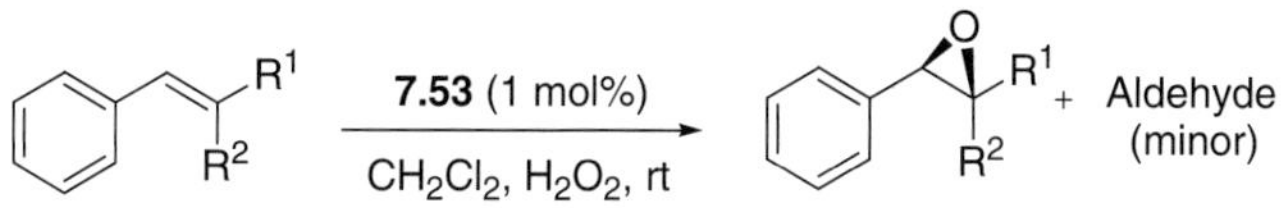

7.54

7.55

42% ee
$R^1 = R^2 = H$

The hydrogen peroxide and pentafluorophenyl Pt(II) complexes **7.56** can be used for the epoxidation of unfunctionalized terminal alkenes[56].

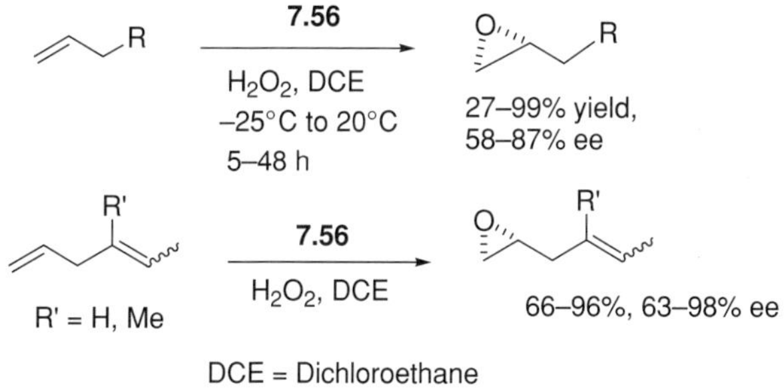

27–99% yield,
58–87% ee

66–96%, 63–98% ee

DCE = Dichloroethane

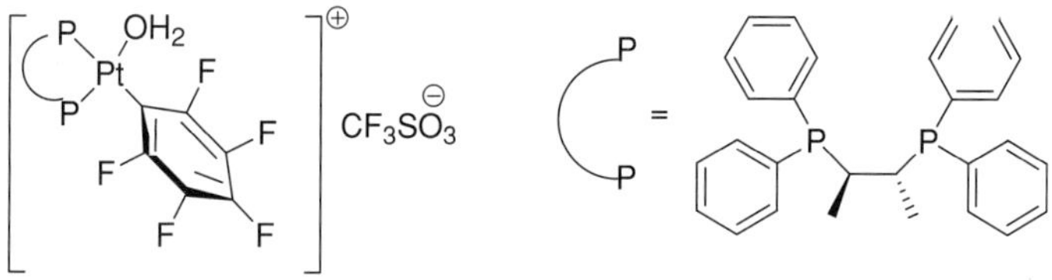

7.56

Dioxiranes are extremely useful reagents for the epoxidation of alkenes under neutral conditions. Since the oxygen atom transfer to the alkene regenerates the initial ketone, this epoxidation is catalytic. Dioxiranes are easily generated by the action of an oxone (potassium persulfate) on a ketone (usually acetone) either in a biphasic mixture or in a homogeneous aqueous organic solution.

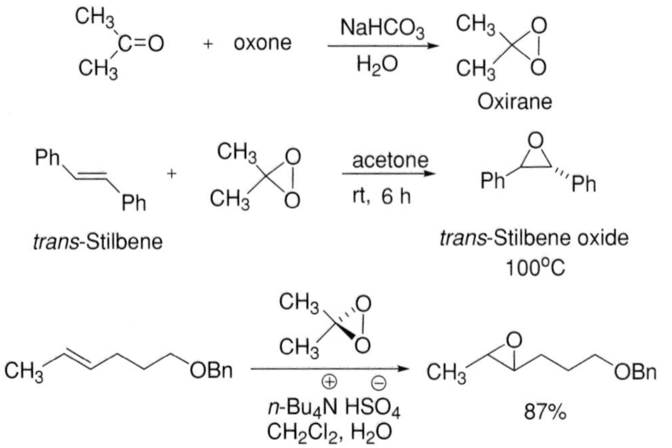

trans-Stilbene

trans-Stilbene oxide
100°C

In the **Shi epoxidation**, an oxone (potassium persulfate, $KOSO_2OOH$) in the presence of a fructose-derived catalyst, **7.57**, generates epoxides with high enantiomeric excess[57,58]; oxone is best used to oxidize aldehydes to carboxylic acids in the presence of DMF.

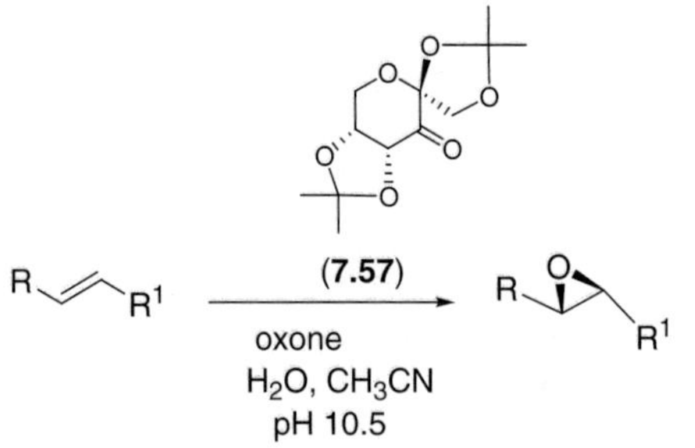

Mechanism: The mechanism of the epoxidation of alkenes with an oxone in the presence of **7.57** is given in Scheme 7.21.

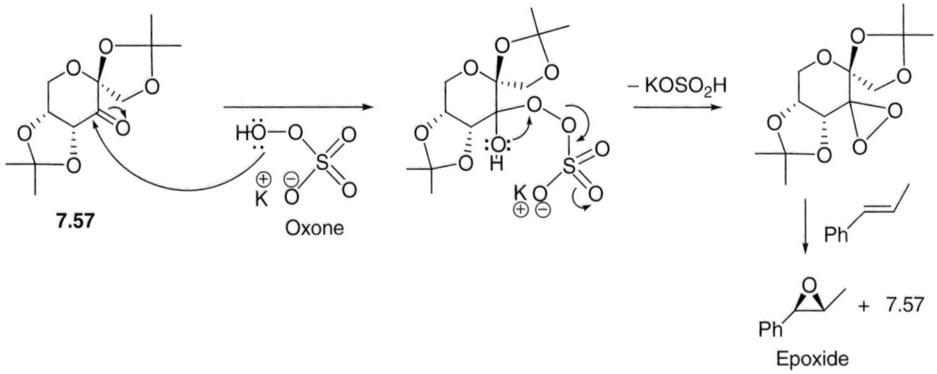

Scheme 7.21

7.5 Dihydroxylation

The introduction of oxygen atoms into unsaturated organic molecules via dihydroxylation reactions leads to 1,2-diols. 1,2-Diols can be synthesized by the reaction of alkenes either with peracids via corresponding epoxides and subsequent hydrolysis or with OsO_4, $KMnO_4$, RuO_4 and $Cr(VI)$ compounds.

Alkenes with cold dilute potassium permanganate (neutral or alkaline media) or osmium tetroxide and sodium bisulfite ($NaHSO_3$) give dihydroxylated products (glycols) with *syn*-stereoselectivity. Although the higher yields of glycols are obtained with OsO_4 than $KMnO_4$ dihydroxylation, OsO_4 is very expensive and toxic. The purple colour or $KMnO_4$ is converted into brown precipitates of MnO_2 during the reaction which forms the basis of **Baeyer test** for the presence of alkenes.

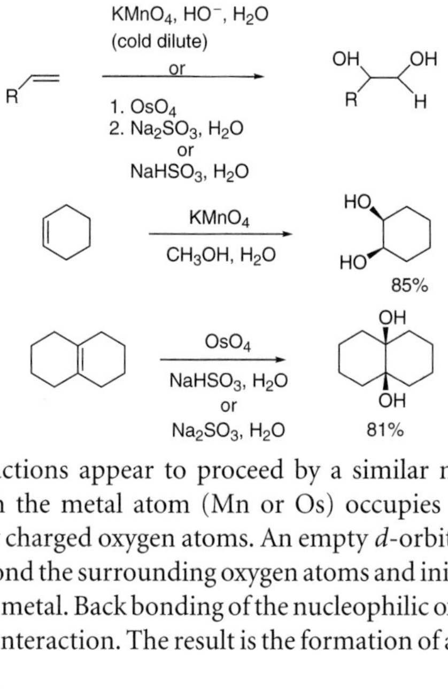

Mechanism: Both reactions appear to proceed by a similar mechanism involving cyclic intermediate in which the metal atom (Mn or Os) occupies the centre of a tetrahedral grouping of negatively charged oxygen atoms. An empty *d*-orbital of the electrophilic metal atom extends well beyond the surrounding oxygen atoms and initiates electron transfer from the double bond to the metal. Back bonding of the nucleophilic oxygen to the antibonding π-orbital completes this interaction. The result is the formation of a metallocyclic intermediate **A** or **B** (Scheme 7.22).

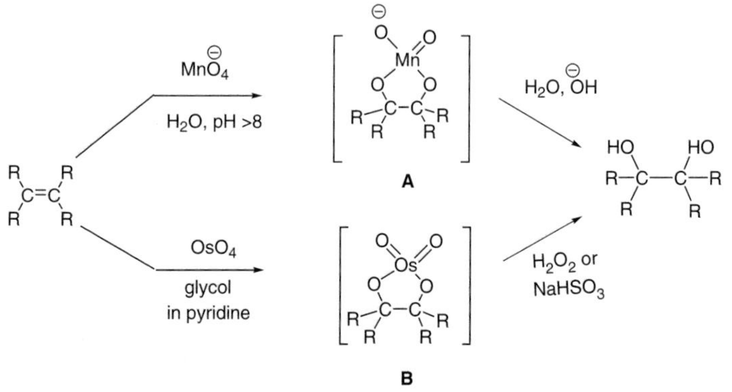

Scheme 7.22

The mechanism of the oxidation of alkenes with KMnO$_4$ is given in Scheme 7.23.

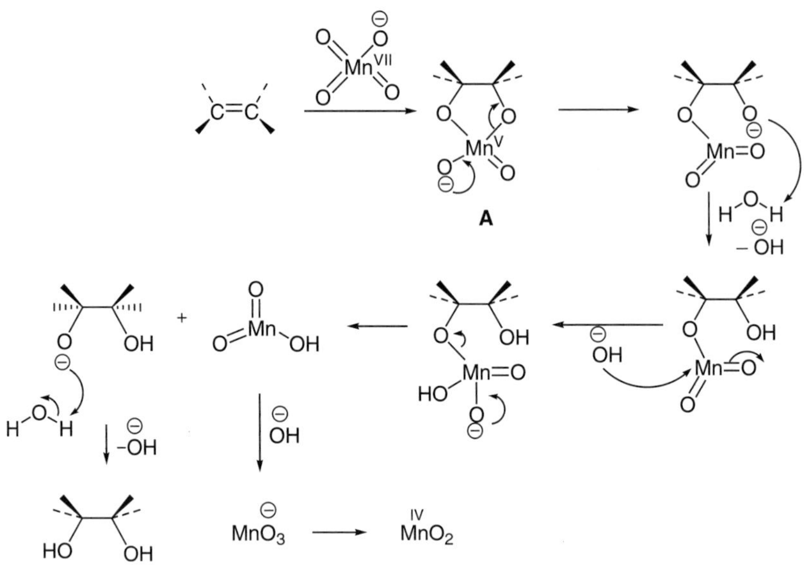

Scheme 7.23

From the mechanism shown in Scheme 7.23, we would expect the dihydroxylation with *syn*-selectivity. The cyclic intermediate may be isolated in the osmium reaction, which is formed by the cycloaddition of OsO$_4$ to the alkene. Since osmium tetroxide is highly toxic and very expensive, the reaction is performed using a catalytic amount of osmium tetroxide and an *oxidizing agent* such as TBHP, sodium chlorate, potassium ferricyanide or NMO, which regenerates osmium tetroxide. For example, **Upjohn dihydroxylation** allows the *syn*-selective preparation of 1,2-diols from alkenes by the use of catalytic amount of OsO$_4$ and a stoichiometric amount of an oxidant such as NMO.

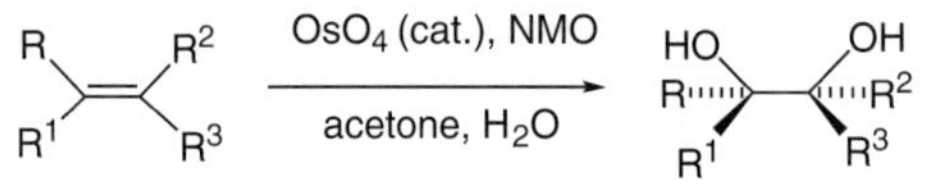

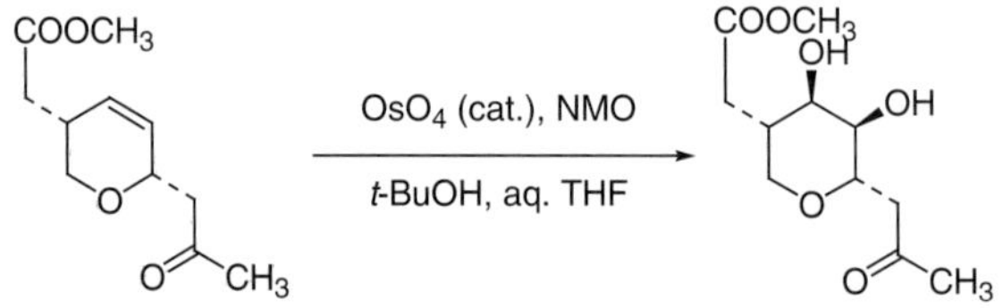

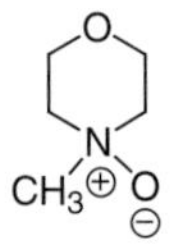

N-Methylmorpholine-*N*-oxide (NMO)

The [3+2]-cycloaddition or [2+2]-cycloaddition of OsO_4 and alkene followed by rearrangement forms the cyclic osmate ester. Tertiary amines (achiral ligand) such as 4-dimethylaminopyridine (DMAP) or pyridine (Py) accelerate the addition reaction (Scheme 7.24).

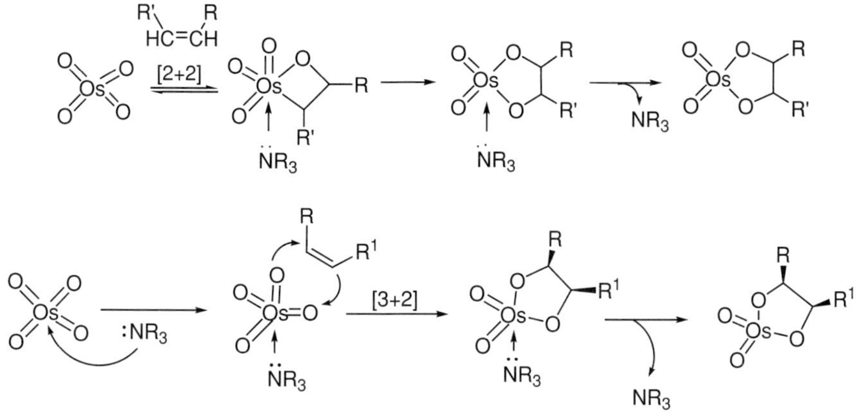

Scheme 7.24

The catalytic cycle for OsO_4-catalyzed alkene dihydroxylation in the presence of the co-oxidant NMO is shown in Scheme 7.25.

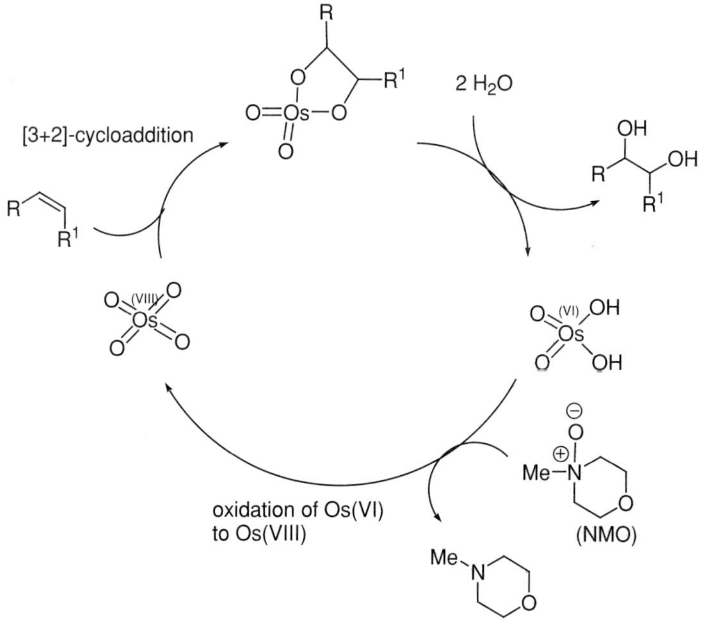

Scheme 7.25

Quantum chemical calculations have shown an initial [3+2]-addition of OsO_4 to alkene to be energetically more favourable.

Alkenes can be oxidized to diols by the **Prevost method**, which involves the reaction of an alkene with iodine and silver acetate. The *trans*-1,2-diacetate formed first on hydrolysis gives *trans*-1,2-diols (Scheme 7.26). In the Woodward variation of the Prevost reaction, the monoester is formed under aqueous conditions, which on hydrolysis gives *cis*-diol.

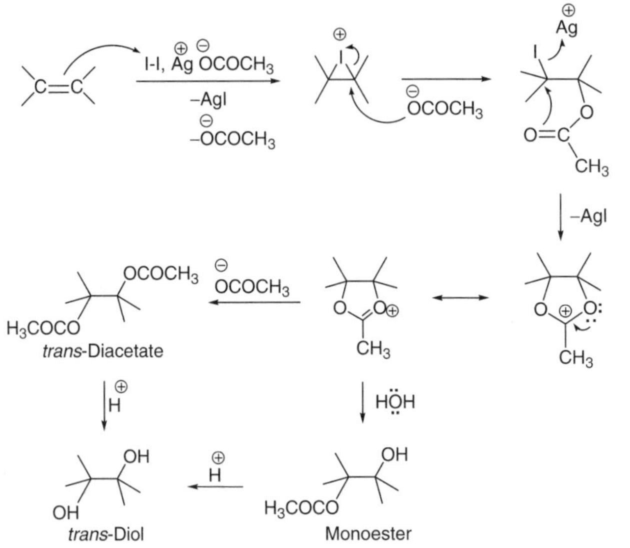

Scheme 7.26

Asymmetric dihydroxylation: Sharpless developed a catalytic system (AD-mix-β or AD-mix-α) that incorporates a chiral ligand into the oxidizing mixture which can be used for the asymmetric dihydroxylation of alkenes. The chiral ligands used in Sharpless asymmetric dihydroxylation are quinoline alkaloids, usually dihydroquinidine (DHQD) or dihydroquinine (DHQ) linked by a variety of heterocyclic rings such as 1,4-phthalhydrazine (PHAL) or pyridazine (PYR) (see section 1.6, reference 32 of Chapter 1).

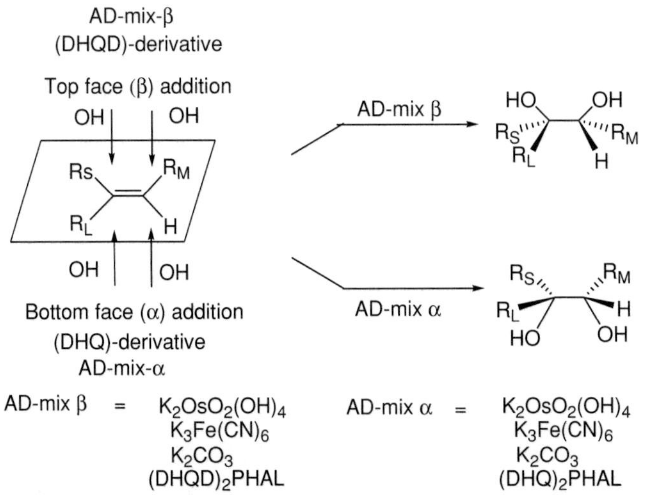

AD-mix-β
(DHQD)-derivative

Top face (β) addition

Bottom face (α) addition
(DHQ)-derivative
AD-mix-α

AD-mix β	=	$K_2OsO_2(OH)_4$
		$K_3Fe(CN)_6$
		K_2CO_3
		$(DHQD)_2PHAL$

AD-mix α	=	$K_2OsO_2(OH)_4$
		$K_3Fe(CN)_6$
		K_2CO_3
		$(DHQ)_2PHAL$

Examples of asymmetric dihydroxylation are shown below and in section 1.6.

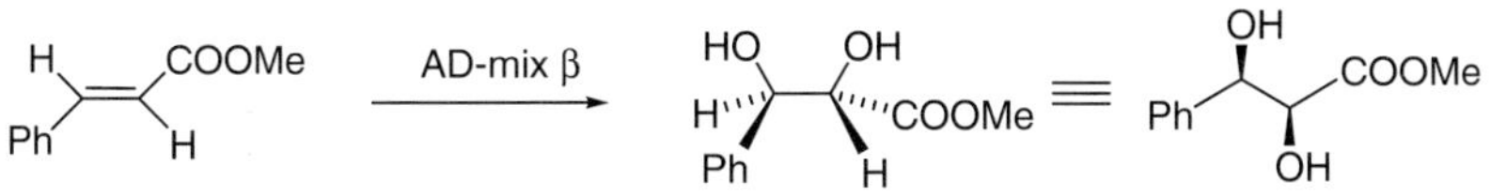

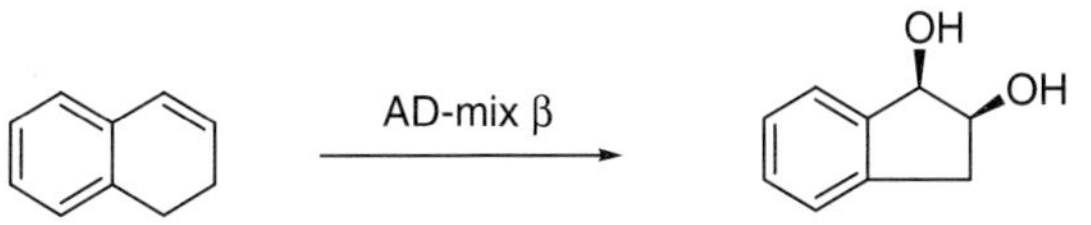

65% yield, 41% ee

Dihydroxylation of styrene to the corresponding vicinal diol was achieved in 73% yield and 96% ee with AD-mix-β[59].

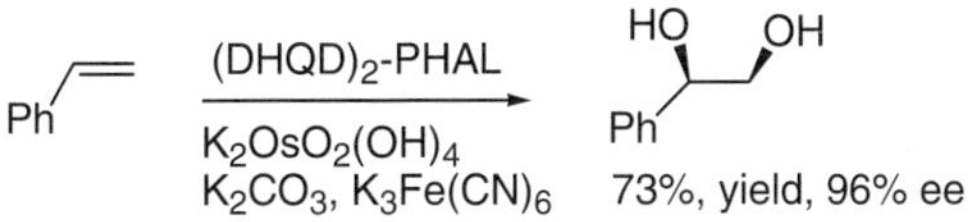

73%, yield, 96% ee

Anti-dihydroxylation: Epoxides may be cleaved by an aqueous acid to give glycols. Proton transfer from the acid catalyst generates the conjugate acid of the epoxide, which is attacked by nucleophiles such as water. The result is **anti-dihydroxylation** of the double bond.

In Scheme 7.27, it is shown that a *cis*-disubstituted epoxide, which, of course, could be prepared from the corresponding *cis*-alkene, is converted into *trans*-diol by an acid or a base.

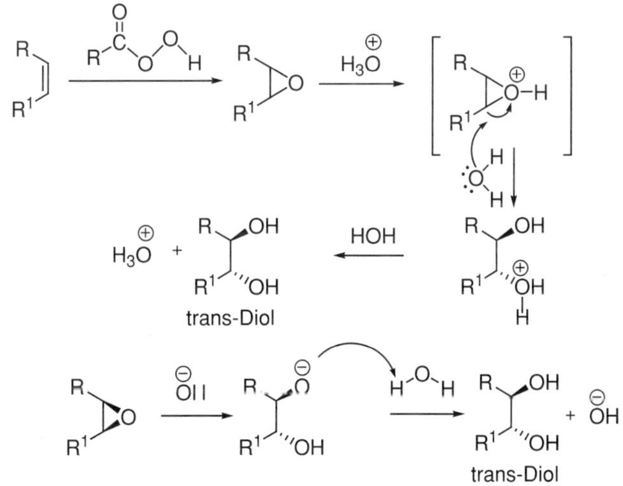

Scheme 7.27

7.6 Aminohydroxylation

Aminohydroxylation of unsymmetrically substituted alkenes, in contrast to dihydroxylation, may give two possible regioisomers of aminoalcohol derivatives; but asymmetric amino-hydroxylation, by using the same catalytic system as that used for Sharpless asymmetric dihydroxylation, can be highly regioselective as well as enantioselective.

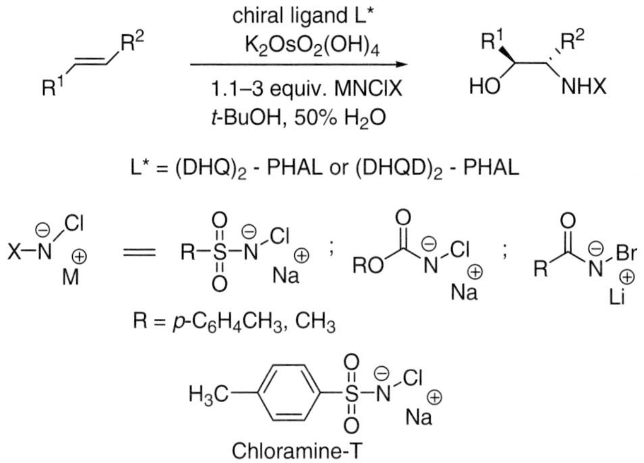

chiral ligand L*
K$_2$OsO$_2$(OH)$_4$

1.1–3 equiv. MNCIX
t-BuOH, 50% H$_2$O

L* = (DHQ)$_2$ - PHAL or (DHQD)$_2$ - PHAL

R = p-C$_6$H$_4$CH$_3$, CH$_3$

Chloramine-T
(sodium N-chloro-p-toluenesulfonamide)

Examples of asymmetric aminohydoxylation are given below and in section 1.6 (see references 25–27 of Chapter 1).

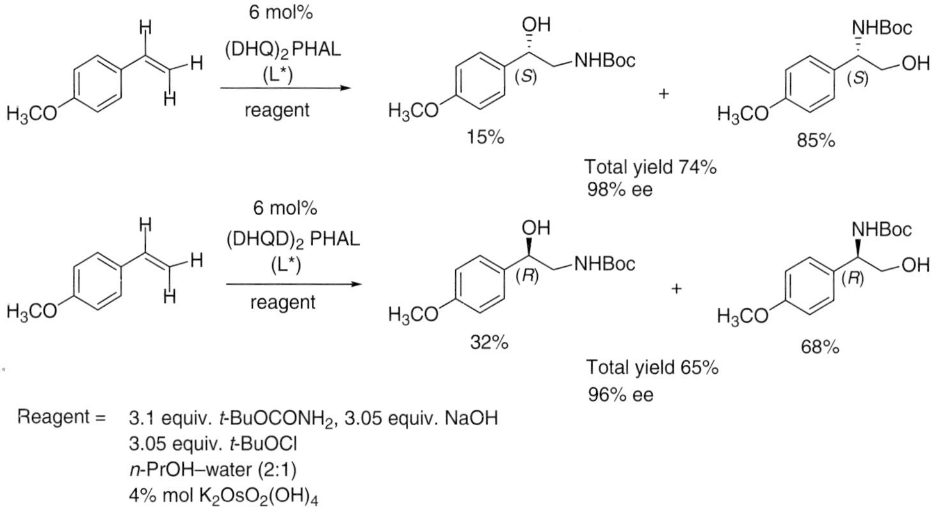

6 mol%
(DHQ)$_2$ PHAL
(L*)
reagent

15%

Total yield 74%
98% ee

85%

6 mol%
(DHQD)$_2$ PHAL
(L*)
reagent

32%

Total yield 65%
96% ee

68%

Reagent = 3.1 equiv. t-BuOCONH$_2$, 3.05 equiv. NaOH
3.05 equiv. t-BuOCl
n-PrOH–water (2:1)
4% mol K$_2$OsO$_2$(OH)$_4$

7.7 Oxidative cleavage of C–C double bonds

Some oxidation reagents actually cleave the carbon–carbon double bonds. The most important reagents for the oxidative cleavage of alkenes to carboxylic acids and/or ketones include ozone, potassium permanganate and ruthenium tetroxide. Hot concentrated KMnO$_4$ (acidic or alkaline) cleaves double bonds, and if it is a terminal alkene then bubbles of CO$_2$ gas are evolved. However, cold KMnO$_4$ gives *syn*-addition, which leads to diol formation (see section 7.5). The purple colour of KMnO$_4$ disappears and brown precipitates of MnO$_2$ are formed (**Baeyer test**). However, there are several methods available for the oxidative cleavage of alkene via 1,2-diol or epoxide.

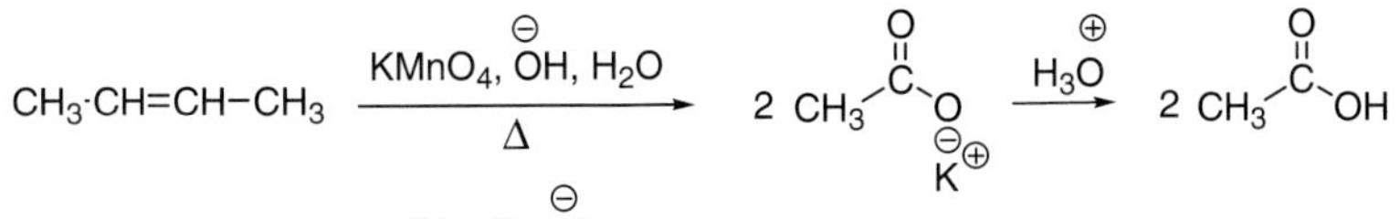

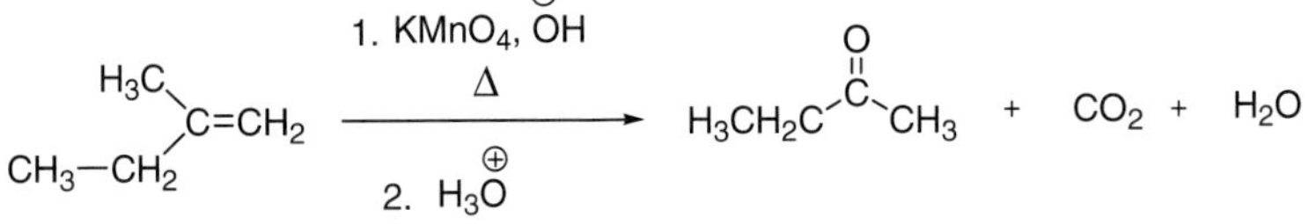

7.7.1 Ozonolysis

Ozone, O_3, is an *allotrope* of oxygen that adds rapidly to carbon–carbon double bonds, which is followed by reductive or oxidative cleavage.

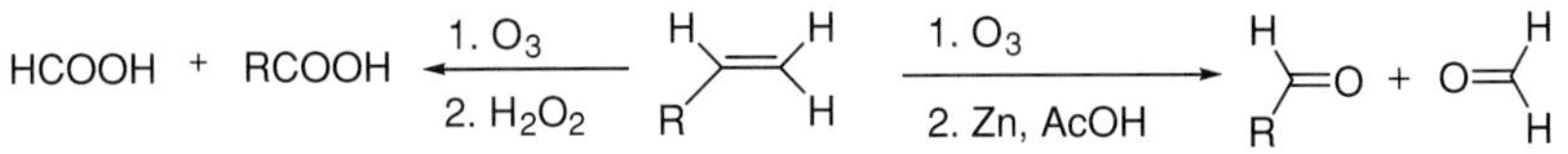

Mechanism: The mechanism of **ozonolysis** suggested by **Criegee** has been extensively studied. *syn*-Addition of ozone to alkene first gives the primary addition product molozonide (**A**), which on rearrangement through dipolar species **B** and **C** gives reactive intermediate called ozonides (**D**). The decomposition of the final ozonide (**D**) by a reductive work up (with reducing agents such as Me_2S, Zn and AcOH, H_2 and Pd–C or Ph_3P) gives aldehydes or ketones whereas an oxidative work up (with H_2O_2 and H_2O or H_2O_2 and AcOH) gives ketone and carboxylic acid (Scheme 7.28).

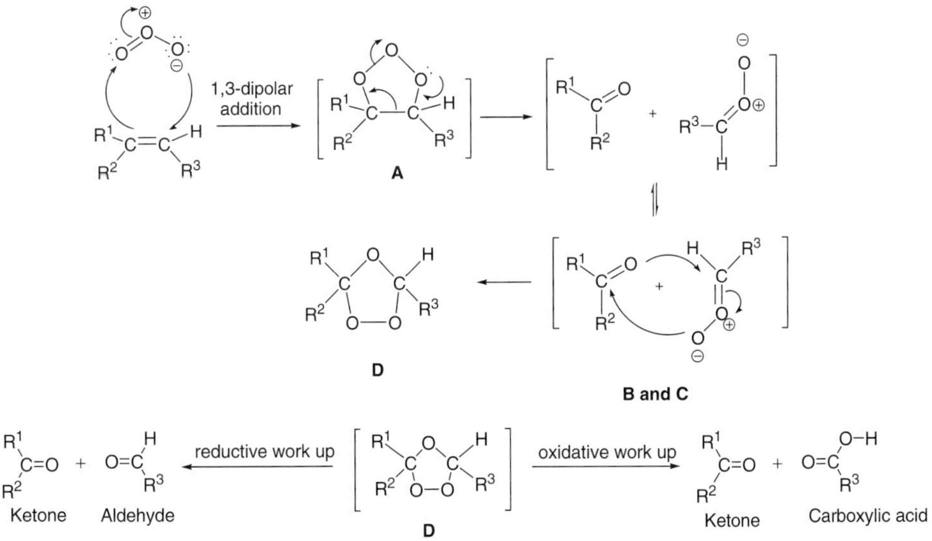

Scheme 7.28

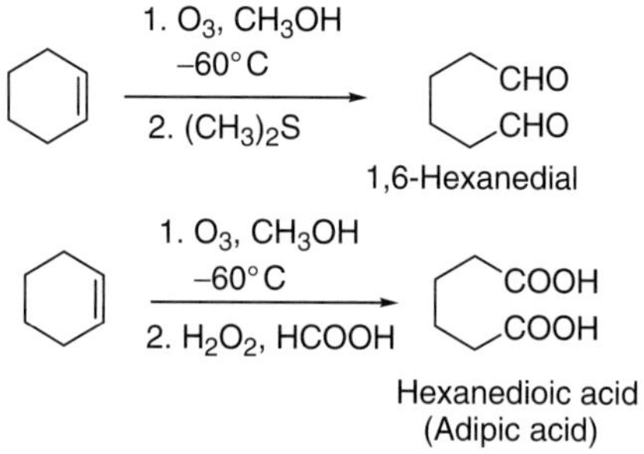

1,6-Hexanedial

Hexanedioic acid
(Adipic acid)

From the analysis of products obtained in an ozonolysis experiment, it is possible to deduce structural formulas for the starting alkenes.

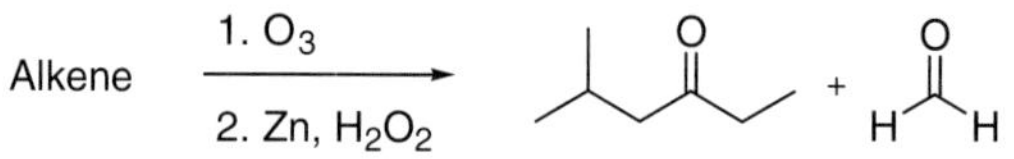

Thus, the structure of the starting alkene in the above reaction is as follows:

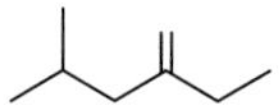

7.7.2 Glycol cleavage

The vicinal glycols prepared by dihydroxylation of alkenes are converted into aldehydes and ketones in high yield by the action of **lead tetraacetate** $[Pb(OAc)_4]$[60] or **periodic acid**[61] (HIO_4 in H_2O and THF). This oxidative cleavage of a carbon–carbon double bond provides a two-step, high-yield alternative to ozonolysis, that is often preferred for small-scale work involving precious compounds. As a rule, *cis*-glycols react more rapidly than *trans*-glycols.

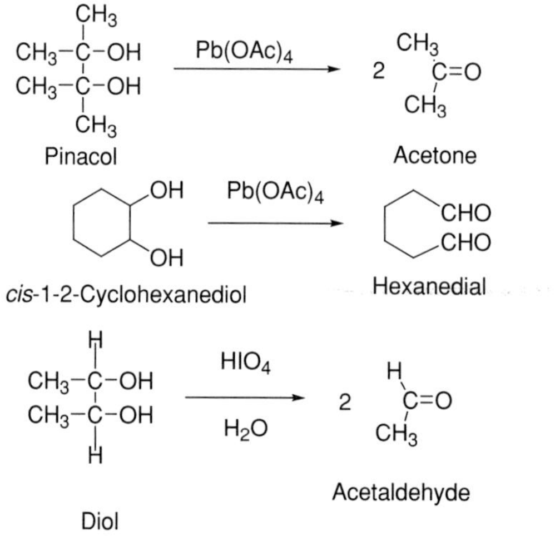

Pinacol

Acetone

cis-1-2-Cyclohexanediol

Hexanedial

Diol

Acetaldehyde

Mechanism: The cleavage of diols with lead tetraacetate $[Pb(OAc)_4]$ proposed by **Criegee** is shown in Scheme 7.29.

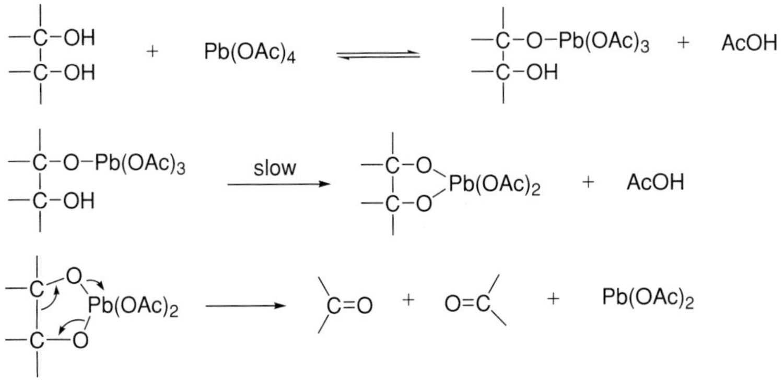

Scheme 7.29

The oxidation with periodic acid also occurs via cyclic intermediate periodate ester. Periodate ester undergoes rearrangement of the electrons, which leads to the formation of a C–C bond and two C=O bonds (Scheme 7.30).

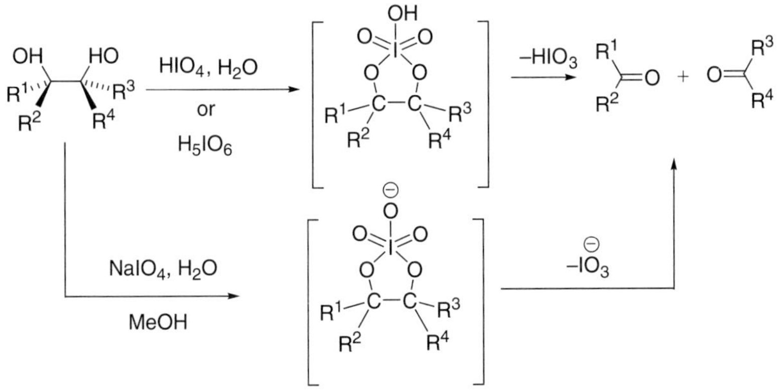

Scheme 7.30

MnO$_2$ is also employed for the glycol cleavage.

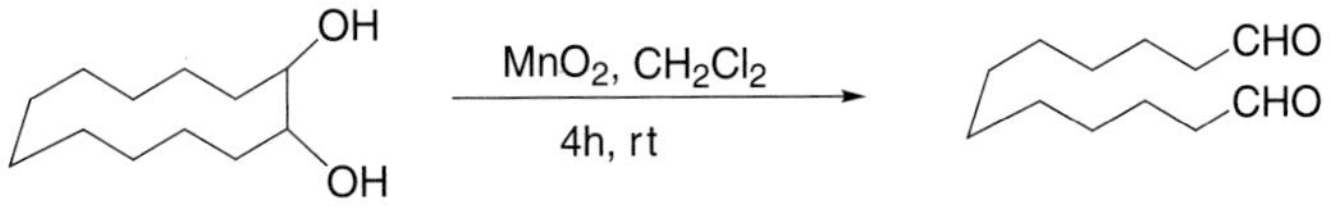

When catalytic amount of OsO$_4$ is used in combination with stoichiometric Jones reagent in acetone[62], alkenes are converted into acids or ketones. The reaction involves the formation of osmate ester, which is cleaved by the chromic acid.

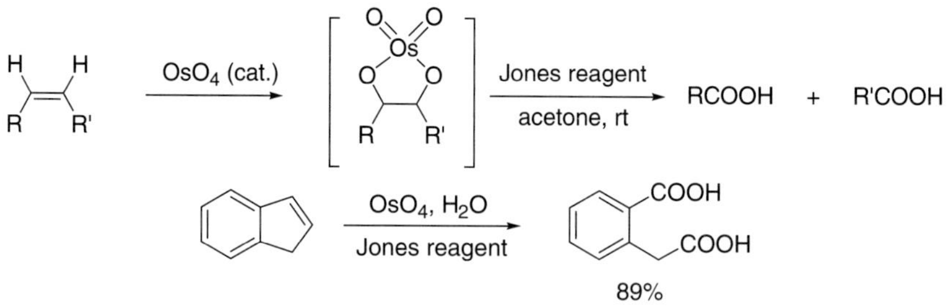

Similarly, the use of OsO_4 (catalytic amount) or $KMnO_4$ (catalytic amount) and $NaIO_4$ (stoichiometric amount) cleaves the carbon–carbon double bond.

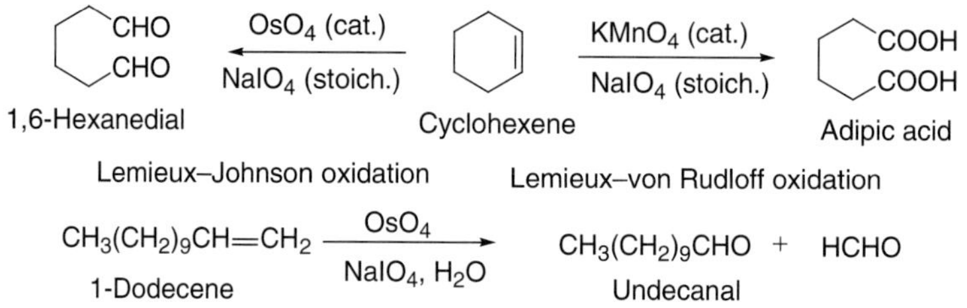

The oxidative cleavage of the carbon–carbon double bond can also be effected by RuO_2 and $NaIO_4$.

7.8 Oxidation of anilines

Oxidation of anilines with alkaline potassium persulfate ($K_2S_2O_8$) followed by hydrolysis to give *ortho*-hydroxyl anilines is known as the **Boyland–Sims oxidation**[63].

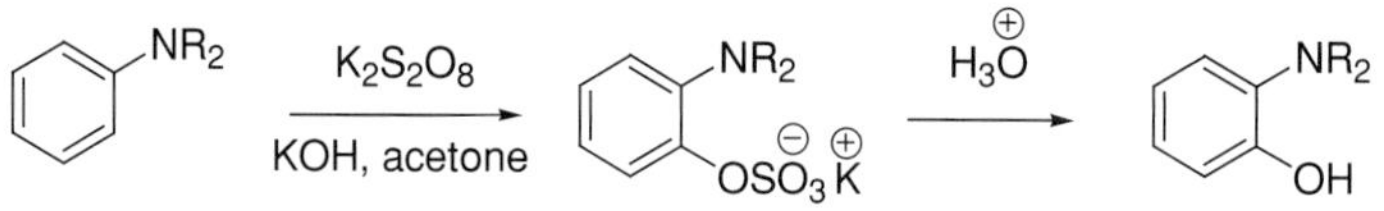

The *ortho*-isomer is formed predominantly. However, the *para*-sulfate is formed in small amounts with certain anilines. The intermediate **A** formed first in the Boyland–Sims oxidation on rearrangement gives both the *ortho*- and *para*-amino aryl sulfates **B** and **C** (Scheme 7.31)[64].

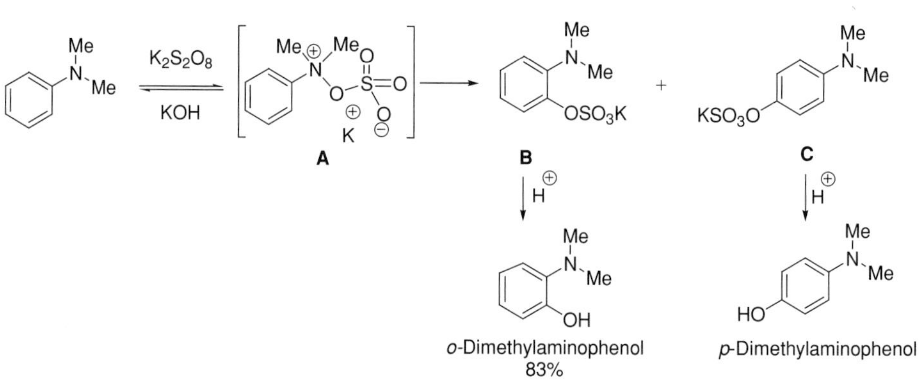

Scheme 7.31

7.9 Dehydrogenation

Dehydrogentaion is removal of two hydrogens to form a multiple bond. Generally, catalysts such as sulfur, selenium, Pd–C, Pd–C$_6$H$_6$, Pt–C, Pt–C$_6$H$_6$ and several quinones (such as **7.58** and **7.59**) remove the hydrogen from an aromatic system. The fewer the number of hydrogens to be removed, the milder the conditions.

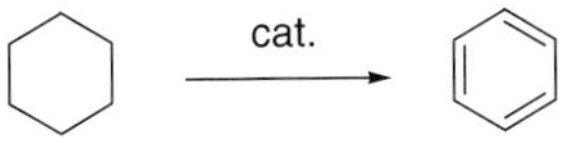

cat.

cat. = Pd–C, Pd–C$_6$H$_6$, Pt–C, Pt–C$_6$H$_6$, S or Se

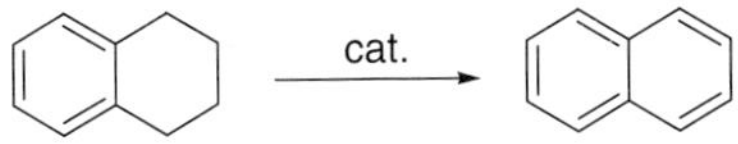

cat.

cat. = Pd–C, Pd–C$_6$H$_6$, Pt–C, Pt–C$_6$H$_6$, S, Se, DDQ or chloranil

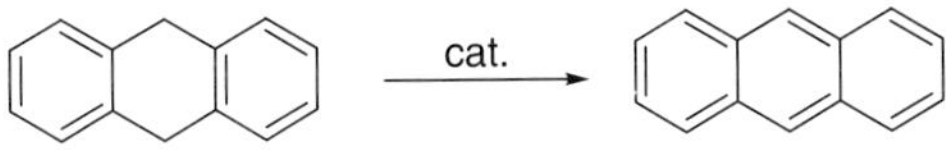

cat.

cat. = Pd–C, Pd–C$_6$H$_6$, SeO$_2$ or Cu

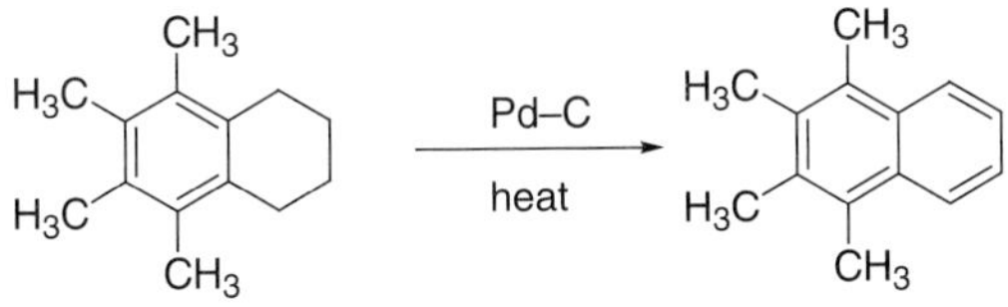

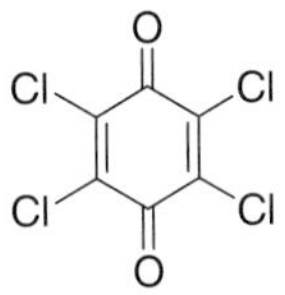

Pd–C

heat

7.58

Chloranil

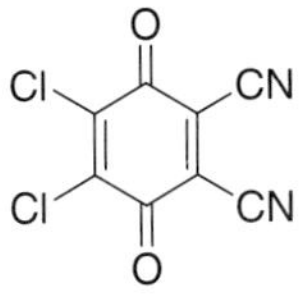

7.59

Dichlorodicyanoquinone
(DDQ)

The mechanism of dehydrogenation of aromatic compounds with dichlorodicyanoquinone (DDQ) is shown in Scheme 7.32.

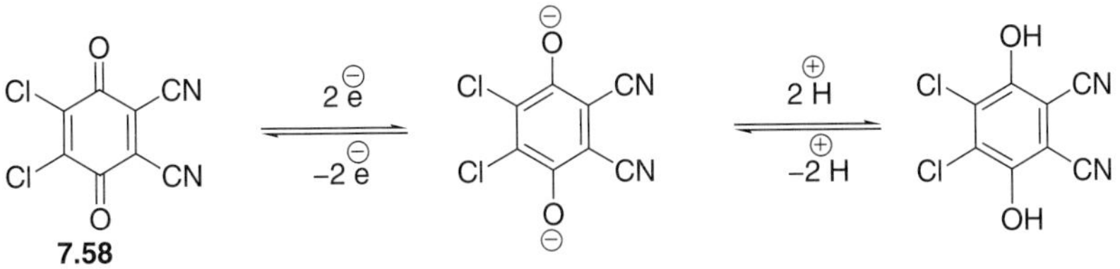

7.58

Scheme 7.32

7.10   Allylic or benzylic oxidation

SeO_2 oxidizes an allylic or benzylic C–H fragment to the corresponding allylic or benzylic alcohols.

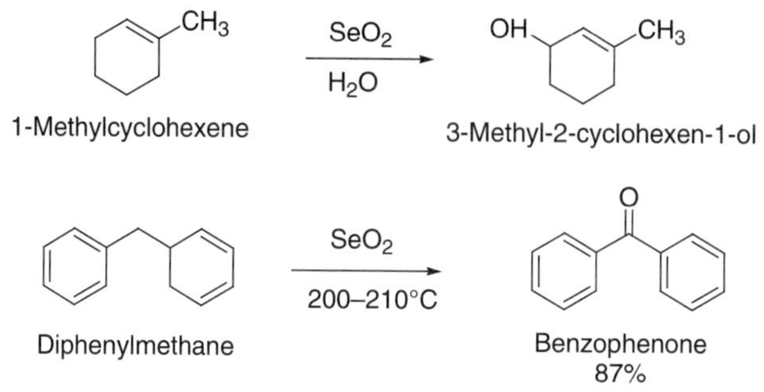

1-Methylcyclohexene SeO_2 / H_2O 3-Methyl-2-cyclohexen-1-ol

Diphenylmethane SeO_2 / 200–210°C   Benzophenone 87%

The mechanism of SeO_2 oxidation involves the formation of seleninic acid **A**. The [2,3]-sigmatropic rearrangement of **A** gives **B**, which is then converted into allylic alcohol and Se(II) by-products (Scheme 7.33).

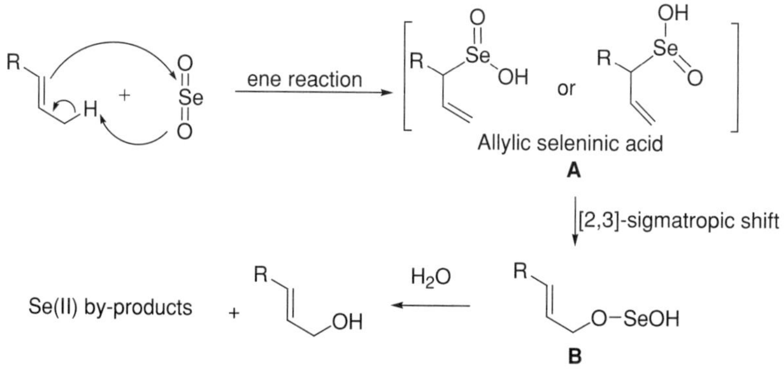

Scheme 7.33

7.11   Oxidation of sulfides

Asymmetric oxidation of a sulfide to a chiral sulfoxide involves the use of a chiral ligand with a transition metal, such as titanium, vanadium or manganese, in the presence of a hydrogen peroxide or a hydrogen peroxide adduct as the oxygen source. The chiral ligands that have been successfully used include bidentate diethyl tartrate (**7.45** and **7.46**), diol **7.60**, BINOL (**7.61**), tridentate Schiff's base ligands **7.62** and tetradentate salen-type ligands **7.63**.

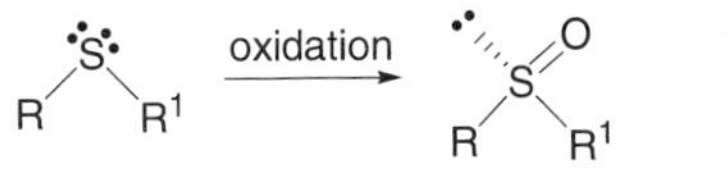

Sulfide → Chiral sulfoxide

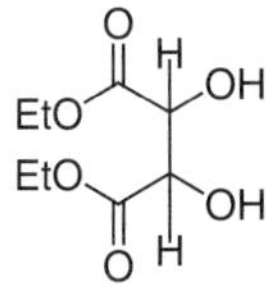

7.45 and 7.46 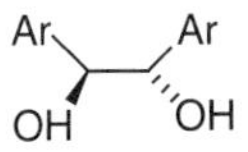   **7.60** 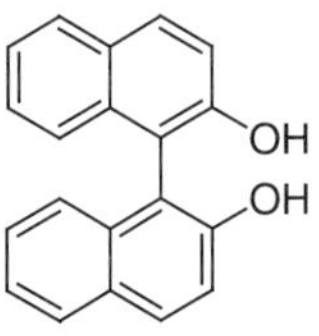   **7.61**

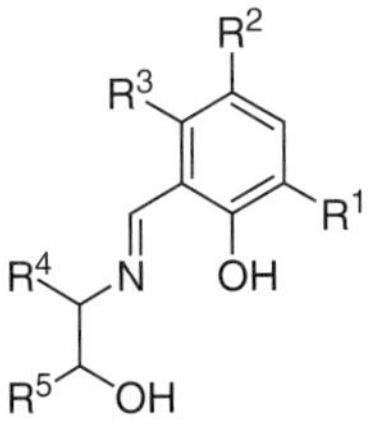

7.62 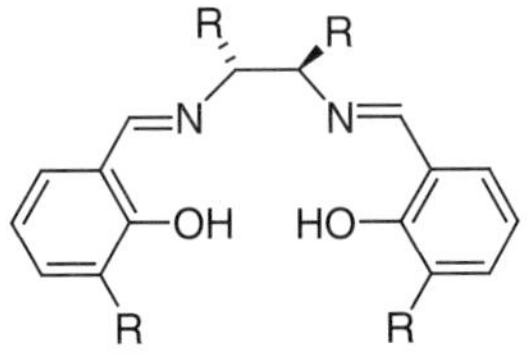   **7.63**

Vanadium(IV) Schiff's base complexes[65–68] derived from β-aminoalcohols **7.62** and **vanadyl acetylacetonate** have been used to oxidize different substrates to chiral sulfoxides.

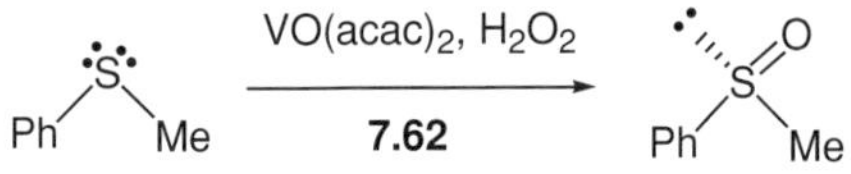

The intermediate formed in this oxidation is 2:1 complex of Schiff's base ligand **7.62** and vanadium. This intermediate then reacts with hydrogen peroxide, eliminating one of the ligands to give a vanadium hydroperoxide complex, which then oxidizes the sulfide to sulfoxide.

Environmentally benign oxidation of sulfides to sulfoxides was reported by Kita and co-workers[69] by using iodine(III) reagents such as iodosobenzene (PhIO) or phenyliodine diacetate (PIDA) with KBr in water.

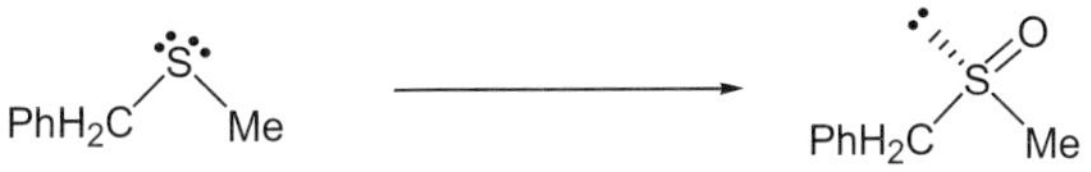

PhIO–KBr (1.5 equiv.: 1.0 equiv.)–H$_2$O = 95%
PIDA–KBr (1.1 equiv.: 0.1 equiv.)–H$_2$O = 85%

However, when recyclable poly(diacetoxyiodo)styrene in water was used for the oxidation of sulfides, the corresponding sulfoxides were obtained in excellent yields without the formation of sulfones[69].

7.12 Oxidation of aliphatic side chains attached to aromatic ring

It is very easy to oxidize aliphatic side chains attached to aromatic rings than the side ring themselves. Alkaline KMnO$_4$ and dichromate convert the alkyl group attached to the aromatic ring to the carboxylic group.

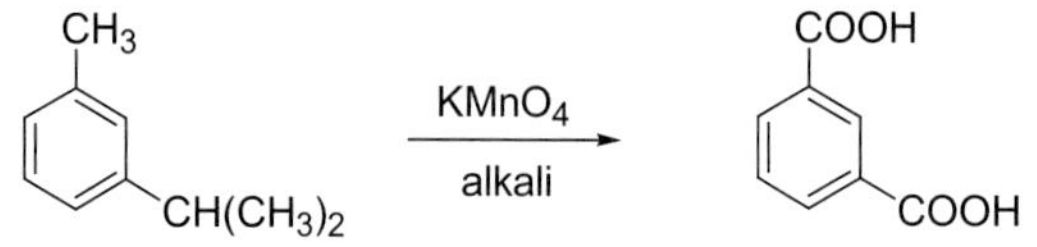

3-Isopropylmethylbenzene   Isophthalic acid

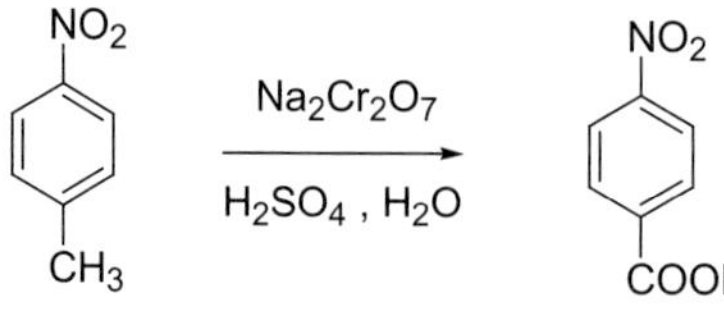

4-Nitrotoluene   4-Nitrobenzoic acid

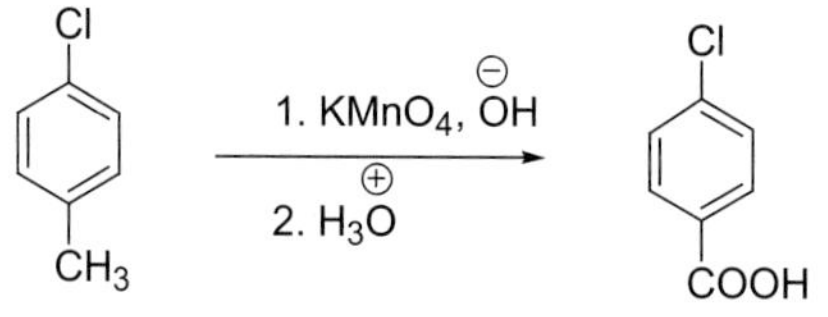

4-Chlorotoluene   4-Chlorobenzoic acd

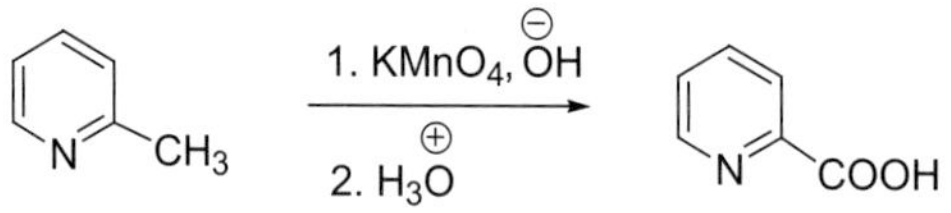

2-Methylpyridine   2-Pyridinecarboxylic acid
                                                        50–51%

Mechanism:[70] The mechanism of the aliphatic side-chain oxidation as shown in Scheme 7.34 probably involves manganese–oxygen–carbon bonding with subsequent reduction of manganese and oxidation of carbon to yield aldehyde, which is much more vulnerable to oxidation with permanganate.

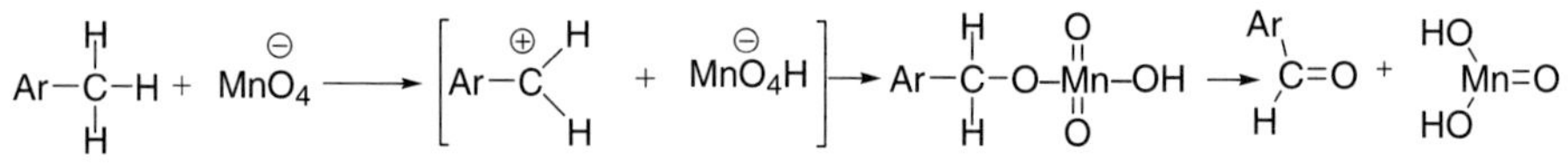

Scheme 7.34

Chromyl chloride converts the aromatic methyl group into an aldehyde; this reaction is known as the **Etard reaction**[71]. For example, toluene can be oxidized to benzaldehyde.

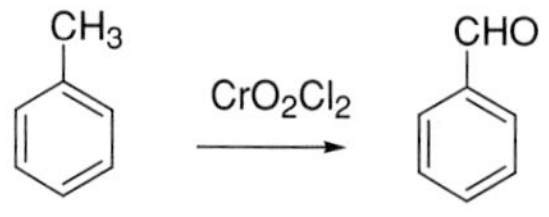

Mechanism: The mechanism of the Etard reaction involves the formation of dichromium species, which on hydrolysis gives an aldehyde (Scheme 7.35).

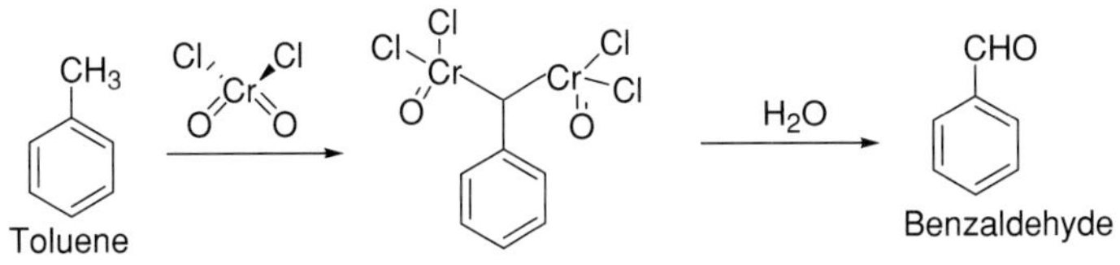

Scheme 7.35

References

1. Bowden, K., Heilbron, I. M., Jones, E. R. H. and Weedon, B. C. L., *J. Chem. Soc.*, **1946**, 39.
2. Ali, M. H. and Wiggin, C., *Synth. Commun.*, **2001**, *31*(9), 1389, and references therein.
3. Zibuck, R. and Streiber, J. M., *Org. Synth. Coll.*, **1998**, *9*, 432.
4. Zibuck, R. and Streiber, J. M., *Org. Synth. Coll.*, **1993**, *71*, 236.
5. Poos, G. I., Arth, G. E., Beyler, R. E. and Sarett, L. H., *J. Am. Chem. Soc.*, **1953**, *75*, 422.
6. Collins, J. C., Hess, W. W. and Frank, F. J., *Tetrahedron Lett.*, **1968**, *9*(30), 3363.
7. Corey, E. J. and Suggs, J. W., *Tetrahedron Lett.*, **1975**, *16*(31), 2647.
8. Coates, W. M. and Corrigan, J. R. *Chem. Ind.*, **1969**, *54*, 1594.
9. Corey, E. J. and Schmidt, G., *Tetrahedron Lett.*, **1979**, *20*(5), 399.
10. Dauben, W. G., *J. Org. Chem.*, **1977**, *42*, 682.
11a. Goldman, I. M., *J. Org. Chem.*, **1969**, *34*, 3289.
11b. Kitamura, M., Tokunaga, M. and Noyori, R. *J. Am. Chem. Soc.* **1995**, *117*, 2931.
12. Quesada, E. and Taylor, R. J. K., *Tetrahedron Lett.*, **2005**, *46*, 6473.
13. Hall, T. K. and Story, P. R., *J. Am. Chem. Soc.*, **1967**, *89*, 6759.
14. Bolm, C., Magnus, A. S. and Hildebrand, J. P., *Org. Lett.*, **2000**, *2*, 1173.
15. Pfitzner, K. E. and Moffatt, J. G., *J. Am. Chem. Soc.*, **1963**, *85*, 3027.
16. Pfitzner, K. E. and Moffatt J. G., *J. Am. Chem. Soc.*, **1965**, *87*, 5661.
17. Fenselau, A. H. and Moffatt J. G., *J. Am. Chem. Soc.*, **1966**, *88*, 1762.
18. Mancuso, A. J. and Swern, D., *Synthesis*, **1981**, 165.
19. Omura, K. and Swern, D., *Tetrahedron*, **1978**, *34*, 1651.
20. Corey, E. J. and Kim, C. U., *Tetrahedron Lett.*, **1974**, *15*(3), 287.
21. Corey, E. J. and Kim, C. U., *J. Am. Chem. Soc.*, **1972**, *94*(21), 7586.
22. Crich, D. and Neelamkavil, S., *Tetrahedron*, **2002**, *58*, 3865.
23. Ohsugi, S., Nishidea, K., Oonob, K., Okuyamab, K., Fudesakaa, M., Kodamaa, S. and Node, M., *Tetrahedron*, **2003**, *59*, 8393.
24. Li, C., Xu, Y., Lu, M., Zhao, Z., Liu, L., Zhao, Z., Cui, Y., Zheng, P., Ji, X. and Gao, G., *Synlett*, **2002**, 2041.
25. Dess, D. B. and Martin, J. C., *J. Am. Chem. Soc.*, **1991**, *113*(19), 7277.
26. Nicolaou, K. C., Zhong, Y.-L. and Baran, P. S., *J. Am. Chem. Soc.*, **2000**, *122*, 7596.
27. Hadfield, J. A., McGown, A. T. and Butler, J., *Molecules*, **2000**, *5*, 82.
28. Lenz, R. and Ley, S. V., *J. Chem. Soc., Perkin Trans. 1*, **1997**, 3291.
29. Ley, S. V., Norman, J., Griffith, W. P. and Marsden, S. P., *Synthesis*, **1994**, 639.
30. Kubias, J., *Collect. Czech. Chem. Commun.*, **1966**, *31*, 1666.
31. Fétizon, M. and Golfier, M. C. R., *Acad. Sci. Ser. C*, **1968**, *267*, 900.
32. Fétizon, M., Golfier, M., Morgues, P., *Tetrahedron Lett.*, **1972**, 4445.
33. Campaigne, E. and Lesuer, W. M., *Org. Synth. Coll.*, **1963**, *4*, 919.
34. Corey, E. J., Gilman, N. W. and Ganem, B. E., *J. Am. Chem. Soc.*, **1968**, *90*, 5616.
35. Riley, H. L., *J. Chem. Soc.*, **1932**, 1875.
36. Sharpless, K. B., *J. Am. Chem. Soc.*, **1976**, *98*, 300.
37. Baldwin, J. E., Adlington, R. M. and Ramcharitar, S. H., *Tetrahedron*, **1992**, *48*(14), 2957.
38. Krow, G. R., *Tetrahedron*, **1981**, *37*, 2697.
39. Bolm, C., Shingloff, G. and Weickhardt, K., *Angew. Chem., Int. Ed. Engl.*, **1994**, *33*, 1848.
40. Gusso, A., Baccin, C., Pinna, F. and Strukul, G., *Organometallics*, **1994**, *13*, 3442.
41. Woodward, S. S., Finn, M. G. and Sharpless, K. B., *J. Am. Chem. Soc.*, **1991**, *113*, 106.
42. Finn, M. G. and Sharpless, K. B., *J. Am. Chem. Soc.*, **1991**, *113*, 113.
43. Lopp, M., Paju, A., Kanger, T. and Pehk, T., *Tetrahedron Lett.*, **1996**, *37*(42), 7583.
44. Kanger, T., Kriis, K., Paju, A., Pehk, T. and Lopp, M., *Tetrahedron: Asymmetry*, **1998**, *9*(24), 4475.
45. Teuber, H. J. and Rau, W., *Chem. Ber.*, **1953**, *86*, 1036.
46. Zimmer, H., Lankin, D. C. and Horgan, S. W., *Chem. Rev.*, **1971**, *71*(2), 229.

47. Teuber, H. J., *Org. Synth. Coll.*, **1988**, *6*, 480.
48. Teuber, H. J., *Org. Synth. Coll.*, **1972**, *52*, 88.
49. Elbs, K. J., *Prakt. Chem.*, **1893**, *48*, 179.
50. Behrman, E. J., *Org. React.*, **1988**, *35*, 421.
51. Kang, B., Kim, M., Lee, J., Do, Y. and Chang, S., *J. Org. Chem.*, **2006**, *71*(18), 6721.
52. Porter, M. J. and Skidmore, J., *Chem. Commun.*, **2000**, *14*, 1215.
53. Zhang, W., Loebach, J. L., Wilson, S. R. and Jacobsen, E. N., *J. Am. Chem. Soc.*, **1990**, *112*(7), 2801.
54. Jacobsen, E. N., Zhang, W., Muci, A. R., Ecker, J. R. and Deng, L., *J. Am. Chem. Soc.*, **1991**, *113*, 7063.
55. Stoop, R. M., Ph.D. Dissertation, ETH to Swiss Federal Institute of Technology Zurich; 2000.
56. Colladon, M., Scarso, A., Sgarbossa, P., Michelin, R. A. and Strukul, G., *J. Am. Chem. Soc.*, **2006**, *128*, 14006.
57. Wang, Z. X., Tu, Y., Frohn, M., Zhang, J. R. and Shi, Y., *J. Am. Chem. Soc.*, **1997**, *119*(46), 1224.
58. Frohn, M. and Shi, Y., *Synthesis*, **2000**, *14*, 1979.
59. Junttila, M. H. and Hormi, O. E. O., *J. Org. Chem.*, **2004**, *69*, 4816.
60. Criegee, R., *Angew. Chem., Int. Ed.*, **1958**, *70*, 173.
61. Corey, E. J. and Ensley, H. E., *J. Am. Chem. Soc.*, **1975**, *97*, 6908.
62. Henry, J. R. and Weinreb, S. M., *J. Org. Chem.*, **1993**, *58*, 4745.
63. Boyland, E. and Sims, P., *J. Chem. Soc.*, **1954**, 980.
64. Behrman, E. J., *J. Org. Chem.*, **1992**, *57*, 2266.
65. Bolm, C. and Bienewald, F., *Angew. Chem., Int. Ed.*, **1995**, *34*, 2640.
66. Tang, T. P., Volkman, S. K. and Ellman, J. A., *J. Org. Chem.*, **2001**, *66*, 8772.
67. Yuste, F., Ortiz, B., Carrasco, A., Peralta, M., Quintero, L., Sanchez-Obregon, R., Walls, F. and Ruano, J. L. G., *Tetrahedron: Asymmetry*, **2000**, *11*, 3097.
68. Gama, A., Flores-Lopez, L. Z., Aguirre, G., Parra-Hake, M., Hellberg, L. H. and Somanathan, R., *ARKIVOC*, **2003**, *11*, 4.
69. Tohma, H., Maegawa, T. and Kita, Y., *ARKIVOC*, **2003** 6, 62.
70. Ladbury, J. W. L. and Cullis, C. F., *Chem. Rev.*, **1958**, 403.; Waters, W. A., *Rev. Chem. Soc.*, **1958**, *12*, 277.
71. Etard, A. L., *Compt. Rend*, **1880**, *90*, 534.

Chapter 8
Pericyclic Reactions

Pericyclic reactions are a class of reactions that include some of the most powerful synthetically useful reactions such as the **Diels–Alder reaction**. Pericyclic reactions often proceed with simultaneous reorganization of bonding electron pairs and involve a cyclic delocalized transition state. They differ from ionic or free radical reactions as there are no ionic or free radical intermediates formed during the course of the reaction. They proceed by one-step concerted mechanisms and have certain characteristic properties (although there are some exceptions to all these rules).

1. Pericyclic reactions often proceed with a high degree of stereospecificity.
2. Although some pericyclic reactions occur spontaneously, most reactions can be frequently promoted by light as well as heat. Normally, the stereochemistry under the two sets of conditions is different. Thus, there may be two main reaction conditions, thermal (in ground state) and photochemical (in excited state).
3. Pericyclic reactions are relatively unaffected by solvent changes and can occur in the gas phase with no solvent. Normally, they are unaffected by the presence of electrophilic and nucleophilic catalysts.
4. Normally, no catalyst is needed to promote the reactions. But Lewis acids may catalyze many forms of pericyclic reactions, either directly or by changing the mechanism of the reaction so that it becomes a stepwise process and hence no longer a true pericyclic reaction.

8.1 Important classes of pericyclic reactions

There are four major classes of pericyclic reactions: cycloaddition, electrocyclic, sigmatropic and ene reactions. All these reactions are potentially reversible. A general illustration of each class is given below.

8.1.1 Cycloaddition reactions

A cycloaddition reaction involves the concerted formation of two σ-bonds between the termini of two π-systems. The reverse reaction involves the concerted cleavage of two σ-bonds to produce two π-systems. The simplest example being the hypothetical combination of two ethene molecules to give cyclobutane. This does not occur under normal heating, but the cycloaddition of 1,3-butadiene to ethene does, and this is an example of the **Diels–Alder reaction**.

8.1.2 Electrocyclic reactions

An electrocyclic reaction involves the concerted formation of a σ-bond between the two ends of a conjugated π-system (**ring forming process**), or the reverse reaction (**ring opening process**) in which the σ-bond is broken to produce a conjugated system.

8.1.3 Sigmatropic rearrangements

A sigmatropic rearrangement involves the concerted migration of an atom or group from one point of attachment to a conjugated system to another point of attachment, during which one σ-bond is broken and another σ-bond is formed. A sigmatropic rearrangement can be classified according to the length of the group that migrates, and the length of the backbone along which it migrates. Thus, sigmatropic rearrangements are given by two numbers in brackets [i,j].

[1,5]-Sigmatropic rearrangement   [3,3]-Sigmatropic rearrangement

8.1.4 Ene reactions

An ene reaction involves the formation and cleavage of unequal numbers of σ-bonds in a concerted cyclic transition state.

8.1.5 Other classes of pericyclic reactions

Many diverse reactions of synthetic interest could be classified as pericyclic reactions.

Cheletropic reactions

A reaction in which formation of two σ-bonds occurs at a single atom (monocentric) is called a cheletropic reaction. The addition[1-4] of SO_2 to diene is a well-known example of cheletropic reaction.

The reverse of this process is known as **cheletropic extrusion (or cheletropic elimination)**. The cheletropic elimination of nitrogen from diazenes **8.1** and **8.2** is a stereospecific reaction (for definition of a stereospecific reaction, see section 1.5).

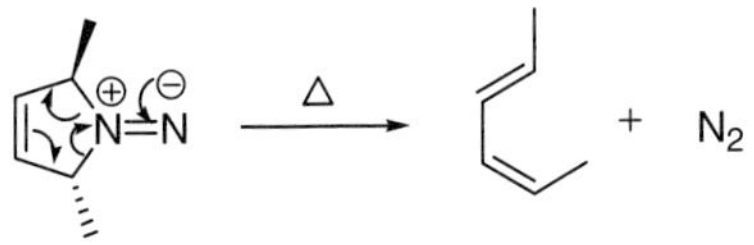

8.1

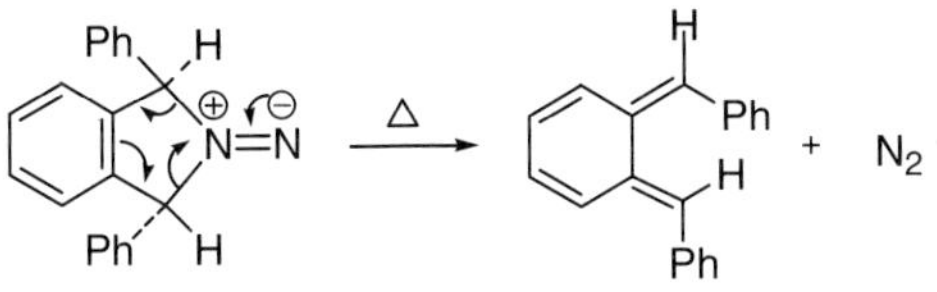

8.2

Decomposition of chromate ester

Chromium(VI) oxidations of alcohols involve the decomposition of chromate ester **8.3**, which proceeds via cyclic transition states (see Scheme 7.1).

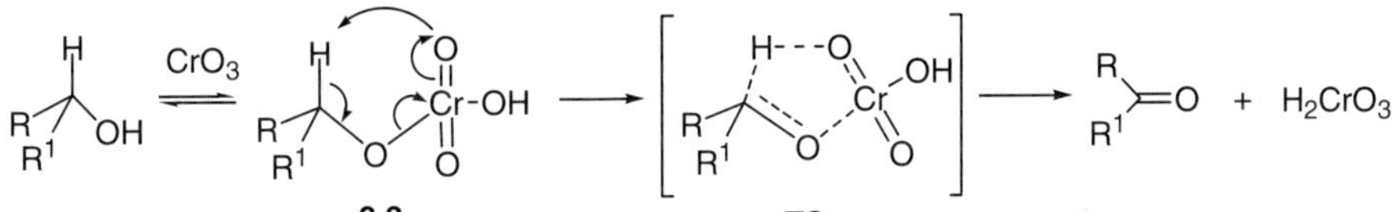

8.3   **TS**

Co-arctic reactions

A co-arctic reaction involves the formation or cleavage of four σ-bonds at a single atomic centre.

In-between class of pericyclic reactions

In some reactions, a cyclopropane ring acts as an alkene and participates in pericyclic reactions.

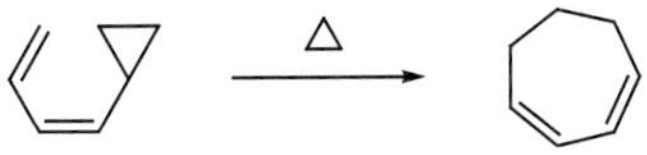

Pseudopericyclic reactions

In a pseudopericyclic reaction[5,6] there is no continuous orbital overlap around the ring of breaking and forming bonds. All pseudopericyclic reactions are allowed; there are no anti-aromatic transition states.

8.2   Theoretical explanation of pericyclic reactions

Although Otto Diels and Kurt Alder won the 1950 Nobel Prize in Chemistry for the Diels–Alder reaction, almost 20 years later R. Hoffmann and R. B. Woodward[7] gave the explanation of this reaction. They published a classical textbook, *The Conservation of Orbital Symmetry*. K. Fukui[8] (the co-recipient with R. Hoffmann of the 1981 Nobel Prize in Chemistry) gave the '**Frontier molecular orbital**' (FMO) theory, which also explains pericyclic reactions. Both theories allow us to predict the conditions under which a pericyclic reaction will occur and what the stereochemical outcome will be. Between these two fundamental approaches to pericyclic reactions, the FMO approach is simpler because it is based on a pictorial approach. Another method similar to the FMO approach of analyzing pericyclic reactions is the '**transition state aromaticity**' approach.

All these theories are based on the molecular orbital (MO) theory.

8.2.1   MOs and their symmetry properties

Since pericyclic reactions proceed with a cyclic reorganization of bonding electron pairs, it is necessary to evaluate changes in the associated MOs that take place in going from reactants to products. These orbitals may be classified by two independent symmetry operations: a mirror plane (m) perpendicular to the functional plane and bisecting the molecule, and a twofold axis of rotation (C_2).

Let us first define the symmetry properties of $1s$- and $2p$-orbitals (Fig. 8.1) and MOs formed by the overlap of two or more atomic orbitals with respect to a plane of symmetry (m) or an axis of symmetry (C_2) (Figs. 8.2–8.4).

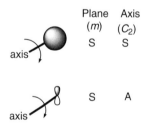

Figure 8.1   Symmetry properties of $1s$- and $2p$-orbitals.

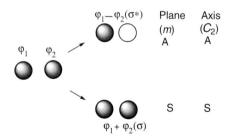

Figure 8.2   Symmetry properties of MOs formed by the overlap of two s-orbitals.

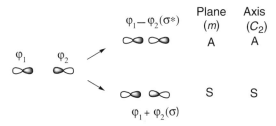

Figure 8.3 Symmetry properties of MOs formed by the head-on overlap of two *p*-orbitals.

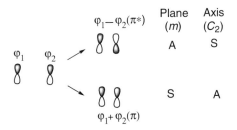

Figure 8.4 Symmetry properties of MOs formed by the side-on overlap of two *p*-orbitals.

MOs of linear polyenes and their symmetry properties

The spatial distribution of electron density for most occupied MOs is discontinuous, with regions of high density separated by regions of zero density (e.g. a **nodal plane**). The wave functions that describe MOs undergo a change in phase (or sign) at nodal surfaces. This phase change is sometimes designated by plus and minus signs associated with discrete regions of the orbital, but this notation may sometimes be confused for an electric charge. In this book, regions having one phase sign are shown as dark regions, while those having an opposite sign are blank.

Schematic, qualitatively correct π-MOs can be easily generated from a linear array of n *p*-atomic orbitals by following a few simple rules:

1. The lowest energy MO always has all *p* orbitals in phase, making it symmetric (S) with respect to **end-for-end reflection**. This MO has no vertical nodes.

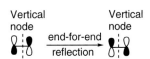

2. The next MO, which has one nodal plane (a single vertical node), is antisymmetric (A) with respect to end-for-end reflection. An example is the π^*-MO of ethylene.

3. The third MO has two vertical nodes, and is symmetric (S) with respect to end-for-end reflection.

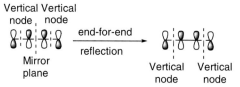

4. The MOs continue to add vertical nodes until the highest energy MO is reached, which has a vertical node between each atom. When combining n p-orbitals, the highest energy π^*-MO will have $n - 1$ vertical nodes.

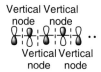

5. As a rule, higher energy MOs have a larger number of nodal surfaces or nodes.
6. A MO is bonding if the number of bonding interactions is greater than the number of nodes between the nuclei, and a MO is antibonding if the number of bonding interactions is less than the number of nodes between the nuclei.

MOs of ethene: The carbon atoms in ethene are sp^2 hybridized. The double bond between two carbon atoms comprises a σ- and a π-bond. The carbon–carbon σ-bond is formed by the head-on overlap of two sp^2-orbitals. The overlapping results in two MOs: a bonding (σ) and an antibonding (σ^*). The π-bond is formed by the side-on overlap of p-orbitals. The bonding orbital is formed by the overlap of in-phase p-orbitals, and the antibonding orbital arises from the interference between two p-orbitals of opposite phases and has a node (or a region of minimum electron density between the nuclei). These orbitals are designated as π and π^*, respectively (Fig. 8.5).

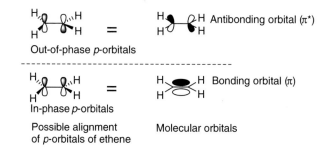

Figure 8.5

The comparison of the energies of σ-, σ^*-, π- and π^*-orbitals of ethene is shown in Fig. 8.6. The bonding electrons are placed in the two orbitals with lowest energies, σ and π, in the ground state of ethene. The π-orbital is the **highest occupied molecular orbital (HOMO)** and π^*-orbital is the **lowest unoccupied molecular orbital (LUMO)**. Both HOMO and LUMO are referred to as **frontier orbitals** (see 'FMO theory') and are used in analyzing pericyclic reactions.

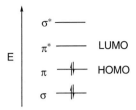

Figure 8.6 Energies of σ-, σ*-, π- and π*-orbitals of ethene.

MOs of butadiene: In butadiene, four p-orbitals combine to give four π-MOs having wave functions ψ_1, ψ_2, ψ_3 and ψ_4 with different energies. Of these, ψ_1 and ψ_2 MOs are bonding orbitals and ψ_3 and ψ_4 are antibonding. These orbitals are arranged in the order of increasing energy, as shown in Fig. 8.7.

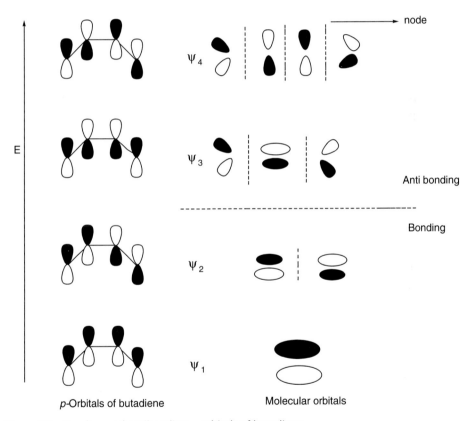

Figure 8.7 Bonding and antibonding π-orbitals of butadiene.

Of the four MOs in the ground state, two, namely ψ_1 and ψ_2, are of lowest energy; ψ_2 is the **HOMO** and ψ_3 is the **LUMO** (see 'FMO theory'). On absorption of a photon of proper wavelength (in the UV range), an electron is promoted from ψ_2 to ψ_3 ($\psi_2 \rightarrow \psi_3$), which then becomes a new HOMO (Fig. 8.8).

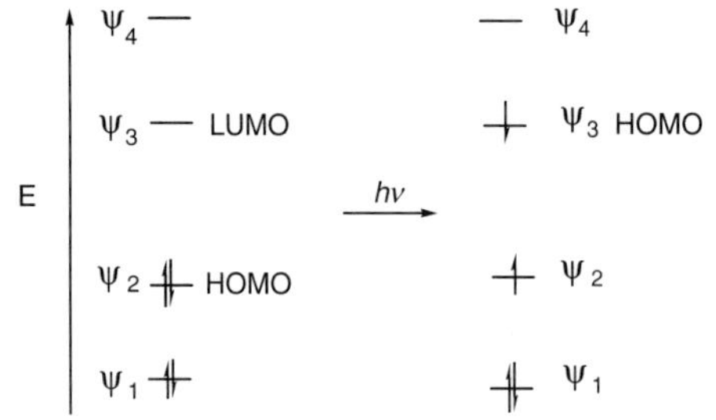

Figure 8.8 Energies of ψ_1-, ψ_2-, ψ_3- and ψ_4-orbitals of butadiene.

MOs of 1,3,5-hexatriene: Six *p*-orbitals of 1,3,5-hexatriene overlap to give six π-MOs (ψ_1 to ψ_6). The six π-electrons of 1,3,5-hexatriene are placed in the first three π-MOs (ψ_1, ψ_2 and ψ_3), which are bonding orbitals. The remaining three higher energy π^*-MOs (ψ_3, ψ_4 and ψ_6) are antibonding orbitals, which remain unoccupied in the ground state (Fig. 8.9).

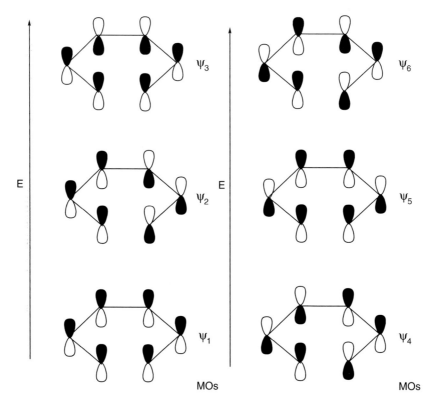

Figure 8.9 Six π-MOs (ψ_1 to ψ_6) are formed by the overlapping of six *p*-orbitals of 1,3,5-hexatriene.

Symmetry properties of reactant and product orbitals
Various MOs of reactants and products are classified according to two independent symmetry operations, either a plane of symmetry (*m*) or a twofold axis of symmetry (C_2). For

example, the symmetry properties of π-orbitals (bonding and antibonding) of ethene are classified in terms of whether they are symmetric (S) or antisymmetric (A) with respect to both the m and the C_2 axis (Fig. 8.10).

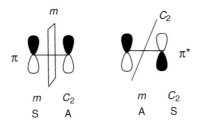

Figure 8.10 Symmetry properties of π- and π^*-MOs of ethene.

The symmetry properties of the π-MOs of the butadiene and a few σ- and π-MOs of cyclobutene are described in terms of whether they are symmetric (S) or antisymmetric (A) with respect to both the mirror plane m and the C_2 axis in Fig. 8.11.

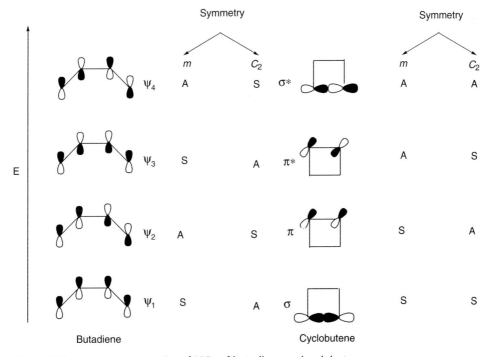

Figure 8.11 Symmetry properties of **MOs** of butadiene and cyclobutene.

Symmetry properties of all the π-MOs of hexatriene (ψ_1 to ψ_6) and a few σ- and π-MOs of cyclohexadiene are described in Fig. 8.12.

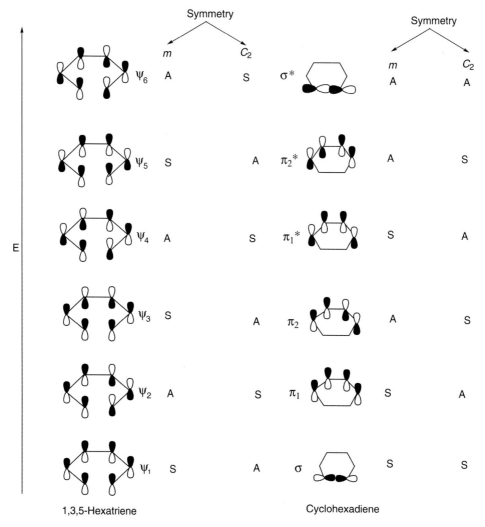

Figure 8.12   Symmetry properties of MOs of 1,3,5-hexatriene and cyclohexadiene.

Such symmetry characteristics play an important role in creating the orbital diagrams used by Woodward and Hoffmann to rationalize pericyclic reactions.

8.2.2   Suprafacial and antarafacial

A *suprafacial* process is one in which the bonds made or broken lie on the same face of the system undergoing reaction. When the newly formed or broken bonds lie on the opposite faces of the reacting systems, it is known as an *antarafacial* process.

In the course of a pericyclic cycloaddition, the interacting terminal lobes of each component may overlap either in a *suprafacial* mode or in an *antarafacial* mode. If both the new bonds form from the same face of the molecule it is known as a *suprafacial* mode (also known as *supra–supra*). It is *antarafacial* if one bond forms from one surface and the other bond forms from the other surface (also known as *supra–antara*) (Fig. 8.13).

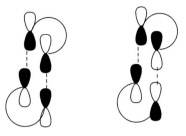

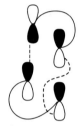

Figure 8.13 *Suprafacial* and *antarafacial* modes in a cycloaddition reaction.

The bracketed numbers that designate reactions of this kind sometimes carry subscripts (s or a) that specify their configuration. Thus, the Diels–Alder reaction (section 8.3.1) may be termed as $[4_s+2_s]$ process under thermal conditions.

In **electrocyclic reactions**, the *suprafacial* mode involves each component of the new σ-bond being formed from the same face of the reactant π-system (this is equivalent to *disrotation*). The *antarafacial* mode involves twisting of the orbitals so that the two components of the new σ-bond form from the opposite face of the reactant π-system (this is equivalent to *conrotation*) (Fig. 8.14).

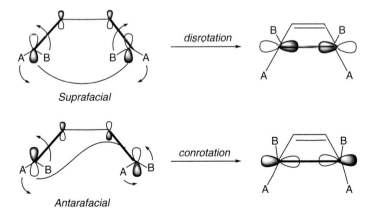

Figure 8.14 *Suprafacial (disrotation)* and *antarafacial (conrotation)* modes in an electrocyclic reaction.

In the course of sigmatropic rearrangement if the migrating group remains on the same face of the π-system, the rearrangement is *suprafacial*. If the migrating group moves to the opposite face of the π-system, the rearrangement is *antarafacial* (Fig. 8.15).

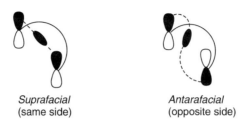

Suprafacial   Antarafacial
(same side)   (opposite side)

Figure 8.15   *Suprafacial* and *antarafacial* modes in a sigmatropic reaction.

8.2.3   Conservation of orbital symmetry

According to orbital symmetry theory, in any concerted process the reactant orbitals must be transformed into product orbitals of the same symmetry. Thus, the symmetry of the orbitals of the reactants must be conserved as they are transformed into the orbitals of the product.

The original explanation of Woodward and Hoffmann involved construction of an orbital correlation diagram for the reaction under consideration, and then carrying out the reaction in such a manner so that the symmetries of the reactant and product orbitals matched exactly. If the correlation diagram indicates that the reaction may occur without encountering a symmetry-imposed barrier, it is termed **symmetry-allowed**. If a symmetry is present, the reaction is designated **symmetry-forbidden**.

Before we move on, it is worthwhile to clarify the implications of the words *allowed* and *forbidden*. An allowed reaction is simply one with a low activation energy relative to some other pathway, while a forbidden reaction is a process for which there is a significant activation energy.

Orbital correlation diagrams

The term *orbital correlation diagram* describes the theoretical device that Woodward and Hoffmann developed to interpret pericyclic reactions. The Woodward–Hoffmann method for correlating reactant orbitals with product orbitals includes the following:

1. Select an appropriate symmetry element, e.g. mirror plane (m) or rotation axis (C_2), which passes through at least one bond that is breaking or forming in order to give useful information.
2. Classify the reactant orbitals and the product orbitals with respect to this element: S = symmetric, A = antisymmetric.
3. Create an energy level diagram that connects orbitals of like symmetry.
4. The reaction will be thermally allowed if there is no crossing to antibonding levels, whereas it will be photochemically allowed if there is crossing to antibonding levels.

Such an approach is not readily applicable to the majority of more complex reactions.

FMO theory

The FMO approach involves using the principles of quantum mechanics to generate the **HOMO** and the **LUMO** for the reactant(s). Then the interaction(s) of the FMO(s) determines whether the reactions are thermally and photochemically allowed or forbidden. The FMOs for the ethene, butadiene and hexatriene are shown in Fig. 8.16.

HOMO and LUMO of ethene

HOMO and LUMO of butadiene

HOMO and LUMO of hexatriene

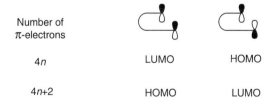

Number of π-electrons		
4n	LUMO	HOMO
4n+2	HOMO	LUMO

A general pattern for HOMO and LUMO orbitals

Figure 8.16 Frontier molecular orbitals (HOMO and LUMO) of a reactant molecule.

The bonding or antibonding interaction(s) of the FMO(s) determine whether the reactions are thermally and photochemically allowed or forbidden.

The formation of the new σ-bond(s) must occur by an appropriate overlap of the same phases of these orbitals. If only one σ-bond is forming, as in electrocyclic reactions, then only the overlap of the HOMO of the open chain reactant is considered. Such an overlap can occur in one of the two fundamental ways: *suprafacial* mode or *antarafacial* mode (see Fig. 8.14). If two or more σ-bonds are formed during the reaction, as in cycloaddition reactions, then the overlap of the HOMO of one reactant with the LUMO of the second reactant must be considered (see section 8.3).

For simple systems, the form of the HOMO and LUMO is not difficult to remember. For more complex systems, explicit calculations have to be made and the FMO method becomes more difficult to apply.

The advantage of the FMO method is that it can be expressed quantitatively in terms of the magnitude of the coefficients involved, and hence can be used to predict regioselectivity (see Fig. 8.33).

Transition state aromaticity (Hückel and Möbius topologies)

Zimmerman[9–15] gave the idea that the favoured transition states for pericyclic reactions are those containing an aromatic number of electrons. What constitutes an aromatic number of electrons depends on the topology of the system of orbitals.

Hückel aromatics: The benzene molecule has the *suprafacial* topology; this means that the π-electron density in benzene is continuous along the top or bottom face of the molecule. If the transition state for the pericyclic reaction has the same topology, it is said to resemble Hückel topology (Fig. 8.17).

Figure 8.17 Hückel topology.

Suprafacial or Hückel transition state in a pericyclic reaction is associated with a plane of symmetry and is particularly favourable if the number of cyclically conjugated π-electrons in the transition state equals $[4n + 2]$ (the Hückel rule, where $n = 0, 1, 2, \ldots$).

Hückel was also able to show that if a cyclic conjugated π-system is irradiated with light so that it goes into the first excited singlet or triplet electronic state, it is especially stable if the number of cyclically conjugated electrons equals $[4n]$. Hence, **photochemically** activated pericyclic reactions will proceed *suprafacially* via a Hückel transition state if the electron count corresponds to $[4n]$.

Möbius aromatics: The *antarafacial* mode is said to resemble Möbius topology (after August Ferdinand Möbius). An *antarafacial* mode can be formed by taking a cyclic alkene 'strip' and giving the π-system a 180° twist, and in doing so for a molecule a twofold axis of symmetry is created in the resulting object. A Möbius aromatic system has an extra node, introduced by twisting the set of orbitals so that each one forms an angle, θ, with its neighbours (Fig. 8.18).

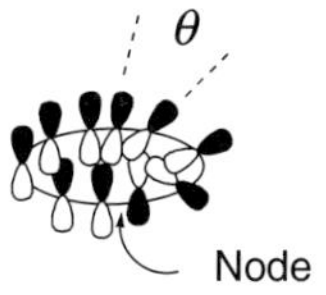

Figure 8.18 Mobius topology.

Edgar Heilbronner in 1964 worked out that such a twisted system (Möbius system) would be aromatic if it contained $4n$ conjugated π-electrons. Curiously, no actual examples were identified until relatively recently, when it was proposed that [12], [16] and [20] annulenes have several higher energy conformations which adopt this mode. In 2003, the first crystal structure of a true stable Möbius [16] annulene was completed, and various heteroannulenes were identified as existing in Möbius form.

Just as with excited state Hückel aromatics, Möbius molecules in the excited singlet or triplet state are thought to be aromatic if they contain $[4n + 2]$ rather than $[4n]$ π-electrons.

The example of Möbius aromatic is $C_8F_8^{2-}$ (perfluorocyclo-octatetraene dianion, a $[4n + 2]$ π-electrons system) in triplet state.

In terms of transition states, an allowed reaction proceeds via an aromatic transition state, while a forbidden reaction does not occur because the transition state would be anti-aromatic.

A summary of selection rules based on Hückel and Möbius topologies is given below:

1. A **thermally** activated pericyclic reaction will proceed via a **Hückel topology** containing only *suprafacial* components if the cyclically conjugated π-electrons equal $[4n + 2]$ ($n = 0, 1, 2, \ldots$).
2. A **photochemically** activated pericyclic reaction may proceed via a **Hückel topology** containing only *suprafacial* components if the cyclically conjugated π-electrons equal $[4n]$ ($n = 0, 1, 2, \ldots$).
3. A **thermally** activated pericyclic reaction will proceed via a **Möbius topology** containing one *antarafacial* component if the cyclically conjugated π-electrons equal $[4n]$ ($n = 0, 1, 2, \ldots$).
4. A **photochemically** activated pericyclic reaction will proceed via a **Möbius topology** containing one *antarafacial* component if the cyclically conjugated π-electrons equal $[4n + 2]$ ($n = 0, 1, 2, \ldots$).

8.3 Cycloaddition reactions

Cycloaddition reactions are a very important class of pericyclic reactions in which two unsaturated molecules join by converting two π-bonds into two new σ-bonds between their termini. Although cycloaddition reactions are concerted (no intermediate species are formed), the two new bonds in a few cases may be formed in an asynchronous fashion. Depending on partial charge distribution in both reactants, the formation of one bond may lead to the development of the other.

8.3.1   [4+2]-Cycloaddition reactions

A classical example of cycloaddition reaction is the **Diels–Alder reaction**[16–19], which allowed the formation of six-membered rings. An example of the Diels–Alder reaction is the reaction between 1,3-butadiene and ethene to give cyclohexene. The 1,3-butadiene is a conjugated π-system with $[4\pi]$ electrons and the ethene is a $[2\pi]$-electrons system. Thus, the Diels–Alder reaction between 1,3-butadiene and ethene to give cyclohexene is described as a [4+2]-cycloaddition reaction.

Butadiene   Ethene   Cyclohexene
(a diene)   (a dienophile)

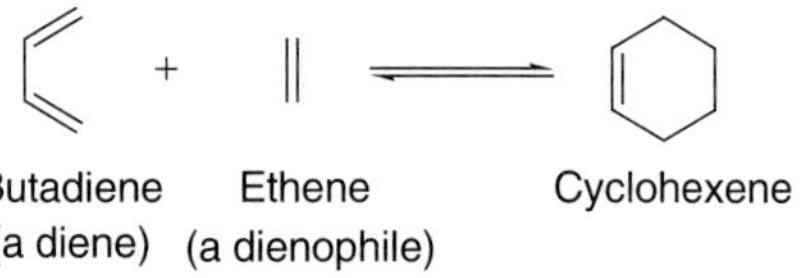

In the Diels–Alder reaction, the $[4\pi]$-electrons system is referred to as the **diene** and the 2π-electrons system as the **dienophile** (diene lover). These terms are used in related [4+2]-reaction systems even when the functional groups are not actually dienes or alkenes.

   Retrocycloaddition: Many cycloaddition reactions require moderate heating to overcome the activation energy, but if it is heated too much the equilibrium will favour **cycloreversion** or **retrocycloaddition**. For example, cyclopentadiene slowly undergoes cycloaddition with itself: one molecule of cyclopentadiene acts as a $[4\pi]$-electrons diene and the other as a $[2\pi]$-electrons dienophile. The product is an *endo*-tricyclo[5.2.1.0]deca-3,8-diene (**8.4**), often called dicyclopentadiene. The product **8.4** gives back cyclopentadiene on heating at 150°C for an hour.

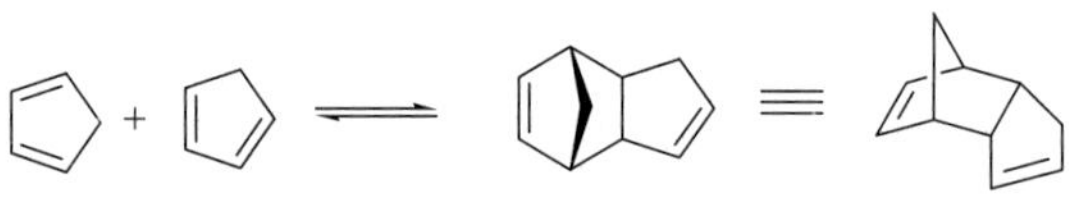

8.4

Dienes and dienophiles: The Diels–Alder cycloaddition reactions proceed more efficiently if the diene is electron rich and the dienophile is electron poor. Steric hindrance at the bonding sites may inhibit or prevent the reaction. Electron-donating groups on the diene facilitate the reaction. The way to make the dienophile electron poor is to add electron-withdrawing groups, such as CN, C=O and NO_2.

   The Diels–Alder reaction is a single-step process, so the diene component must be able to assume an s-*cis* conformation (the s refers to the single bond connecting the two double bonds) in order for the end carbon atoms (i.e. C-1 and C-4) to bond simultaneously to the dienophile. For many acyclic dienes the s-*trans* conformer is more stable than the s-*cis* conformer (due to steric crowding of the end groups), but the two are generally in rapid equilibrium as shown below:

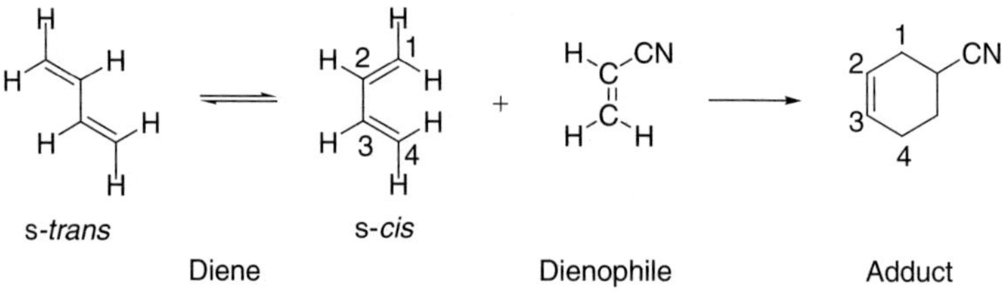

s-*trans*   s-*cis*

Diene   Dienophile   Adduct

The dienes which cannot adopt an s-*cis* orientation cannot be used as reactants in Diels–Alder reactions.

Endo **versus** *exo* **geometry in the Diels–Alder reaction:** When the Diels–Alder reaction forms bridged bicyclic adducts and an unsaturated constituent is located on this bicyclic structure, the chief product is normally the kinetically favoured *endo*-isomer, **Alder's endo rule**.

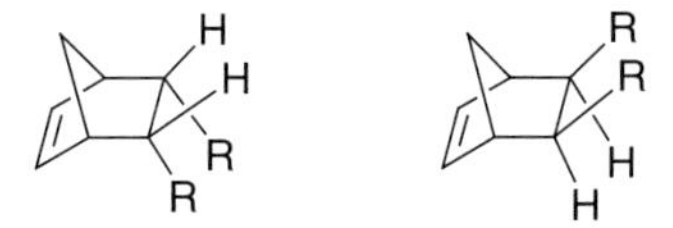

endo
(favoured)

exo

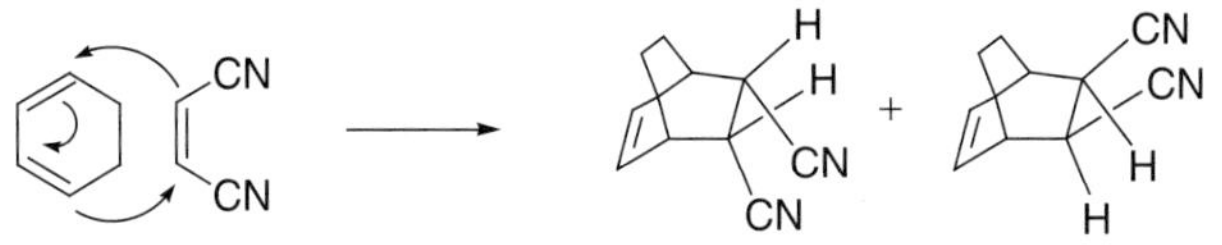

endo
(favoured)

exo

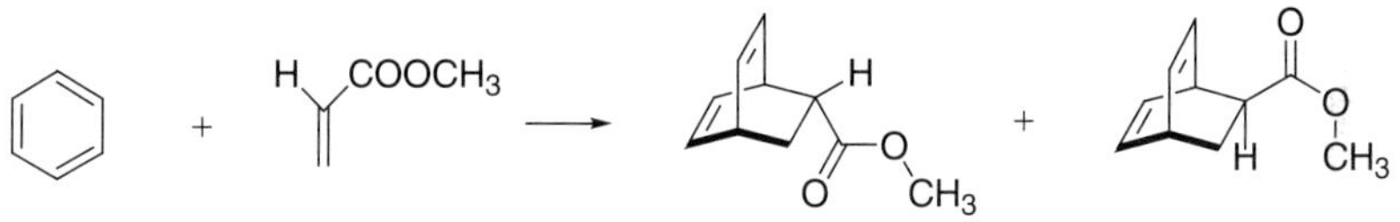

endo
(favoured)

exo

The overall Diels–Alder reaction between maleic anhydride and cyclopentadiene gives Diels–Alder adduct **8.5** with *endo* conformation. On heating at 190°C, the *endo* conformation adduct **8.5** adopts the thermodynamically more stable *exo*-adduct **8.6**.

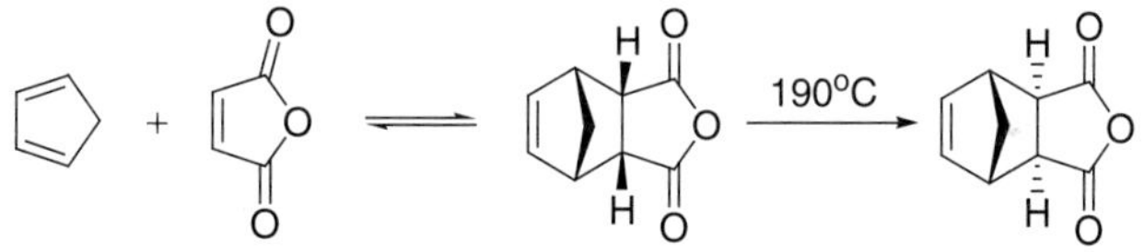

endo **8.5**

exo **8.6**

Endo and *exo* additions require different approach and transition state geometries as shown below:

endo **8.5**

exo **8.6**

Solvent: Diels–Alder reactions do not require a solvent that dissolves both the reagents. Hydrocarbons are the common solvent used. When water[20,21] is used as a solvent, the rate of reaction accelerate and the *endo* selectivity of these reactions increase. This is because the water acts as an anti-solvent, the organic reagents that are not soluble in water are clumped together in oily droplet by water and forced into close proximity.

Regiochemistry and stereochemistry: When both components of a cycloaddition reaction are unsymmetrically substituted, two regioisomeric cycloadducts are possible. In the case of Diels–Alder reactions, these are shown in the reactions of both C-1 and C-2 substituted dienes and monosubstituted dienophiles. Isomeric adducts can be referred as *ortho*, *meta* and *para* in reference to similar disubstitution isomers of benzene.

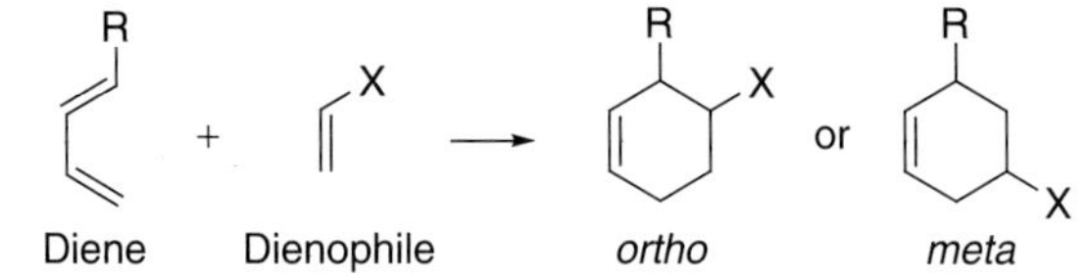

| Diene | Dienophile | *ortho* | *meta* |

| Diene | Dienophile | *meta* | *para* |

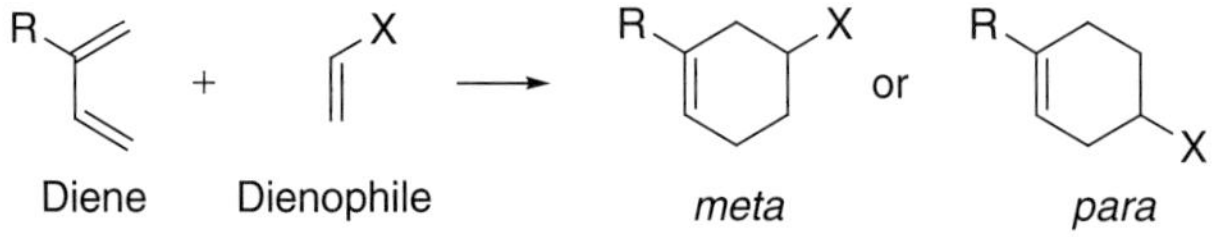

As a rule, C-1 substituted dienes form *ortho*-adducts predominantly; for example, 1-methoxy-1,3-butadiene reacts with acrylonitrile to give 3-methoxy-4-cyanocyclohexene (**8.7**) in a major amount.

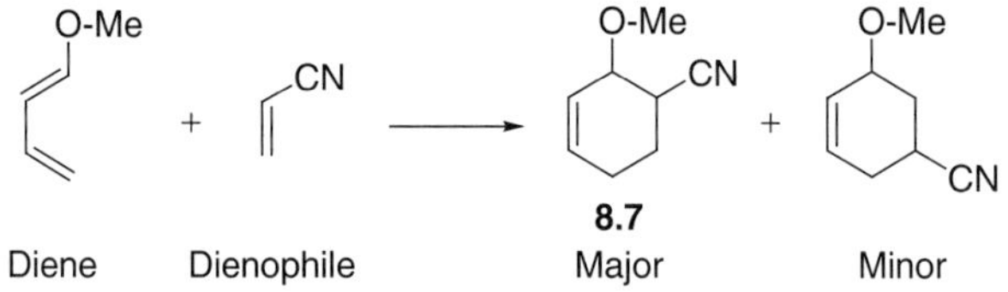

| Diene | Dienophile | **8.7** Major | Minor |

On the other hand, C-2 substituted dienes produce *para*-adducts as the major product. For example, 2-methyl-1,3-butadiene reacts with methylacrylate to give **8.8** in a major amount. Diels–Alder reactions can be catalyzed by Lewis acids and there may be improvement in yield of **8.8** product (Scheme 8.1).

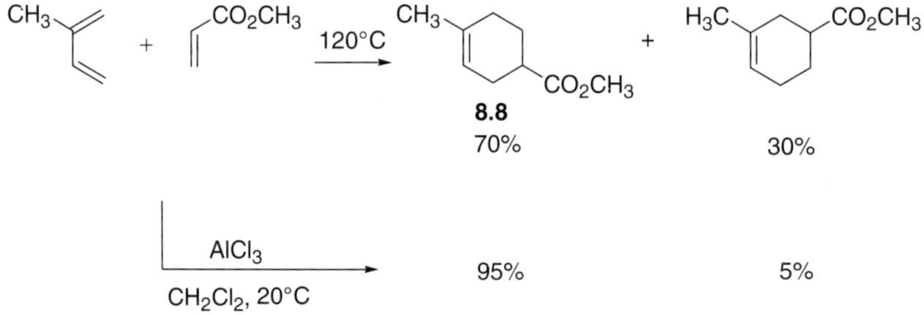

Scheme 8.1

In some cases, Lewis acid catalysis may change the regioselectivity of a Diels–Alder reaction.

The cycloaddition reaction is stereospecific with respect to substituent configuration in both the dienophile and the diene. The relative configurations of both the reactants are preserved in the product (the adduct). In the following reaction of 1,3-butadiene with *cis*-dinitroethene and *trans*-dinitroethene, the *cis* and *trans* relationships of the nitro groups in the dienophiles are preserved in the six-membered ring of the adducts, **8.9** and **8.10**, respectively.

cis-Dinitroethene   *cis* **8.9**

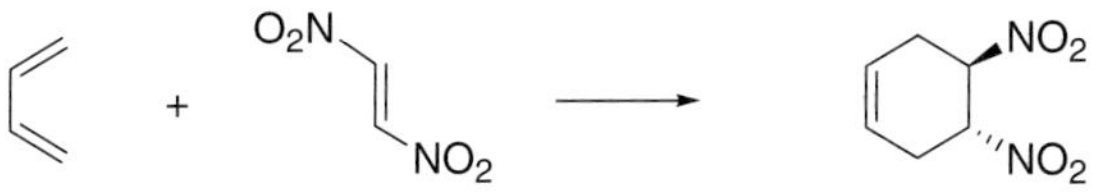

trans-Dinitroethene   *trans* **8.10**

Using the earlier terminology, we could say that bonding to both the diene and the dienophile is *syn*. An alternative description, however, refers to the planar nature of both reactants and terms the bonding in each case to be *suprafacial*. This stereospecificity also confirms the synchronous nature of the 1,4-bonding that takes place (Fig. 8.19).

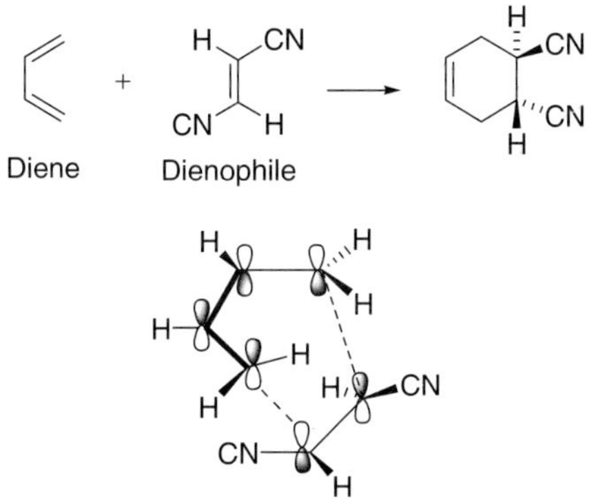

Diene   Dienophile

Figure 8.19

8.3.2 [2+2]-Cycloaddition reactions

Dimerization of ethene to cyclobutane, described as a [2+2]-cycloaddition reaction, does not proceed in either the forward or reverse directions under thermal conditions. But the [2+2]-cycloaddition reaction does occur under photochemical conditions. For example, ethene and maleic anhydride when exposed to UV light at $-65°$C gave cyclobutane diacid anhydride (**8.11**) in 77% yield.

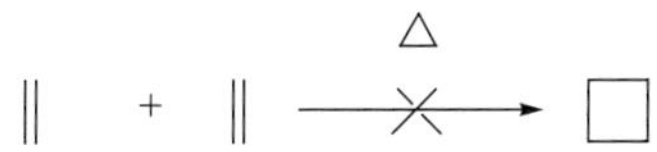

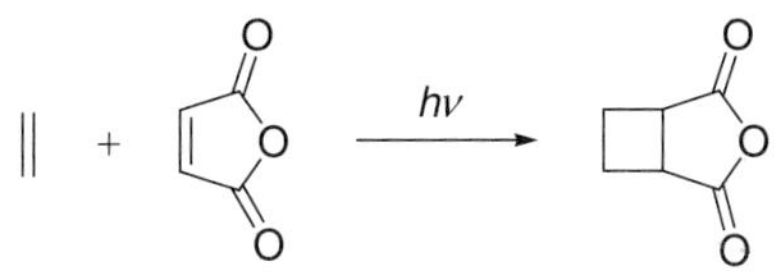

8.11

8.3.3 1,3-Dipolar additions

[4_s+2_s]-Cycloaddition reactions leading to five-membered heterocyclic adducts are classified as dipolar cycloadditions[22,23]. 1,3-Dipolar cycloaddition involves two components just like the Diels–Alder reaction, one dipolar heteroatom compound analogous to diene and the other dipolarophile analogous to dienophile. The dipolar heteroatom compound (1,3-dipole) may be exemplified by ozone and diazomethane.

Ozone

8.3.4 Theoretical explanation

Cycloaddition reactions can be explained in terms of the π-MOs of the reactants.

Correlation diagram

Cycloaddition reactions can be explained by using correlation diagrams. According to the orbital symmetry theory, the symmetry of the orbitals of the reactants must be conserved as they are transformed into the orbitals of the product.

Consider a simple example of a cycloaddition reaction of two molecules of ethene to form cyclobutane. Let us classify all the MOs of reactants and the product as symmetric (S) or antisymmetric (A) with respect to symmetry planes m and C_2. Once these symmetries are noted, correlations of reactants and product orbitals may be drawn so that orbitals of like symmetry are connected. It is assumed that ethene molecules attack each other in parallel planes (i.e. vertically). There are two symmetry planes (the mirror planes), one bisecting the π-system of the molecules (plane 1, vertical) and the other between the interacting molecules (plane 2, horizontal), as shown in Fig. 8.20.

There are two π-MOs of each ethene molecule, π and π^* (bonding and antibonding, respectively). Orbitals of one ethene molecule can interact with orbitals of another ethene in four ways leading to cyclobutane.

In the course of [$2 + 2$]-cycloaddition the four π-orbitals of the two ethene molecules are transformed into four σ-orbitals of cyclobutane (Fig. 8.20).

Let us consider the orbitals of the cyclobutane (product) and analyze them with respect to the same planes used for the ethenes interaction. This is shown in Fig. 8.21.

After discussing the symmetry properties of the four π-orbitals of two ethene molecules and four σ-orbitals of cyclobutane, a correlation diagram can be constructed for the ethene–cyclobutane conversion, as shown in Fig. 8.22.

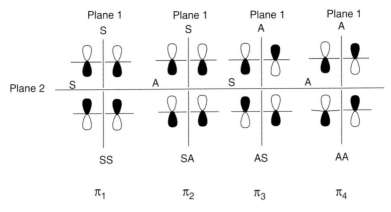

Figure 8.20 The possible interactions of the orbitals of the two ethene molecules with respect to two symmetry planes (1 and 2).

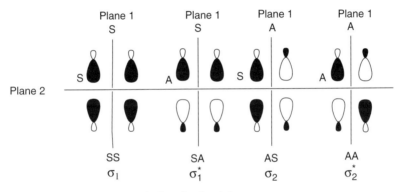

Figure 8.21 Symmetry properties of σ-bonds of cyclobutane.

Let us examine the correlation of the orbitals of reactant with those of the product (Fig. 8.22). There is a correlation of a bonding reactant level with an antibonding product level, and vice versa. But if orbital symmetry is to be conserved, two ground state ethene molecules cannot combine in a concerted reaction to give ground state cyclobutane nor can cyclobutane be decomposed in a concerted fashion to two ethene molecules.

The correlation diagram shown in Fig. 8.22 clearly illustrates that the [2+2]-cycloaddition reaction is **thermally forbidden**.

But if an electron of one ethene molecule is promoted to the antibonding orbital, then there is no symmetry-imposed barrier to the reaction. Thus, **photochemically** it is **symmetric allowed**.

[4+2]-**Cycloaddition:** The transition state for this addition is considered to reasonably have the diene and dienophile in parallel planes with the reacting termini as close together as possible. Thus, reacting molecules are arranged as shown in Fig. 8.23. The most reasonable symmetric approach is characterized by a single plane of symmetry bisecting the two components.

In this case, every bonding level of reactants correlates with a bonding product level, and there is no correlation which crosses the large energy gap between the bonding and

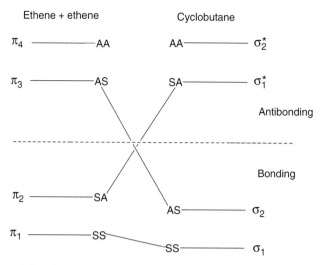

Figure 8.22 Correlation diagram for the formation of cyclobutane from two molecules of ethene.

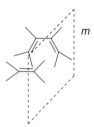

Figure 8.23 Symmetry approach of butadiene and ethene in the [4+2]-cycloaddition reaction.

antibonding levels. The plane of symmetry (*m*) bisects both the diene and the dienophile. The correlation diagram for $[4_s+2_s]$ process is shown in Fig. 8.24.

It is clear from the correlation diagram that the $[4n+2]$-cycloaddition is thermally allowed but photochemically forbidden.

Since orbital correlation diagrams for more cycloaddition reactions are more complex, we will not deal further with this approach.

FMO theory[24]

According to the FMO theory, for the cycloaddition reactions to occur there are two requirements: (i) a molecule must donate electrons from its HOMO to the LUMO of the other; and (ii) the interacting orbitals must have identical symmetry, or more simply, the phases of the terminal *p*-orbitals of each MO must match. Thus, in the case of the cycloaddition reactions, the FMO theory requires us to select the HOMO of one molecule and LUMO of the other to interact so that electron flow can occur from the HOMO of one molecule to the LUMO of the other.

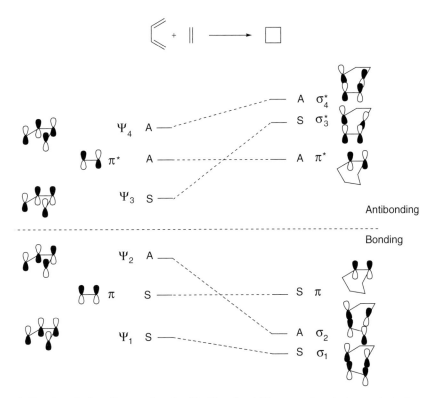

Figure 8.24 Correlation diagram for the [4+2]-cycloaddition reaction between butadiene and ethene.

[2+2]-Cycloaddition

Four stereocombinations of two molecules in the [2+2]-cycloaddition reaction are possible (Fig. 8.25): *supra–supra* (**A**), *antara–antara* (**B**), *supra–antara* (**C**) and *antara–supra* (**D**). **C** and **D** have the same allowedness, but may give different product stereochemistries depending on the nature of the reactants.

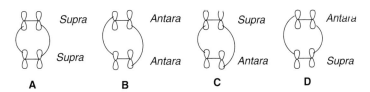

Figure 8.25 Stereocombinations of two molecules in cycloaddition.

The FMO argument for the [2+2]-cycloaddition in the ground state (i.e. under thermal conditions) is that the LUMO (π^*) of one ethene and the HOMO (π) of another ethene are phase mismatched for the *supra–supra* [2_s+2_s]-cycloaddition (Fig. 8.26). Thus, it is **symmetry forbidden** under thermal conditions (two molecules in the ground state). Similarly, the *antara–antara* mode is symmetry forbidden. The *anatra–supra or supra–anatara* mode of cycloaddition is, however, ***symmetry allowed***, but it is not feasible due to small size of the ring. However, a photon promotes one of the electrons from the HOMO into the LUMO. Now, the HOMO of the excited alkene (π^*) is phase matched to combine with the LUMO of the ground state alkene (also π^*) so that the *supra–supra* $[2_s+2_s]$ is ***photochemically symmetry allowed***.

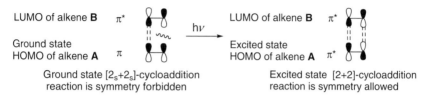

Figure 8.26   FMOs for the [2+2]-cycloaddition reaction.

The concerted [2+2]-cycloadditions in the ground state are known only when one of the components is an allene or ketene that is sterically less demanding than the ordinary alkenes.

[4+2]-Cycloaddition

The alkene (dienophile) component has two electrons in a π-bond; thus, the FMO theory identifies both HOMO and LUMO (Fig. 8.16). Likewise, the diene which has four electrons in conjugated π-system can have its HOMO and LUMO (Fig. 8.16). The most common situation finds electron-withdrawing substituents (X) on the dienophilic double bond and electron-donating substituents (R) on the diene. The bonding interaction in a normal electron demand will therefore have electrons flowing from the HOMO of the diene (Ψ_2) to the LUMO of the dienophile (π^*). Various possible ways for the [4+2]-cycloaddition reaction to occur are shown in Fig. 8.27.

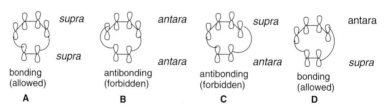

Figure 8.27   Stereocombinations of butadiene and ethene in the [4+2]-cycloaddition reaction.

If we examine the phases at the ends (termini) of the diene and dienophile (Fig. 8.28), we find that the HOMO and LUMO *supra–supra* interactions are phase matched. Thus, the $[4_s+2_s]$-*supra–supra* cycloaddition is **symmetry allowed** under thermal conditions.

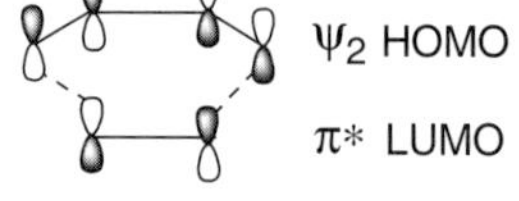

Ψ_2 HOMO

π^* LUMO

Figure 8.28   Thermal $[4_s+2_s]$-cycloaddition is symmetry matched for normal electron demand.

The *antarafacial–antarafacial* combination, although **allowed**, is sterically difficult to attain, and hence is unknown in practice. The *antara–supra* or *supra–antara* is **symmetry forbidden**.

Note that for the inverse electron demand, the LUMO of the diene (Ψ_3) and HOMO of the ethene (π) are also symmetry matched for *supra–supra* interaction, the **symmetry allowed** process (Fig. 8.29).

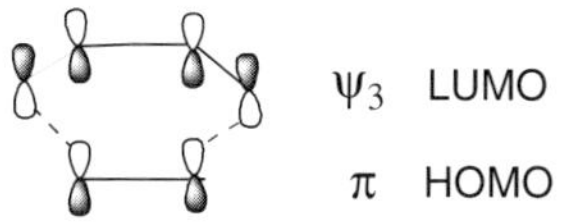

Ψ_3 LUMO

π HOMO

Figure 8.29   Thermal [4_s+2_s]-cycloaddition is symmetry matched for the inverse electron demand.

Photochemically, the [4_s+2_s] process for normal electron demand is **symmetry forbidden** because the LUMO of dienophile (π^*) does not symmetry matched with the excited state HOMO (i.e. the ground state LUMO) of diene (Ψ_3) (Fig. 8.30).

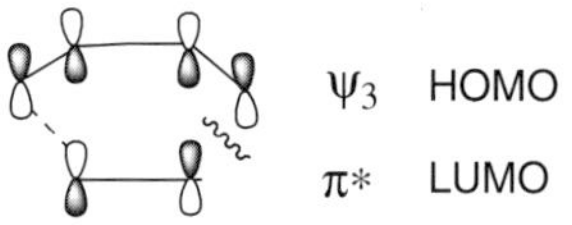

Ψ_3 HOMO

π^* LUMO

Figure 8.30   Photochemical [4_s+2_s]-cycloaddition is symmetry mismatched for normal electron demand.

Similarly, it is clear from the analysis of FMO for [6_s+2_s] *suprafacial* cycloaddition that *supra–supra* cycloaddition is **symmetry forbidden** under **thermal conditions** as shown in Fig. 8.31.

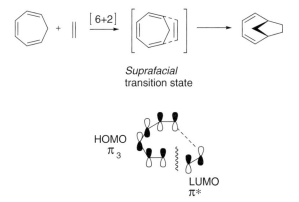

Suprafacial
transition state

HOMO
π_3

LUMO
π^*

Figure 8.31   [6+2]-*Suprafacial* cycloaddition is symmetry forbidden under thermal conditions.

In the FMO diagrams shown above (Figs. 8.28–8.31) for cycloaddition reactions the orbital lobes are shown with the same sizes (or coefficients), but usually they are of different sizes (Fig. 8.32). Coefficients are derived from the wave equations for the π-orbitals. The unsymmetrical substitution of a diene or dienophile perturbs the orbital coefficients in an unsymmetrical fashion. Calculations of orbital coefficients in such cases lead to an attractive explanation of the regioselectivity that characterizes their Diels–Alder chemistry. Thus, not only phase but also coefficients must match as shown in Fig. 8.32.

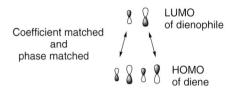

Figure 8.32 [4+2]-Cycloaddition.

If the coefficients are calculated for 1-methoxy-1,3-butadiene the termini are +0.3 and −0.58 (or −0.3 and +0.58). For acrylonitrile the coefficients are +0.2 and −0.66 (or −0.2 and +0.66). The cycloaddition reaction proceeds so that the coefficients match, in terms of both phase (essential) and coefficient magnitude: +0.3 with +0.2 and −0.58 with −0.66. Thus, the preferred orientation of reactants (diene and dienophile) for the initial bonding interaction is displayed and this orientation agrees with the regioselectivity reported in Fig. 8.33.

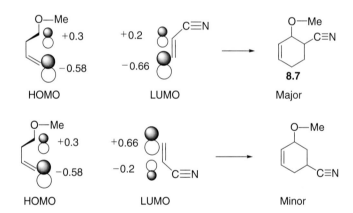

Figure 8.33 [4+2]-Cycloaddition of 1-methoxy-1,3-butadiene and acrylonitrile.

MO calculations that give π-orbital coefficients for dienes and dienophiles are beyond the scope of this book. However, there is a simple mnemonic trick that will predict regioselectivity in many cases. It involves drawing the four possible diradical intermediates that can be formed by homolytic bonding at one end of each reactant. Always remember, this is just a mnemonic trick; most Diels–Alder reactions are concerted and do not proceed through a diradical intermediate.

The FMO theory also explains the effect of substituent on rates of the Diels–Alder reaction. A donor substituent (R) on the diene raises the energy of its HOMO and an electron-withdrawing substituent (X) on the dienophile lowers the energy of its LUMO, bringing the two orbitals closer together in energy (Fig. 8.34). Orbitals close in energy interact more strongly than those far apart because the energy of the transition state is lower than it was without the substituents. Thus, the reaction occurs faster if the dienophile has electron-withdrawing groups on it and electron-donating groups on the diene.

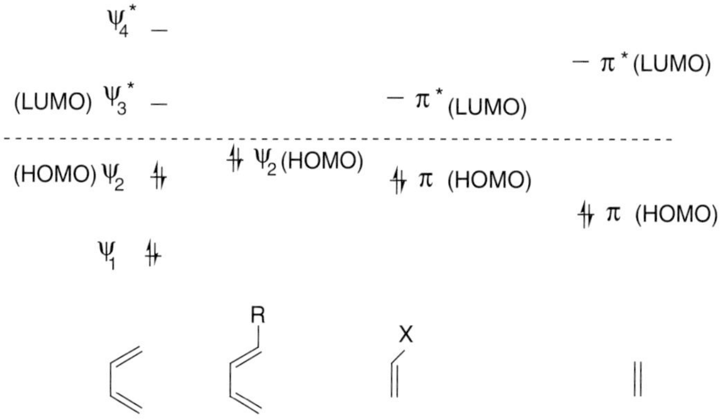

$\psi_4^* \; -$

$-\pi^*(\text{LUMO})$

$(\text{LUMO}) \; \psi_3^* \; -$ $\qquad$ $-\pi^*(\text{LUMO})$

- -

$(\text{HOMO}) \; \psi_2 \; \text{⇅}$ $\qquad$ $\text{⇅} \; \psi_2 \,(\text{HOMO})$ $\quad$ $\text{⇅} \; \pi \; (\text{HOMO})$

$\qquad\qquad\qquad\qquad\qquad\qquad\qquad\qquad\qquad\qquad\qquad \text{⇅} \; \pi \; (\text{HOMO})$

$\psi_1 \; \text{⇅}$

Figure 8.34 Energy of the FMO changes when a substituent is present on diene or dienophile.

Aromatic transition state theory

An aromatic molecule, according to Hückel's rule, contains a cyclic array of orbitals in which there are $[4n + 2]$ electrons. Since the $[4+2]$-cycloaddition reaction involves the four π-electrons of 1,3-butadiene and the two π-electrons of ethene, there must be six electrons involved in the transition state. Therefore, the $[4+2]$-cycloaddition reaction involves an aromatic transition state.

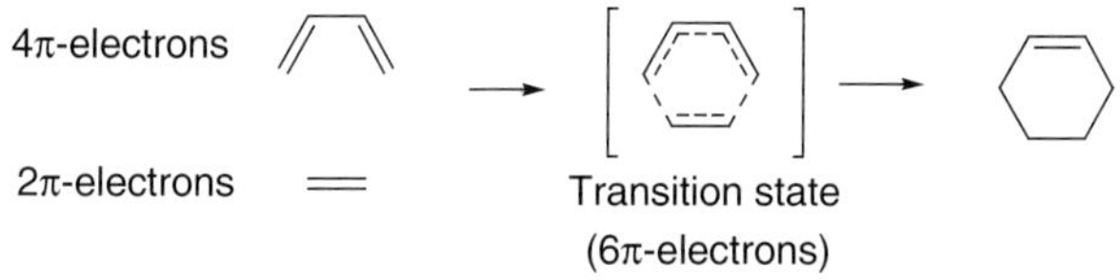

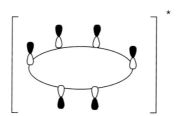

The orbitals overlap *suprafacially* on both π-systems (Fig. 8.35); thus, the $[4_s+2_s]$-cycloaddition reaction is **thermally allowed** (see the rules summarized in Table 8.1). However, the rules are reversed when the reaction is **photochemically induced**.

Figure 8.35

In the case of the $[2+2]$-cycloaddition reaction, the transition state contains four π-electrons; thus, the Hückel transition state (*suprafacial*) is anti-aromatic (Fig. 8.36). Therefore, it is **thermally forbidden**.

Table 8.1 Rules for cycloaddition reactions

Designation	Thermal	Photochemical
$\pi2_s + \pi2_s$	Forbidden	Allowed
$\pi2_s + \pi4_s$	Allowed	Forbidden
$\pi2_s + \pi2_a$	Allowed	Forbidden
$\pi2_s + \pi4_a$	Forbidden	Allowed
$\pi2_a + \pi2_a$	Forbidden	Allowed
$\pi2_a + \pi4_a$	Allowed	Forbidden

Figure 8.36

8.4   Electrocyclic reactions

Electrocyclic reactions are the concerted cyclization of a conjugated π-electron system by converting one π-bond to a ring-forming σ-bond. The reverse reaction may be called electrocyclic ring opening. For example, 1,3,5-hexatriene undergoes an electrocyclic ring closure to give 1,3-cyclohexadiene on heating. The reverse or retroelectrocyclic reaction is the ring opening of 1,3-cyclohexadiene to 1,3,5-hexatriene.

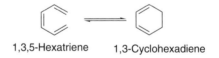

1,3,5-Hexatriene   1,3-Cyclohexadiene

Similarly, thermal (at 150°C) electrocyclic opening of cyclobutenes forms conjugated butadienes; this mode of reaction is favoured by relief of ring strain. However, the reverse ring closure is not normally observed. Photochemical ring closure can be affected, but the stereospecificity is opposite to that of thermal ring opening.

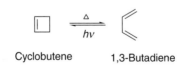

Cyclobutene   1,3-Butadiene

Stereochemistry: Electrocyclic reactions, like all pericyclic processes, exhibit great stereospecificity. The stereospecificity of such reactions is demonstrated by thermal closure of *trans,cis,trans*-2,4,6-octatriene (**8.12**) to *cis*-5,6-dimethyl-1,3-cyclohexadiene (**8.13**) and of the isomeric *trans,cis,cis*-octatriene (**8.14**) to *trans*-5,6-dimethyl-1,3-cyclohexadiene (**8.15**).

8.12　　　　　　　**8.13**

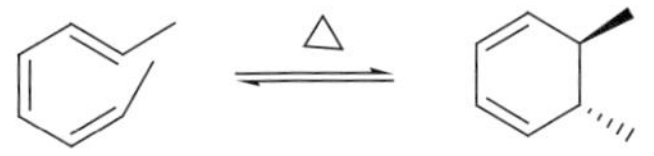

8.14　　　　　　　**8.15**

In a similar manner, stereochemistry for the ring-opening reactions of *trans-* and *cis-*3,4-dimethylcyclobutenes (**8.16 and 8.19**) is shown below.

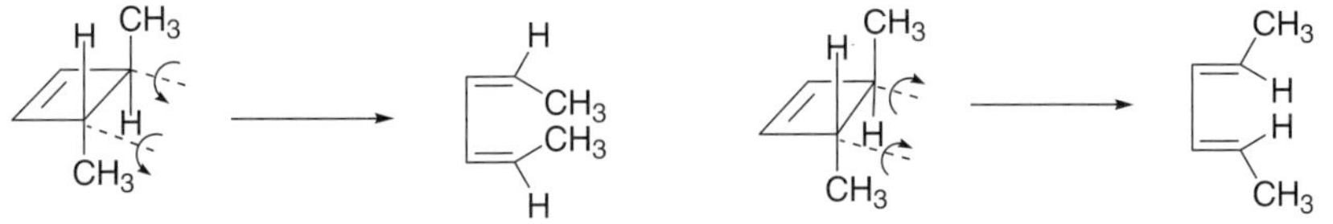

8.16　　　　**8.17**　　　　**8.16**　　　　**8.18**

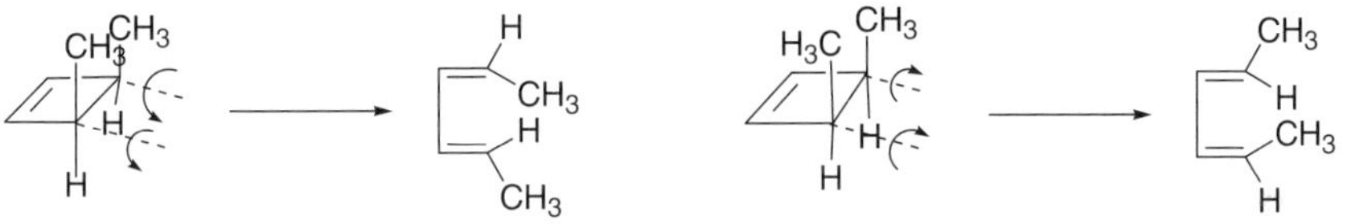

8.19　　　　**8.20**　　　　**8.19**　　　　**8.21**

*trans-*3,4-Dimethylcyclobutene (**8.16**) will give rise to either *cis,cis-*2,4-hexadiene (**8.17**) or *trans,trans-*2,4-hexadiene (**8.18**). Formation of *cis,trans-* or *trans,cis-*isomer is forbidden. *cis-*3,4-Dimethylcyclobutene (**8.19**) will form butadiene derivatives **8.20** or **8.21** with *cis,trans or trans,cis* geometry, respectively. The formation of *cis,cis-* and *trans,trans-*isomer being forbidden by orbital symmetry rules.

Disrotatory and conrotatory rotation: The concerted rotation around two bonds in the same direction, either clockwise or counterclockwise, is described as *conrotatory*. In the electrocyclic ring opening, the terminal *p*-orbitals rotate (or twist, roughly 90°) in the same direction known as *conrotation* (comparable to *antarafacial*) to form a new σ-bond. In *disrotatory* cyclization (comparable to *suprafacial*) the terminal *p*-orbitals rotate in opposite directions. These two modes of the electrocyclic reaction are shown in Fig. 8.37.

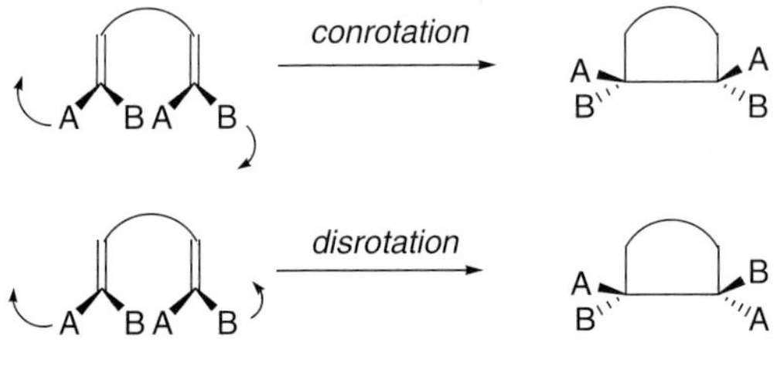

Figure 8.37

The bond breaking and bond making associated with these motions and transformations are symmetrical with respect to one of the symmetry elements of the molecule. For example, the symmetry element preserved during a *disrotatory* ring closing is the mirror plane (*m*) and for *conrotatory* ring closing it is C_2 (Fig. 8.38).

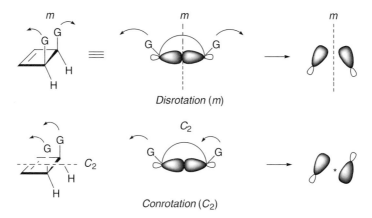

Disrotation (*m*)

Conrotation (C_2)

Figure 8.38

8.4.1 Theoretical explanation

There are several equivalent ways to rationalize the observed stereospecificity in the electrocyclic reactions.

Correlation diagram

Let us first examine how the relevant MOs for two interconverting molecules, for example, cyclobutene and butadiene, change when one molecule is converted into the other. Note particularly that we need to examine only the MOs explicitly involved in the reaction; most of the σ framework remains unchanged and no orbitals derived from this need to be considered.

In order to convert cyclobutene into butadiene, the four MOs labelled σ, π, π* and σ* must be converted into ψ_1, ψ_2, ψ_3 and ψ_4. There are two stereochemically distinct ways in which this might be accomplished: *conrotation* and *disrotation*. As mentioned earlier, the symmetry element preserved during a *disrotatory* ring closing is the mirror plane (*m*) and for *conrotation* it is C_2.

The orbitals are classified whether they are symmetric (S) or antisymmetric (A) with respect to both mirror plane (*m*) and (C_2) rotation axis. Figure 8.39 shows the symmetries of the orbitals of cyclobutene and butadiene.

The correlation diagrams for the *conrotatory* process show that there is a good correlation between the bonding orbitals ψ_1 and ψ_2 of butadiene and σ- and π-orbitals of cyclobutene (Fig. 8.40). Thus, the ring opening of cyclobutene to butadiene or the reverse reaction is **thermally allowed** and occurs by the *conrotatory* process. The reaction proceeds with conservation of orbital symmetry. The **photochemical** *conrotatory* process in this case will be **symmetry forbidden**.

Let us consider a *disrotatory* ring opening of cyclobutene to butadiene in which a mirror plane symmetry (*m*) is maintained. In the ground state, the orbitals of the reactant do not

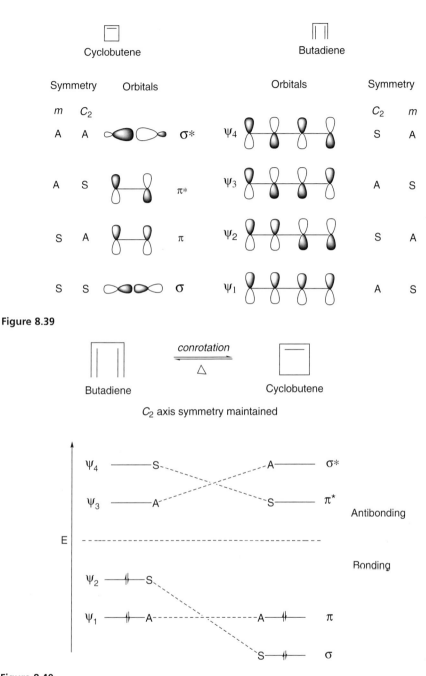

Figure 8.39

Butadiene →[conrotation][Δ] Cyclobutene

C_2 axis symmetry maintained

Figure 8.40

correlate with the ground state orbitals of the product (Fig. 8.41). Thus, the **thermally** *disrotatory* process is **symmetry forbidden**. However, irradiation of butadiene (photochemical reaction) promotes electrons from the ground state to the excited state ($\psi_2 \to \psi_3$). In this case, σ-, π- and π^*-orbitals of cyclobutene correlate with ψ_1-, ψ_2- and ψ_3-orbitals of butadiene. Hence, the *disrotatory* ring opening of cyclobutene to butadiene and the reverse reaction is **photochemically** a **symmetry-allowed reaction**.

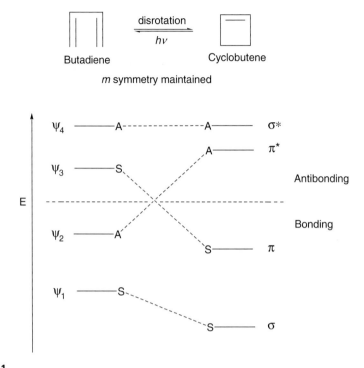

Figure 8.41

After studying the correlation diagrams for both *conrotatory* and *disrotatory* processes, it is concluded that the interconversion of butadiene to cyclobutene **thermally** proceeds in a *conrotatory* fashion while **photochemically** it proceeds in a *disrotatory* fashion.

Working on similar lines, one can draw correlation diagrams for $[4n + 2]$ π-electrons system. An example of $[4n + 2]$ π-electrons system is interconversion of hexatriene to cyclohexadiene (Fig. 8.42).

The orbitals are classified as symmetric (S) or antisymmetric (A) with respect to both mirror plan (m) and (C_2) rotation axis. Figure 8.42 shows the symmetries of the orbitals of cyclohexadiene and hexatriene.

The correlation diagram for the *disrotatory* process (m symmetry is maintained) clearly indicates that the bonding orbitals of the reactant correlate with the bonding orbitals of the product. Thus, this process is **thermally allowed** (Fig. 8.43).

On the other hand, the *conrotatory* process (C_2 axis symmetry is maintained) is **thermally not allowed** as the bonding orbitals (in the ground state) of hexatriene do not correlate with the bonding orbitals (in the ground state) of cyclohexadiene (Fig. 8.44).

However, on irradiation (i.e. photochemically) an electron is promoted from ψ_3 to ψ_4 in hexatriene. Now all the orbitals of hexatriene (in the excited state) correlate with the orbitals of cyclobutadiene.

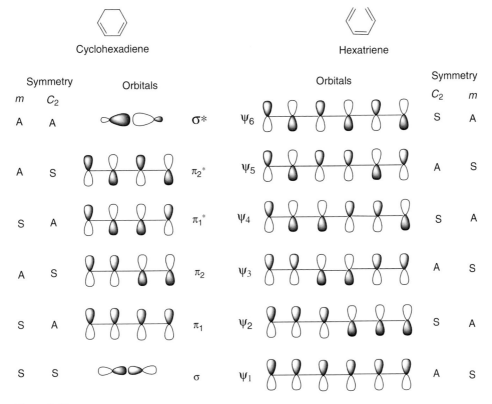

Figure 8.42

FMO theory

The easiest way to rationalize the stereospecificity in electrocyclic reactions is by examining the symmetry of the HOMO of the open (non-cyclic) molecule, regardless of whether it is the reactant or the product. For example, the HOMO of hexatriene is Ψ_3, in which orbital lobes (terminal) that overlap to make the new σ-bond have the same phase (sign of the wave function). Thus, in this case, the new σ-bond between these two terminal orbital lobes can be formed only by the *disrotation* (*suprafacial* overlap) (Fig. 8.45). If the terminal orbital lobes of HOMO of hexatriene were to close in a *conrotatory* (*antarafacial* overlap) fashion, an antibonding interaction would result.

Thus, *trans,cis,trans*-2,4,6-octatriene (**8.12**) on heating gives *cis*-5,6-dimethylcyclo-hexadiene (**8.13**) by the *disrotatory* process of ring closure.

Me ⌄⌃ Me $\xrightarrow[\text{process}]{\triangle\ \ \text{disrotatory}}$ H H / Me Me

8.12   **8.13**

However, when photoactivated, hexatriene absorbs a photon and an electron moves from the HOMO, Ψ_3, to the next orbital, the LUMO, Ψ_4 (now this orbital contains an electron;

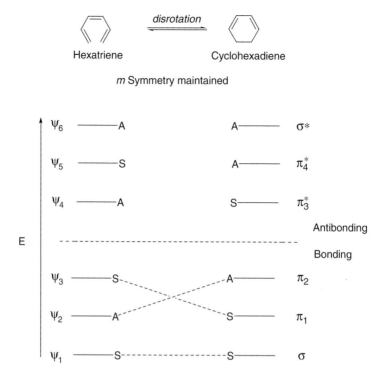

Figure 8.43

it is no longer unoccupied; it is either a SOMO (*singly occupied molecular orbital*) or an excited state HOMO). This photoexcited system will ring close in an opposite manner to the thermal system and the triene undergoes ring closure in a *conrotatory* fashion (Fig. 8.46).

On exposure to light, *trans,cis,trans*-2,4,6-octatriene (**8.12**) gives *trans*-5,6-dimethyl-1,3-cyclohexadiene (**8.15**) by *conrotatory* mode of ring closure (*antarafacial* overlap).

However, the HOMO of butadiene is Ψ_2 (in the ground state), in which the orbital lobes (terminal) that overlap to make the new σ-bond have the opposite phase (sign of the wave function). Thus, in this case, the new σ-bond between these two terminal orbital lobes cannot be formed by the *disrotation*. Thus, *disrotatory* ring closing of a diene to cyclobutene is **thermally forbidden**. If the terminal orbital lobes of the HOMO of butadiene were to close, it would be in a *conrotatory* fashion (Fig. 8.47).

Thus, *conrotatory* ring closing of butadiene to cyclobutene is **symmetry allowed** under **thermal** conditions. However, under **photochemical** conditions the LUMO becomes HOMO and the *disrotatory* ring opening is **symmetry allowed** (Fig. 8.48).

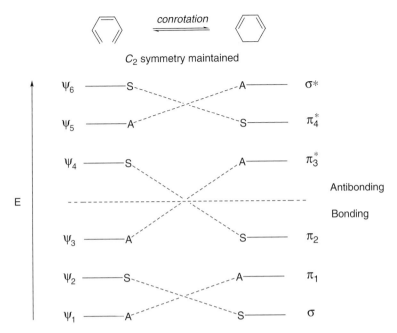

conrotation

C_2 symmetry maintained

E

Ψ_6	—S—	—A—	$\sigma*$
Ψ_5	—A—	—S—	π_4^*
Ψ_4	—S—	—A—	π_3^*

Antibonding

Bonding

Ψ_3	—A—	—S—	π_2
Ψ_2	—S—	—A—	π_1
Ψ_1	—A—	—S—	σ

Figure 8.44

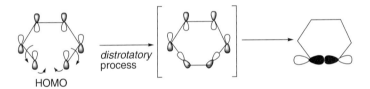

distrotatory process

HOMO

Figure 8.45

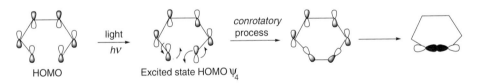

HOMO

light
$h\nu$

Excited state HOMO Ψ_4

conrotatory process

Figure 8.46

Aromatic transition state theory

The electrocyclic reaction shown below (i.e. ring opening of cyclobutene) involves two electron arrows and hence is [4n] system ($n = 1$).

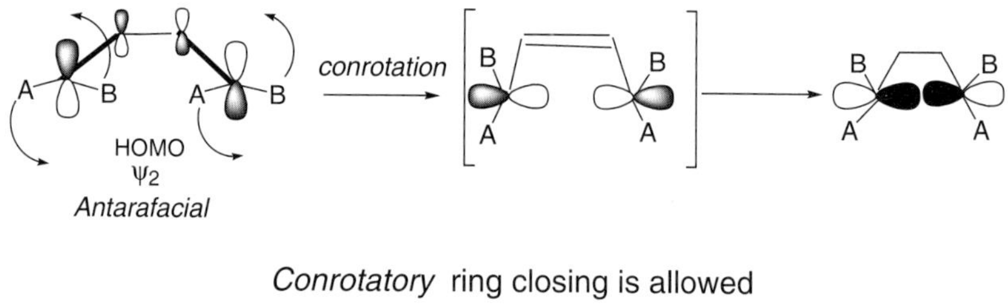

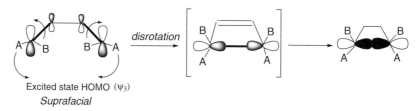

Conrotatory ring closing is allowed

Figure 8.47

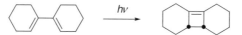

Figure 8.48

Thermally, it will proceed via a Möbius topology involving one *antarafacial* component. Both methyl groups will rotate in the same direction (*conrotation*) leaving one endocyclic and one exocyclic. However, under photochemical conditions, a $[4n]$ π-reaction is predicted to proceed via a Hückel topology with *suprafacial* (*disrotation*) bond formation.

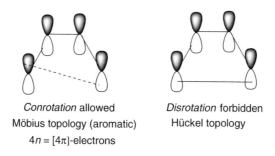

An analysis of reaction possibilities in the Hückel–Möbius theory requires us to construct a fully interacting basis set, which is simply a sketch of the transition state, drawn with the maximum possible bonding character (the fewest nodes possible). Next, draw a line to connect the reactant orbitals as they would be connected by *conrotatory* and *disrotatory* processes. This gives a complete model of the transition state for butadiene as shown in Fig. 8.49.

Figure 8.49

Now, inspect the transition state model in both the cases. If the number of nodes (apart from the nodal plane of the *p*-orbitals) is zero or any even number, the transition state has a **Hückel topology**, and will be aromatic if $[4n + 2]$ π-electrons are present (Fig. 8.50). If the number of node is one or any odd number, the transition state has a **Möbius topology**, and will be aromatic if $[4n]$ π-electrons are present (Fig. 8.49).

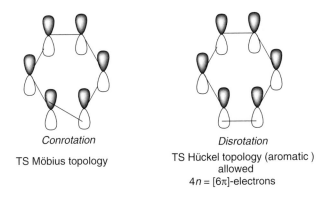

Conrotation

TS Möbius topology

Disrotation

TS Hückel topology (aromatic)
allowed

$4n = [6\pi]$-electrons

Figure 8.50

In the butadiene example, the *conrotatory* transition state has a Möbius topology, and will be aromatic (*four* electrons are present). Thus, the *conrotatory* process is allowed. The *disrotatory* transition state is a Hückel system, and will require *two* or *six* electrons for aromaticity. Since in butadiene only four electrons are present, the *disrotation* pathway will be forbidden.

Similarly, for hexatriene, the Möbius and Hückel topologies are shown in Fig. 8.50. It is clear that in a $[6\pi]$-electrons system, the Hückel topology is aromatic and the reaction is symmetry allowed by *disrotation* cyclization.

8.4.2 General rules for electrocyclic reactions

For thermal reactions, with $4n$ electrons in the transition state the *conrotatory* process is allowed, and with $[4n + 2]$ electrons in the transition state the *disrotatory* process is allowed. For photochemical reactions, rules are usually reversed.

8.5 Sigmatropic rearrangements

Sigmatropic rearrangements, like electrocyclic reactions, are unimolecular processes. They involve the migration of a σ-bond from one site to another along a π-system, with the simultaneous reorganization of the π-bonds. The total number of σ-bonds and π-bonds remain unchanged. There are two different types of sigmatropic reactions: (i) those that involve the migration of a hydrogen atom or one of its isotopes and (ii) those that involve a carbon or other atom. These rearrangements are described by two numbers set in brackets $[i,j]$, which refer to the relative distance (in atoms) each end of the σ-bond has moved. The following examples illustrate this:

$$\underset{R}{C=C-C=C-\overset{|}{C}} \quad \xleftarrow{\text{[1,5]-shift}} \quad \underset{R}{\overset{|}{C}-C=C-C=C} \quad \xrightarrow{\text{[1,3]-shift}} \quad \underset{R}{C=C-\overset{|}{C}-C=C}$$

Examples of widely encountered sigmatropic rearrangements are [1,3]- and [1,5]-shifts.

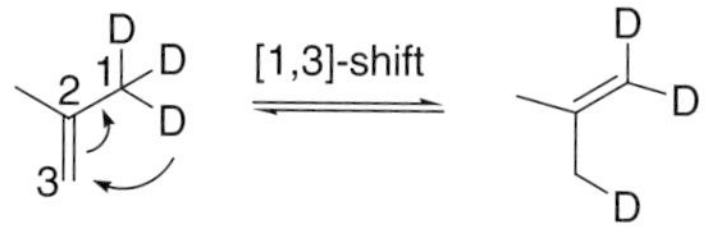

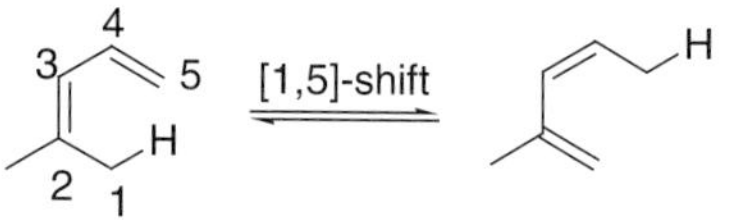

The [1,5]-hydrogen shift in cyclopentadiene can be observed at room temperature. At 60°C, migration is so fast that only one peak for all hydrogens appears in ^{1}H-NMR (Scheme 8.2).

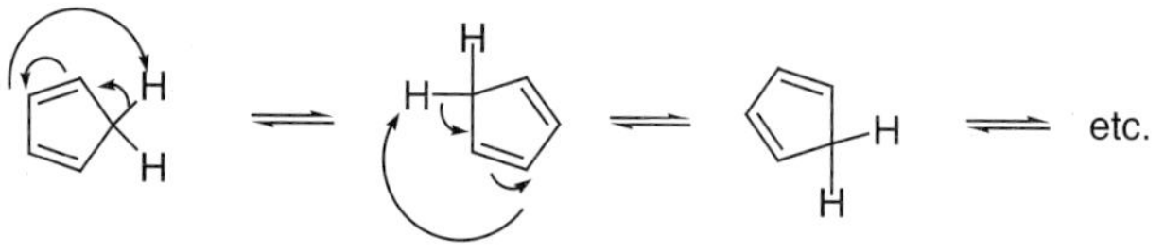

Scheme 8.2

The biosynthesis of vitamin D$_3$ (**8.24**) involves the thermal [1,7]-sigmatropic rearrangement of pre-vitamin **8.23**, which is obtained by conrotatory electrocyclic ring opening of 7-dehydrocholesterol (**8.22**) (Scheme 8.3).

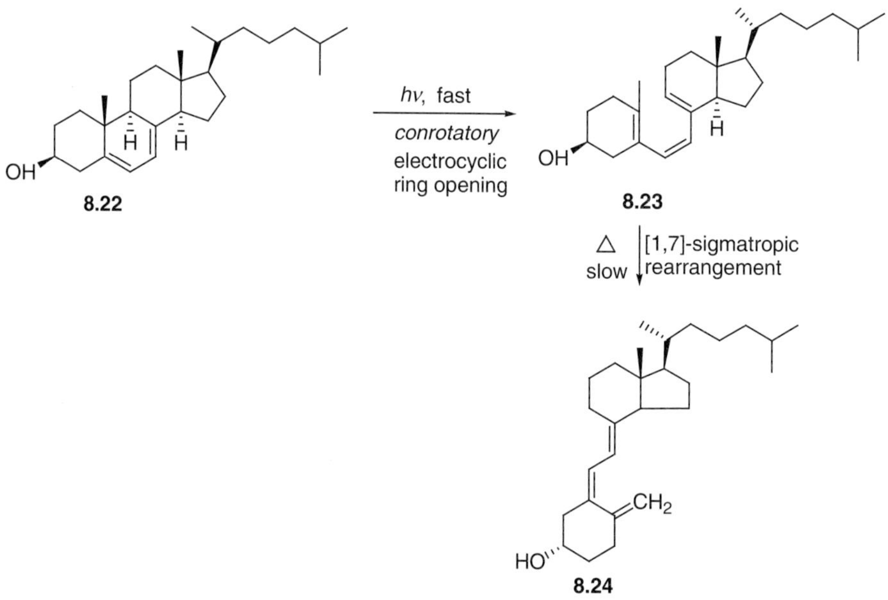

Scheme 8.3

The most common varieties of [*i,j*]-sigmatropic reactions are the [3,3] and [5,5]. These rearrangements have been widely utilized for the synthesis of structurally complex organic molecules because of the ease with which carbon–carbon bonds are formed in a regio- and stereo-controlled manner.

[3,3]-Sigmatropic rearrangement

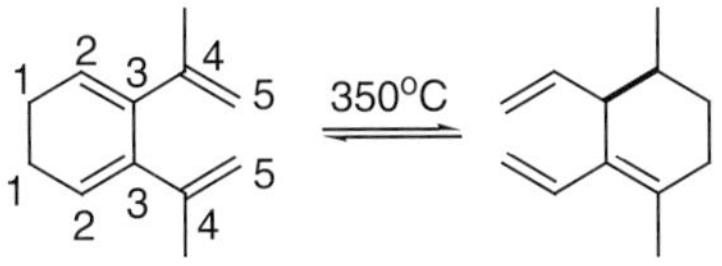

[5,5]-Sigmatropic rearrangement

The benzidine rearrangement[25–28] is an important example of the [5,5]-sigmatropic rearrangement (Scheme 8.4).

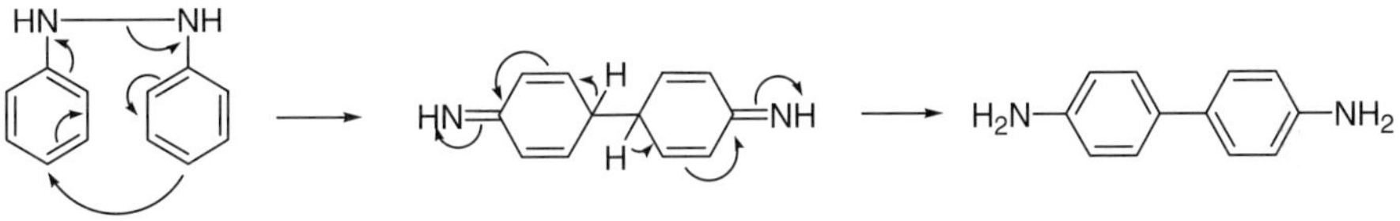

Scheme 8.4

Rearrangement of 1,5-dienes is known as the **Cope rearrangement**[29,30], and rearrangement of allyl phenyl ethers and allyl vinyl ethers is known as the **Claisen rearrangement**[31–33]. The Cope and Claisen rearrangements are the [3,3]-sigmatropic rearrangements and are among the most commonly used sigmatropic reactions. They have been the subject of numerous reviews.

Cope rearrangement

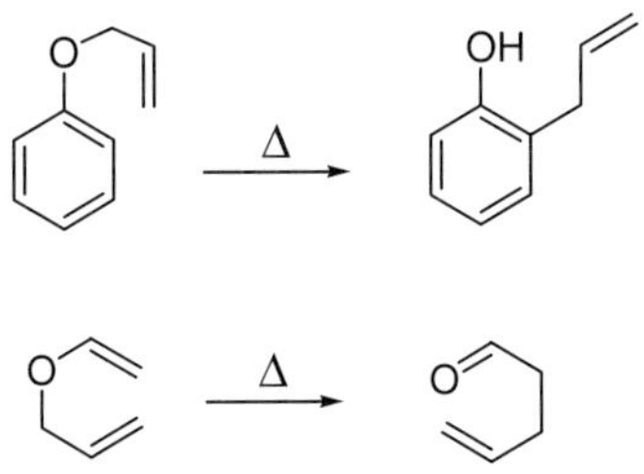

Claisen rearrangement

These rearrangements involve a six-membered transition state[34] (Scheme 8.5). It is found that a chair-like transition state is usually preferred for the Cope and Claisen rearrangements.

Chair-like transition state

Boat-like transition sate

X = — CH$_2$ — Cope rearrangement
X = — O — Claisen rearrangement

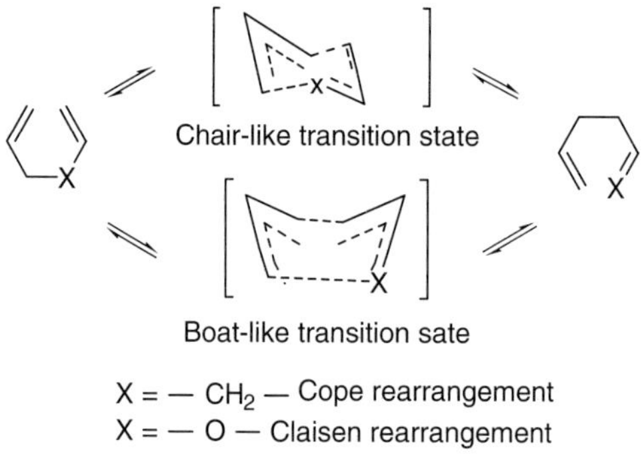

Scheme 8.5

The majority of these rearrangements require high temperatures (100–350°C), although numerous examples of catalytic syntheses are also known[35].

Important preparative examples of the [3,3]-sigmatropic rearrangements include oxy-Cope rearrangements. When a hydroxyl substituent is present at C-3 of a 1,5-diene, the Cope rearrangement is known as the **oxy-Cope rearrangement** (Scheme 8.6).

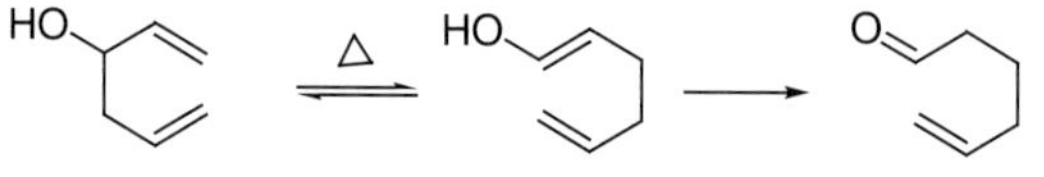

Scheme 8.6

In 1975, Golob and Evans used the alkoxide instead of the traditional alcohol[36]. The reaction is carried out at low temperature, thus making the methodology more versatile and minimizing the competing thermal retro-ene side reaction.

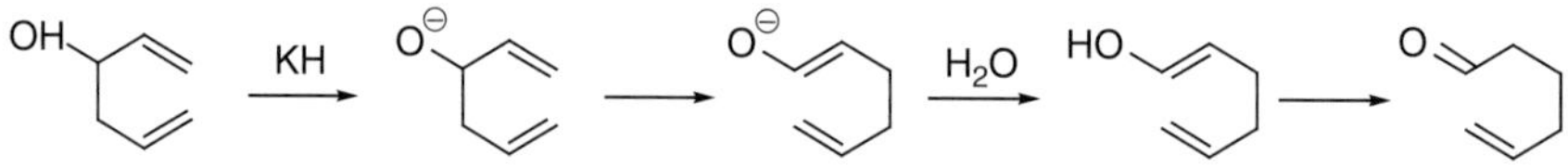

In the presence of a base, the alcohol is converted into the corresponding alkoxide and the rate of rearrangement is accelerated and occurs at room temperature or lower temperature. Thus, the **anionic oxy-Cope rearrangement** is often performed at or near room temperature and exhibits a considerable tolerance towards functional groups present in the original reactant.

The anionic oxy-Cope rearrangement is used for the ring expansions (Schemes 8.7 and 8.8). The anionic oxy-Cope rearrangement was found applicable in the synthesis[37] of muscone (**8.29**), a valuable compound with the odour of musk from an unsaturated alcohol **8.25**. Thus, alcohol **8.25** forms alkoxide **8.26** on treatment with the base KH, which undergoes the oxy-Cope rearrangement to give **8.27**. Rearrangement of **8.27** forms **8.28**, which on reduction gives muscone (**8.29**) (Scheme 8.7).

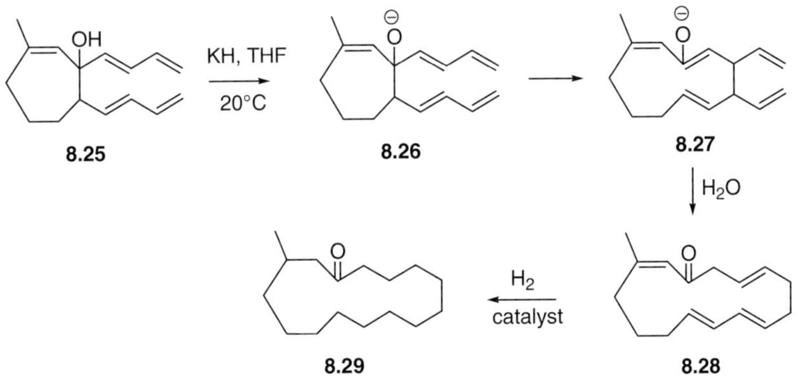

Scheme 8.7

In another example, a six-membered ring is expanded to a ten-membered ring by the anionic oxy-Cope rearrangement (Scheme 8.8). The formation of an enolate provides the driving force to overcome the increase in strain energy on going from a six- to ten-membered ring.

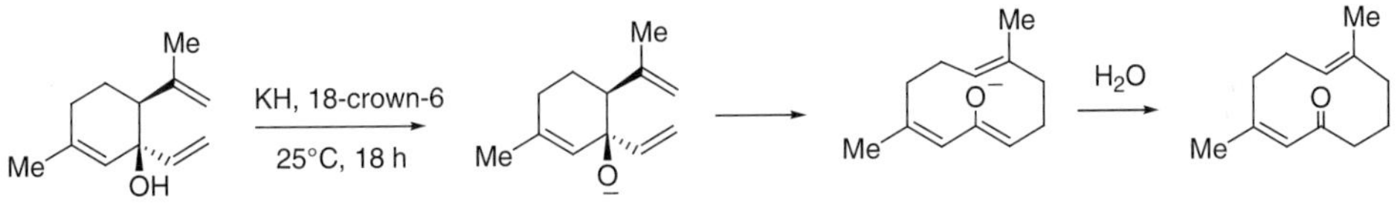

Scheme 8.8

Potassium hydride and potassium hexamethyldisilazide are the most commonly used metal sources to generate the alkoxide[38]. But recently, the indium(I)-mediated tandem carbonyl addition-oxy-Cope rearrangement of γ-pentadienyl anions to cyclohexenones and conjugated aromatic ketones has been reported. For example, indium alkoxide **8.32**, obtained after the addition of 5-bromopenta-1,3-diene (**8.31**) to the aromatic conjugated ketones **8.30**, undergoes a spontaneous oxy-Cope rearrangement[39] to give **8.33** in 55% yield (Scheme 8.9).

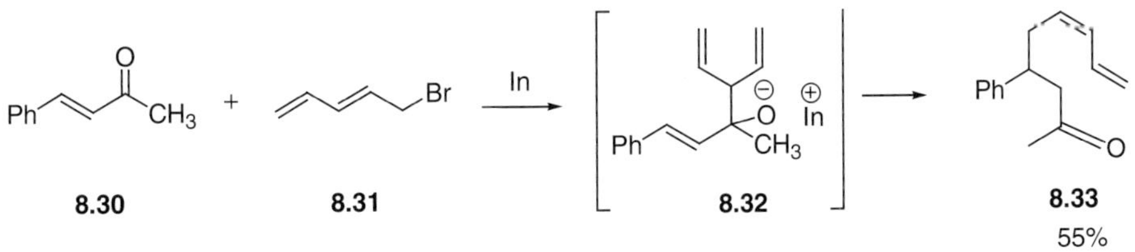

Scheme 8.9

The product of the Claisen rearrangement of allyl aryl ethers is a ketone which immediately tautomerizes to its enol form (Scheme 8.10).

The Claisen rearrangements of allyl vinyl ethers yield γ,δ-unsaturated carbonyl compounds. When $R^1 = NR_2$ the rearrangement is known as the **Eschenmoser–Claisen**

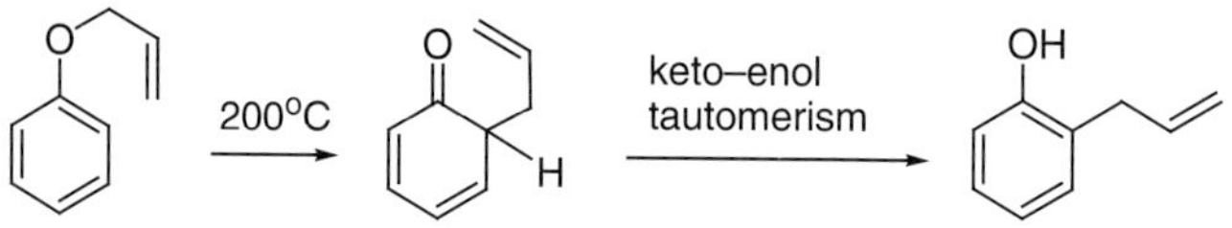

Scheme 8.10

rearrangement[40,41], and when R^1 = OR it is known as the **Johnson–Claisen rearrangement**[42] (Scheme 8.11).

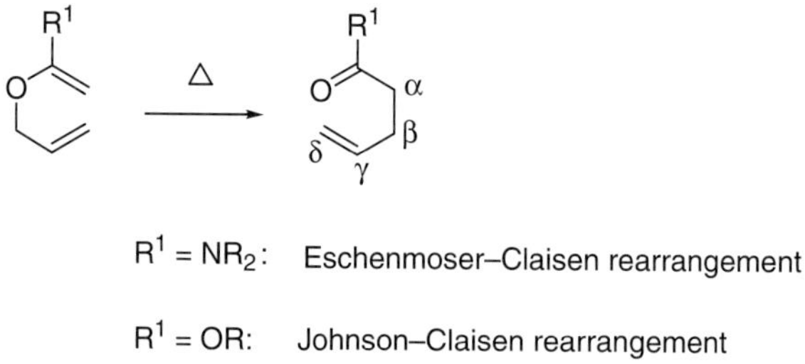

$R^1 = NR_2$:   Eschenmoser–Claisen rearrangement

$R^1 = OR$:   Johnson–Claisen rearrangement

Scheme 8.11

When $R^1 = OSiR_3$ or OLi, the reaction is referred to as the **Ireland–Claisen rearrangement**[43] (Scheme 8.12).

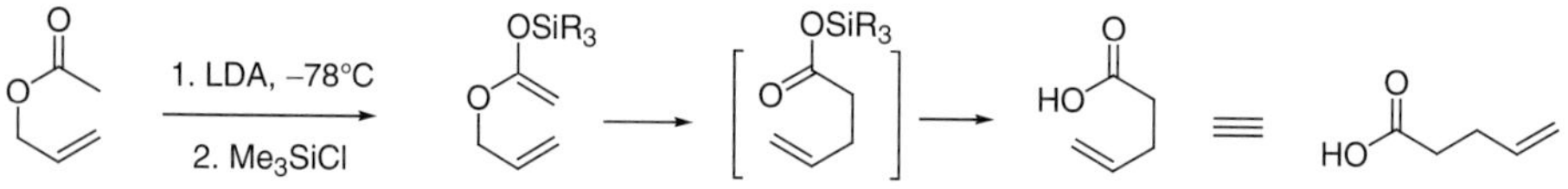

Scheme 8.12

By the **Carroll rearrangement**[44], 2-methyl-2-hepten-6-one (**8.34**) from commercially available materials such as acetone, acetylene and ethyl acetoacetate is synthesized (Scheme 8.13).

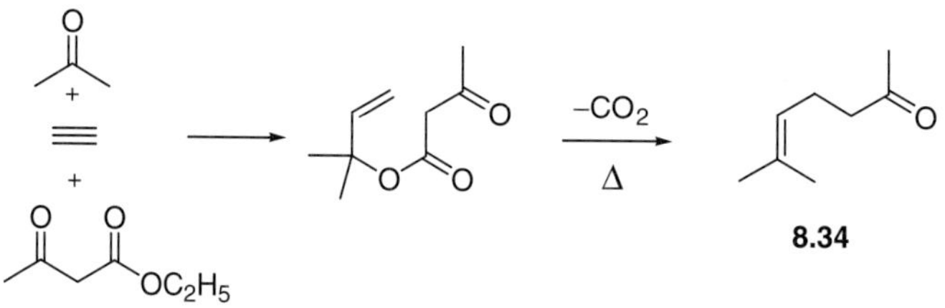

Scheme 8.13

Hetero-Claisen rearrangement. The rearrangement of allylic trichloroacetimidates (**8.35**) to allylic trichloroacetamides (**8.36**) is an example of the hetero-Claisen rearrangement[45,46].

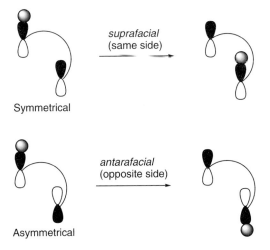

Another example of the hetero-Claisen rearrangement is the thermal and catalyzed [3,3]-sigmatropic rearrangement of allylic phosphorimidates (**8.37**) to phosphramidates[47,48] (**8.38**), which are converted into alylic amines of diverse structure.

8.5.1 Analysis of sigmatropic rearrangements

There are two topologically distinct ways (Fig. 8.51) by which hydrogen atom can migrate, i.e. *suprafacial* or *antarafacial* migration. When the hydrogen atom remains on the same face of the π-electron system (top or bottom), the rearrangement is termed as *suprafacial*. But when the migrating hydrogen atom is passed from the top face of one carbon terminus to the bottom of the other, the rearrangement is known as *antarafacial*. Sigmatropic rearrangements involve cyclic transition states. If the transition state has six or fewer atoms in the ring, the migration must be *suprafacial* because of the geometric constraints of small rings. The *antarafacial* shift can occur only with large rings.

Figure 8.51

Correlation diagram

For the analysis of these sigmatropic reactions, correlation diagrams are not relevant since it is only the transition state and not the reactants or products which may possess molecular symmetry elements.

FMO Theory

For a sigmatropic rearrangement to take place, the involved frontier orbitals (i.e. HOMO and LUMO of migrating group and polyene component, respectively) must be able to overlap during the reaction. For example, in the [1,3]-sigmatropic rearrangement of a hydrogen atom, the hydrogen HOMO and the allyl LUMO are considered. The four electrons involved are considered to occupy the hydrogen HOMO (two) and the lowest orbital of allyl Ψ_2 (two). The only way for this shift to occur is by *antarafacial* process, which is **geometrically not allowed** (Fig. 8.52).

LUMO (Ψ_2)

HOMO

Figure 8.52

However, the [1,5]-sigmatropic rearrangement is a concerted *suprafacial* reaction because the HOMO and LUMO (of migrating group, which is H, and of polyene component, pentadienyl) can interact in a *suprafacial* process (Fig. 8.53). The six electrons involved are considered to occupy the hydrogen HOMO (two) and the Ψ_1 (two) and Ψ_2 (two) of pentadienyl. The LUMO of pentadienyl is Ψ_3.

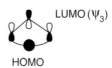

LUMO (Ψ_3)

HOMO

Figure 8.53

However, the easiest way to analyze these rearrangements is to assume that the migrating σ-bond undergoes homolytic cleavage to yield a pair of radicals. For example, in the case of a [1,5]-sigmatropic rearrangement, the products of a hypothetical cleavage are a hydrogen atom and pentadienyl radical. The latter contains five π-electrons and therefore there are five π-MOs, as shown in Fig. 8.54.

Because in the ground state the HOMO is ψ_3, the hydrogen shift is controlled by the symmetry of the ψ_3 of the pentadienyl radical. The ψ_3 has similar signs on the terminal lobes; i.e. it is symmetrical. Thus, the [1,5]-hydrogen shift is **thermally allowed** and occurs in a *suprafacial* process. This involves a transition state in which the C-1 and C-5 orbitals overlap with 1s hydrogen orbital. This shift is both **symmetry allowed** and **geometrically favourable**, as shown in Fig. 8.55.

In an excited state (i.e. photochemically) the HOMO is now ψ_4 and it is antisymmetrical. Therefore, [1,5]-*suprafacial* migration of H is no longer possible; i.e. it is **symmetry forbidden**. The photochemical reaction will follow the *antarafacial* route, which is

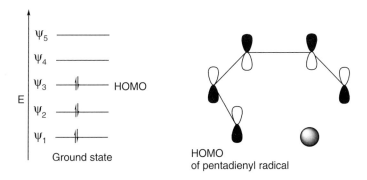

Figure 8.54

Figure 8.55 *Suprafacial* [1,5]-H shift.

symmetry allowed. However, this cannot be accomplished due to its steric problem in this type of system.

In contrast to [1,5]-H-atom shift, in a *suprafacial* [1,3]-H-atom shift, the $1s$-orbital of the H-atom cannot bond simultaneously to both terminals of the allyl radical. This is because the HOMO for a thermal reaction (i.e. in a ground state) of allyl radical is ψ_2, which is antisymmetrical; thus [1,3]-*suprafacial* migration is **symmetry forbidden**. The $1s$ hydrogen atom must be transferred from one end of the HOMO to the other by overlapping between like signs if bonding is to result. However, for this shift to be accomplished in an allyl system the H-atom must be transferred from the top face to the bottom face of the HOMO, i.e. by an *antarafacial* process. The *antarafacial* [1,3]-H-atom shift is **symmetry allowed** but **geometrically unfavourable**; hence, the reaction fails (Fig. 8.56).

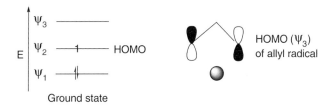

Figure 8.56

However, [1,3]–H-atom shift takes place photochemically. The HOMO of allyl system for an excited state is ψ_3 ($\psi_1^2\,\psi_2^1\,\psi_3 \rightarrow \psi_1^2\,\psi_2\,\psi_3^1$), which is symmetrical; i.e. now it has terminal lobes of the same phase. Therefore [1,3]-H shift is **photochemically symmetry allowed** and occurs in a *suprafacial* process (Fig. 8.57).

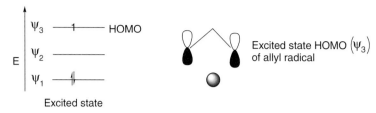

Figure 8.57

8.5.2   Carbon shift

Unlike [1,3]-sigmatropic migrations of H-atom, which cannot occur under thermal conditions, the [1,3]-sigmatropic migrations of carbon can occur under thermal conditions. This is because carbon has two-lobed *p*-orbital. Therefore, carbon can simultaneously interact with the migrating source and the migrating terminus using either one of its lobes or both of its lobes, as shown in Fig. 8.58.

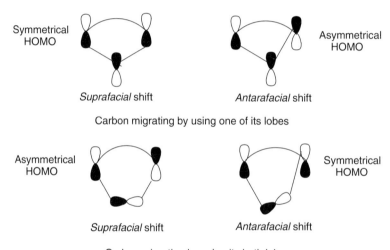

Figure 8.58   Shift of carbon.

Thus, in a *suprafacial* carbon shift if the HOMO is symmetrical, carbon will migrate using one of its lobes, but if the HOMO is antisymmetrical, it will migrate using both of its lobes. However, when carbon uses only one of its lobes, the migrating group retains its configuration. When carbon migrates by using both of its lobes, there is an inversion of configuration (Fig. 8.59). In the case of [1,3]-carbon shift, the HOMO is antisymmetrical; therefore, carbon can migrate using both of its lobes in a *suprafacial* process.

Figure 8.59

The simplest sigmatropic reaction, **1,2-shift (2-electron system)**, in carbocations is the well-known 1,2-alkyl shift (Schemes 2.9 and 2.10). This shift can be concerted **Wagner–Meerwein rearrangement** (see section 2.1.3) and *suprafacial* in carbocations. The 1,2-methyl shift involves three carbons held together by a three-centre two-electron bond at the transition state, representing the smallest and simple system (Scheme 8.14).

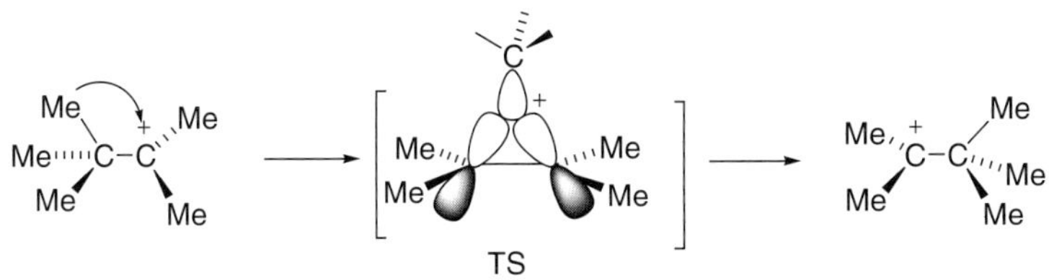

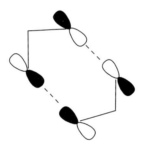

Scheme 8.14

The transition state for $[3_s,3_s]$-shifts can be seen as composed of two allyl radicals (Fig. 8.60).

Figure 8.60

Stereospecificity is a characteristic of sigmatropic rearrangements. For example, the [3,3]-sigmatropic reaction shown in Scheme 8.15 is stereospecific; i.e. the stereoisomer of the starting material produces exclusively the stereoisomer of the product shown.

Scheme 8.15

The orbital symmetry theory rationalizes the result shown above in terms of a *disrotatory* motion of the orbitals that are involved in the reaction. As shown in Fig. 8.61, the reaction involves simultaneous (concerted) rotation about the C2–C3, C3–C4, C5–C6 and C6–C7 bonds. Rotation about the C2–C3 bond proceeds in a clockwise direction, opposite to that around the C6–C7 bond. This is called a *disrotatory* motion.

For sigmatropic rearrangement of order (i,j) in which both i and j are greater than 1, the migrating group of the above analysis consists of more than one atom, and appropriate

Figure 8.61 *Disrotation* is allowed.

Table 8.2 Selection rules for sigmatropic rearrangements

Number of electron pairs in the starting compound	Allowed	Excited state
Even	*Antara* (only with large rings)	Thermal
	Supra	Photochemical
Odd	*Supra*	Thermal
	Antara (only with large rings)	Photochemical

topological distinction must be made in moving to both π-systems through which the σ-bond is moving.

The Woodward–Hoffmann selection rules for sigmatropic rearrangements are summarized in Table 8.2.

8.6 Ene reactions

The joining of a double or triple bond to an alkene reactant having a transferable allylic hydrogen is called an **ene reaction**[49–52]. The reverse process is called a **retro-ene reaction**. Like Diels–Alder reactions, Lewis acids such as $AlCl_3$ or BF_3 can catalyze ene reactions.

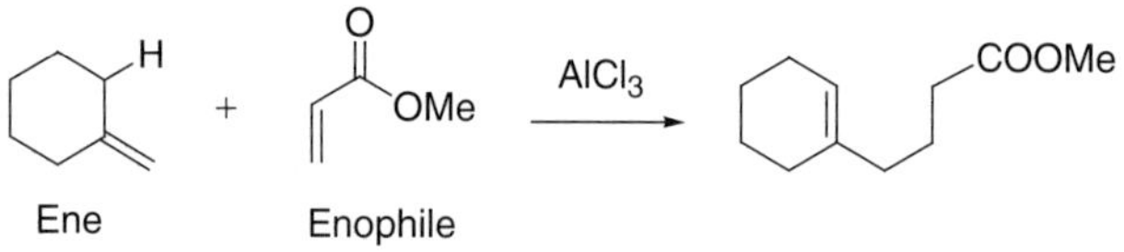

Reduction of alkenes and alkynes with diimides is also an example of an ene reaction.

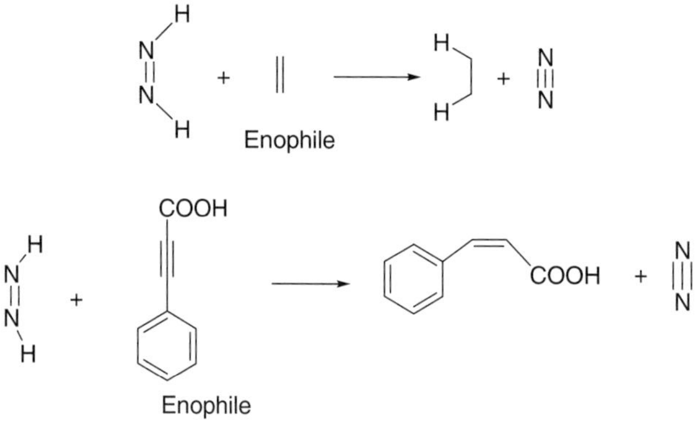

In the case of an intramolecular ene reaction, a ring is formed or broken. For example, alicyclic alcohols **8.40** and **8.42** were obtained from linalool (**8.39**) and dehydrolinalool (**8.41**), respectively.

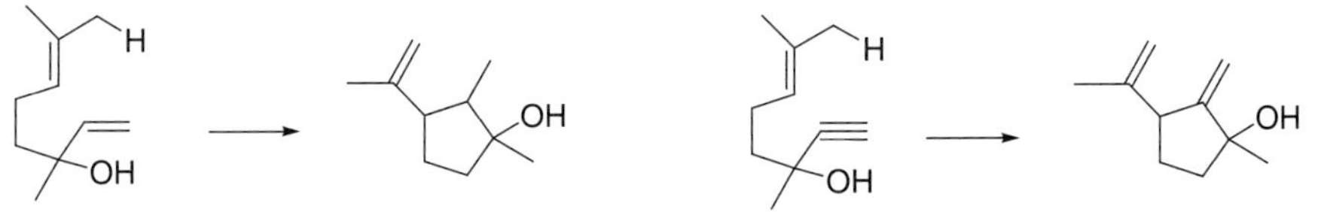

8.39	**8.40**	**8.41**	**8.42**

The McLafferty rearrangement is a **retro-ene reaction** and so is ester pyrolysis (Scheme 8.16) and similar reactions.

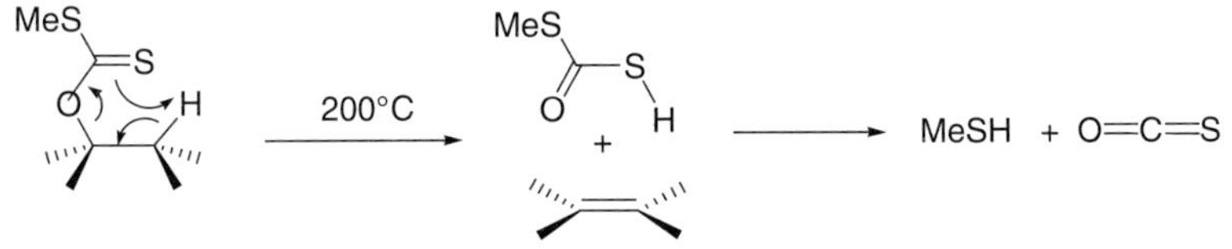

Scheme 8.16

The **retro-ene reaction** cleaves an unsaturated compound into two unsaturated fragments. A common example of a **retro-ene reaction** in organic synthesis is the acid-catalyzed decarboxylation of a β-ketoester. The ester is hydrolyzed to the β-ketoacid by the aqueous acid, which rapidly loses carbon dioxide to form enol. The loss of CO_2 drives the reaction to the right-hand side. The enol rapidly tautomerizes to the methyl ketone (Scheme 8.17).

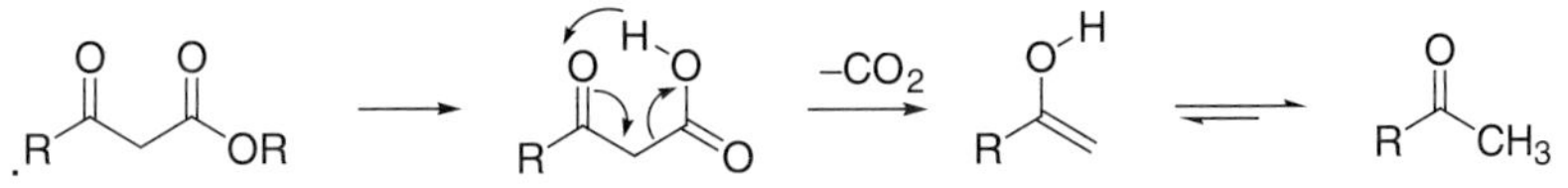

Scheme 8.17

Enophiles may include carbonyls, thiocarbonyls, imines, alkenes and alkynes. When the carbonyl is an enophile, the reaction is called a carbonyl ene reaction. The enol form of an unsaturated ketone may serve as an ene in an intermolecular ene reaction known as the **Conia-ene reaction**[53]. Ene reactions proceed best when the enophilic double bond is electron deficient. Hydrogen is the most common atom transferred in an ene reaction. Other atoms or groups may, however, participate in ene-like transformations.

Since **ene reactions** usually involve coupled bond-making and bond-breaking operations associated with short π-electron systems (2 or 3 carbons), their stereospecificity is almost always *suprafacial* with respect to both components. Note that bond-breaking and bond-making steps take place in a *suprafacial* orientation with respect to each π-electron system. This reaction is facilitated by the relief of small ring strain.

A frontier orbital analysis of the ene reaction is shown in Fig. 8.62. It displays many features of the [3,3]-sigmatropic shift. Thus, the hydrogen atom transfer from an allylic site to a double bond is seen to be **symmetry allowed** with respect to the HOMO of an

Table 8.3   Selection rules for pericyclic reactions

	Transition state class	Configurational preference
Thermal reaction	$[4n + 2]$ (aromatic) $[4n]$ (anti-aromatic)	*Suprafacial* or *disrotatory* *Antarafacial* or *conrotatory*
Photochemical reaction	$[4n + 2]$ (aromatic) $[4n]$ (anti-aromatic)	*Antarafacial* or *conrotatory* *Suprafacial* or *disrotatory*

allyl radical and the LUMO of an alkene. There is also symmetry correlation between the HOMO and LUMO terminal sites of the ene and enophile reactants.

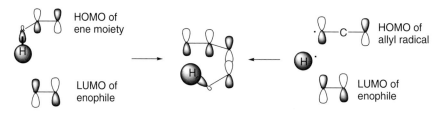

Figure 8.62

8.7   Selection rules

Before pericyclic reactions can be put to use in a predictable and controlled manner, a broad mechanistic understanding of the factors that influence these concerted transformations must be formulated. The simplest method for predicting the configurational path favoured by a proposed pericyclic reaction is based on a **transition state electron count**. In most of the earlier examples, pericyclic reactions were described by a cycle of curved arrows, each representing a pair of bonding electrons. The total number of electrons undergoing reorganization is always even, and is either a $[4n + 2]$ or $[4n]$ number (where n is an integer). Once this electron count is made, Table 8.3 may be used for predictions.

A simple summary of rules for pericyclic reactions is as follows:

1. Count the number of electron pairs in the delocalized transition state: even number of electron pairs = $[4n]$ electrons; odd number of electron pairs = $[4n + 2]$ electrons.
2. Even/*antara*/*con* and odd/*supra*/*dis* are **thermally allowed** pathways.
3. It is reverse for photochemical reactions.
4. Inversion of stereochemistry counts like *antarafacial* and two *antarafacial* components count like *suprafacial*.

References

1. Fleming, I., *Frontier Orbitals and Organic Chemical Reactions*, Wiley, London, **1976**, p. 96.
2. Gilchrist, T. L. and Storr, R. C., *Organic Reactions and Orbital Symmetry*, 2nd edn, Cambridge University Press, Cambridge, **1979**, pp. 81, 159, 229.

3. Baker, H. and Botema, J. A., *Recl. Trav. Chim. Pays-Bas*, **1932**, *51*, 294.
4. Staudinger, H. and Ritzenthaler, B., *Chem. Ber.*, **1935**, *68*, 455.
5. Wei, H.-X., Zhou, C., Ham, S., White, J. M. and Birney, D. M., *Org. Lett.*, **2004**, *6*, 4289.
6. Birney, D. M., *Org. Lett.*, **2004**, *6*, 851.
7. Woodward, R. B. and Hoffmann, R., *The Conservation of Orbital Symmetry*, Academic, New York, **1971**.
8. Fukui, K., *Tetrahedron Lett.*, **1965**, *2009*, 2427.
9. Zimmerman, H. E., *J. Am. Chem. Soc.*, **1966**, *88*, 1564.
10. Zimmerman, H. E., *Tetrahedron*, **1982**, *38*, 753–758.
11. Heilbronner, E., *Tetrahedron Lett.*, **1964**, 1923–1928.
12. Dewar, M. J. S., *Angew. Chem., Int. Ed.*, **1971**, *83*, 859.
13. Shen, K. W., *J. Chem. Educ.*, **1973**, *50*, 238.
14. Houk, K. N., Li, Y. and Evanseck, J. D., *Angew. Chem., Int. Ed.*, **1992**, *104*, 711.
15. Bernardi, F., Olivucci, M. and Robb, M. A., *Acc. Chem. Res.*, **1990**, *23*, 405.
16. Diels, O. and Alder, K. *Justus Liebigs Ann. Chem.*, **1928**, *460*(1), 98–122.
17. Diels, O. and Alder, K., *Chem. Ann.*, **1929**, *470*, 62.
18. Diels, O. and Alder, K., *Chem. Ber.*, **1929**, *62*, 208.
19. Fringuelli, F. and Taticchi, A., *Dienes in the Diels–Alder Reaction*, 1st edn, Wiley-Interscience, New York, **1990**.
20. Kumar, A., Deshpande, S. S. and Pawar, S. S., *Sci. Lett.*, **2003**, *26*, 232.
21. Breslow, R. and Guo, T., *J. Am. Chem. Soc.*, **1988**, *110*, 5612.
22. Huisgen, R., *Angew. Chem., Int. Ed.*, **1963**, *2*(11), 633.
23. Huisgen, R., *Angew. Chem., Int. Ed.*, **1963**, *2*(10), 565.
24. Houk, K. M., *Acc. of Chem. Res.*, **1975**, *8*(11), 361.
25. Hammond, G. S. and Shine, H. J., *J. Am. Chem. Soc.*, **1950**, *72*, 220.
26. Hughes, E. D. and Ingold, C. K., *Q. Rev.*, **1952**, *6*, 53.
27. Wittig, G. and Grolig, J. E., *Chem. Ber.*, **1961**, *94*, 2148.
28. Shine, H. J. and Chamness, T., *J. Org. Chem.*, **1963**, *28*, 1232.
29. Cope, A. C., *J. Am. Chem. Soc.*, **1940**, *62*, 441.
30. Wilson, S. R., *Org. React.*, **1993**, *43*, 93.
31. Hiersemann, M. and Nubbemeyer, U., *The Claisen Rearrangement*, Wiley-VCH, Weinheim, **2007**.
32. Rhoads, S. J. and Raulins, N. R., *Org. React.*, **1975**, *22*, 1–252.
33. Ziegler, F. E., *Chem. Rev.*, **1988**, *88*, 1423.
34. Jiao, H. P., Schleyer, V. R. and Von, R., *J. Phys. Org. Chem.*, **1998**, *11*, 655.
35. Lutz. R. P., *Chem. Rev.*, **1984**, *84*, 205.
36. Evans, D. A. and Golob, A. M., *J. Am. Chem. Soc.*, **1975**, *97*, 4795.
37. Tsui, J., *Tetrahedron Lett.*, **1979**, *24*, 2257.
38. Paquette, L. A., *Tetrahedron*, **1997**, *53*, 13971.
39. Villalva-Servin, N. P., Melekov, A. and Fallis, A. G., *Synthesis*, **2003**, *5*, 790.
40. Wick, A. E., *Helv. Chim. Acta*, **1964**, *47*, 2425.
41. Lautens, M., Huboux, A. H., Chin, B. and Downer, J., *Tetrahedron Lett.*, **1990**, *31*, 5829.
42. Johnson, S., Werthemann, L., Bartlett, W. R., Brocksom, T. J., Li, T. T., Faulkner, D. J. and Petersen, M. R., *J. Am. Chem. Soc.*, **1970**, *92*, 741.
43. Ireland, R. E. and Muller, R. H., *J. Am. Chem. Soc.*, **1972**, *94*, 5897.
44. Carrol, K. F., *J. Chem. Soc.*, **1940**, 704.
45. Overman, L. E., *J. Am. Chem. Soc.*, **1974**, *96*, 597.
46. Overman, L. E., *J. Am. Chem. Soc.*, **1976**, *98*, 2901.
47. Chen, B. and Mapp, A. K., *J. Am. Chem. Soc.*, **2005**, *127*, 6712.
48. Challis, B. C. and Frenkel, A. D., *J. Chem. Soc., Chem. Commun.*, **1972**, 303.
49. Alder, K., Pascher, F. and Schmitz, A., *Chem. Ber.*, **1943**, *76*, 27.
50. Mikami, K. and Shimizsu, M., *Chem. Rev.*, **1992**, *92*, 1021.
51. Mikami, K., Terada, M., Narisawa, S. and Nakai, T., *Synlett*, **1992**, 255.
52. Rouessac, F., Beslin, P. and Conia, J. M., *Tetrahedron Lett.*, **1965**, *6*(37), 3319.
53. Conia, J. M. and Perchec, P. L., *Synthesis*, **1975**, 1.

Index